A COMPREHENSIVE ASSESSMENT OF THE ROLE OF RISK IN U.S. AGRICULTURE

NATURAL RESOURCE MANAGEMENT AND POLICY

EDITORIAL STATEMENT

There is a growing awareness to the role that natural resources such as water, land, forests and environmental amenities play in our lives. There are many competing uses for natural resources, and society is challenged to manage them for improving social well being. Furthermore, there may be dire consequences to natural resources mismanagement. Renewable resources such as water, land and the environment are linked, and decisions made with regard to one may affect the others. Policy and management of natural resources now require interdisciplinary approach including natural and social sciences to correctly address our society preferences.

This series provides a collection of works containing most recent findings on economics, management and policy of renewable biological resources such as water, land, crop protection, sustainable agriculture, technology, and environmental health. It incorporates modern thinking and techniques of economics and management. Books in this series will incorporate knowledge and models of natural phenomena with economics and managerial decision frameworks to assess alternative options for managing natural resources and environment.

Agricultural producers are exposed to risk, both in terms of crop prices and crop yields, as a result of changing global institutions and global environment. Although agricultural policies in the U.S. and elsewhere shift continuously from income-support to risk-related help, the treatment of agricultural risk is not satisfactory at all. This book defines and critically evaluates the current state of the literature on economic risk in agriculture. The book sets a research agenda that is expected to meet future needs and prospects on agricultural production risk.

The Series Editors

Recently Published Books in the Series

Casey, Frank, Schmitz, Andrew, Swinton, Scott, and Zilberman, David:
Flexible Incentives for the Adoption of Environmental Technologies in Agriculture
Feitelson, Eran and Haddad, Marwan
Management of Shared Groundwater Resources: the Israeli-Palestinian Case with an International Perspective
Wolf, Steven and Zilberman, David
Knowledge Generation and Technical Change: Institutional Innovation in Agriculture
Moss, Charles B., Rausser, Gordon C., Schmitz, Andrew, Taylor, Timothy G., and Zilberman, David
Agricultural Globalization, Trade, and the Environment
Haddadin, Munther J.
Diplomacy on the Jordan: International Conflict and Negotiated Resolution
Renzetti, Steven
The Economics of Water Demands

A COMPREHENSIVE ASSESSMENT OF THE ROLE OF RISK IN U.S. AGRICULTURE

edited by

Richard E. Just
University of Maryland, College Park

and

Rulon D. Pope
Brigham Young University

KLUWER ACADEMIC PUBLISHERS
Boston / Dordrecht / London

Distributors for North, Central and South America:
Kluwer Academic Publishers
101 Philip Drive
Assinippi Park
Norwell, Massachusetts 02061 USA
Telephone (781) 871-6600
Fax (781) 871-6528
E-Mail <kluwer@wkap.com>

Distributors for all other countries:
Kluwer Academic Publishers Group
Distribution Centre
Post Office Box 322
3300 AH Dordrecht, THE NETHERLANDS
Telephone 31 78 6392 392
Fax 31 78 6546 474
E-Mail <services@wkap.nl>

Electronic Services <http://www.wkap.nl>

Library of Congress Cataloging-in-Publication Data

A C.I.P. Catalogue record for this book is available from the Library of Congress

Printed on acid-free paper.

Printed in the United States of America

Contents

PART III
ADEQUACY OF GENERAL METHODOLOGICAL APPROACHES FOR RISK ANALYSIS

PART IV
SOURCES AND CONSEQUENCES OF AGRICULTURAL RISK: HOW FARMERS MANAGE RISK

PART V
POLICY ISSUES RELATING TO RISK IN AGRICULTURE

List of Figures

List of Tables

Preface

This book has grown out of a request to us to co-chair the annual meetings of the Regional Agricultural Experiment Station Project SERA-IEG-31, which is an exchange group of the leading economists working on agricultural risk in the United States. It was inspired by the two papers that were presented at the annual meetings of this group last year, one which assessed the past twenty-five years of research on agricultural risk since that group was first formed by Professor Arne Hallam of Iowa State University, and the other which assessed possibilities for the next twenty-five years of research on agricultural risk presented by Professor Richard Just of the University of Maryland. Feeling that the extent of those two papers could hardly be comprehensive in their efforts, the program for the March 22-24, 2001, meetings of SERA-IEG-31 was organized with the intent of preparing a thorough assessment of the importance of risk in agriculture and preparing from the proceedings a comprehensive reference book on the economics of agricultural risk. A great need exists for this kind of book because risk received relatively little treatment in the new *Handbook of Agricultural Economics* currently in preparation due to the wide class of issues it covers, and because there has been no comprehensive book on agricultural risk since the well-known book edited by James A. Roumasset, Jean-Marc Boussard, and Inderjit Singh on *Risk and Uncertainty in Agricultural Development* published in 1979, which had a clear emphasis on agriculture in developing countries.

This book is intended to (i) define the current state of the literature on agricultural risk research, (ii) provide a critical evaluation of economic risk research on U.S. agriculture to date, and (iii) set a research agenda that will meet future needs and prospects. This type of research promises to become of increasing importance because agricultural policy in the United States and elsewhere has decidedly shifted from explicit income support objectives to risk-related motivations of helping farmers deal with risk. Beginning with the 1996 Farm Bill, the primary set of policy instruments for U.S. agriculture shifted from target prices and set aside acreages to agricultural crop insurance. The related risk motivation is likely to be at the center of the upcoming farm bill debate, which will consider continuation or alteration of this dramatic legislation, as well as subsequent agricultural policy required under free trade agreements. Because this book is intended to have specific implications for U.S. agricultural policy, it has a decidedly domestic scope, but clearly many of the issues will have application abroad. Our hope is that this volume will inspire further research on agricultural risk and serve as a guide to where that work may be most productive. Our final thoughts along this line are offered in the concluding chapter.

We wish to thank all those authors who have contributed excellent papers to this work and have gathered to discuss their contributions at the annual meetings of Regional Agricultural Experiment Station Project SERA-IEG-31. We thank collectively the discussants who provided the excellent critical discussions at those meetings, which have inspired further refinements in the papers, including the following: **Brian D. Wright**, University of California, Berkeley; **Loren W. Tauer**, Cornell University; **Matthew T. Holt**, North Carolina State University; **Shiva Makki**, Economic Research Service; **Paul D. Mitchell**, Texas A&M University; **Meredith Soule**, Economic Research Service; **Bruce L. Ahrendsen**, University of Arkansas; and **Agapi Somwaru**, Economic Research Service. We especially want to thank Ms. Liesl Koch of the University of Maryland who provided the excellent technical editing of this volume. Without her work the quality would be far from what it is.

Richard E. Just, University of Maryland
Rulon D. Pope, Brigham Young University
July, 2001

Part 1

BEHAVIOR UNDER RISK: GENERAL CONCEPTS AND THEIR SIGNIFICANCE FOR AGRICULTURE

Chapter 1

EXPECTED UTILITY AS A PARADIGM FOR DECISION MAKING IN AGRICULTURE

Jack Meyer
Michigan State University

INTRODUCTION

The assumption that a decision maker maximizes expected utility has been, and still is, a frequently employed model specification. This is true in economics and in agricultural economics. Decision models, where maximization of expected utility is the goal of the decision maker, have developed significantly during the forty or so years they have been in use. There now exists a substantial set of definitions, theorems, and empirical procedures available to those applying this paradigm. The goal of this chapter is to briefly describe the development of this expected utility (EU) decision model and to describe in some detail its current state.

The discussion begins with an overview that applies to economics in general, but has interspersed a number of comments specific to agricultural economics. Following this overview is a more detailed presentation of specific details of EU decision modeling as it is practiced today. This latter discussion is divided into three segments, each focusing on a different component of the EU model. The chapter concludes with discussion of an important and often overlooked fact that is essential to recognize when using an EU decision model in applied analysis.

AN OVERVIEW

The traditional date used as the beginning of modern EU decision theory is 1944, when von Neumann and Morgenstern's *Theory of Games and Economic Behavior* was published. During the twenty-year period following this event, 1944-1963, the hypothesis that individuals maximized expected utility was extensively discussed and debated by economists, but used surprisingly little in the modeling of economic decisions. Significant progress was made in developing and understanding the axiomatic basis for the expected utility maximization hypothesis, but progress in using this hypothesis in modeling economic behavior was much more limited.

During this period the implications of EU for the measurability of utility were intensely debated, and a few hypotheses concerning the shape of the utility function were proposed. Allais (1953), and a few others, were vocal skeptics, pointing to examples where the EU hypothesis is violated, but for the most part the hypothesis was accepted. As it is now, the independence axiom was the focus of this discussion. Also during this period Savage (1954) combined into a single set, axioms yielding both probability and expected utility as ways to represent and choose among random alternatives. Finally, it was also during this period that Tobin (1958) popularized the mean-variance (MV) approach in his work concerning the portfolio aspects of holding money.

During the 1950s and 1960s, EU was used little in agricultural economics. The modeling and analysis of decisions in agricultural economics is typically oriented toward application, and the EU decision model of the time was not very well suited to applied work. The risk analysis procedures used in agricultural economics instead were quadratic programming and other computer-based analytical tools. These latter decision models usually lead to a specific recommended course of action, or at least identify only a very small number of possible alternatives. It is also the case that empirical implementation of risk analysis was better developed for ranking procedures which use means, variances, and correlations to represent risky alternatives. For many of these same reasons, quadratic programming and other computer-based decision models are still in use today.

During the period 1963-1971, EU changed from being a term used to describe a ranking criterion, to one that instead represents a complete modeling framework. The term EU now refers to the analysis of an economic model under the assumption that expected utility is maximized. This change in terminology began to occur gradually in the mid and early 1960s and was partially instigated by Arrow's (1965) three foundational lectures on risk. These lectures provided definitions for various measures of risk aversion. In these lectures, Arrow also illustrated the use of the expected utility hypothesis in the analysis of economic models. He showed that EU could be used to predict and explain economic decisions. Arrow's analysis involved portfolio and insurance decisions and these models and his results are still being extended in work carried out today. Contemporaneous with Arrow, Pratt proposed measures of absolute and relative risk aversion that are similar to those of Arrow, and these definitions, and extensions of them, are still an important aspect of EU model analysis. The Pratt-Arrow measures of risk aversion are in such wide use that they have become part of instruction to first-year doctoral students studying economics.

The contributions of Arrow and Pratt led to a large number of mainly theoretical papers concerning economic decisions involving risk. During the ensuing eight or so years, that research firmly established the expected utility ranking criterion as a workable feature of an economic model. The many papers published from 1965-1971 include those of Hadar and Russell (1969) and Hanoch and Levy (1969) defining stochastic dominance, and those of

Rothschild and Stiglitz (1970, 1971) defining increases in risk. Also during this time period, papers by Sandmo (1971), Leland (1972), and Baron (1970) examine the behavior of the firm under risk. These latter papers serve to illustrate the EU approach, but also drew the attention of agricultural economics toward EU because of their focus on the firm. Maximization of expected utility of profit is still the most commonly assumed goal for the agricultural firm. With this theoretical base in place, EU began to be used as an applied risk analysis tool, especially among agricultural economists working and studying in Australia. This work culminated in an important book summarizing the use of EU to model decision making by agricultural producers, Anderson, Dillon, and Hardaker's *Agricultural Decision Analysis*, published in 1977.

By 1971, EU was clearly regarded as a decision model rather than just a ranking criterion. EU decision models displayed many common features. Although multi-argument utility functions were sometimes used, most often utility was assumed to depend on a single outcome variable. This outcome variable in turn was modeled as depending on random parameters and on choices made by the decision maker. The probability distributions describing these sources of risk were viewed as known to the decision maker. This total framework, in addition to the assumption that expected utility was maximized, was referred to as an EU decision model.

From 1972 to approximately 1981, economists mostly refined and extended the theoretical framework for analysis of EU decision models, while in agricultural economics the applied/empirical aspects of EU modeling were extensively examined. When the definitions of stochastic dominance were extended in ways that made application more feasible, these extensions were used extensively in agricultural economics. The issues of hypothesis testing and estimation error within the EU framework of analysis were a major concern. Jointly estimating production functions and producer risk aversion measures became standard analysis procedures in agricultural economics.[1] Even though EU was gaining wide acceptance and use, agricultural economics also continued to refine and use the mean-variance (MV) approach. In part this is due to its simplicity, but also because MV is better suited for dealing with multiple sources of risk.

During this same time period the focus in the analysis of an EU model shifted from the ranking of alternatives to comparative static analysis. Special types of increases in risk were identified that were particularly suitable for determining their effect on choices made by decision makers. Hypotheses concerning risk-aversion measures were proposed and defended on the basis of comparative static theorems whose proofs depended on those hypotheses. In order to use the Pratt and Arrow risk-aversion measures, the multi-argument form for the utility function was replaced with one that combines all

---

[1] Meyer (1977), King and Robison (1981), Pope and Ziemer (1984), and Just and Pope (1978) are early references on these topics.

arguments into a single outcome. Within economics, MV decision models were mostly pushed aside and are used very little even today. Finally, it was also during this period of the 1970s that carefully obtained experimental evidence clearly indicated that the expected utility hypothesis, especially the independence axiom, was violated systematically and under a variety of conditions.

Beginning in 1981, Machina published a series of papers describing and illustrating the tractability of non-expected utility alternatives to EU decision models. This work suggested that EU could be replaced by a more general and yet tractable alternative. Possible alternatives, including rank-dependent preferences, are ranking criteria that clearly could resolve some of the experimentally obtained violations of EU.[2] Other chapters in this volume discuss this work in more detail. A consensus was developing that EU was dying if not already dead, and this perception continued for several years.

Beginning in the late 1980s, two distinct developments changed the perception that EU was dead. One development was instigated by Pratt and Zeckhauser (1987) and added to by many others. This work suggested that rather than being overly restrictive in imposing the independence axiom, EU decision modeling was not restrictive enough in that more assumptions concerning the utility function should be considered. This possibility of imposing additional assumptions on utility led to a renaissance of analysis of EU models of both general and specific forms. Numerous additional definitions, theorems, and comparative static findings were presented, each suggesting and making use of additional rather than fewer restrictions on the EU model.

Concurrent with this resurgence of EU analysis, a number of review papers were published evaluating the EU decision model and its potential non-EU replacements. Primary among these is a paper by Machina (1987) in the *Journal of Economic Literature*. In short, these reviews indicate that EU was far from dead, and that despite its flaws, EU was likely the best currently available model for analysis of decisions under risk. This is still the view among many if not most doing applied research concerning decision making under risk. The next section discusses the recent refinements to the EU modeling framework and attempts to specify the current state of EU decision modeling.

THE CURRENT STATE OF EXPECTED UTILITY MODELING

The EU decision model has three main components and the discussion concerning its current state is divided into sections focusing on each of them. The first component requires only brief discussion and concerns the nature of the argument or arguments of utility. The second component is the utility function, and deals with assumptions concerning risk-aversion measures, and

[2] Quiggin (1982) defines rank-dependent preferences.

this discussion is considerably longer. Finally, the section concludes with a description of restrictions on changes in the random parameter that allow the effects of the change to be evaluated. For each of the three components, the restrictions that are listed should be viewed as ones whose purpose is to allow sharper conclusions to be drawn from the analysis of the EU model. Of course, the cost of these sharper conclusions is that the model is applicable to fewer decision makers and situations.

In the 1960s and early 1970s, it was not unusual to assume that the utility function used to calculate expected utility depended on a vector of variables, as is the case for the usual consumer utility function. For instance, Rothschild and Stiglitz (1971) use the general utility function $u(x, \alpha)$ in their analysis of the effect of an increase in risk. In this model, x is a random parameter and α is chosen by the decision maker, and the decision maker's expected utility depends on each. At the same time, however, a less general but more convenient formulation was also used. In this formulation, utility depends only on a single argument denoted z, which in turn depends on random variable x and choice variable α. This assumption concerning utility implies that the random parameter and the choice variable interact in the same way for each decision maker, and that this interaction does not depend on the decision maker's risk preferences. Thus, any differences in behavior between individuals are the result of differences in risk preferences. The outcome variable $z = z(x, \alpha)$ is the same for all, although utility from that outcome, $u(z) = u[z(x, \alpha)]$, does differ across decision makers.

From the mid 1970s on, this less general but more convenient single argument assumption has been used in the vast majority of EU decision models. As mentioned earlier, one reason for this is that the Pratt-Arrow definitions of measures of risk aversion are for single argument utility functions, and are difficult to extend to the multi-variable case. The restriction to a single argument is a sensible one when indeed all decision makers receive the same single dimension payoff. This is the case for portfolio models where wealth is that payoff, or for models of the firm where the payoff is profit. In cases where a single dimension payoff is more difficult to identify, the restriction can be interpreted as defining $z(x, \alpha)$ to be the utility for the decision maker who is risk neutral. This implies that $u(z)$ then represents the extent to which the decision maker's risk preference deviates from those risk-neutral preferences.

The $z(x, \alpha)$ function is often restricted in ways that facilitate analysis. The discussion that follows is for the case of one random parameter and one decision variable, although usually the restrictions have a natural extension to cases where there is more than one of each. The outcome variable is commonly assumed to be monotonic in the random parameter and concave in the decision variable; that is, $z_x(x, \alpha) \geq 0$ and $z_{\alpha\alpha}(x, \alpha) \leq 0$ for all x and α. For some random parameters such as rainfall, monotonicity of preference is not a useful assumption, but for many parameters such as price or rate of return, it is. Being monotonic in x is an important property when one attempts to determine the effect of altering this random parameter. Concavity in the choice variable is important for second-order conditions and interior solutions. This assumption is almost always an acceptable one because the EU model is typically formulated by adding randomness to an existing model where concavity in α is already assumed.

Additional assumptions sometimes made concerning $z(x, \alpha)$ include $z_{\alpha x} \geq 0$, $z_{\alpha xx} \leq 0$, and $z_{xx} \leq 0$. For the most part these assumptions are made when the proof of a comparative static theorem is facilitated by the assumption. Most of the specific economic models that are analyzed, such as the portfolio model, the firm under output price risk, and the insurance demand model, satisfy each of these assumptions. In fact, for each of these three decision models, the outcome is linear in x and concave in α. For agricultural economics, with its focus on agricultural producers, the assumption of a single argument utility function is a standard one. Most often profit is that outcome variable, although wealth is also frequently used. Which of these to use and how to measure it is discussed further in the last section of this chapter.

The second component of an EU model that is often restricted in order to yield sharper conclusions is the utility function itself. The discussion presented next is for the case where utility depends on a single argument. Because the utility function is unique to a positive linear transformation, from the very beginning, restrictions on utility have specified either the sign of various derivatives of utility, or the magnitude of ratios of these derivatives. It should be recognized, however, that many of the restrictions could be stated without assuming that $u(\cdot)$ is differentiable to the required degree.

The first obvious restriction placed on $u(z)$ results from the practice in economics to focus on goods rather than bads. This yields $u'(z) \geq 0$ as a natural assumption and is consistent with z representing such outcomes as wealth or profit. Models where smaller values of the outcome variable are preferred to larger ones are easily transformed to meet this restriction by redefining the payoff variable. Situations where preferences are not monotonic in the outcome variable are less easily handled, and this restriction eliminates dealing with those cases. The assumption that $u'(z) \geq 0$ is imposed in virtually all EU modeling.

The property of risk aversion is also almost always imposed, and requires that $u(z)$ be concave, that is, $u''(z) \leq 0$. This assumption is commonly imposed for two very different reasons. First, risk aversion is necessary to explain commonly observed behavior in situations involving risk. Diversification of a portfolio or purchase of insurance would not occur without risk aversion, nor would there be the positive correlation between average rate of return and risk that data suggest. The second reason for assuming that $u(z)$ is concave is more pragmatic. This assumption is important and necessary when solving and analyzing the decision model using traditional differential calculus methods for finding an interior solution. While it is clearly the case that some decision makers do behave in ways that are not consistent with risk aversion, for the most part the profession has paid little attention to modeling or explaining this behavior, and $u''(z) \leq 0$ is almost always assumed.

The assumptions concerning $u(\cdot)$ and its derivatives, which remain, are not ones that are universally imposed. Interestingly and perhaps surprisingly, the main and often only support for several of these assumptions comes from examination of the conclusions they allow to be drawn in the theoretical analysis of a model. This "proof by comparative statics" is unusually common in testing hypotheses in EU decision models. For some utility properties, there is also modest empirical support for the assumption. This empirical support takes the form of observations consistent with the comparative static theorems derived in theory, or sometimes by direct estimation of the utility function itself. The various assumptions on utility are discussed approximately in the order in which they were introduced, although this is usually also from least to most restrictive and in order of acceptability.

Arrow (1965) and Pratt (1964) each discuss the assumption of decreasing absolute risk aversion (DARA) as a sensible restriction to impose on $u(z)$. They propose this restriction for the situation where variable z represents the wealth of the decision maker. This assumption, that $A(z) = -u''(z)/u'(z)$ is monotonically decreasing in z, has been demonstrated to lead to sensible conclusions in a variety of decision models. For instance, DARA implies that increases in wealth lead to increases in the holding of the risky asset in the two-asset portfolio model. In contrast, the reverse assumption, that $A(z)$ is increasing, implies the opposite. For the competitive firm, Sandmo (1971) shows that DARA for utility of profit is required if an increase in fixed cost is to lead to lower rather than higher output levels.

In the almost forty years since DARA was first proposed, at least a hundred theorems have used this assumption to derive conclusions that appear to be consistent with observation. While empirical testing of this and other properties of utility is very limited and many times difficult to interpret, consistent evidence against DARA has not been found, and some evidence does support DARA. As a consequence of these two types of evidence, it is very acceptable to assume that $u(z)$ displays DARA. This is certainly true when z represents wealth, but is also the case for many other outcome variables as well.

Arrow (1965) also proposed the assumption of increasing relative risk aversion (IRRA); that is, he assumed that $R(z) = -u''(z)\cdot z/u'(z)$ is an increasing function. In proposing this, Arrow clearly indicates that the outcome variable z represents wealth. This assumption also has some support on the basis of implied comparative static results, but far fewer theorems use IRRA than is the case for DARA. In the portfolio model, for instance, IRRA implies that as wealth increases, the portfolio contains proportionally less of the risky asset, a finding that is often but not always supported in empirical work. Perhaps the most useful and important feature of the IRRA assumption is that it is a way to limit the rate at which the absolute risk aversion measure can decrease. If the choice is to assume IRRA or its opposite, the more sensible comparative static findings support IRRA. Most researchers, however, would expend considerable effort and make other assumptions in order to not employ this assumption. The acceptability of IRRA for outcome variables other than wealth is even more in doubt.

Another assumption concerning relative risk aversion proposed by Arrow (1965) is that its magnitude be less than one, $R(z) \leq 1$, where z represents wealth. This assumption has proven to be a powerful and useful one in comparative static analysis in a variety of EU decision models. For state preferences models satisfying EU, state contingent commodities are gross substitutes (complements) if and only if $R(z) \leq (\geq)\ 1$. Equivalently, own price of elasticity of demand for state contingent commodities exceeds one if and only if $R(z) < 1$. Although the usefulness and sensible comparative static implications of the $R(z) \leq 1$ assumption is not in question, empirical evidence does not seem to support the assumption. This may be due to measurement problems that are discussed in the last section of the chapter. Certainly no strong empirical evidence for $R(z) \leq 1$ exists, and there is some evidence to the contrary. As with IRRA, in theoretical analysis, the assumption that $R(z) \leq 1$ is more acceptable than its opposite, but making no assumption at all is best of all. For outcome variables that are not wealth, the assumption has little support even on the theoretical side.

Kimball (1990) introduced the concept of "prudence," defined by $u'''(z) \geq 0$, and a measure of absolute prudence defined by $P(z) = -u'''(z)/u''(z)$. Prudence is a weaker restriction than DARA and is in many cases equally useful in demonstrating comparative static theorems because the concavity or convexity of $u'(z)$ is an important feature of that analysis. Prudence is a very acceptable and supported assumption simply because DARA is. The magnitude of absolute prudence and also the assumption of decreasing absolute prudence are also sometimes useful in obtaining sharper predictions. Decreasing prudence combined with DARA is referred to as standard risk aversion. These conditions on risk preferences are sufficiently new, however, to have little evidence supporting them other than a limited set of comparative static theorems that do support the hypothesis.

Kimball (1990) has also defined the word "temperance" by $u''''(z) \leq 0$. This assumption also is useful in some comparative static analyses but is too unexplored to have support at this time. Both prudence and temperance are consistent with an assumption suggested many years ago that the derivatives of utility should alternate in sign with the first being positive, an assumption which has not been employed in economic analysis to any great extent.

The terms "proper risk aversion" and "risk vulnerability" also deserve mention (see Pratt and Zeckhauser 1987 and Gollier and Pratt 1996). Each of these restrictions on utility was proposed in order to prevent combinations of independent risks from being improvements on the individual components.

Proper risk aversion eliminates the possibility that two undesirable and independent risks, when added together, yield a risk that is desirable. Risk vulnerability ensures that the addition of an independent risk with zero mean cannot change an existing undesirable risk into a desirable one. Neither definition was proposed with comparative static analysis as its main purpose, and neither has been used very much in model analysis.

Finally, other measures of risk aversion, such as the partial risk aversion measure, have been proposed. These measures are a one-to-one transformation of either the absolute or relative risk aversion measures of Pratt and Arrow, and hence little seems to be gained from the additional terminology. As a consequence, measures other than absolute and relative risk aversion have been used very little.

One of the main comparative static questions asked in EU model analysis is: What is the result of a change in the random parameter or a change in the random outcome? Focusing on only this question allows discussion of the third and final component of an EU decision model. For a change in the random outcome z, the effect on the expected utility of the decision maker is of main concern. On the other hand, for a change in the random parameter x, the effect on the decision that is made is primary. Before describing restrictions on changes in random variables, a few comments are needed concerning their representation and some commonly assumed restrictions.

It is typical in EU models in economics to describe a random variable by giving its cumulative distribution function (CDF), and $F(x)$ and $G(x)$ are commonly used to denote these. This practice of using CDFs rather than probability functions allows both discrete and continuous probability to be described by the same notation. In addition, to avoid the possibility of infinite or undefined levels of expected utility, the support of the random variable is usually assumed to be bounded and denoted either $[a, b]$ or sometimes $[0, 1]$. When the random variable is discrete, the number of outcomes is typically assumed to be finite. The assumptions of a bounded support or a finite number of outcomes can often be relaxed, but doing so complicates the analysis.

The first restrictions proposed for changes in CDFs are the stochastic dominance definitions that place conditions on the difference between a pair of CDFs. Letting $F(x)$ and $G(x)$ denote two CDFs, first-degree stochastic dominance (FSD) of $F(x)$ over $G(x)$ is defined by $[G(x) - F(x)] \geq 0$ for all x in $[a, b]$. FSD expresses the fact that under $F(x)$ the random variable is "bigger" than under $G(x)$. Similarly, second-degree stochastic dominance (SSD) of $F(x)$ over $G(x)$ is defined by $\int_a^x [G(s) - F(s)]\, ds \geq 0$ for all x in $[a, b]$, and SSD indicates that the variable is bigger and/or less risky under $F(x)$ than $G(x)$. In order to focus on only the risk associated with the random variable, Rothschild and Stiglitz (1970) define $F(x)$ to be less risky than $G(x)$ if their means are equal, and if $\int_a^x [G(s) - F(s)]\, ds \geq 0$ for all x in $[a, b]$. Thus, their definition involves SSD plus the equal means restriction.

These three definitions, all from the period 1969-1970, were formulated to answer the question of how a change in the random outcome z affects expected utility. In each instance, the definition gives a restriction that is necessary and sufficient for expected utility to change in a predictable way for a broad group of EU decision makers defined by assumptions on the first two derivatives of their utility functions. Additional stochastic dominance definitions have been proposed and used. Third-degree stochastic dominance adds a restriction concerning the third derivative of utility. Another form of stochastic dominance, used more often in agricultural economics than in economics, restricts the set of decision makers by specifying upper and lower bounds on their measures of absolute risk aversion (Whitmore 1976 and Meyer 1977, respectively). This particular definition makes the stochastic dominance ranking procedure better suited for applied research and explains its use in agricultural economics. All in all, the question concerning the effects of changes in random outcomes on the level of expected utility was mainly resolved by the late 1970s and little has been published on the topic since then.

Beginning in the 1970s and continuing to the present, the focus in EU model analysis has shifted to the question of how a change in the random parameter x affects the decision made by an economic agent. This question was addressed by Rothschild and Stiglitz (1971). They asked how an increase in the riskiness of the random parameter alters the choice made by the decision maker. This same question has since been asked for FSD and SSD changes as well.

The answers to these general questions concerning the effect of FSD, SSD, or increases in risk for x on the choice of α are not very encouraging. Research has made it clear that many additional restrictions on utility are needed if one is to ensure that α is to change in a predictable way. The restrictions on utility described in earlier paragraphs are some of these conditions, but often even stronger assumptions on utility are needed. In order to avoid imposing unacceptably strong conditions on utility, restrictions have been proposed for the change in the random parameter instead. In most cases, the practice is to further restrict the original FSD, SSD, or increase in risk definitions.

A number of special increases in risk have been identified as yielding sharp comparative static results. Each further restricts the Rothschild and Stiglitz definition. Most focus on specifying additional conditions on $[G(x) - F(x)]$, the difference between the pair of CDFs. Terms such as "strong" and "relatively strong" increases in risk have been introduced to indicate that the increase in risk under consideration is more narrowly defined than that of Rothschild and Stiglitz. Recently, Gollier (1995) consolidated and completed a portion of this literature by determining a necessary and sufficient condition for a change in a random parameter to have a determinate effect on the choice made by all risk-averse decision makers. He used the term "greater central riskiness" for the special case where the decision model is linear in the random parameter. Although there are still many unanswered questions, much progress has been made toward the continuing goal of determining for which increases in risk and for which decision makers does the change in risk cause the decision variable to be adjusted in a predictable direction.

For FSD changes in the random parameter, the literature has followed a similar path. The question of interest is for which special FSD changes and for which decision makers is the change such that the decision variable is adjusted in a predictable direction. The answers given for this question, however, have a somewhat different formulation. Rather than further restricting the difference between $F(x)$ and $G(x)$, a restriction is placed on ratios instead. The monotone likelihood ratio condition (MLR) restricts F'/G', and the monotone probability ratio condition (MPR) restricts F/G. In each case, a special FSD change is identified as yielding determinate comparative static conclusions for a rather broad group of decision makers (Landsberger and Meilijson 1990 and Eeckhoudt and Gollier 1995). While this literature is less complete than that for increases in risk, a substantial and growing body of work does exist to help answer the question concerning the effect of an FSD change in a random parameter. Since SSD changes are a combination of an FSD change and a decrease in risk, they are rarely dealt with separately.

Finally, changes in risk and FSD improvements have been modeled for more than forty years using a transformation of the random parameter to specify the change that occurs. Sandmo (1971) does this in his analysis of the competitive firm, and his procedure has since been generalized to include more general changes in random parameters than those that he considered (Meyer and Ormiston 1989). In this transformation approach, random variable x is replaced by $[x + \theta \cdot k(x)]$ for some function $k(x)$. The effect of changing θ from zero to a positive value is then determined. When $k(x)$ is positive, increases in θ are an FSD improvement in x, and for other appropriately defined $k(x)$ functions, the change is an increase in the riskiness of x instead. Using this approach, special FSD improvements and increases in risk have also been identified so that sharp comparative static conclusions can be drawn.

AN APPLICATION ISSUE AND CONCLUSIONS

In this section a fact that is very important to recognize when using the EU decision model framework in an applied setting is pointed out and discussed. Although the discussion is quite brief, its purpose is to partially explain why it is that empirical evidence concerning utility functions and risk aversion measures is so limited, hard to interpret, and seems to vary widely from study to study. That is, the discussion attempts to offer a reason why, after nearly thirty years of attempting to determine the risk-taking characteristics of agricultural producers, we seem to know so little. The same discussion also serves as a partial explanation of why restrictions on utility are more often supported by the comparative static theorems they yield than by direct empirical tests.

The fact that must be recognized when specifying preferences in an EU decision model is that those preferences depend critically on the precise definition of, or measure used for, the argument of the utility function. It is common in economics and agricultural economics to use words such as "wealth," "consumption," or "profit" to describe the argument of utility, but it is less common to clearly indicate what those words represent or how they are measured. The general question addressed here is when and how this matters. The specific question that is addressed is: How does the way in which the variable entering the utility function is defined or measured affect the properties of utility?

The discussion begins with a brief derivation that is well understood but serves to illustrate the relationships that form the basis for the analysis. When an EU decision model specifies utility to be $u(z)$ and defines $z = z(x, \alpha)$ to be the outcome variable, if $z(x, \alpha)$ is not linear in x, then even the risk-neutral person's preferences for x, the random parameter, depend on more than its mean value. This fact has been recognized in many contexts and has been used to explain why such things as a bankruptcy provision can lead to seemingly risk-loving behavior even for risk-neutral or risk-averse decision makers. As indicated earlier, the function $z(x, \alpha)$ is the utility for the risk-neutral decision maker, and hence its properties determine that decision maker's preferences for x.

Another implication that is derived from this same specification, and seems to be less recognized, is the fact that when two different outcome variables are specified for the same decision maker, then the two utility functions for that person must also be different. Moreover, the relationship between these two utility functions representing one person's risk preferences is determined by the relationship between the arguments of utility. This particular fact is illustrated with three examples, each of which has particular relevance when applying EU decision models to agricultural producers. Throughout the discussion of the three examples, the implications for applied EU analysis are emphasized.

The first example involves two different EU representations of the same agricultural firm, where the difference arises because of a differing definition or measure of profit. To introduce notation, suppose the first EU model is for a firm-maximizing expected utility from profit denoted π, while the second uses a different measure of profit π_1 which is larger than π by an amount c. Thus, $\pi = \pi_1 - c$. This constant term c could represent an opportunity cost that accountants do not include in defining profit, while economists do, so π is economic profit and π_1 is accounting profit. Alternatively, π and π_1 can simply differ because one empirical study has more complete data concerning costs than another. An example of a cost that illustrates either situation is the cost of owner-supplied farm labor. For accountants this is not a cost if it is unpaid, and also this cost is frequently noted as specifically excluded from the calculation of profit because the data are not available.

The firm can be modeled as maximizing expected utility either from $U(\pi)$ or from $u(\pi_1)$, and exactly the same behavior is represented as long as the two utility functions satisfy the relationship $u(\pi_1) = U(\pi_1 - c)$. This says that if cost c is not subtracted from revenue in determining profit, this fact can be adjusted for by replacing utility function $U(\cdot)$ with $u(\cdot)$. This point is rather obvious, but what may be less clear is that this adjustment requires that both the level of risk aversion and the functional form for utility be adjusted. This is demonstrated by comparing the risk-aversion measures for $U(\pi)$ with those for $u(\pi_1)$. These risk-aversion measures are obtained by differentiating the identity $u(\pi_1) = U(\pi_1 - c)$. Space considerations prevent displaying the derivations so only the results are listed.

In this analysis, the same agricultural producer is represented in two models whose only difference is the measures of profit. In these two models, it can be demonstrated that $u(\cdot)$ is increasing, and/or concave, and/or displays DARA if and only if $U(\cdot)$ has those same properties. Thus, when only these restrictions on utility are of interest, the fact that profit is measured in two different ways is inconsequential. For the majority of theoretical analyses of the firm these are the assumptions on utility, and thus how profit is defined is unimportant. For relative risk aversion measures, and for choice of the functional form for utility, however, the variation in the way profit is measured is of much greater consequence, and this is discussed next.

For empirical studies of the agricultural producer a functional form for utility is often chosen and then parameters in that form are estimated. In addition, measures of relative risk aversion are usually reported because these measures are unit free. It can easily be shown that the relative risk aversion measure for $u(\pi_1)$ is not the same as that for $U(\pi)$, in fact is equal to the relative risk aversion measure for $U(\pi)$ multiplied by the factor $[\pi_1/\pi]$. This extra factor implies that the magnitudes of the two relative risk aversion measures are not the same, and also that the slopes for the two measures are different. Relative risk aversion for utility from the more inclusive measure π_1 is larger than that for utility from the less inclusive π. Similarly, since the multiplicative factor $[\pi_1/\pi]$ is decreasing in π_1, the relative risk aversion measure for $u(\pi_1)$ is less steeply sloped than that for $U(\pi)$.

Suppose it is known that a cost is omitted when profit is reported or calculated in the data being used in an application of the EU decision model. This discussion indicates that the form chosen for the utility function must reflect the fact that the slope of the relative risk aversion is less steeply sloped compared with what would be the case if the cost were not omitted. Thus, if CRRA is the appropriate form to estimate for utility from economic profit, as some would argue, then a decreasing relative risk averse form must be chosen when estimating $u(\pi_1)$ from accounting profit data. In addition, the estimate for the magnitude of relative risk aversion is larger than would be the case for correctly measured profit.

It is the case that estimates of the magnitude of relative risk aversion for utility from profit determined in agricultural economics are almost always significantly larger than one, and often between three and five. Even if omitted costs c are not precisely known, sometimes an approximate magnitude can be determined, and this information can be used to determine the degree to which the omission impacts the measure of relative risk aversion. For instance, if omitted family labor costs comprise half to two-thirds of an agricultural producer's accounting profit, this is sufficient to yield a relative risk aversion magnitude near one for utility from economic profit. Of course, this same discussion is relevant when comparing results across studies where different measures of profit may be used. Adjusting for these differences could well reduce the variation in the reported estimates.

The second example where one decision maker's risk preferences are represented by two different utility functions comes from a multi-period consumption model. While this model is not directly related to agricultural producers, a suggestion near the end of the discussion indicates that perhaps it should be. In this multi-period consumption model, the relationship between utility for wealth and utility for consumption is a well-known one so only a brief description is provided.

Utility from wealth is a derived or indirect utility, defined to be the maximum utility from the consumption that the wealth allows. When utility is of the additive separable form, this implies that $u(W) = v(C_1) + \Sigma_{j=2}^{\infty} v(C_j)$, where the consumption levels are those that maximize utility. One can apply the envelope theorem to this equation, and the additive separable nature of utility from consumption implies that the two utility functions, $u(W)$ and $v(C_1)$, have the same slope at the optimum; that is, $u'(W) = v'(C_1(W))$ holds as an identity. $C_1(W)$ is the optimal consumption level in the first period. The relationship between the risk-aversion measures for these two utility functions follows directly from this identity. The absolute and relative risk aversion measures for $u(W)$ and $v(C_1)$ are related by the equations $A_u(W) = A_v(C_1)[dC_1/dW]$ and $R_u(W) = R_v(C_1)[dC_1/dW][W/C_1]$.

Many different things can be demonstrated in this model and a few of those results are mentioned. Some things are general. For instance, $u(W)$ is less risk averse than $v(C_1)$, but $u(W)$ does display risk aversion if $v(C_1)$ does. It is also the case that the two utility functions each display constant absolute risk aversion (CARA) if and only if optimal consumption is linear in wealth, and each displays constant relative risk aversion (CRRA) if and only if optimal consumption is proportional to wealth. Finally, it is possible that $u(W)$ displays CRRA even though $v(C)$ does not.

Other relationships between the risk-aversion measures for the two utility functions depend on the exact specification of the model. In one specification, where a portion of consumption comes from wealth and the remainder from labor income, it can be shown that the magnitude of the relative risk aversion measures for utility from wealth is smaller than that for utility from consumption. In fact, when 80-90 percent of consumption comes from labor income and not wealth, the one is approximately ten times larger than the other. This may explain why macroeconomists using asset pricing models based on consumption over time use measures of relative risk aversion for consumption on the order of ten, while researchers in finance examining portfolio issues using expected utility from wealth, use relative risk aversion measures near one.

The reason this example is discussed is twofold. First, it is another case where confusion concerning the argument of utility has led to seemingly very disparate empirical findings. In this case it is not how the variable is measured that differs across studies, but which variable is viewed as the argument of utility. It is likely that models of the agricultural producer would benefit from discussion of why profit is the argument of utility and how utility for profit relates to utility from consumption and/or utility from wealth. When the relationship between profit, consumption, and wealth is correctly modeled, it may well be that the risk-aversion measures for agricultural producers do not differ from those of consumers in general.

The third and final example of multiple, but related, utility functions for a single decision maker arises simply because the decision being modeled is repeated in successive time periods, a circumstance faced by most agricultural producers. The example used is a portfolio model. Assume an investor begins with wealth W_1 which can be invested in the risky or riskless asset to yield wealth $W_2 = W_1[\alpha r + (1-\alpha)\rho]$. In this notation, r is the risky return, ρ the riskless return, and α the proportion of wealth invested in the risky asset.

In this model, utility for wealth at the beginning of a period is the expected utility this wealth can generate at the conclusion of the period, $v(W_1) = \mathrm{E}u(W_2) = \mathrm{E}u(W_1[\alpha r + (1-\alpha)\rho])$. If this model is extended by adding a time period before period 1, $v(W_1)$ is the utility function whose expected value is maximized in that earlier period. Hence comparison of the risk-aversion properties of $v(W_1)$ and $u(W_2)$ is relevant to the question of how the representation of risk preferences is altered by the fact that the decision is to be repeated.

Using $v(W_1) = \mathrm{E}u(W_1[\alpha r + (1-\alpha)\rho])$, one can show that the two utility functions are not generally the same. It is the case that if $u(\cdot)$ displays CARA, then so does $v(\cdot)$, but even then the magnitude of the absolute risk aversion coefficient need not be the same. It is also the case that when $u(\cdot)$ displays CRRA, then so does $v(\cdot)$.

The point to be extracted from these three examples is a simple one. The agricultural producer has but one set of risk preferences. How these preferences translate into attributes of the utility function, however, depends critically on what is selected as the argument of utility and how it is measured.

One can derive information concerning those risk preferences for any argument of utility that is convenient, and with appropriate adjustment translate that information so that it can be compared with similar information on risk preferences determined for utility with a different argument. In symbols, if $u(w)$ and $v(z)$ represent the utility functions for the same decision maker when w and z are the arguments, and z and w are related by $z = z(w)$, then $u(w)$ must equal $v(z(w))$. With this formulation, information concerning any two of the three functions involved is sufficient to determine the third. In the three examples given here, $z(w)$ is determined by the model, and as a consequence once either $u(w)$ or $v(z)$ is specified or estimated, the other is also determined.

Hopefully, this chapter has provided a sense of how the EU decision model has developed through time and how it is used today. Particular emphasis was given to description of how EU is used to describe the behavior of agricultural producers. In addition, an important fact to recognize when applying EU modeling assumptions to agricultural producers was discussed in some detail. Many other areas where the EU model can be improved are not discussed for space reasons, and further work can make the EU decision model even more suitable for the representation of the agricultural producer.

REFERENCES

Allais, M. 1953. "Le Comportement de l'Homme Rationnel Devant le Risque: Critique des Postulats et Axiomes de l'Ecole Américaine." *Econometrica* 21: 503-546.

Anderson, J.R., J.L. Dillon, and J.B. Hardaker. 1977. *Agricultural Decision Analysis*. Ames: Iowa State University Press.

Arrow, K.J. 1965. *Aspects of the Theory of Risk Bearing*. Helsinki: Yrjo Jahnssonin Saatio.

___. 1971. *Essays in the Theory o Risk Bearing*. Chicago: Markham.

Baron, D. 1970. "Price Uncertainty, Utility, and Industry Equilibrium in Pure Competition." *International Economic Review* 11: 463-480.

Eeckhoudt, L., and C. Gollier. 1995. "Demand for Risky Assets and the Monotone Probability Ratio." *Journal of Risk and Uncertainty* 11: 113-122.

Gollier, C. 1995. "The Comparative Statics of Changes in Risk Revisited." *Journal of Economic Theory* 66: 522-535.

Gollier, C., and J.W. Pratt. 1996. "Risk Vulnerability and the Tempering Effect of Background Risk." *Econometrica* 64: 1109-1123.

Hadar, J., and W.R. Russell. 1969. "Rules for Ordering Uncertain Prospects." *American Economic Review* 59: 25-34.

Hanoch, G., and H. Levy. 1969. "Efficiency Analysis of Choices Involving Risk." *Review of Economic Studies* 38: 335-346.

Just, R.E., and R.D. Pope. 1978. "Stochastic Specification of Production Functions and Economic Implications." *Journal of Econometrics* 7: 67-86.

Kimball, M. 1990. "Precautionary Saving in the Small and in the Large." *Econometrica* 58: 53-74.

___. 1993. "Standard Risk Aversion." *Econometrica* 61: 589-611.

King, R.P., and L.J. Robison. 1981. "An Interval Approach to the Measurement of Decision Maker Preferences." *American Journal of Agricultural Economics* 63: 510-520.

Landsberger, M., and I. Meilijson. 1990. "Demand for Risky Financial Assets: A Portfolio Analysis." *Journal of Economic Theory* 50: 204-213.

Leland, H. "Theory of the Firm Facing Uncertain Demand." *American Economic Review* 62: 278-291.
Machina, M. 1982. "'Expected Utility' Analysis Without the Independence Axiom." *Econometrica* 50: 277-323.
___. 1987. "Choice Under Uncertainty: Problems Solved and Unsolved." *Journal of Economic Perspectives* 1: 121-154.
Meyer, J. 1977. "Choice Among Distributions." *Journal of Economic Theory* 14: 326-336.
Meyer, J., and M.B. Ormiston. 1989. "Deterministic Transformations of Random Variables and the Comparative Statics of Risk." *Journal of Risk and Uncertainty* 2: 179-188.
Pope, R.D., and R.F. Ziemer. 1984. "Stochastic Efficiency, Normality, and Sampling Errors in Agricultural Risk Analysis." *American Journal of Agricultural Economics* 66: 31-40.
Pratt, J.W. 1964. "Risk Aversion in the Small and in the Large." *Econometrica* 32: 122-136.
Pratt, J., and R.J. Zeckhauser. 1987. "Proper Risk Aversion." *Econometrica* 55: 143-154.
Quiggin, J. 1982. "A Theory of Anticipated Utility." *Journal of Economic Behavior and Organization* 3: 323-343.
Rothschild, M., and J.E. Stiglitz. 1970. "Increasing Risk I: A Definition." *Journal of Economic Theory* 2: 225-243.
___. 1971. "Increasing Risk II: Its Consequences." *Journal of Economic Theory* 3: 66-84.
Sandmo, A. 1971. "On the Theory of the Competitive Firm Under Price Uncertainty." *American Economic Review* 61: 65-73.
Savage, L. 1954. *The Foundations of Statistics*. New York: John Wiley and Sons.
Tobin, J. 1958. "Liquidity Preference as Behavior Towards Risk." *Review of Economic Studies* 25: 65-86.
von Neumann, J., and O. Morgenstern. 1944. *Theory of Games and Economic Behavior*. Princeton: Princeton University Press.
Whitmore, G.A. 1970. "Third-Degree Stochastic Dominance." *American Economic Review* 60: 457-459.

Chapter 2

NON-EXPECTED UTILITY: WHAT DO THE ANOMALIES MEAN FOR RISK IN AGRICULTURE?

David E. Buschena
Montana State University

INTRODUCTION

After more than a quarter century of analysis into its predictive value, the validity of the expected utility model (EU) is seriously called into question. These questions are particularly critical for agricultural economists since we have long relied on EU to assess behavior under the pervasive environment of risk in agricultural and natural resource issues. In this chapter, I will review some of the primary violations of EU, assess their implications, and consider various responses put forward in light of them. Questions addressed include:

- What is the nature of behavior violating EU as commonly applied?
- What do these anomalies tell us about modeling behavior under risk and what testable implications can be drawn from models incorporating them?
- Can we still use EU for some risky choices in light of these anomalies; i.e., how robust are the anomalies to (1) the design of the risky questions, (2) real payoffs, and (3) experimental vs. non-experimental risky questions?

THE EXPECTED UTILITY MODEL, LINEAR IN PROBABILITIES AND NON-LINEAR IN WEALTH

Expected utility has been a success story in economics generally, and in agricultural and resource economics in particular. With its very early inception in the 1700s and the development of its formal axiomatic structure by von Neumann and Morgenstern in 1953, EU forms the basis for most predictions for behavior under risk. Expected utility is pervasive in models of insurance purchases, investment pricing, technology adoption, finance, marketing, economic development, resource policy, storage decisions, and for many other issues.

Since its inception, EU has arisen as a structure to define optimizing behavior that is consistent with observation – that is, a positive model. This observed behavior differed from the predictions of an alternative model viewed as normative up to that time. Mathematicians Gabriel Cramer in 1728 and Daniel Bernoulli in 1738 were the first on record to discuss expected utility concepts, which combine non-linear transformations of monetary outcomes in a linear manner with the outcome's probabilities of occurrence to construct a weighted sum providing a cardinal valuation of a risky enterprise. Cramer and Bernoulli's contributions were initiated by a puzzle known as the "St. Petersburg Paradox," devised by Nicholas Bernoulli in 1713. In this puzzle, most people would be willing to sell a risky enterprise that has infinite expected value (with very small and declining probabilities of increasing and very high payments) for a relatively small sum (see Fishburn 1988 for a summary). The "paradox" was the discrepancy between the finite willingness to sell the gamble and its infinite expected value, where expected value maximization was the erstwhile normative model. Although the St. Petersburg paradox dealt with games of chance, its solution had relevance for understanding everyday decisions by merchants, governments, and others making risky decisions in that age.

Expected utility holds that decision makers evaluate a risky discrete event outcome vector $\boldsymbol{x}$ through its expected utility,

$$U(\boldsymbol{x}, \boldsymbol{p}) = \sum_{i=1}^{n} u(x_i) \cdot p_i \,. \qquad (1)$$

In equation (1), the elements of the vector $\boldsymbol{p}$ define the probability of every possible event x_i.[1] While the utility function $u(\cdot)$ over outcomes is generally non-linear – its most common form is concave, reflecting risk aversion – probabilities always enter the EU valuation in a linear fashion. The outcomes $\boldsymbol{x}$ are generally taken to be final wealth, but other outcomes, including changes from a given reference level of wealth, can be specified instead. For example, Friedman and Savage (1948) treat $\boldsymbol{x}$ as gains or losses in income from current levels, allowing simultaneous modeling of (a) insurance purchases over large amounts and (b) gambling over small amounts by the same agent.

It is clear that EU in equation (1) relies on a person's "taste" for risk, an approach that differs from most economic modeling approaches (e.g., Stigler and Becker 1997). EU provides a framework to mathematically characterize

[1] Extensions of EU to continuous probability distributions are straightforward.

and differentiate the willingness of different farmers to take risks. The degree of curvature of $u(\cdot)$, defined as the size of the second derivative of $u(\cdot)$ relative to its first, indicates whether or not an individual will accept a particular risky enterprise (Pratt 1964 and Arrow 1970). For example, farmers with low degrees of relative risk aversion are more likely to plant new and untried crops, forgo insurance, hold unprotected grain stocks or livestock inventories, and in general have portfolios weighted toward risky undertakings. Expected utility rests on a necessary and sufficient axiomatic framework widely recognized as normatively compelling. This axiomatic framework also provides structure for discussing the nature of EU violations. The three axioms in Jensen's (1967) framework are defined for a preference ordering ($\succ$ denoting strict preference, $\sim$ denoting indifference, and $\succeq$ denoting weak preference) for all distributions $\{\boldsymbol{p}, \boldsymbol{q}, \boldsymbol{r}\} \in \boldsymbol{P}$ that are defined over an outcome vector $\boldsymbol{x}$:[2]

Order. $\succ$ is a weak ordering on $\boldsymbol{P}$. This implies that $\succ$, $\succeq$, and $\sim$ are transitive, and also that $[\boldsymbol{p} \sim \boldsymbol{q}, \boldsymbol{q} \succ \boldsymbol{r}] \Rightarrow \mathbf{p} \succ \mathbf{r}$ and $[\boldsymbol{p} \succ \boldsymbol{q}, \boldsymbol{q} \sim \boldsymbol{r}] \Rightarrow \boldsymbol{p} \succ \boldsymbol{r}$.

Independence. $\boldsymbol{p} \succ \boldsymbol{q} \Rightarrow \alpha\boldsymbol{p} + (1\text{-}\alpha)\boldsymbol{r} \succ \alpha\boldsymbol{q} + (1\text{-}\alpha)\mathbf{r}$ for any $\alpha \in (0,1)$.

Continuity. $[\boldsymbol{p} \succ \boldsymbol{q}, \boldsymbol{q} \succ \boldsymbol{r}] \Rightarrow [\alpha\boldsymbol{p} + (1\text{-}\alpha)\boldsymbol{r} \succ \boldsymbol{q}$, and $\boldsymbol{q} \succ \beta\boldsymbol{p} + (1\text{-}\beta)\boldsymbol{r}$ for some $\alpha\ \beta \in (0,1)]$.

The *order* axiom provides consistency for multiple lottery comparisons, particularly through its transitivity requirement. Further, Fishburn (1988) holds that the order axiom provides the basis for economic rationality across risky choices, and that violations of it are generally seen as aberrations. The *independence* axiom is necessary for the linear treatment of probabilities and closely relates to substitution principles, cancellation conditions, additively axioms, and sure-thing principles. Violations of the independence axiom have been well documented, and are discussed further below. The *continuity* axiom guarantees that no one gamble is infinitely preferred to another, a condition called into question by behavior modeled by safety rules (Roy 1952, Telser 1955-56, Katoka 1963).

Expected utility's linearity in probabilities offers useful predictions that are evidenced by Meyer's and other chapters in this volume. For example, if risk increases through a mean-preserving spread, a risk averter (with a concave utility function) is made worse off. This risk averter would be willing to pay some positive amount for actuarially fair insurance that decreases an enterprise's variance while keeping its mean constant (Newbery and Stiglitz 1981). This same risk averter will prefer second-degree stochastically dominant distributions. Risk-averse price-taking producers will increase input use and thus production as output risk decreases (Sandmo 1971), a result used to

[2] See also Fishburn 1988 for descriptions of this and other axiomatic formulations of EU.

support government agricultural price intervention and crop insurance provision. With well-functioning futures markets, the (deterministic) production decision can be separated under EU from the risk-averse producer's subjective price distribution (Feder, Just, and Schmitz 1980).

ANOMALOUS BEHAVIOR INCONSISTENT WITH EXPECTED UTILITY DEFINED OVER FINAL WEALTH

There are four categories of behavior that call into question the assumptions underlying EU, with most of the evidence coming from controlled experiments. There is some evidence that decision makers do not treat objective probabilities linearly and thus violate the independence axiom (Allais 1953, 1979). This evidence has received the greatest amount of attention. Behavior violating transitivity over these subjective probabilities has also been documented in experimental settings, but has received considerably less attention and is difficult to reconcile with optimizing behavior in social settings. Questions regarding whether risky outcomes are best treated as final wealth or as changes from a reference level of wealth do not necessarily entail violations of EU, but do raise questions about most common EU applications. Finally, respondents in some settings exhibit probability miscalibration, exhibiting some bias in their estimation of subjective probabilities.

I describe the nature of these four types of anomalies below, and further discuss their implications for the use of EU in the remainder of the chapter.

Non-Linear Probabilities Violating Independence

Psychologists and economists have been actively designing experiments designed to reveal behavior violating EU independence for a number of years. Additional effort has been devoted toward devising models that alter the EU preference structure to permit such choice patterns within an optimization framework. Independence violations involve respondents selecting between gamble pairs generally similar to those introduced by Allais (1953, 1979) or by Kahneman and Tversky's (1979) pairs using more modest outcomes. These pairs are provided below.

Camerer and Hogarth (1999) review a number of risky choice and related experiments using real payoffs, and find that offering real payoff in some cases decreases, but does not eliminate, violations of normative models in replicated studies. Harrison (1994) has also questioned the saliency of the real payoffs used in experiments, a concern that generally remains a valid criticism (see, however, Kachelmeier and Shehata 1992). Consistency with EU for choice in non-experimental risky settings has shown mixed qualitative results and will be discussed in greater detail below.

Allais' Risky Pairs (outcomes are 1950s French francs):

Pair 1: Choose A or B
A: 1 million with probability 1.0
B: 3 million with probability .98, 0 with probability .02

Pair 2: Choose C or D
C: 1 million with probability .05, 0 with probability .95
D: 3 million with probability .049, 0 with probability .951

Most respondents select A over B and then D over C, indicating their preference for the certain very large payoffs of A but also their willingness to accept alternative D's slightly larger probability of a zero payoff for its higher positive payments. The combined choice pattern A and D violate EU independence because C and D are formed by combining A and B linearly with a common alternative O that gives a payoff of zero: C = (1/20)*A + (19/20)*O, D = (1/20)*B + (19/20)*O. Respondents are thought to overweight the probability for the low outcome (zero payoff) event provided in alternative B, and/or to underweight alternative D's probabilities for receiving the zero payoff.

Kahneman and Tversky's Certainty Effect Pairs (outcomes are 1970s Israeli shekels):

Pair 1: Select between E and F
E: 3000 with probability 1.0
F: 4000 with probability .80, 0 with probability .20

Pair 2: Select between G and H
G: 3000 with probability .25, 0 with probability .75
H: 4000 with probability .20, 0 with probability .80

The majority of Kahneman and Tversky's respondents selected E over F and also H over G. This selection of both E and H violates independence because G = (1/4)*E + (3/4)*O and H= (1/4)*F + (3/4)*O. Independence under EU calls for consistency of choice through either selection of both E and G or selection of both F and H. The common pattern of choices for Kahneman and Tversky's "certainty effect" experimental pair and of Allais' pairs above are viewed by Kahneman and Tversky as including a discrete certainty effect. This certainty effect holds that experimental subjects treat a lottery with a certain choice of a positive payoff (p = 1.0) qualitatively differently than they treat a lottery with less than a certain probability of a positive payoff. In an alternative view, choices over the certainty effect pairs reflect the overweighting of the probability of the zero payoff in alternative F, or an underweighting of this zero outcome's probability in H, or both.

Kahneman and Tversky's Common Ratio Effect (outcomes are 1970s Israeli shekels):

Pair 1: Select between R and S
R: 3000 with probability .9, 0 with probability .1
S: 6000 with probability .45, 0 with probability .55

Pair 2: Select between T and V
T: 3000 with probability .002, 0 with probability .998
V: 6000 with probability .001, 0 with probability .999

In these common ratio effect pairs, the majority of respondents select R over S and also V over T. The selection of both R and V violates independence because T = (1/450)*R + (449/450)*O and V= (1/450)*S + (449/450)*O. EU independence calls for consistency of choice through either the selection of both R and T or the selection of both S and V. Here, the probabilities for the non-zero outcome events are thought to be overweighted in R or underweighted for V, or both.

These three risky choice examples, plus a host of additional work, has led many to view respondents' choices as consistent with overweighting small probabilities and underweighting large probabilities, through the probability transformation g(·) illustrated in Figure 1. The transformation function g(·) is first above the 45-degree line with low probabilities overweighted, and then crosses and is below the 45-degree line with probabilities underweighted. Depending on the model, g(·) may or may not take a value of 1.0 for a probability p = 1.0. In some of the earliest of this work, Preston and Baratta (1948) find accurate valuation at around .2 (see also Edwards 1954). In most of this work, probability weighting is thought to occur for each gamble independently – it is invariant to the other gamble in the pair. This non-linear weighting is very closely tied with psychological studies of human perception and the biases and distortions in information processing.

Positive models for choice that adjust preferences to allow for independence violations, while generally retaining transitivity and continuity, are still the dominant approach to these violations. These axiomatic choice generalized-EU (GEU) models offer a reduced-form model of choice that reflects both the underlying (normative) preference structure and the imperfect and costly information processing of the stated probabilities and outcomes. Recent summaries and comparative tests of these models are given in Hey and Orme (1994), Harless and Camerer (1994), and Buschena and Zilberman (2000). Fishburn (1988) also provides an extremely comprehensive evaluation of these models. These GEU models introduce various forms of a non-linear probability weighting function, g(·), that weights properties in a manner that is consistent with the experimental risk research showing violations of EU and as illustrated in Figure 1. The decision maker is taken to maximize:

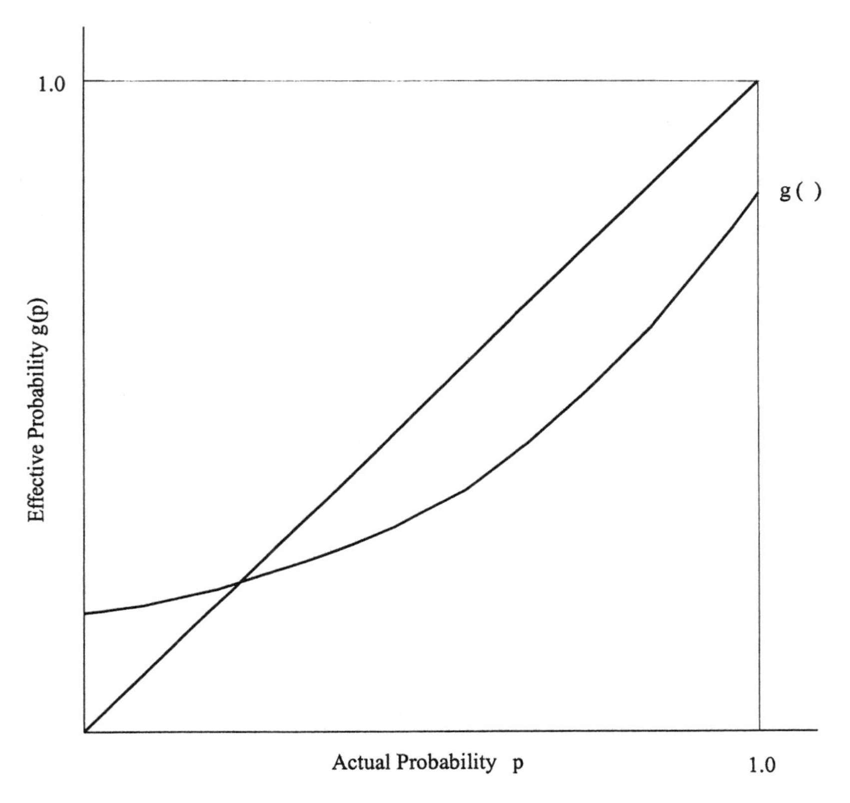

Figure 1. Probability Weighting Function

$$U(\boldsymbol{x},\boldsymbol{p})=\sum_{i=1}^{n}u(x_i)\cdot g_i(\boldsymbol{p})\,. \qquad (2)$$

These GEU models reflect heuristics used to evaluate the decision maker's underlying preferences. Note that the probability weighting function depends on the entire probability vector of a risky alternative in some of the GEU models. However, these weights depend only on the probability vector for one of the two risky alternatives. For example, the weights would only depend on alternative R in Kahneman and Tversky's common ratio effect pair RS above. These heuristics are relied on because the decision maker faces evaluation costs and imperfections in their decision making.

Hey and Orme's (1994) and Buschena and Zilberman's (2000) comparative tests of GEU model performance show relative predictive support for Quiggin's (1982) rank-dependent EU model (also related to Tversky and Kahneman's [1992] cumulative prospect theory) and Viscusi's (1989) prospective reference theory model. It is notable that Quiggin's and Viscusi's models rely on substantial additions beyond their axiomatic framework to achieve their predictions for risky choice. They, like Machina's (1982) smooth utility model with fanning out, are empirically successful because they incorporate behavioral regularities that are not limited to their axiomatic framework.

Choice Patterns Violating Transitivity

Although it is unlikely that an economic agent would become substantially and directly disadvantaged due to non-linear probability weighting reflecting independence violations, rigid violations of transitivity point to the potential for an agent to face serious monetary losses. Consider intransitivities of the type A preferred to B, B preferred to C, and C preferred to A. Suppose an agent with these intransitive preferences first possesses C. He or she would accept a trade giving the agent B-̣ for C, then A-̣ for B, and then C-̣ for A, with " . " representing an arbitrarily small amount. The agent finishes this cycle with C-3̣, but hopefully revises his or her intransitive preferences. In the event that the agent is no wiser after the first cycle of this fleecing, the cycle could be repeated infinitely and the agent becomes a "money pump" (Machina 1989).

There is some experimental evidence revealing intransitive choice patterns, evidenced by the results from an experiment reported in Tversky (1969). Tversky presented subjects with pairs of gambles, with the pairs given in random order to each respondent. The gambles are denoted by pairs whose first element denotes the probability of winning and the second denotes the outcome: gamble A (7/24, 5.00), gamble B (8/24, 4.75), gamble C (9/24, 4.50), gamble D (10/24, 4.25), and gamble E (11/24, 4.00). For example, gamble A provides a 7/24 chance of winning $5 and a 17/24 chance of winning $0. Note that this list of gambles exhibits increasing expected values. Tversky's respondents faced a choice between gambles A and B, a choice between gambles B and C, a choice between gambles C and D, a choice between gambles D and E, and a choice between gambles A and E. Tversky presented respondents with these pairs in multiple sessions and found a significant occurrence of transitivity violations. The most common pattern of choice was A over B, B over C, C over D, D over E, but also E over A. There are two aspects to this pattern. First, the gamble with the higher outcome, higher variance, and lower expected value was favored when the gamble pairs were quite similar (A vs. B, B vs. C, C vs. D, and D vs. E). Second, the gamble with the higher expected value and lower variance was favored when the gambles were dissimilar (E vs. A).

Despite the understandable nature of the intransitivity pattern shown by Tversky, the treatment by risk researchers of these transitivity violations contrasts substantially with that afforded independence violations. Transitivity violations are generally not allowed in non-EU risky choice models (Loomes and Sudgen's [1982] regret theory, Bell's [1982] disappointment aversion models, and Fishburn's [1981, 1988] SSB utility model are the exceptions). Machina (1989) relates that the reluctance of economists to incorporate such intransitive behavior into choice models reflects the potential for agents with such choice patterns to be taken advantage of by others in a social setting.[3]

Reference Points: Final Wealth Versus Changes from Current Wealth

From a research standpoint, the clearest way to model decision making under risk is to define the outcomes $\boldsymbol{x}$ as final wealth. However, there exists a body of research, both experimental and non-experimental, pointing to the importance of the individual's point of reference in understanding his or her behavior under risk. One individual may view a risky enterprise as providing either losses or gains from a reference point of current wealth. Another individual may have a reference point representing an expected rate of return over current wealth, with a risky venture offering outcomes above or below that rate, a reference point greater than current wealth. Another individual facing almost certain losses due to bankruptcy could have a reference point reflecting the losses from financial transactions costs, a reference point less than current wealth.

Let the reference wealth level be s. Then, risky alternatives with an outcome vector $\boldsymbol{x}$ can be defined as giving changes in wealth from s through the elements: $x_i = s + \Delta w_i$. The agent then selects between risky alternatives to maximize

$$\sum_{i=1}^{n} p_i \cdot v(\Delta w_i + s) . \tag{3}$$

Note that the linear treatment of the probabilities in equation (3) is consistent with EU. All of the EU axioms can apply to these outcomes defined as changes (w_i) from a reference level in the same way that they apply to outcomes as final wealth (x_i). The critical issue for agricultural economists applying this model is the dependence of predictions on the specification of the reference wealth level (s), how this reference wealth is determined, and how robust this reference level is to alternative presentation of the risky alternatives. These issues present challenges for defining clear testable implications for behavior under risk using this reference point approach.

Reference points appear to vary between gambles viewed as single plays (such as experimental lotteries) and those viewed as part of a larger set of risky decisions within a larger time frame such as production decisions made by farmers (see Roberts 1994 and Hershey and Schoemaker 1981). In a

---

[3] Cherry, Crocker, and Shogren (1999) report that experimental subjects quickly learned to adjust their responses to avoid monetary losses in light of market discipline that extracts a monetary cost for behavior somewhat like transitivity violations.

sense, reference points introduce another element of subjectivity into modeling risk, with the behavioral predictions differing substantially with different reference points.

Friedman and Savage (1948) refer to anecdotal knowledge of simultaneous gambling and insurance purchase that supports reference points. Kahneman and Tversky (1979) offer evidence from a number of experiments showing the importance of reference points, with their discussion of reference points comprising a larger part of that paper's coverage than is their discussion of probability weighting that has arguably received more attention in the literature.

Collins, Musser, and Mason (1991) found evidence of reference points in the (non-experimental) behavior of grass seed producers in Oregon. Recently, Jullien and Salanié (2000) found significant evidence for reference points in betting decisions over British horse races, reflecting Friedman and Savage's (1948) model of risk aversion for losses and slight risk preference for gains on race bets. Both of these studies will be addressed in more detail later in a discussion of nonexperimental results addressing the validity of EU.

Differences Between Subject's and Population Probabilities: Error or Supplemental Information?

In addition to the challenges of modeling risk described above, there is some evidence questioning the ability of decision makers to estimate the probabilities of risky events. This evidence stems from experiments, but its implications have been extended to risky decisions in business, medicine, environmental policy, and other fields. These experiments find that respondents appear to violate the laws of probability in their construction of subjective probabilities.

Ellsberg (1961) reports experimental results viewed as showing an inconsistency in the estimation of subjective probabilities. A single ball is to be drawn from an urn containing 90 unobserved balls. The respondent is told that 30 of the balls are red, while the remaining 60 are either black or yellow. The respondent is allowed to choose between the following:

Alternative 1: a payment of $100 if a red ball is drawn, $0 if black or yellow is drawn

Alternative 2: a payment of $0 if red, $100 if black, and $0 if a yellow ball is drawn. Most respondents select Alternative 1, avoiding the probability ambiguity of Alternative 2.

The respondent then faces a draw from a similar urn in another choice. Ellsberg asks the respondent to "...consider the following two actions, under the same circumstances" (as above for Alternatives 1 and 2):

Alternative 3: a payment of $100 if a red or a yellow ball is drawn, $0 if black is drawn.

Alternative 4: a payment of $0 if red, $100 if black or yellow ball is drawn.

Ellsberg reports that Alternative 3 is the most frequently selected alternative in this pair. The paradox arises if the urn is indeed viewed by the respondent as the same for both pairs because the respondent's subjective estimate of the proportion of yellow to black balls should be unchanged across the two draws.

Ellsberg's discussion of this experiment is a bit unclear, however. It is difficult to determine if both stages of the experiment are clearly using the same urn. If the experimental respondents thought that the urns were different, the most common pattern of choices described by Ellsberg appears to be rational if the respondent distrusts the experimenter. That is, a devious experimenter could be thought to place a high proportion of yellow balls in the first urn and a high proportion of black balls in the second. Replication of this experiment clarifying this issue is feasible, but to my knowledge has not been undertaken.

Work by Lichtenstein et al. (1978) shows significant biases in subjects' estimation of the probabilities of relative health risks. Lichtenstein et al.'s respondents were asked for their estimates of the frequency of over one hundred causes of death for the total U.S. population. These respondents tended to underestimate the occurrence of high-frequency causes of death – such as heart disease, strokes, and cancer – and to overestimate the occurrence of low-frequency causes of death, such as lightning and firearms. Lichtenstein et al.'s respondents included two groups, undergraduate students from the University of Oregon and members of the Eugene, Oregon, chapter of the League of Women Voters. The bias shown by the Lichtenstein et al. study, and subsequent replication, has been used to justify educational programs and additional study of health and environmental risk. Perhaps the strongest call for use of this body of evidence is Smith (1992), who argues that although they might be biased, individuals' subjective probabilities of health risks should be used in addition to expert judgments in evaluating the welfare implications of policy analysis. Smith argues that because the actual probabilities of health risks are unknown and may differ among individuals, their subjective beliefs should be incorporated with expert judgments for policy decisions. A key issue with respect to the use of these subjective judgments is the nature of the "biases" reported in Lichtenstein et al. If these biases reflect only psychological factors relating to agents' decision making processes, then policy should not make use of them (Machina 1990). Alternatively, if these biases reflect the agent's own subjective probability estimation (as Smith claims), they have clear relevance for policy if expert judgments are imperfect.

WHAT ARE THE IMPORTANT NEW DEVELOPMENTS, AND WHAT SHOULD BE THE RESPONSE TO THE ANOMALIES?

With the exception of reference point concerns and perhaps differences between subjects' and experts' probability estimates, most economists view models adjusted to permit anomalies as reflecting positive, rather than normative, behavior. That is, these "violations" of EU are unlikely to show that people are truly irrational, because we know very little about decision errors and the costs of the decision itself (Plous 1993). Therefore, models adjusting the EU preference structure to allow for anomalous behavior should be viewed as in part reflecting the decision process of imperfect decision makers facing decision costs, rather than as only reflecting the decision makers' underlying preferences. In this light, even intransitive behavior, social setting and robustness issues aside, may have a place in choice models; that is, it is problematic to exclude intransitive behavior on normative grounds, since these adjusted models are not thought to be normative. At issue for risk researchers are the modeling tradeoffs between a reduced-form generalized-EU (GEU) model for choice and a more detailed structural model of the choice process that captures relevant aspects of the benefits and costs of decision effort. Another issue to be addressed involves policy approaches. Should we strive to "correct" anomalies through learning, or should we work toward reducing decision costs, for example through information provision?

Modeling Developments

Deriving Implications of GEU Models. A longstanding criticism of GEU models reflects the desire to retain the EU model's clear and testable implications with respect to risky behavior. Quiggin (1991), Quiggin and Chambers (1998), and Machina (1989) have evaluated extensively the implications of GEU models for risky choice. Their innovations have taken the GEU approach beyond theoretical development into achieving precise implications for application and testing. They have successfully taken separate models, rank-dependent EU for Quiggin and Quiggin and Chambers and smooth utility for Machina, and evaluated them for well-known results such as comparative static effects of risk increases, stochastic dominance, differentiability, risk-aversion coefficients, and other topics. Their work provides a framework for empirical estimation such as in Jullien and Salanié (2000) of both the non-linear utility portion of risk attitudes over income, $u(\cdot)$, and the non-linear probability weighting portion $g(\cdot)$ in equation (2).

Similarity and Choice Error Models. An alternative to GEU models is to introduce instruments to reflect the importance and costs of risky decision making into the analysis. One approach is to differentiate between the importance of the various gambles through their similarity (Rubinstein 1988, Leland

1994, Buschena and Zilberman 1995, 1999a, 1999b, 2001). Respondents are shown to devote less effort to selection between very similar and also very dissimilar gambles, and choice patterns over these gambles reflect these differences in effort allocation (Buschena and Zilberman 2001). This similarity approach is not inferior in fit to the best of the GEU models, and has superior fit to many other GEU models (Leland 1994, Buschena and Zilberman 1995, 1999a). These similarity models attempt to model the constrained choice process more directly than in GEU models, and to reconsider the nature of the empirical evidence supporting them. In essence, they suggest that EU should provide useful predictions when the risky pairs are dissimilar (distinct). Additionally, these similarity approaches provide an explanation for intransitive choice pattern such as Tversky's (1977) (Leland 1994, Buschena and Zilberman 1995, 1999a).

A related approach to modeling the choice process is to allow for heteroscedastic error for multiple choices from a respondent (Hey 1995, Buschena and Zilberman 2000). Heteroscedastic error and similarity approaches are related when the error variance is allowed to increase as the gambles become more alike (similar). This heteroscedastic error approach has considerable econometric appeal, and EU with heteroscedastic error performs well relative to GEU models (Hey 1995, Buschena and Zilberman 2000). Additionally, EU with heteroscedastic error (where the error variance increases with similarity) compares very favorably to GEU models with heteroscedastic error; that is, heteroscedastic error structures appear to substitute for GEU formulations in terms of predictive power (Buschena and Zilberman 2001). Carbonne and Hey (2000) show that error varies by choice pairs for some individuals and that the correct error formulation depends on the choice model. This error/model relationship makes joint estimation of the choice models' axiomatic form and the error structure complicated, providing additional challenge to making operational this error approach to modeling risk.

Real World (Non-Experimental) Data, EU, and Reference Points. The bottom line for agricultural and resource economists is whether or not violations of EU that appear in experiments are also present in non-experimental settings. There have been relatively few attempts to assess the performance of EU and GEU models using non-experimental data. It is difficult to find comprehensive data sets that provide econometric identification of both the relevant probabilities and then the parametric forms describing the non-linearity in probabilities under GEU models. Whether or not EU violations actually occur beyond experimental settings is a critical question for economists. This question goes beyond real vs. hypothetical payoff effects: at issue is whether or not social and market forces provide information, signals, and experience that reduce the degree of potentially costly EU violations.

The importance of choice context has been shown experimentally by Hershey and Schoemaker (1981), Heath and Tversky (1991), and Roberts (1994). In Hershey and Schoemaker, individuals facing the same probabilities and outcomes are more likely to accept small losses in order to avoid large but improbable losses when the hypothetical question is posed as an insurance problem instead of as a simple lottery. More generally in Roberts, individuals are more likely to take risks and less likely to violate EU when hypothetical risky questions are posed to them as familiar decisions (insurance, purchasing a used car, part-time jobs, etc.) instead of being posed as simple gambles. These findings of contextual effects point to the potential importance of studying the relative performance of EU for risky decisions made in everyday, non-experimental contexts.

Bar-Shira (1992) uses time-series/cross-section data of Israeli farmers to determine if their decisions can be consistent with EU. Conceptually, Bar-Shira searches for a hyperplane (linearity) in the probability space that is consistent with crop choice, under both adaptive and rational expectations frameworks. Bar-Shira finds such a hyperplane for the majority of farmers in his sample, so in effect EU is consistent for crop choice by these farmers. There may, however, be GEU models that offer superior fit to EU models, a potential extension of Bar-Shira's work.

Harless and Peterson (1998) evaluate enrollment and liquidation decisions of mutual fund members. They find that investors rely on heuristics in estimating the probabilities for future fund performance, and that their use of these heuristics depends on whether or not they have equity in the plan.

Collins, Musser, and Mason (1991) estimate risk attitudes for Oregon grass seed producers using production decisions from 1973 and 1974 crop years. They find (consistent with Kahneman and Tversky's [1979] prospect theory) that for farmers whose income declined, the change in income from 1973 to 1974 is significantly related to the farmers' estimated risk attitude level. They also point to a potential link between farmers' financial condition and increased underlying risk arising from debt increases.

Jullien and Salanié (2000) evaluate the predictive power of EU versus a GEU model (rank-dependent EU) for racetrack betting patterns in Britain. For tractability, they assume a constant absolute risk aversion functional form for utility. They also assess reference points. Jullien and Salanié find no economically significant improvements in model fit when GEU models are considered as alternatives to EU for the entire set of payoffs.[4] They do find empirical support for reference point effects, where the probabilities of losses are treated differently from the probabilities for gains, and bettors do not treat these probabilities over losses linearly as required in EU.

Farm finance provides a wealth of information for consideration of reference points. Final wealth levels often imply discrete and important changes in well-being, such as bankruptcy or liquidation for consumers, small business owners, or farmers. Robison and Barry (1987) provide comprehensive models for risky decisions under the EU model with such discontinuities. Atwood,

---

[4] Jullien and Salanié (2000) find statistically significant but small non-linearity for positive payoffs under one of two specifications of the rank-dependent model.

Watts, and Baquet (1996) carry out econometric estimates of these effects. Mahul (2000) provides a comprehensive model for the effects of such financial transactions costs on decision makers that maximize long-term discounted EU.

Calibration Problems for Subjective Probabilities. The usefulness of models assuming rationality is seriously called into question if individuals routinely violate probability laws when assessing risky choices, as in Lichtenstein et al. (1978) discussed above. Also at issue is the optimal response to these biased or miscalibrated subjective probabilities – should policymakers accept them at face value and combine them with expert judgments, as suggested in Smith (1992)? Alternatively, should policymakers attempt to correct for these biases in their decisions and provide additional information to combat these biases?

Just how bad are individuals at assessing their relevant probabilities of health risks, and why? Recall that Lichtenstein et al. (1978) required undergraduate students and League of Women Voters members in Oregon to assess death rates for the entire U.S. population. Particularly for students, these death rates are unlikely to be very relevant for their own immediate health risks. Given information gathering and processing costs, why should we expect students to be very good estimators of health risks for the entire population? Put another way, might the biases inferred by Lichtenstein et al. reflect rationality in light of the costliness of gathering and assessing health information?

Recent efforts by Benjamin and Dougan (1997) and Benjamin, Dougan, and Buschena (2001) have shed new light on the evidence of biases in risk assessment from Lichtenstein et al. (1978). This new work has found that although respondents show similar patterns of probability bias for population risk assessment, they exhibit no significant bias when asked to judge risks more clearly relevant to them (within their age group). This recent work supports the view that rational decision makers facing information acquisition and processing costs should only possess workable information for health risks they face, not necessarily for health risks they likely do not. This research also suggests targeting risk information efforts toward subsets of the population that are in the near future likely to make decisions influencing their likelihood of facing these specific risks.

CONCLUSIONS

This chapter opened with three questions:

- What is the nature of behavior violating EU as commonly applied?
- What do these anomalies tell us about modeling behavior under risk and what testable implications can be drawn from models incorporating them?

- Can we still use EU for some risky choices in light of these anomalies; i.e., how robust are the anomalies to (1) the design of the risky questions, (2) real payoffs, and (3) experimental vs. non-experimental risky questions?

The Nature of Behavior Violating Expected Utility as Commonly Applied

There are four types of behavior violating the standard application of EU. Two direct violations of EU include violations of independence and violations of transitivity. Violations of independence question the EU model's linear treatment of the probabilities of risky gambles and have been addressed by generalizations of EU. Violations of transitivity are not generally accepted in models for economic agents and appear to rely on decision makers reacting to a series of gambles with disparate similarity.

Another departure from EU as commonly applied is for decision makers to make risky decisions providing changes from some reference level of wealth, often taken as their current wealth level. Kahneman and Tversky (1979) coined the term "prospects" for these wealth changes, but these changes from a reference level can be incorporated into the EU framework. The challenge in using this reference-level approach is in understanding how this reference level is determined and observed.

A final departure from EU maximization arises when decision makers exhibit significant bias in estimating the probabilities of risky events. Much of the experimental evidence of this bias comes from questions of health risks, but these findings have been expanded to other areas, including environmental risks.

What the Anomalies Tell Us About Modeling Behavior Under Risk And Testable Implications Drawn from Models Incorporating Them

Independence and transitivity violations provide an interesting contrast in approaches to addressing these experimental regularities. Allowance for independence violations has been incorporated into a number of axiomatic choice models because it is difficult to see a decision maker with those choice patterns being taken advantage of by another agent. Rigid transitivity violations clearly lend themselves to potentially costly interaction with others. Therefore, these positive models of risky choice are limited in the type of observed behavior they allow.

Quiggin and Machina have been particularly active in developing the behavioral implications of their respective models. They have shown that many important risk topics – such as comparative static effects of risk increases, stochastic dominance, differentiability, and risk-aversion coefficients – can be applied to some of the GEU models (Quiggin 1991, Quiggin and Chambers 1998, Machina 1989). Their efforts in these developments have generated additional interest in the GEU models among general economists.

As discussed in the body of this chapter, recent authors have successfully estimated reference point effects using instruments related to underlying factors thought to affect them. These efforts require identification of the relevant factors describing reference points, but may provide additional information regarding factors underlying risk attitudes.

The long-held belief that many people were severely biasing their estimation of health risks and other events has been called into serious question by recent research reviewed above. These results point toward a need for less simplistic views of subjective probability estimation, and a tailoring of models for such issues to reflect the risks people actually face.

The Use of Expected Utility for a Subset of Risky Choices

Many questions have been answered regarding the robustness of the anomalies to experimental design and to offering real payoffs. Similarity arguments, including the incorporation of these effects into heteroscedastic error structures, show that common patterns of independence and transitivity violations can be successfully addressed using instruments that reflect the underlying difficulty and importance of the risky choice. This success has been shown in both absolute terms and relative to GEU models. These similarity findings suggest that EU has considerable predictive power for decisions between distinct, and clearly important, risky alternatives. Offering real payments in experimental settings has not eliminated, but has occasionally reduced, the significant occurrence of EU independence violations (Camerer and Hogarth 1999).

There appears to be some evidence that choice patterns violating EU are mitigated in non-experimental settings. Putting risky choices into familiar contexts, and providing mechanisms for market discipline, have successfully reduced the incidence of some EU violations.

Finally, what does this large body of literature in EU violations mean for agricultural and resource economists? We certainly have many options in addressing these violation patterns. The GEU models offer reduced-form methods to model choice, and it is possible to draw useful predictions from them. Similarity approaches model the decision process more directly, and suggest ways to construct elicitation methods to avoid common patterns of EU violations.

Studies showing the importance of context reveal the importance of couching risk elicitation in ways familiar to decision makers. To the extent that new risk enterprises are quite unfamiliar to farmers and resource managers, these context effects call for relating these new problems to familiar ones. Reference points also appear to be influenced by the presentation of the alternatives (a fact clearly known by pollsters). Care must be made to present risky choice problems in ways valid to the decision maker's problem. Reference points are arguably the greatest challenge facing us in modeling risky decisions.

It may be that there is no universal model for understanding risky choice. As in almost all productive economic research, it is important to understand various costs and constraints, such as (1) heteroscedastic error structures, (2) information processing, and (3) reference points (including their underlying causes). A richer understanding of these issues would move us from "simple" tests of EU to a more accurate and interesting understanding of the underlying factors affecting risky choice.

REFERENCES

Allais, M. 1953. "Le Comportement de l'Homme Rationnel Devant le Risque: Critique des Postulats et Axiomes de l'Ecole Américaine." *Econometrica* 21: 503-546.

___. 1979. "The So-Called Allais Paradox and Rational Decisions Under Uncertainty." In M. Allais and O. Hagen, eds., *Expected Utility Hypotheses and the Allais Paradox*. Dordrecht, Holland: Reidel.

Arrow, K. 1970. *Essays in the Theory of Risk Bearing*. Amsterdam: North Holland.

Atwood, J., and D.E. Buschena. 2000. "Evaluating the Magnitudes of Financial Transactions Costs on Risk Behavior." Working Paper, Department of Agricultural Economics and Economics, Montana State University.

Atwood, J.A., M.J. Watts, and A. Baquet. 1996. "An Examination of the Effects of Price Supports and Federal Crop Insurance Upon the Economic Growth, Capital Structure, and Financial Survival of Wheat Growers in the Northern High Plains." *American Journal of Agricultural Economics* 78: 212-224.

Bar-Shira, Z. 1992. "Nonparametric Tests of The Expected Utility Hypothesis." *American Journal of Agricultural Economics* 74: 523-533.

Bell, D.E. 1982. "Regret in Decision Making Under Uncertainty." *Operations Research* 30: 961-981.

Benjamin, D.K., and W.R. Dougan. 1997. "Individual's Estimates of the Risk of Death: Part I - A Reassessment of the Previous Evidence." *Journal of Risk and Uncertainty* 15: 115-134.

Benjamin, D., W. Dougan, and D.E. Buschena. 2001. "Individuals' Estimates of the Risks of Death: Part II – New Evidence." *Journal of Risk and Uncertainty* 22: 35-57.

Buschena, D.E., and D. Zilberman. 1995. "Performance of the Similarity Hypothesis Relative to Existing Models of Risky Choice." *Journal of Risk and Uncertainty* 11: 233-262.

___. 1999a. "Testing the Effects of Similarity and Real Payoffs on Choice. In M. Machina and B. Munier, eds., *Beliefs, Interactions, and Preferences in Decision Making*. Boston: Kluwer Academic Publishers.

___. 1999b. "Testing the Effects of Similarity on Risky Choice: Implications for Violations of Expected Utility." *Theory and Decision* 46: 253-280.

___. 2000. "Generalized Expected Utility, Heteroscedastic Error, and Path Dependence in Risky Choice." *Journal of Risk and Uncertainty* 20: 67-88.

___. 2001. "Predictive Value of Incentives, Decision Difficulty, and Expected Utility Theory for Risky Choices." Working Paper, Department of Agricultural Economics and Economics. Montana State University.

Camerer, C.F., and R. Hogarth. 1999. "The Effect of Financial Incentives in Experiments: A Review and Capital-Labor-Production Framework." *Journal of Risk and Uncertainty* 19: 7-42.

Carbonne, E., and J.D. Hey. 2000. "Which Error Story is Best?" *Journal of Risk and Uncertainty* 20: 161-176.

Cherry, T.L., T.D. Crocker, and J.F. Shogren. 1999. "Rationality Spillovers." Working Paper, Department of Economics and Finance, University of Central Florida.

Collins, A., W.N. Musser, and R. Mason. 1991. "Prospect Theory and Risk Preferences of Oregon Seed Producers." *American Journal of Agricultural Economics* 73: 429-435.

Edwards, W. 1954. "The Theory of Decision Making." *Psychological Bulletin* 51: 380-417.

Ellsberg, D. 1961. "Risk, Ambiguity, and the Savage Axioms." *Quarterly Journal of Economics* 75: 643-669.

Feder, G., R.E. Just, and A. Schmitz. 1980. "Futures Markets and the Theory of the Firm Under Price Uncertainty." *Quarterly Journal of Economics* 94: 317-328.

Fishburn, P.C. 1981. "An Axiomatic Characterization of Skew-Symmetric Bilinear Functionals, With Applications to Utility Theory." *Economics Letters* 8: 311-313.

___. 1988. *Nonlinear Preference and Utility Theory*. Baltimore: Johns Hopkins University Press.

Friedman, M., and L.J. Savage. 1948. "The Utility Analysis of Choices Involving Risk." *Journal of Political Economy* 56: 279-304.

Harless, D., and C. Camerer. 1994. "The Predictive Utility of Generalized Expected Utility Theory." *Econometrica* 62: 1251-1290.

Harless, D.W., and S.P. Peterson. 1998. "Investor Behavior and the Persistence of Poorly-Performing Mutual Funds." *Journal of Economic Behavior and Organization* 37: 257-276.

Harrison, G.W. 1994. "Expected Utility Theory and the Experimentalists." *Empirical Economics* 19: 223-253.

Heath, C., and A. Tversky. 1991. "Preference and Belief: Ambiguity and Competence in Choice Under Uncertainty." *Journal of Risk and Uncertainty* 4: 5-28.

Hershey, J.C., and P. J.H. Schoemaker. "Risk Taking and Problem Context in the Domain of Losses: An Expected Utility Analysis." *Journal of Risk and Insurance* 47: 111-132.

Hey, J.D. 1995. "Experimental Investigations of Errors in Decision Making Under Risk." *European Economic Review* 39: 633-640.

Hey, J.D., and C. Orme. 1994. "Investigating Generalizations of Expected Utility Theory Using Experimental Data." *Econometrica* 62: 1291–1326.

Jensen, N. 1967. "An Introduction to Bernoullian Utility Theory: Utility Functions." *Swedish Journal of Economics* 69: 163-183.

Jullien, B., and B. Salanié. 2000. "Estimating Preferences Under Risk: The Case of Racetrack Bettors." *Journal of Political Economy* 108: 503-530.

Kachelmeier, S.J., and M. Shehata. 1992. "Examining Risk Preferences Under High Monetary Incentives: Experimental Evidence From the Peoples' Republic of China." *American Economic Review* 82: 1120-1141.

Kahneman, D., and A. Tversky. 1979. "Prospect Theory: An Analysis of Decision Under Risk." *Econometrica* 47: 263-291.

Katoka, S. 1963. "A Stochastic Programming Model." *Econometrica* 31: 181-196.

Leland, J.W. 1994. "Generalized Similarity Judgments: An Alternative Explanation for Choice Anomalies." *Journal of Risk and Uncertainty* 9: 151-172.

Lichtenstein, S., P. Slovic, B. Fischoff, M. Lyman, and B. Combs. 1978. "Judged Frequency of Lethal Events." *Journal of Experimental Psychology: Human Learning and Memory* 4: 551-578.

Loomes, G., and R. Sudgen. 1982. "Regret Theory: An Alternative Theory of Rational Choice Under Uncertainty." *Economic Journal* 92: 805-824.

Machina, Mark J. 1982. "'Expected Utility' Analysis Without the Independence Axiom." *Econometrica* 50: 277-323.

___. 1989. "Comparative Statics and Non-Expected Utility Preferences." *Journal of Economic Theory* 47: 393-405.

___. 1990. "Choice Under Uncertainty: Problems Solved and Unsolved." In R.B. Hammon and R. Coppock, eds., *Valuing Health Risk, Costs and Benefits for Environmental Decisionmaking.* Washington, D.C.: National Academy Press.

Mahul, O. 2000. "The Output Decision of a Risk-Neutral Producer under Risk of Liquidation." *American Journal of Agricultural Economics* 82: 49-58

Meyer, J. 2001. "Expected Utility as a Paradigm for Decision Making in Agriculture." In R.E. Just and R.D. Pope, eds., *A Comprehensive Assessment of the Role of Risk in U.S. Agriculture.* Boston, MA: Kluwer Academic Publishers.

Newbery, D.M.G., and J.E. Stiglitz. 1981. *The Theory of Commodity Price Stabilization.* Oxford: Oxford University Press.

Plous, S. 1993. *The Psychology of Judgment and Decision Making.* Philadelphia: Temple University Press: Philadelphia.

Pratt, J.W. 1964. "Risk Aversion in the Small and in the Large." *Econometrica* 32: 122-136.

Preston, M.G., and P. Baratta. 1948. "An Experimental Study of the Auction Value of an Uncertain Outcome." *American Journal of Psychology* 61: 183-193.

Quiggin, J. 1982. "A Theory of Anticipated Utility." *Journal of Economic Behavior and Organization* 3: 323–343.

___. 1991. "Comparative Statics for Rank-Dependent Expected Utility Theory." *Journal of Risk and Uncertainty* 4: 339-350.

Quiggin, J., and R.G. Chambers. 1998. "Risk Premiums and Benefit Measures for Generalized-Expected Utility Theories." *Journal of Risk and Uncertainty* 17: 121-138.

Roberts, M. 1994. "The Sensitivity of Expected Utility Violations to the Experimental Design: How Context Affects Risky Choice." Master's thesis, Department of Agricultural Economics and Economics, Montana State University.

Robison, L.J., and P.J. Barry. 1987. *The Competitive Firm's Response to Risk.* New York: MacMillan.

Roy, A.D. 1952. "Safety First and the Holding of Assets." *Econometrica* 20: 431-449.

Rubinstein, A. 1988. "Similarity and Decision Making Under Risk: Is There a Utility Theory Resolution to the Allais Paradox?" *Journal of Economic Theory* 46: 145-153.

Sandmo, A. 1971. "On the Theory of the Competitive Firm Under Price Uncertainty." *American Economic Review* 61: 65-73.

Smith, V.K. 1992. "Environmental Risk Perception and Valuation: Conventional Versus Prospective Reference Theory." In D.W. Bromley and K. Segerson, eds., *The Social Response to Environmental Risk: Policy Formulation in an Age of Uncertainty.* Boston: Kluwer Academic Publishers.

Stigler, J.E., and G.S. Becker. 1977. "De Gustibus Non est Disputandum." *American Economic Review* 67: 76-90.

Telser, L.G. 1955-56. "Safety First and Hedging." *Review of Economic Studies* 23: 1-16.

Tversky, A. 1969. "Intransitivity of Preferences." *Psychological Review* 76: 31-48.

___. 1977. "Features of Similarity." *Psychological Review* 84: 327-352.

Tversky, A., and D. Kahneman. 1992. "Advances in Prospect Theory: Cumulative Representation of Uncertainty." *Journal of Risk and Uncertainty* 5: 297-323.

Viscusi, W.K. 1989. "Prospective Reference Theory: Toward an Explanation of the Paradoxes." *Journal of Risk and Uncertainty* 2: 235-264.

von Neumann, J., and O. Morgenstern. 1953. *Theory of Games and Economic Behavior* (3rd ed.). Princeton, NJ: Princeton University Press.

Chapter 3

ORDERING RISKY CHOICES

Lindon J. Robison and Robert J. Myers
Michigan State University

INTRODUCTION

This chapter connects assumptions about probability distributions, decision makers' risk attitudes, and the resulting methods used to order risky choices. The methods we discuss are consistent with the expected utility hypothesis because up to this point in time no alternative for decision making under risk has gained widespread acceptance.

This chapter emphasizes two themes. The first theme is that different methods for ordering choices whose outcomes are described by probability distributions have different assumptions with varying degrees of strength. These assumptions define decision makers' preferences and probability distributions describing possible outcomes.

The second theme is that methods used to order choices with uncertain outcomes depend on the strength or weakness of assumptions applied to both decision makers' preferences and probability distributions. Complete ordering (ranking all choices according to their preference) requires strong assumptions about both decision makers' preferences and probability distributions. Weaker assumptions about decision makers' preferences and probability distributions preclude complete orderings and instead order probability distributions into efficient and inefficient sets.

STRONG AND WEAK ASSUMPTIONS

Strong and Weak Assumptions About Decision Makers' Preferences. Strong assumptions about decision makers' preferences are represented by a single-valued absolute risk aversion function defined over income and wealth. Weaker assumptions about decision makers' preferences are represented by an interval of possible absolute risk aversion values for each level of wealth. In other words, a strong assumption about risk preferences describes a single decision maker. A weaker assumption about decision makers' preferences describes a set of decision makers and is less restrictive than assumptions describing a single individual.

Strong assumptions about decision makers' preferences are capable of completely ordering well-defined probability distributions from most preferred to least preferred. Weak assumptions about decision makers' preferences are capable of less complete orderings. As the assumptions about decision makers' preferences are weakened, the number of decision makers whose preferences are characterized by the assumptions increases, as does the proportion of choices included in the efficient set.

The history of ordering choices under risk can be characterized by efforts to first weaken assumptions about risk preferences and then, in response to increasing numbers of choices included in the efficient set, to strengthen assumptions about risk preferences.

Strong and Weak Assumptions About Probability Distributions. Strong assumptions about probability density functions assume that the population density functions are known completely. Weaker assumptions about probability distributions assume some information about the population distribution is known, but not enough to describe it completely. In such cases, an estimated distribution can be used to describe the population distribution, but with error. Another weak assumption about probability distributions is to claim that the family of the population density function (e.g., normal, chi square, gamma, etc.) is known, or that its first and second moments are known. These weaker assumptions stop short of claiming to know the complete distribution.

When ordering choices under risk, researchers have mostly employed strong assumptions about probability distributions. Although some efforts have been made to describe the consequences of weak assumptions about probability distributions, these have been mostly ignored (Pope and Ziemer 1984). One concern is that the evidence does not support strong probability assumptions and it is unlikely that studies employing them will be using appropriate methods for ordering probability distributions. For example, in an important study, Meyer and Rasche (1992) employed a weak assumption about probability distributions (that their sample distributions were not the complete distribution) and found that, once adopted, this assumption leads to renewed confidence in an expected value-variance (EV) criterion, a method largely rejected under strong probability assumptions.

Empirical Realities. Weak assumptions applied to decision makers' preferences and probability distributions have altered the methods used for ordering risky choices and the size of efficient sets (Meyer 1977). In part, the substitution of weak assumptions for strong ones has been motivated by empirical realities that have left us with reduced confidence in both measures. In the case of risk attitudes, we are now more likely to rely on theoretical deductions to describe them than on empirical measures, despite considerable efforts to measure risk attitudes (Meyer and Meyer 2001). In the case of probability distribution measures, we now recognize that most of them are only estimates created by draws from some population that may be time-indexed.

INDICES AND CERTAINTY EQUIVALENTS

In the beginning was certainty. Once a choice was made, outcomes measured in terms of income or income plus wealth were considered to be known. Furthermore, because all decision makers preferred more income plus wealth to less, when faced with the same opportunities and constraints, all chose alike. As a result, beginning production economic theory typically makes no reference to decision makers' preferences and instead focuses on maximizing profits. More refined preference information is not needed when the world is assumed to be certain.

Then came uncertainty. Uncertainty means that choices have more than one possible outcome and the likelihoods of these outcomes are described using probabilities. The challenge of decision makers facing uncertainty was how to order choices. This more complicated ordering of choices with multiple possible outcomes and probabilities required that the value of each choice be reduced to an index that could be compared to other choices. Moreover, ordering choices under uncertainty required that the index be capable of providing a ranking consistent with decision makers' preferences.

One index that satisfies the ranking requirement is the certainty equivalent, or some monotonic transformation of the certainty equivalent. When each choice's possible outcomes are reduced to their certainty equivalent, a complete ordering of choices described by probability distributions can be obtained.

The solution for a complete ordering of choices according to preferences was to specify with certainty a decision maker's preferences in the form of a utility function, unique up to a linear transformation. A decision maker's utility function provides the basis for a complete ordering of choices because it reflects all that is known about decision makers' preferences for monetary outcomes. Let $U(y)$ be a utility function that reflects a decision maker's preferences for income and let $f_1(y), f_2(y), ..., f_n(y)$ represent probability distributions describing the likelihood of income level y for n risky choices. According to the expected utility hypothesis (EUH), a decision maker whose preferences are consistent with certain axioms can have his or her preferences reflected by the indices $EU(y)f_1(y)$, $EU(y)f_2(y)$, ..., $EU(y)f_n(y)$. These indices, when known, provide a complete ordering of choices.

When a complete ordering of distributions is required, a strong assumption is required about preferences represented by a single-valued utility function that can be mapped into an absolute risk aversion function. The advantage of such an explicit ordering is also a disadvantage – it applies only to a single individual whose preferences can be described completely by $U(y)$. For theoretical purposes, though, $U(y)$ need not be specified beyond some continuity conditions and derivative signs. Still, the vagueness of the function often limits even theoretical results, especially if choices depend on more than one random variable.

If the argument can be made convincingly that many decision makers share the same preference function or that the function describes a representative agent, then an index employing a specific $U(y)$ can be argued to have more generally applicable results than for a single individual. Early on, Bernoulli argued that decision makers' preferences could be described using logarithmic functions. The implication was that probability distributions could be ordered according to their geometric means. Some solid theoretical reasons exist for the support of the logarithmic function, including the fact that it displays decreasing absolute risk aversion. But the inability of the function to handle negative values has limited the enthusiasm that might otherwise have existed for it.

A linear utility function, on the other hand, orders distributions according to their arithmetic means (expected values). At different times, ordering choices according to their arithmetic means has had its supporters, particularly when linear programming was the rage. But for many, the problem of the St. Petersburg gamble with its infinite mean and finite value was sufficient to reject expected values as a sufficient criterion for ordering distributions. And for those left unconvinced, the practice by successful insurance companies of charging customers for eliminating risk and reducing expected wealth casts further doubt on the linear preference function.

Then there were, for a time, those who argued that when an individual's or firm's survival was at risk, preferences produced a criterion that ordered choices according to the maximum probability of survival. However, after some reflection, thoughtful scholars recognized that such a criterion for ordering choices could only be consistent with a utility function that was constant for income levels above and below the survival threshold. As a result of the implausible nature of this assumption, this restricted preference function was generally rejected, although occasionally pockets of support for it still manifest themselves in the literature.

EFFICIENT AND INEFFICIENT SETS

Wanting results more general than those provided by a single-valued utility function, assumptions about the form of decision makers' preferences were relaxed or weakened. Instead of assuming we knew with certainty the complete utility function, we assumed we knew the functional form of decision makers' preferences. In effect, this approach allowed us to order preferences for a class of decision makers instead of a single decision maker. However, ordering distributions for a class of decision makers generally resulted in failure to make complete orderings. Instead, probability distributions were ordered into inefficient sets and efficient sets that contained the expected utility maximizing choice for the class of decision makers corresponding to the criterion. For a distribution to be a member of the inefficient set required that there existed a choice in the efficient set such that all decision makers would prefer it to the ones in the inefficient set.

Extremely popular in the past has been the class of decision makers whose preferences could be represented by a quadratic utility function. Significant support for this function focused on two of its properties. First, it can be argued that any concave down function $U(y)$ can be approximated to the second degree by a quadratic function. And second, it naturally led to one of our most enduring efficiency criteria, the expected value-variance (EV) efficient set. In effect, this criterion relaxes the assumption that we know $U(y)$ and adopts a weaker assumption that all we know about $U(y)$ is that it can be represented by the quadratic function: $U(y) = y - by^2$, where $b > 0$.

Taking the expectation of a quadratic utility function leads to an equivalent expression containing measures of the expected value and variance parameters of probability density functions to be ordered. For example, for constant levels of variance, the probability distribution with the highest expected value is preferred. Applied generally, if one assumes that preferences are quadratic, then it follows that the popular EV frontier contains the expected utility maximizing choices. Points interior to the EV frontier are considered to be inefficient.

As has been so often the case, decision theorists identified reasons for rejecting quadratic utility as an appropriate description of decision makers' risk preferences. One reason was that it implied increasing absolute risk aversion. For these decision makers, the risk premium for a gamble of fixed dimensions increased with decision makers' wealth, a counterintuitive result. And a worse implication of quadratic utility is that beyond some level of income and wealth, marginal utility of income turns negative. Furthermore, to prevent negative marginal utilities of income and wealth over relevant ranges often required such a small coefficient "b" that in effect the function was linear. So much for preferences defined by quadratic functions.

So what next? What next was to substitute a weaker assumption about decision makers' preferences for the stronger one that assumes we could describe decision makers' preferences with a function or a functional form. This substitution of a weaker assumption for a stronger one led to a less restrictive description of decision makers' preferences. It turns out that we could still order probability distributions into efficient and inefficient sets without specifying the preference function of the decision makers.

Hanoch and Levy (1969) and Hadar and Russell (1969), who published their independent work simultaneously, asked and answered the question: What kind of criterion would follow if the most we knew about decision makers' preferences was that they preferred more income and wealth to less? Thoughtful reflections produced an answer. Begin with two choices and assume both have the same two possible outcomes, say 0 and $1,000, but with different probabilities. The choice with the larger probability for earning $1,000 would be preferred by all who prefer more to less, regardless of the shape of their preference function. So, the two choices could be ordered by knowing only that decision makers preferred more to less.

This example, generalized to the case of multiple outcomes, became known as first-degree stochastic dominance (FSD). In effect, the criterion demands the following. When comparing any two distributions, the first is preferred to the second by all decision makers who prefer more to less if the following condition exists. For any common outcome, the probability of earning the outcome or less for the first distribution is the same or smaller than for the second distribution. When this condition is satisfied, then the first distribution is included in the efficient set while the second is excluded and becomes a member of the inefficient set.

The reader might already suspect a problem with this criterion. If all decision makers who prefer more to less must agree that distribution one is preferred to distribution two before distribution two can be consigned to the inefficient set, then rarely will distribution two or any other choice be considered inefficient. Indeed, a very large efficient set is often the outcome of applying the FSD criterion, especially if the procedure for generating probability distributions relies on some experiment that generates large numbers of distributions.

Another way to describe FSD is that it represents a procedure very unlikely to reject preferred distributions. The cost of FSD is that it is unlikely to reject distributions that most risk-averse agents would eliminate from the efficient set.

Another problem with FSD, and for many of the criteria that followed, is what has been referred to as the left-hand tail problem. The left-hand tail problem is that the distribution which first accumulates probability can never dominate or be preferred to other distributions that accumulate probability later by all decision makers who prefer more to less, no matter what happens later on. In other words, suppose for cumulative distributions F and G that $G(y_{min}) > F(y_{min})$ for $(y_{min} < y < y_{max})$. Then, G can never be preferred to F even if $G(y) < F(y)$ for all other values of y. The reason for the left-hand tail problem is evident on reflection. Recall that the FSD criterion applies to all decision makers who prefer more to less. Obviously, this is a very large set of decision makers and all of these (universally) must prefer G to F before we can say that G is preferred or dominates F or vice versa. But getting all decision makers to agree is not an easy task because included in the set of decision makers who prefer more to less are the most risk averse and the most risk preferring of all decision makers. And the most risk averse decision makers are those who want to minimize the probability of the worst possible event from occurring.

The limitations of the FSD criterion soon became evident and a criterion that applied to a reduced set of decision makers followed. The second criterion posed and answered the question: What criterion would order choices for the set of all risk-averse decision makers? While the set of decision makers subject to the FSD criterion included both risk-averse and risk-preferring decision makers, in effect the entire space of possible values of an absolute risk averse function, the second set of decision makers in search of a criterion was limited to only risk averters whose utility functions satisfied $U = (y) > 0$ and $U''(y) < 0$. For these, their risk preferences were limited to the positive values of absolute risk aversion functions.

The criterion that applied to risk-averse decision makers also had an intuitive appeal. Create a new distribution from an existing one by moving probabilities to the tails of the distribution while preserving the mean, a mean-preserving spread. The new distribution with probability spread to the tails will be less preferred for all risk averters because of diminishing marginal utility. Diminishing marginal utility implies that increasing the probability of lower-valued outcomes is a greater reduction in expected utility than an equal increase in the probability of outcomes greater than the mean. Of course, the efficient set for risk averters was smaller than the efficient set for those who preferred more to less, and the criterion that assured that one distribution could be obtained from another by shifting probabilities became known as second-degree stochastic dominance (SSD).

In general, spreading probability from the center of a probability density function to its tails makes the transformed density function less preferred by risk averters than the original function and the shift in probability need not always preserve the mean. All that is required is that probability shifted from the center of the distribution to higher outcomes never exceed the amount of probability shifted an equal distance to lower-valued outcomes. Later Rothschild and Stiglitz (1971) gave other intuitive meaning to SSD, that of obtaining an inefficient distribution from an efficient one by adding white noise.

While SSD restricted preference representations measured using absolute risk aversion functions to positive values, instead of all possible absolute risk aversion values as under FSD, the efficient set associated with SSD is still not very discriminating and it still suffers from the left-hand tail problem. SSD suffers from the left-hand tail problem because the most risk averse decision maker whose sole concern was the least favorable outcome is still included in the risk-averse class of decision makers. And identifying a distribution as inefficient still requires unanimous agreement between all risk-averse decision makers.

Some enterprising economists, noting deductive similarities in the methods used to produce FSD and SSD, continued the effort and produced third-degree stochastic dominance and even fourth-degree stochastic dominance that applied restrictions to the third and fourth derivatives of the $U(y)$. However, these later dominance criteria have not been generally accepted, mostly because they are difficult to defend intuitively. This may change in the future as new measures of risk attitudes, such as prudence, that do depend on higher derivatives are considered (Kimball 1993).

SSD, with a more restrictive assumption on the probability distributions to be ordered, produced an interesting result. Suppose that the probability distributions to be ordered are normally distributed and completely described by their first two moments. Furthermore, suppose that any mean-preserving shift in probability to the tails preserves the mean and increases the variance of the distribution. As a result, all SSD-efficient members can be fully described by an EV-efficient set. So, without imposing additional assumptions about decision makers' preferences besides risk aversion, it is possible to argue that all risk averters facing normal distributions can find their expected utility maximizing choices in an EV-efficient set.

The EV criterion is particularly appealing to those choosing between portfolios consisting of many risky investments. A distribution describing a portfolio's outcomes should, according to probability laws applied to central limits, approach normality. Furthermore, even when choices could not be characterized by a large number of risky choices, transformations on initial outcome distributions produced by such risk management strategies as insurance, storage, diversification, holding reserves, and others, tend to shift probabilities from the tails to the center of distributions, and transform the final outcome distributions into what may more nearly approximate normal ones. While these tendencies support the assumption that final outcome distributions are likely to approach normal distributions, this hypothesis is in need of further development and testing before it is likely to produce increased support for the EV criterion.

For example, initial outcome distributions may describe profits from farm operations that depend on production levels and stochastic prices. These might also be used as interrogations of the decision maker to infer his or her risk preferences. But, if insurance programs also exist that subsidize incomes below a certain level and otherwise charge premiums, then the final outcome distribution is different than the initial one used to derive risk attitude measures and has more probability in the center of the distribution than does the initial outcome distribution.

As was bound to happen, careful scholars found reasons to distrust the orderings produced by FSD and SSD (and EV analysis). What happens to our orderings using FSD and SSD if, instead of knowing the true distributions, we really know only that they are empirical estimates subject to sampling error? Because the results of FSD and SSD depend on knowing the crossing points of cumulative density functions, not knowing these crossing points with certainty may lead to inconsistent orderings. Now FSD and SSD suffer not only from left-hand tail problems but also from requirements of knowing probability distributions with certainty, a requirement that is rarely satisfied.

For a brief moment, there was hope that convex set stochastic dominance could solve the problems inherent in FSD and SSD. Alas, hope was short-lived. A convex combination of portfolios that include a choice with a left-hand tail problem continues to have a left-hand tail problem.

STOCHASTIC DOMINANCE WITH RESPECT TO A FUNCTION

A significant breakthrough in methods for ordering probability distributions was Meyer's (1977) work on choices among distributions or, as it has become known, stochastic dominance with respect to a function (SDWRF). Meyer's important breakthrough demonstrated how a "bang-bang" optimal control program could order probability distributions for any class of decision makers defined by an interval measured in absolute risk aversion space. Let the absolute risk aversion function $r(y)$ be defined as $-U''(y)/U'(y)$ and define a class of decision makers whose preferences can be included in an interval around $r(y)$, bounded below by the absolute risk aversion function $r_L(y)$ and bounded above by the function $r_U(y)$. With this class of decision makers, available probability distributions can then be separated into efficient and inefficient sets.

Meyer's SDWRF was a gift that continued to give. Kramer and Pope (1981) demonstrated its use, employing the program that Meyer provided. Others developed program refinements. However, one of the major contributions of SDWRF was that it provided the means for eliminating the left-hand tail problem. Indeed, the method was so flexible that it could be used to order distributions for any well-defined class of decision makers. Indeed, a single function $r(y)$ was simply a special case of SDWRF.

SDWRF's strength, its flexibility to define classes of decision makers, produced an unexpected outcome. It returned attention to the measurement of decision makers' preferences. Early efforts had focused on measuring single-valued preference functions that employed various certainty equivalent interrogations that had generally been regarded as unreliable. FSD, SSD, and EV required no explicit risk preference measures, which was part of their popularity. Yet SDWRF required stronger assumptions about risk preferences than either FSD or SSD. SDWRF required that decision makers' preferences be defined by intervals, perhaps something less than knowing a unique function, $r(y)$, but still a stronger risk assumption than was required by either FSD, SSD, or EV analysis.

The solution for obtaining the interval that described decision makers' preferences was found in the method. King and Robison (1981) described how SDWRF and carefully defined probability distributions could be used to find decision makers' risk preferences. Repeating the procedures for several individuals could produce some interval describing risk preferences for a class of decision makers. So finally there was a method for measuring risk preferences and for ordering distributions for the class of decision makers flexibly defined.

Alas, our enthusiasm for SDWRF was not to be long-lived because SDWRF, like FSD and SSD, assumed we knew more about the distributions being ordered than could be justified by the data. In other words, just as our confidence in FSD and SSD to order distributions was reduced by having only empirical distributions as opposed to true distributions, the same could be said for SDWRF. One approach for dealing with this problem was to recognize our uncertainty about probability distributions by reducing the certainty applied to our measures of risk preferences, in effect increasing the interval around $r(y)$. While this may have some logical appeal, the empirical benefits have yet to be demonstrated.

Finally, our confidence in our ability to measure risk preferences using SDWRF was shaken by another empirical reality. Efforts to measure risk preferences assumed that the probability choices were final outcome distributions, i.e., distributions describing what would ultimately be experienced by respondents.

But the distributions used to measure preferences rarely qualified. Existing risky holdings and nonlinear functions associated with insurance and other risk responses acted to transform responses to pre-selected probability distributions used to measure risk preferences into something quite different than final outcome distributions. In effect, our efforts to elicit risk attitude measures were in fact identifying a transformed preference function – something like risk preferences inferred from a football coach's choice of plays at the end of the game and with few opportunities left to score and win the game (Robison and Lev 1986). And so efforts to measure risk attitudes have lately been on vacation, or at least they should be until we resolve some of the problems inherent in the process.

LOCATION AND SCALE

Most recent efforts to order probability distributions have returned attention to assumptions about probability distributions. In their book, Robison and Barry (1987) note that if decision makers were facing a single distribution, and choices were between varying amounts of the risky asset and a safe asset, then all distributions would lie on an EV-efficient frontier. Sinn (1983) and later Meyer (1987) improved upon this observation. They demonstrated that if distributions were related to each other by location-scale, then expected utility choices would be included in EV or, equivalently, mean-standard deviation sets. Expected utility functions defined over means and standard deviations were attractive and allowed economists to represent income and substitution effects graphically, something we had not generally been able to do in expected utility models.

Then came a major result. Relaxing the assumption that we knew the exact form of the probability distribution, Meyer and Rasche (1992) demonstrated using a Kolmogorov-Smirnov test that for most portfolios comprised of financial data, even with surprisingly large numbers of sample observations, one could not reject the hypothesis that the distributions were related to each other by location-scale. Furthermore, if one could not reject the hypothesis that the distributions were related to each other by location-scale, neither could one reject the hypothesis that the efficient set for risk-averse agents was the EV-efficient set.

Despite the importance of Meyer and Rasche's results, there appears to be little evidence that their contributions have been applied. Application of stochastic dominance with their strong assumptions about probability density functions still dominates among efforts to identify efficient sets. Some considerable resistance to EV methods generally remains, and perhaps with some good reasons. Meyer and Rasche's results have not been replicated in other settings besides the financial markets.

CONCLUSIONS

In a somewhat surprising turn of events, substituting weaker assumptions about probability distributions for strong ones returned us to the EV criterion. But this is not the end of the story either. Can we define an EV frontier if what we have are only estimates of expected values and standard deviations? The answer is no. And instead of a single EV frontier, what we have is a confidence interval around our estimates of expected values and standard deviations.

Nevertheless, the renewed focus on EV sets because of their empirical properties is likely to yield some possible theoretical dividends. The reason is that it will likely facilitate the next set of advances in developing the theory of the firm facing risk. The reason is simple. Generalized expected utility models are cumbersome and generally lead to ambiguous theoretical results when decision makers are facing multiple risk choices. This is not the case for EV analysis that can sum over risky choices to obtain combined measures of risk and expected income and wealth.

REFERENCES

Hadar, J., and W.R. Russell. 1969. "Rules for Ordering Uncertain Prospects." *American Economic Review* 59: 25-34.

Hanoch, G., and H. Levy. 1969. "Efficiency Analysis of Choices Involving Risk." *Review of Economic Studies* 38: 335-346.

Kimball, M. 1993. "Standard Risk Aversion." *Econometrica* 61: 589-611.

King, R., and L. Robison. 1981. "An Interval Approach to Measuring Decision Maker Preferences." *American Journal of Agricultural Economics* 63: 510-520.

Kramer, R.A., and R.D. Pope. 1981. "Participation in Farm Commodity Programs: A Stochastic Dominance Analysis." *American Journal of Agricultural Economics* 63: 119-128.

Meyer, D.J., and J. Meyer. 2001. "Habit Formation Utility, the Equity Premium Puzzle, and Risk Preferences in Multi-Period Consumption Models." Draft Working Paper (January).

Meyer, J. 1977. "Choice Among Distributions." *Journal of Economic Theory* 14: 326-336.

___. 1987. "Two Moment Decision Models and Expected Utility Maximization." *American Economic Review* 77: 421-430.

Meyer, J., and R.H. Rasche. 1992. "Sufficient Conditions for Expected Utility to Imply Mean-Standard Deviation Rankings: Empirical Evidence Concerning the Location and Scale Condition." *The Economic Journal* 102: 91-106.

Pope, R.D., and R.F. Ziemer. 1984. "Stochastic Efficiency, Normality, and Sampling Errors in Agricultural Risk Analysis." *American Journal of Agricultural Economics* 66: 31-40.

Robison, L.J., and P.J. Barry. 1987. *The Competitive Firm's Response to Risk.* New York: Macmillan Publishing Company, Inc.

Robison, L.J., and L. Lev. 1986. "Distinguishing Between Indirect and Direct Outcome Variables to Predict Choices Under Risk or Why Woody Chip Went to the Air." *North Central Journal of Agricultural Economics* 8: 59-68.

Rothschild, M., and J.E. Stiglitz. 1971. "Increasing Risk II: Its Economic Consequences." *Journal of Economic Theory* 2: 66-84.

Sinn, H.-W. 1983. *Economic Decisions Under Uncertainty.* New York: North-Holland Publishing Company.

Chapter 4

CONCEPTUAL FOUNDATIONS OF EXPECTATIONS AND IMPLICATIONS FOR ESTIMATION OF RISK BEHAVIOR

Richard E. Just and Gordon C. Rausser
University of Maryland and University of California, Berkeley

INTRODUCTION

Many prominent agricultural economists suggest that risk has only a second-order effect in the agricultural economy and that attention should rather be focused on understanding first-order effects. To determine whether risk effects are of first- or second-order importance, however, producers cannot be modeled in isolation. The linkages among input suppliers and downstream processing, wholesaling, and distribution are also critical. Many of the other agents in agriculture and food systems are keenly concerned not only about market risk but also about basis, credit, and output risk.

Presumably, no prominent agricultural economist would argue that market price, revenue, or cost expectations are of second-order importance. In fact, conventional methodologies require an assessment and measurement of such expectations before quantifying and measuring the effects of risk. Typically, expectation formation patterns are imposed as part of the maintained hypothesis in any measurement and causal analysis of the impact of risk and/or assessment of risk aversion.

The purpose of this chapter is to relax the maintained hypothesis of such analyses, which typically imposes an expectation mechanism arbitrarily, and determine the resulting implications for modeling behavior under risk. This focus requires consideration of the literature on modeling risk and expectation formation patterns. An assessment of this literature leads us to the introduction of a more generic framework of rational expectations and rational determination of the effects of risk than in existing literature. We show that under certain circumstances, adaptive and even naïve expectations can, in fact, be rational.[1] The driving force determining this result is nothing more than the

[1] In some related conceptual work, we note that Brock and Hommes (1997) use the term "rational expectation" to refer to decision makers' choices to use naïve versus adaptive

cost of collecting and processing information. Few if any rational expectation models that appear in the literature balance the benefits and costs associated with such information processing and collection.

After reviewing the relevant literature, we frame the fundamental problem. We demonstrate that, for problems where risk is material, the efforts in collecting and processing information are based on two dimensions: accuracy of expectations and assessment of risk. An obvious intuitive result emerges: heterogeneity of risk preferences implies heterogeneity of information collection and processing. Regardless of such heterogeneity, unforeseen possibilities arise. The realization of low-probability–high-stakes events can often result in unpredictable shifts in confidence, sometimes resulting in significantly biased expectations driven by fear. As Alan Greenspan (2001) recently noted, "our economic models have never been particularly successful in capturing a process driven in large part by non-rational behavior" (p. 4).

The main paradigm that has been offered in an attempt to systematize non-rational behavior has been characterized as the behavioral branch of economics. It has gained increasing acceptance over the past few decades. The documented empirical results from this paradigm emphasize the biases in human judgment under uncertainty. We compare and contrast the non-rational paradigm with the cost-consistent rational expectation formulation presented here. We note that cost-consistent rational expectations as well as other formal approaches can apparently achieve considerable observational equivalence to paradigms based on systematic non-rational behavior and other non-expected utility formulations.

The final section of this chapter goes beyond individual decision making to evaluate competitive market interactions. In this setting, market discipline can force some convergence toward a cost-consistent rational expectation equilibrium. However, existence of such an equilibrium is not clear and, if it exists, it is possibly not locally stable. Convergence properties, as in any risky environment, cannot be formally analyzed without recognizing market learning. Here, as in much of the literature, we draw a distinction between rational versus irrational learning. Based on this analysis, we suggest that one explanation for the emergence of the non-rational behavioral literature is failure of conventional approaches to incorporate cost of information and/or learning appropriately.

DEPENDENCE OF BEHAVIORAL ESTIMATION ON EXPECTATIONS SPECIFICATIONS

A vast literature on agricultural production attempts to estimate farmers' response to risk and farmers' attitudes toward risk. While a few studies have

---

expectations based on performance. However, in their work these are the only two expectation choices available.

attempted to assess farmers' attitudes toward risk by survey approaches related to hypothetical risk choices (e.g., Lin, Dean, and Moore 1974), and the innovative study by Binswanger (1980) attempted to estimate risk attitudes by offering real risky choices with controlled objective probabilities, most studies have attempted to estimate risk preferences and response using revealed preference data on actual farm decision making (real data). A critical problem with using real data is that estimating response to risk depends on correctly representing the level of risk at each data point. In turn, representing the level of risk depends on properly characterizing the measure of central tendency, i.e., the expectation. Clearly, the estimated response to risk is flawed if the level of risk is characterized incorrectly. Examination of the literature, however, reveals that the profession has been extremely cavalier about this critical dependence of risk measures on the expectations mechanism and on the specification and measurement of the expectations mechanism.

Risk is typically represented in terms of the probabilities of (squared) deviations from an expectation. If the wrong expectation is used to assess deviations, then risk facing agents can be either over- or under-estimated. As a result, response to risk is either under- or over-estimated, respectively. As Pesaran (1987) concludes,

> "In the absence of direct observations on expectations, empirical analysis of the expectations formation process can be carried out only indirectly, and conditional on the behavioural model which embodies the expectational variables. This means that conclusions concerning the expectations formation process will not be invariant to the choice of the underlying behavioral model.... Only when direct observations on expectations are available is it possible to satisfactorily compare and contrast alternative models of expectations formation" (p. 207).

Or as Nerlove and Bessler (2001) state it, "If we reject a particular hypothesis it is not clear whether we are rejecting the underlying behavioral theory or the expectational hypothesis in question" (p. 179).

In spite of this critical dependence on modeling expectations, most empirical studies of risk behavior arbitrarily impose a specific expectation mechanism as part of the maintained hypothesis without properly acknowledging the dependence of results on that choice. Typical practices for modeling expectations include specifying expected price or expected revenue per acre as (1) a naïve expectation equal to price or revenue in a previous period (e.g., Antonovitz and Green 1990 average several recent months), (2) a 3-period moving average (Behrman 1968) or 3-period weighted average (Traill 1978), (3) an adaptive expectation with an estimated geometric expectation parameter (Nerlove 1958, Just 1974, 1977), (4) a polynomial lag model (Lin 1977), (5) an autoregressive process (Holt and Moschini 1992, Chavas and Holt 1996), (6) a futures price (Gardner 1976, Just and Rausser 1981, Rausser et al. 1986), (7) an extrapolative expectation developed by extending the trend of a few recent time periods (Goodwin 1947, Ryan 1977), (8) an implicit expectation developed by finding the expectation that best fits observed behavior (Mills 1955, 1962; see Nerlove and Bessler 2001 for a critical evaluation), (9) a reduced-form estimate derived by regression on exogenous, e.g., pre-harvest, production inputs (Antle 1987, 1989), (10) a structural estimate derived by simultaneous solution of an estimated supply-demand system, sometimes called a "fully rational expectation" (Antonovitz and Green 1990; see also Nerlove and Bessler 2001),[2] and (11) a quasi-rational expectation developed by choosing the best-fitting Autoregressive Integrated Moving Average (ARIMA) model (Eckstein 1984, Goodwin and Sheffrin 1982).

An examination of this literature on estimation of risk behavior reveals that the choice of specification for the expectations mechanism is rarely discussed. Rather, the expectations mechanism is almost always imposed arbitrarily without justification. Yet, as demonstrated by Just and Weninger (1999), the choice of the mean function with which to interpret residual variation has dramatic effects on the characterization of risk including skewness and kurtosis as well as variance.

Proposition 1. *Misspecification of the expectations mechanisms used to characterize risk may be seriously biasing conclusions from the empirical risk behavior literature.*

After imposing an arbitrary expectation mechanism, risk studies typically proceed to model risk response as a response to variation in the induced deviations, most often measured by some type of average of squared deviations from the imposed expectation mechanism. These include using (1) a simple 3-period moving average of squared deviations (Behrman 1968), (2) an arbitrarily weighted 3-period moving average of squared deviations (Chavas and Holt 1996) or a square root of such (Ryan 1977, Lin 1977), (3) a weighted 3-period moving average of absolute deviations with estimated weights (Traill 1978), (4) an adaptive risk specification with an estimated geometric weighting of past squared deviations (Just 1974, 1977), (5) a reduced-form estimate derived by regressing squared deviations on exogenous production inputs (Antle 1987, 1989), (6) a "rational" variance found as the expected squared deviation from a simultaneous solution of an estimated supply-demand system taking into account the usual stochastic disturbances in supply and demand specifications (Antonovitz and Green 1990), (7) a [Generalized] AutoRegressive Conditional Heteroskedasticity framework, i.e., an ARCH [GARCH] model, which estimates heteroskedasticity as a function of lagged squared

[2] A reduced-form approach such as used by Antle (1987, 1989) might be construed as a fully rational expectation with certain qualifications. However, reduced-form estimation is typically less efficient than deriving the implied reduced form based on estimation of a complete structural model (Dhrymes 1973).

deviations [and additional moving average terms] (Aradhyula and Holt 1989, Holt and Moschini 1992) and (8) a nonparametric kernel estimator that averages squared deviations in surrounding years (Holt and Moschini 1992). In the few studies that have characterized risk with more than second moments, the third and higher moments have similarly been characterized by regressions of third and higher powers of deviations on other variables in the model (Antle 1987, 1989).

As for the case of expectation mechanisms, the choice of the empirical measurement of risk for a given expectation mechanism is rarely given much discussion either. Some of the more recent ARCH-GARCH literature is an exception where generalizations are motivated by the need for more flexibility in determining a risk measure. For example, Holt and Moschini (1992, p. 1) criticize "simple fixed-weight extrapolative methods to generate time-varying risk measures" because "they do not make optimal use of available data" (see also Pagan and Ullah 1988). Nevertheless, this literature stops far short of a comprehensive test of applicability of the many approaches that have been proposed. Furthermore, this literature focuses on Box-Jenkins type flexibility as opposed to flexibility in representing relationships grounded in economic theory.

Proposition 2. *Mischaracterization of risk given the expectations mechanism may be seriously biasing conclusions from the empirical risk behavior literature.*

Optimal use of data to form expectations is also a concept that depends on the context of the problem. Rational expectations for empirical work are typically defined in a variety of ways. Antle (1987, 1989), for example, determines expectations by regressing revenue on other variables in the model. This approach has often been regarded as estimation of a reduced-form equation that summarizes the "rational" implications of observable variables for expectations of others. Antonovitz and Green (1990), on the other hand, simultaneously solve supply and demand specifications to obtain the implied "rational" expectation. Holt and Moschini (1992, p. 2) refer to the ARCH-GARCH approach as "a type of rational-expectations model" that includes both first and second moments. However, by using an autoregressive expectations mechanism, this is a Box-Jenkins type of rationality compared to the structural rationality implied by a supply-demand system.

Comparing these approaches, an additional point that follows immediately is that what is defined as the rational expectation depends on the scope in which the decision problem is represented. A broader problem scope, i.e., a more comprehensive rational expectation, requires more information. For example, should a rational expectation be defined by a supply-demand system that determines the particular price facing an agent, or should it also be based on determining the rational implications of related markets, say, for the prices that serve as determinants of the supply and demand in question? In theory, a rational expectation summarizes the implications of all available information, but in practice the information base must be limited in some practical way for tractability.[3]

Proposition 3. *In practice, rational expectations are not uniquely defined. They depend on the scope and information space of the decision problem, which in turn determine the database and information required to model them.*

HETEROGENEITY OF RISK PREFERENCES AND ASSESSMENTS

As emphasized by Just and Pope (1999) and R.E. Just (forthcoming), one of the greatest obstacles to risk research in agriculture is heterogeneity of farms and farmers. Indeed, the empirical work of Binswanger (1980) and Antle (1987, 1989) gives evidence of heterogeneity of risk preferences among farmers. However, empirical work also gives evidence of heterogeneity of expectations (Chavas 1999, 2000, Baak 1999).

In the rare early study by Antonovitz and Green (1990), the applicability of alternative expectations schemes is compared empirically. They consider a variety of the leading expectations mechanisms: (1) naïve expectations, (2) ARIMA expectations, (3) futures price expectations, (4) adaptive expectations, and (5) rational expectations. Their empirical results imply that no one expectations model dominates the rest. The "evidence suggests that expectations are heterogeneous rather than homogeneous" (p. 485). In particular, they find with respect to the root-mean-squared error fit of these models that "the highest value was observed for the rational expectations model perhaps adding additional evidence to reject the hypothesis that expectations are formed rationally" (p. 485). And, as they note, "signs and elasticities may be significantly different depending on which model is used" (p. 486).

Antonovitz and Green's conclusions are in harmony not only with later empirical results but also with conceptual understanding. For example, Schultz (1975) has argued that farmers acquire information at different rates, which leads to heterogeneous expectations. Pingali and Carlson (1985) find that human capital determines the accuracy of expectations associated with technological change. The underlying theme in this literature is that information bases differ among agents. If so, then those differences must be considered in

---

[3] In practice, these considerations often amount to technical considerations such as whether price supports or target prices are used to truncate distributions suggested by adaptive or rational expectations approaches, or whether the disturbances of simultaneous equations systems describing supply and demand are included in describing risk (see, e.g., Antonovitz and Green 1990). More realistically, however, a much broader set of issues could conceivably be incorporated into the scope of rational expectations and risk measurements.

assessing the risk faced by different agents because of the dependence of the risk assessment on the expectations mechanism.

In an early but little-known paper that predates Antonovitz and Green's work, Just and Rausser (1983) consider the case of heterogeneity of expectations mechanisms due to differences in circumstances, the cost of acquiring information, and the associated differences in information used by different agents. They develop a detailed model considering six alternative expectations regimes: (1) rational expectations, (2) adaptive expectations, (3) naïve expectations, (4) futures market prices, (5) normal expectations, and (6) convex combinations of (1)-(5). Both price and production uncertainty as well as risk aversion, storage, and basis risk are formally incorporated into the model representation. A variety of agents are considered in developing the equilibrium dynamics of both futures and spot market prices: (1) producer/hedgers, storer/hedgers, exporter-processor/hedgers, and speculators. Three important points are suggested by this work, which are stated as propositions based on demonstrations in the following section.

Proposition 4. *What a rational expectation for an individual agent is depends on the cost of information relative to its associated benefits, which in turn depends on market volatility. As a result, the optimal expectation mechanism is appropriately determined endogenously.*

Just and Rausser (1983) suggest that naïve and normal expectations may be optimal in periods of great stability because the relative benefits of information-intensive expectations such as traditionally defined rational expectations may be small compared to their information costs. On the other hand, traditionally defined rational expectations or other information-intensive expectations may have benefits that exceed their information costs in periods of volatility. Because greater model complexity requires (1) more information to facilitate specification, (2) more data to facilitate calculation and updating, and (3) more cost of information processing, a clear tradeoff emerges between expectation accuracy and the cost of information. Intuition also suggests a tradeoff between model complexity and the accuracy of forecasts produced. By incorporating the cost of acquiring and processing information, even naïve expectations may be truly rational expectations in certain economic environments or for certain agents after considering the cost of information and information processing. More important, the expectations mechanism of choice becomes endogenous depending on information cost and volatility rather than fixed and exogenous as typically modeled.

Proposition 5. *A tradeoff between information and accuracy of expectations associated with choice of the expectation mechanism induces a tradeoff between cost and accuracy of expectations.*

Taking these arguments to their logical conclusion implies that no truly rational expectation can be defined independent of the cost of information or independent of the volatility of the economic environment. Yet, the existing literature fails to define rational expectations as depending on either (Brock and Hommes 1997 is an exception). Moreover, this endogeneity of information and expectation choices naturally leads to heterogeneity of expectations depending on individual circumstances and preferences.

Proposition 6. *The tradeoff between the cost of information and accuracy of expectation together with the heterogeneity of risk preferences leads to heterogeneity of expectations and risk assessments.*

While the literature has recognized that better information results in more accurate expectations and has even attempted to measure decision makers' willingness to pay for information under price uncertainty (see, e.g., Roe and Antonovitz 1985, Antonovitz and Roe 1986), we are not aware of studies that consider dependence of expectations or assessment of risk on the cost of information in empirical measurement. Thus, we turn to developing a formal framework in the next section that illustrates these three points.

FRAMING THE PROBLEM

For purposes of facilitating analytic discussion of the issues raised above, we introduce a stylized model in which information for forming expectations and associated perceptions of risk is costly. We pose the problem with minimal generality essential to focus on the issues raised by this chapter. For example, the model considers formation of expectations for only the mean and variance (which would represent all two-parameter distributions) even though similar results could be derived in principle for models with higher moments and more general expected utility functions. We also assume constant marginal cost for simplicity, which is easily relaxed by substituting marginal cost for average cost in the equilibrium first-order conditions.

Suppose a single-product, constant-cost producer has profit given by $\pi = pq - cq$, where p is output price, q is output quantity, and c is a constant average cost of production. Suppose output q must be chosen before price uncertainty is resolved. Production uncertainty is not assumed because it is unnecessary to illustrate essential issues. Suppose with full information at decision making time that the expected price is $E(p) = \mu$ and the variance of price is $V(p) = \sigma_p$. Alternatively, with minimal information (for example, under naïve expectations), expected price is $\overline{E}(p) = \mu + \delta$ and the variance of price is $\overline{V}(p) = \sigma_p /(1+\varepsilon)$. In other words, δ represents an error made in forming the price expectation and ε represents an error made in assessing the variance of price relative to the case of full information.

Suppose that the producer maximizes a mean-variance expected utility function. The decision problem under minimal information and absolute risk aversion α is

$$\max_q \bar{E}[U(\pi)] = \bar{E}(pq - cq) - (\alpha/2)\bar{V}(pq - cq)$$
$$= (\mu + \delta - c)q - (\alpha/2)[\sigma_p/(1+\varepsilon)]q^2,$$

for which the first-order condition implies optimal output

$$\tilde{q} = \frac{(\mu + \delta - c)(1+\varepsilon)}{\alpha\sigma_p}.$$

Now consider the effects of unanticipated errors in expectations due to limited information availability. Although these errors are not anticipated, they nevertheless affect ex post utility. Specifically, assume $E(\delta) = E(\varepsilon) = 0$, $V(\delta) = \sigma_\delta$, $V(\varepsilon) = \sigma_\varepsilon$, and that δ, ϵ, and p are mutually independent.[4] Then, correctly considering full information, expected ex post output is $E(\tilde{q}) = (\mu - c)/\alpha\sigma_p$ and expected ex post utility is[5]

$$E[U(\pi)] = E(p\tilde{q} - c\tilde{q}) - (\alpha/2)V(p\tilde{q} - c\tilde{q})$$
$$= \frac{(\mu - c)^2}{\alpha\sigma_p} - \frac{[\sigma_p + (\mu - c)^2][\sigma_\delta + (\mu - c)^2][\sigma_\varepsilon + 1]}{2\alpha\sigma_p^2}. \tag{1}$$

The second right-hand term of (1) is the risk premium. It is increasing in both σ_δ and σ_ε, illustrating how ex post expected utility is adversely impacted by poor information about either the mean or the variance of prices.

Now consider the tradeoff between benefits and costs of information. Suppose the producer can purchase information that improves assessment of both the mean and/or the variance of price resulting in expected price $\bar{E}(p) = \mu + \theta\delta$ and variance of price is $\bar{V}(p) = \sigma_p/(1+\phi\varepsilon)$. That is, suppose

---

[4] Independence of δ and ϵ from p is reasonable since δ and ϵ are not part of the moments of the true p distribution. They affect only the subjective distribution of p under imperfect information. Errors in this distribution would be due to factors that do not ultimately affect p.

[5] Note that

$$V\left[(p-c)\frac{(\mu + \delta - c)(1+\varepsilon)}{\alpha\sigma_p}\right] = E\left[\frac{(p-c)^2(1+\varepsilon)^2(\mu+\delta-c)^2}{\alpha^2\sigma_p^2}\right] - \frac{(\mu-c)^4}{\alpha^2\sigma_p^2},$$

which upon simplification leads to the second right-hand term of (1) aside from the risk aversion factor, $\alpha/2$. Note that we assume unanticipated variation in profit has the same adverse impact on expected utility as anticipated variation.

θ reflects the amount of information obtained on the mean price where $\theta = 1$ represents minimal information, $\theta = 0$ represents full information, and $0 \leq \theta \leq 1$. Similarly, suppose ϕ reflects the amount of information obtained on the variance of price where $\phi = 1$ represents minimal information, $\phi = 0$ represents full information, and $0 \leq \phi \leq 1$. Finally, to represent the tradeoff between benefits and costs of information, suppose the cost function for information follows $\zeta(\theta, \phi)$ and is appropriately decreasing in both arguments (i.e., increasing in information). For intuitive purposes, it may be useful to regard $(\theta, \phi) = (1,1)$ as the case with naïve expectations and $(\theta, \phi) = (0,0)$ as the case of full information rational expectations (using the broadest scope of definition). Each of the other standard expectations mechanisms can be construed as corresponding to intermediate points in the unit square for (θ, ϕ).

By analogy with results above, the producer's ex ante production problem is thus

$$\max_{q} \overline{\mathrm{E}}\,[U(\pi)] = (\mu + \theta\delta - c)q - (\alpha/2)[\sigma_p/(1 + \phi\varepsilon)]\,q^2,$$

which yields the optimal decision choice

$$\tilde{q} = \frac{(\mu + \theta\delta - c)(1 + \phi\varepsilon)}{\alpha\sigma_p}, \tag{2}$$

and expected ex post utility

$$E\,[U(\pi)] = \frac{(\mu - c)^2}{\alpha\sigma_p} - \frac{[\sigma_p + (\mu - c)^2][\theta^2\sigma_\delta + (\mu - c)^2][\phi^2\sigma_\varepsilon + 1]}{2\alpha\sigma_p^2} \equiv W(\theta, \phi).$$

Now consider which tradeoff between benefits and costs of information maximize the producer's expected net benefits. Assuming an internal solution, the information optimization problem,

$$\max_{\theta,\phi} W(\theta, \phi) - \zeta(\theta, \phi),$$

yields first-order conditions,

$$W_\theta - \zeta_\theta = -\frac{\theta\sigma_\delta}{\alpha\sigma_p^2}[\sigma_p + (\mu - c)^2][\phi^2\sigma_\varepsilon + 1] - \zeta_\theta = 0,$$

$$W_\phi - \zeta_\phi = -\frac{\phi\sigma_\varepsilon}{\alpha\sigma_p^2}[\sigma_p + (\mu - c)^2][\theta^2\sigma_\delta + (\mu - c)^2] - \zeta_\phi = 0,$$

where subscripts on W and ζ represent differentiation.[6] The optimal information choices satisfy

$$\theta = -\frac{\alpha\sigma_p^2\zeta_\theta}{\sigma_\delta[\sigma_p + (\mu - c)^2][\phi^2\sigma_\varepsilon + 1]}, \tag{3}$$

$$\phi = -\frac{\alpha\sigma_p^2\zeta_\phi}{\sigma_\varepsilon[\sigma_p + (\mu - c)^2][\theta^2\sigma_\delta + (\mu - c)^2]}. \tag{4}$$

Note that $\zeta_\theta < 0$ and $\zeta_\phi < 0$ so both θ and ϕ will be positive unless information is costless ($\zeta_\theta = \zeta_\phi = 0$), risk aversion is zero ($\alpha = 0$), or price risk is not present ($\sigma_p = 0$), in which case optimal information choices are zero. If ζ_θ and ζ_ϕ are highly negative throughout (information is very expensive), then the results in (3) and (4) will exceed the unit interval constraints and a full Kuhn-Tucker solution would result in $\theta = 1$ and/or $\phi = 1$, implying use of minimal information.

To further examine the implications of (3) and (4) for the case of an internal solution, suppose ζ_θ and ζ_ϕ are constant.[7] Then, by inspection of (3) and (4), the optimal choices of θ and ϕ are characterized by $\partial\theta/\partial\zeta_\theta < 0$, $\partial\theta/\partial\sigma_\delta < 0$, $\partial\theta/\partial\sigma_\varepsilon < 0$, $\partial\theta/\partial\sigma_p > 0$, $\partial\theta/\partial\mu < 0$, $\partial\theta/\partial c > 0$, $\partial\theta/\partial\alpha > 0$, $\partial\phi/\partial\zeta_\phi < 0$, $\partial\phi/\partial\sigma_\delta < 0$, $\partial\phi/\partial\sigma_\varepsilon < 0$, $\partial\phi/\partial\sigma_p > 0$, $\partial\phi/\partial\mu < 0$, $\partial\phi/\partial c > 0$, and $\partial\phi/\partial\alpha > 0$.[8] These results have the following implications regarding the purchase and use of information about both the mean and variance of prices:

---

[6] One can show that joint second-order conditions are satisfied if the sample space for δ and ε is limited by $|\delta| < \mu - c$ and $|\varepsilon| < 1$. This restriction is essentially the same as assuming that $\tilde{q}$ is positive for all possible choices of θ and ϕ over the entire sample space for δ and ε.

[7] The assumption of constant ζ_θ and ζ_ϕ is not necessary but allows simple discussion because comparative static results are evident by visual inspection. Alternatively, assuming $\zeta_{\theta\theta} \geq 0$, $\zeta_{\phi\phi} \geq 0$, and $\zeta_{\theta\theta}\zeta_{\phi\phi} - \zeta_{\theta\phi}^2 \geq 0$, more complicated comparative static methods yield the same qualitative conclusions as when ζ_θ and ζ_ϕ are constant. Note that these assumptions are plausible and correspond to increasing marginal cost of information (recall that increases in information correspond to reductions in θ and ϕ).

[8] These partial results are sufficient to reach global conclusions for a joint solution of θ and ϕ because $\partial\theta/\partial\phi > 0$ and $\partial\phi/\partial\theta > 0$. That is, suppose partial derivatives of (3) and (4) are represented by $\partial\theta/\partial x$ and $\partial\phi/\partial x$, respectively, for $x = \sigma_\delta, \sigma_\varepsilon, \sigma_p, \mu, c, \alpha$. Then complete comparative static results follow from

$$\begin{bmatrix} 1 & -\partial\theta/\partial\phi \\ -\partial\phi/\partial\theta & 1 \end{bmatrix}\begin{bmatrix} d\theta \\ d\phi \end{bmatrix} = \begin{bmatrix} \partial\theta/\partial x \\ \partial\phi/\partial x \end{bmatrix} dx$$

1. Higher marginal cost of information reduces optimal information use.
2. More effectiveness of information in predicting the price mean increases information use.
3. More effectiveness of information in predicting price variance increases information use.
4. Increases in price variability cause reduced information use.
5. Increases in expected price cause increased information use.
6. Increases in the cost of production cause reduced information use.
7. Producers with more risk aversion use less information.

The first of these implications has already been discussed at length above. These are simply confirming results. The second and third of these implications are also highly plausible. The fourth implication is surprising but explained by the contraction in production scale and consequent expected profit caused by increased price variability. As the scale of production is contracted, the volume of output on which the producer benefits from information is reduced, causing less information to be used. The fifth and sixth implications are plausible and follow from two complementary effects. Increases in net revenue per unit of output both increase the expected profit per unit of output and the output by which it is multiplied. Because of increased revenue, the risk premium becomes relatively less important and the costs of information are spread across more output and revenue. The seventh implication is counterintuitive but also has an explanation similar to implications 4 and 6. Increased risk aversion leads to a contraction of output. With a scaling back of output, the information costs cannot be spread over as much output and, therefore, are also reduced.

Of course, in principle, more general utility specifications could be used to show similar results for higher moments. Also, different structures for risk preferences will likely lead to different results. For example, constant relative rather than constant absolute risk aversion will no doubt change the sensitivity of the comparative static results to changes in production scale. Nevertheless, the general implications are clear. For example, in support of Proposition 4, the optimal amount of information depends on the cost of acquiring it, and the optimal amount of information used in forming expectations depends on volatility. Under constant absolute risk aversion, the optimal amount of information is increasing with respect to unanticipated volatility and

---

and yield

$$\begin{bmatrix} d\theta / dx \\ d\phi / dx \end{bmatrix} = \frac{1}{|D|} \begin{bmatrix} 1 & \partial\theta / \partial\phi \\ \partial\phi / \partial\theta & 1 \end{bmatrix} \begin{bmatrix} \partial\theta / \partial x \\ \partial\phi / \partial x \end{bmatrix},$$

where $|D|$ is the determinant of the premultiplying matrix in the first equation, which is positive when second-order conditions are satisfied. Thus, $\partial\theta / \partial\phi > 0$ and $\partial\phi / \partial\theta > 0$ are sufficient for the complete comparative static results to agree in signs with the partial derivatives as long as $\partial\theta / \partial x$ and $\partial\phi / \partial x$ agree in sign for $x = \sigma_\delta, \sigma_\varepsilon, \sigma_p, \mu, c, \alpha$.

decreasing with respect to anticipated volatility. Accordingly, the optimal expectation mechanism endogenously depends on information costs and market volatility.

Also, in support of Proposition 6, because optimal information acquisition depends on risk aversion, heterogeneity of risk preferences necessarily leads to heterogeneity of information choices and, thus, heterogeneity of expectations. This is illustrated by the dependence of the optimal information choice on risk aversion. The comprehensive review of expectations by Nerlove and Bessler (2001) emphasizes heterogeneity of expectations in the determination of aggregate outcomes based on work by Nerlove (1983) and Frydman and Phelps (1983), and notes that heterogeneity is inconsistent with the representative agent assumption. However, they advance no theory explaining the heterogeneity of expectations. Interestingly, the implications of the expectations model here explain the observed heterogeneity of expectations based on the cost of information and the heterogeneity of risk preferences.

Finally, the results of this section suggest a nontrivial definition of information. That is, information is not simply one-dimensional. Some information may improve expectations while other information may also improve risk perceptions. Forecasts that simply give an expected price can offer little toward assessing risk. Similar comments may apply to higher moments as well.

Proposition 7. *For problems where risk is material, the choice of information is nontrivial; it must include dimensions of information that are relevant to anticipating both expected prices and the variation of prices about expectations.*

UNFORESEEN POSSIBILITIES

In early work, Frank Knight (1921) drew a clear conceptual distinction between outcomes that are foreseeable and outcomes that are not foreseeable. In his distinction, outcomes for which probabilities could be characterized by decision makers and used in optimizing expected profits and expected utility were characterized as risky outcomes whereas outcomes that could not be foreseen with probabilities representing their likelihood were characterized as uncertain outcomes. Conceptually, this distinction is intuitively plausible but the literature has failed to build much upon it because the standard paradigm of decision making, von Neumann-Morgenstern expected utility maximization, does not admit problems with outcomes that cannot be foreseen or for which probabilities cannot be determined in some way by decision makers. As a result, more than half a century of research following Knight's work simply ignored Knightian uncertainty.

Interestingly, this approach to economic analysis is quite similar to the practice found by the psychology literature whereby economic agents in reality ignore or underweight outcomes that are unforeseeable, difficult to evaluate, or too remote to quantify easily (see, e.g., Alpert and Raiffa 1982, and Oskamp 1982). Sometimes these remote outcomes, because they are low-probability events far from the mean, are high-stakes outcomes. While the economic literature to date has tended to ignore these cases of Knightian uncertainty, these findings by psychologists provide a basis for quantitative modeling upon considering the cost of information and the cost of processing it.

Low-Probability–High-Stakes Events. To consider a simple model to illustrate the point and in the spirit of the stylized model above, consider a single-product, constant-cost producer with profit given by $\pi = pq - cq$ and other notation as above where output q must be chosen before price uncertainty is resolved. Suppose that the producer is aware of conventional outcomes for prices described by a distribution with mean μ_1 and variance σ_1, but unaware of another set of low-probability–high-stakes (LPHS) price outcomes with mean μ_2 and variance σ_2. Under minimal information, suppose the producer simply ignores the LPHS possibilities due to unawareness. Consistent with LPHS, the true probability ρ of an outcome from the second set of outcomes is assumed small ($\rho < .5$) and the variation of LPHS outcomes is high ($\sigma_2 > \sigma_1$). Also, to focus on adverse outcomes that are likely of more concern in agricultural producer problems, suppose $\mu_2 < \mu_1$. The true mean and variance of price outcomes are, respectively,

$$\begin{aligned}\overline{\mu} &= (1-\rho)\mu_1 + \rho\mu_2\\ \overline{\sigma} &= (1-\rho)(\sigma_1 + \mu_1^2) + \rho(\sigma_2 + \mu_2^2) - \overline{\mu}^2\\ &= (1-\rho)\sigma_1 + \rho\sigma_2 + \rho(1-\rho)(\mu_2 - \mu_1)^2.\end{aligned}$$

To represent partial acquisition of information about LPHS outcomes, suppose the information choice is represented by θ where the producer accordingly weights the LPHS distribution by $\theta\rho$ and the conventional distribution by $(1-\theta\rho)$. Thus, minimal information where the producer is unaware of the LPHS outcomes corresponds to $\theta = 0$ and full information where LPHS outcomes are considered appropriately corresponds to $\theta = 1$. The information decision set is thus $0 \le \theta \le 1$.[9]

Based on information θ, the producer's perceived ex ante price distribution has respective mean and variance

$$\begin{aligned}\mu &= (1-\theta\rho)\mu_1 + \theta\rho\mu_2 = \overline{\mu} + \lambda\rho\delta\\ \sigma &= (1-\theta\rho)\sigma_1 + \theta\rho\sigma_2 + \theta\rho(1-\theta\rho)(\mu_2 - \mu_1)^2\\ &= \overline{\sigma} - (1-\theta)\rho(\omega + \delta^2) + (1-\theta^2)\rho^2\delta^2,\end{aligned}$$

[9] One could also examine possibilities of overshooting due to poor information by considering cases where $\theta > 1$, but for simplicity we only compare the two extremes.

where for convenience $\delta = \mu_2 - \mu_1$ and $\omega = \sigma_2 - \sigma_1$.

With given information θ, the producer's problem is

$$\max_q E_\theta[U(\pi)] = (\mu - c) - (\alpha / 2)\sigma q^2 ,$$

which yields

$$\widetilde{q} = \frac{\mu - c}{\alpha\sigma} .$$

The concentrated decision problem that endogenizes the information choice is thus

$$\max_\theta E_\theta[U(\pi)] = (\mu - c)\,\widetilde{q} - (\alpha / 2)\,\sigma\widetilde{q}^2 - \psi(\theta) \equiv \overline{U} ,$$

where $\psi(\theta)$ represents the cost of information as an increasing function of θ. The first-order condition is

$$\overline{U}_\theta = (\mu - c - \alpha\sigma\widetilde{q})\,\widetilde{q}_\theta - \psi_\theta = 0 ,$$

where subscripts of $\overline{U}$, $\widetilde{q}$, and ψ represent differentiation.

In a related paper, R.E. Just (2001) has analyzed the comparative static properties of this model. Obviously, if the cost of information is sufficiently high, then producers are better off with minimal information. On the other hand, ignoring the cost of information leads to irrational use of full information. For the case where θ has an internal solution, results show that second-order conditions hold $(\overline{U}_{\theta\theta} < 0)$ and that the optimal information choice, denoted by $\widetilde{\theta}$, satisfies

$$\widetilde{\theta}_{\psi_\theta} = -\frac{\overline{U}_{\theta\psi_\theta}}{\overline{U}_{\theta\theta}} < 0 ,$$

$$\widetilde{\theta}_\delta = -\frac{\overline{U}_{\theta\delta}}{\overline{U}_{\theta\theta}} < 0 ,$$

$$\widetilde{\theta}_\omega = -\frac{\overline{U}_{\theta\omega}}{\overline{U}_{\theta\theta}} > 0 ,$$

$$\widetilde{\theta}_{\overline{\sigma}} = -\frac{\overline{U}_{\theta\overline{\sigma}}}{\overline{U}_{\theta\theta}} < 0 ,$$

$$\tilde{\theta}_{\bar{\mu}} = -\frac{\bar{U}_{\theta\bar{\mu}}}{\bar{U}_{\theta\theta}} > 0\,,$$

$$\tilde{\theta}_{c} = -\frac{\bar{U}_{\theta c}}{\bar{U}_{\theta\theta}} < 0\,,$$

$$\tilde{\theta}_{\alpha} = -\frac{\bar{U}_{\theta\alpha}}{\bar{U}_{\theta\theta}} < 0\,.$$

These results have the following implications regarding the purchase and use of information about LPHS outcomes:

1. Higher information costs cause reduced information use.
2. Less adverse LPHS (expected) outcomes cause reduced information use.
3. More variable (unpredictable) LPHS outcomes cause increased information use.
4. Increases in conventional risk cause reduced information use.
5. Improvements in expected price conditions cause increased information use.
6. Increases in production costs cause reduced information use.
7. Increases in risk aversion cause reduced information use.

The first three of these implications are highly plausible and underscore some central themes of this chapter. The fourth implication is surprising but is explained by a reduction in the profit margin adjusted for risk premium. Because this margin falls, production is contracted and the cost of information per unit of output rises. The fifth and sixth implications are plausible by mirror image reasoning because of their impact on the scale of output. The seventh implication may also seem counterintuitive but is largely due to the reduction in output associated with an increase in risk aversion, which reduces the scale over which the marginal benefits of information can be spread.

Again, in principle, more general utility specifications could be used to show similar results with greater notational difficulty and less tractability. Presumably, different qualitative implications would be reached, for example, under constant relative risk aversion. Nevertheless, several central results emerge.

Proposition 8. *When low-probability–high-stakes outcomes are sufficiently difficult to assess (acquiring information and processing it are sufficiently costly), producers are better off to ignore them or obtain only partial information with which to consider them.*

Proposition 9. *More risk-averse producers are better off using less information because the scale of output over which they can enjoy the marginal benefits is less.*

Proposition 10. *Conditions that increase (reduce) marginal profit adjusted for the risk premium generally increase information use because the scale of production over which the marginal benefits of information can be enjoyed increases (declines).*

An important lesson of these results is that information has the characteristics of a fixed input. Whatever information is acquired can be used on all scales of output without affecting cost.

This analysis can be extended in a number of theoretical and empirical directions. For example, non-Gaussian probability distributions with "fat tails" and volatility that fluctuates over time have often been advanced to explain daily commodity market (futures) price movements. Non-Gaussian distributions have also been advanced to explain crop yields (see the review by Just and Weninger 1999). More recently, however, a mixture of Gaussian distributions has been shown to explain the same phenomena (see Bobenrieth 1999, in the case of daily price movements, and Just and Weninger 1999, in the case of crop yields). As in the model advanced here, normal distributions with different means, if combined ignoring the difference in means, will result in a composite distribution that exhibits both skewness and kurtosis, including heavily weighted tails (see Just and Weninger 1999). These results demonstrate that appropriate empirical representations of rational expectations require (1) eliminating misspecifications of expectations mechanisms (correct modeling of their dependence on the cost of information and the associated rational information choice) and (2) careful modeling of rational expectations of higher moments including their dependence on the cost of information and the associated rational information choice. To complicate matters, these choices depend on risk preferences.

ALTERNATIVE PARADIGMS AND OBSERVATIONAL EQUIVALENCE

In a world of risk and/or uncertainty, human judgments can be analyzed from many perspectives. These various perspectives can be integrated into a (non-exhaustive) set of three paradigms. The traditional paradigm and mainstay for analyzing risk behavior is the rational expected utility formation originally advanced by von Neumann and Morgenstern (1944). A second paradigm that has enjoyed much attention within the profession over the past few decades is the behavioral paradigm. Many anomalies in which behavior does not appear to follow conventional von Neumann-Morgenstern expected utility have been identified and various behavior modification (non-expected utility) approaches have been suggested as alternatives. The strongest empirical support for the behavioral paradigm has emerged from the anomalies that have been captured by non-expected utility theory research.

The behavioral paradigm has emerged from a number of empirical results originating with psychological research. The behavioral paradigm focuses on two types of departures from the traditional expected utility paradigm: biases in judgment under uncertainty and non-expected utility maximizing behavior. Much of the interest in this work has grown out of experiments based on observed choices when stated probabilistic alternatives are specified objectively for respondents. The results suggest that people dislike losses significantly more than they like gains (Rabin 1998), the "law of small numbers" (Tversky and Kahneman 1971), belief perseverance and confirmatory bias (Rabin 1998), anchoring and adjustment (Rabin 1998), and related phenomena. Based on these results originating in psychology, experimental economists have accumulated evidence that appears to reject rational expected utility theory (Camerer 1995).

Alpert and Raiffa (1982) and Oskamp (1982) are representative of the judgment bias literature.[10] This literature finds that low-probability events, muddled information, or unusual information tends to be underweighted in probability assessments. The related literature characterizes circumstances where individuals underweight prior information (called representativeness) and others where individuals underweight new information (called conservatism) when updating beliefs. For example, Hogarth and Einhorn (1992) argue that complication of stimuli can lead to excessive primacy or recency bias. In common terms, primacy is related to recall (of primal or early events) and recency relates to learning (modification of beliefs with experience). See Buschena (2001) and D.R. Just (2001) in this volume for further explanation regarding this literature.

In this chapter, we propose a third paradigm described by cost-consistent rational expectation formulation. We suggest that the presumed biased assessments of probability distributions emanating from the behavioral paradigm may in fact be due to nothing more than the cost of collecting and processing information used in generating expectations relative to the associated benefits. The evidence that is offered to reject expected utility maximization presumes that human agents have internalized full-information probability distributions. The formulation of this chapter demonstrates that such a presumption is seriously flawed.

Alternatively, we suggest with sufficiently broad cost-consistent corrections to rational expectation formulations that expected utility may explain many or all of the anomalies noted by behavioralists. Specifically, we have shown that the same underweighting of extreme or low-probability events as noted in the judgment bias literature will occur rationally because of the high

[10] Kahneman, Slovic, and Tversky (1982) also argue that when human agents form expectations they use distributions that are too narrow.

costs of learning about them. High costs of learning about such events as well as high costs of processing information about such events are natural and highly plausible for the very reason that they occur rarely or cause adjustments that go beyond well-understood processes and mechanisms. As a result, expected utility based on cost-consistent rational expectations may well offer a paradigm for studying behavior under risk that is broader than either the conventional expected utility paradigm or the alternative behavioral paradigm. That is, the explanatory power of the conventional von Neumann-Morgenstern expected utility paradigm applies for common cases of low-cost information while uncommon outcomes associated with higher cost of information explain the anomalies upon which behavioralists focus.[11]

In addition, we note that cost-consistent rational expectations is not the only approach that may generate an observational equivalence relative to the behavioral models. D.R. Just (forthcoming) has developed a model incorporating the psychological concepts of recall and learning with which he shows that virtually every documented expected utility violation can be explained by improper reliance on prior information, particularly in complicated lotteries. His model is developed by adding two weighting functions into a standard Bayesian updating framework to represent learning and recall. The resulting limited learning model gives weight to either the prior (recall) or likelihood of observed information (learning) based on information complexity following

$$p_{t+1}(x) = \frac{p_t(x)^{R(l,p)} l(\theta \mid x)^{L(l,p)}}{\int_{-\infty}^{\infty} p_t(x)^{R(l,p)} l(\theta \mid x)^{L(l,p)} dx},$$

where p is the probability distribution indexed on time, x represents possible outcomes, l is the likelihood function, and R and L represent recall and learning as respective functions of the relationship of the prior (in p) to the likelihood of observed phenomena (in l). Using experimental data and estimation methods from Hey and Orme (1994), he shows how this explanation compares favorably to the various alternative models (mostly heuristic decision rules) that have been proposed in the literature.

[11] We note that many of the behavioral developments tend to focus on explaining a particular anomaly rather than explaining general behavior. That is, the alternative models generally have superior performance only on the phenomenon for which they were designed to address. Before an alternative model that can address anomalies is likely to replace the traditional expected utility model in common applications, it will likely need to retain much of the explanatory power of the traditional model for traditional problems.

AGGREGATE MARKET BEHAVIOR

The alternative paradigms of the previous section focus on individual decision problems. When aggregating across individuals and analyzing competitive interactions and market behavior, however, a host of additional complications arise. Intuitively, for example, some agents may be irrational and make serious errors in human judgment, while others who are less prone to such proclivities could be expected to have a competitive advantage. To address such problems requires determining the competitive interactions landscape where some producers are rational expected utility maximizers and other producers are non-expected utility maximizers.

More generally, if the anomalies of expected utility theory are explained by cost-consistent rational expectations, heterogeneity of expectations due to heterogeneity of information choices raises similar fundamental questions about aggregate market behavior. Under what conditions will equilibrium exist and converge to a market-determined "cost-consistent rational expectations equilibrium?" Central to squarely answering this question is understanding the process by which individual agents process information and learn about market conditions and probability distributions.

Based on the individual decision frameworks introduced above, a small step toward addressing this question might involve modeling the distribution of risk aversion among producers, say, by $F(\alpha)$, and then integrating over producers to find aggregate supply. For example, based on the first model presented above, suppose c, ζ_θ, ζ_ϕ, σ_δ, and σ_ε are constant across all producers but that aggregate demand gives the price as a decreasing function of aggregate quantity, $p = a - bQ + \nu$, where $a > 0$, $b < 0$, $E(\nu) = 0$, and $E(\nu^2) = \sigma_\nu$. Aggregate supply is $Q = \int \tilde{q} dF(\alpha)$ where each producer's $\tilde{q}$ is a function of α not only directly but also through optimal choices of θ and ϕ.

To describe a cost-consistent rational expectations equilibrium in this model, note that each producer's output has respective mean and variance

$$E(\tilde{q}) = \frac{\mu - c}{\alpha \sigma_p},$$

$$\sigma_q = \frac{\theta^2 \sigma_\delta + (\mu - c)^2 \phi \sigma_\varepsilon + \theta^2 \phi^2 \sigma_\delta \sigma_\varepsilon}{\alpha^2 \sigma_p^2}.$$

Substituting into the demand equation implies

$$\mu = a - b \int \frac{\mu - c}{\alpha \sigma_p} dF(\alpha), \tag{5}$$

$$\sigma_p = b^2 \int \sigma_q dF(\alpha) + \sigma_\nu \,, \tag{6}$$

assuming that δ and ε are uncorrelated among producers, and that ν is uncorrelated with other random variables in the model. Of course, (6) can be developed under other assumptions such as where δ and ε are identical among producers or, more generally, correlated among producers. The main point is that the variance of prices is endogenous because information choices are endogenous. The cost-consistent rational expectations equilibrium is thus described by (2), (3), and (4) for each producer plus (5) and (6) at the aggregate level.

In general, the existence of an equilibrium described by conditions such as (5) and (6) is not clear. For example, equation (6) is cubic in σ_p. In addition, when such an equilibrium is disturbed, the conditions under which rational actions will return to an equilibrium are not simple to analyze. The problem is that when aggregate demand shifts, each producer must assess both the price mean and price variance for the next cycle of production. Even if the parameters of demand are known, each producer must assess how other producers will react not only in adjusting production but also in adjusting their underlying use of information. This involves assessing not only the effect of information on other producers' expected price, but also how their resulting assessments of variance will alter their risk premiums.

In a much simpler attempt to analyze equilibrium with heterogeneous expectations, Brock and Hommes (1997) consider the case where agents choose rationally between only two alternative expectations mechanisms – naïve and adaptive expectations – depending rationally on performance. They find that when the intensity of switching between expectations mechanisms is high, a complicated dynamical orbit emerges whereby local instability can be part of a fully rational expectations equilibrium. Without doubt, adding further expectation mechanisms into the heterogeneity mix, if at least some are rationally naïve or adaptive given the cost of information, would only complicate rather than simplify the nature of any equilibrium.

The considerations of this section thus suggest the following proposition.

Proposition 11. *What a rational expectation for an individual agent is depends on the expectation mechanisms and risk response behavior in use by other agents including the associated competitive interactions.*

The principle suggested in Proposition 11 is that truly rational expectations must account for the competitive behavior of other agents. The underlying expectations depend on understanding how other agents learn from observed prices and information. While a host of studies have attempted to model learning – most often as a Bayesian process – no models have addressed the considerations raised by endogenizing variance assessments nor the underlying information choices. Thus, for the remainder of this chapter, we dispense with formality and attempt to suggest some concepts in the existing literature that have promise for future theoretical and empirical specifications of agricultural (competitive supply) risk models.

In a Bayesian learning process, an agent efficiently uses information regarding the variables of interest. There are at least three types of Bayesian learning, which may be characterized by how the learning agent interacts with the variables about which learning takes place. First, an agent may seek to learn about a state variable that is unaffected by the agent's actions; this is statistical learning and, under appropriate conditions, may be represented using a Kalman filter. Second, an agent may seek to learn about the behavior of other agents; this type of learning may be represented in a game theoretic model. Third, a competitive market agent may seek to learn about aggregate market relationships (supplies and demands) and their implications for variables such as prices that affect the agent's profits as a function of the agent's decisions. This learning may take place in the context of a market frequently in transition, where the behavior of other producers, as represented by aggregate market behavior, is unknown. While each producer in this environment (rightly) views his or her individual influence on the market outcome as negligible, the relative predominance of various expectations and behaviors among producers determines the realized market outcome that impacts each individual producer. Accordingly, agents face a learning problem similar to the inference problem faced by an econometrician seeking to estimate demand or supply in an environment in which both functions are shifting.

Of course, the third case is not only the most realistic, but also the most challenging. The competitive interactions among producers who face unknown demand functions (or unknown parameters for given functional specifications) and their learning processes can assume a number of different forms. Not only does the usual distinction between passive and active learning apply (Rausser and Hochman 1979), but also a rich literature on rational learning versus ad hoc learning is relevant.

Ideally, conditions are needed under which market discipline can structure a learning process to achieve convergence to a stable market equilibrium. These conditions obviously depend on the underlying model specifications of individual producers, which determine how informed each is. Do producers operate with subjective or objective probability distributions? How are such distributions revised? What signals are used in this revision process? How frequent is updating? In most models, public signals include market prices, while private signals include among other things individual producers' costs of production. If some producers are informed and rational and others are uninformed and/or irrational, how is the familiar infinite regress problem modeled? The available literature provides only partial answers to these issues and related conditions.

The literature on modeling market learning processes began in earnest with Grossman and Stiglitz (1976), who drew a sharp distinction between informed and uninformed traders. Townsend (1978) followed with a rational market learning model in which agents learn about critical demand parameters that are presumed to be realized before the beginning of trading. This work was extended by Blume and Easley (1982), Bray and Kreps (1988), and Frydman (1982) and Townsend (1983). Much of this work is in the spirit of the earlier Cyert and DeGroot (1974) model. In these models, convergence to rational expectations follows directly from the asymptotic properties of Bayesian estimators. Bray (1982) and Jordan and Radner (1982) have designed further conceptual frameworks in the spirit of the Grossman and Stiglitz model. Jordan and Radner's framework requires each agent to attempt determination of every other agent's price formation process, while Bray specifies an ad hoc learning process where informed agents do not take into account uniformed agents' behaviors.

The above literature has been extended by Goodhue, Simon, and Rausser (1998) and Goodhue, Rausser, and Simon (2000). In this work, the earlier study specifies an adaptive expectation mechanism, while the latter introduces rational Bayesian learning. The relationship between market risk and the number of competitive firms is shown to have a number of crucial effects. These effects, of course, depend upon the unknown parameters of the market demand function. One illustrative result derived for the rational learning formulation is that the impact of the number of competitive producers on the variance of price is positive if the demand elasticity is sufficiently large. Accordingly, as markets become more open, an increase in the number of competitive producers can result in increased market risk. This particular result is modified when the "cost-consistent rational expectations" formulation advanced in this chapter is incorporated. The resulting specification does not arbitrarily divide producers into two regimes, those who are informed and those who are not, but rather results in a continuum of information. In stable or policy-controlled markets, results consistent with adaptive expectation mechanisms obtain, while in more volatile markets the tendency is to move toward rational learning with greater risk, again resulting from increases in the number of competitive producers. As with the much simpler model represented by equations (5) and (6), we have not as yet been able to prove existence of a "cost consistent rational expectations equilibrium" in this more general setting.

CONCLUSIONS

In the most recent comprehensive synthesis of the expectation literature, the *Handbook of Agricultural Economics*, Nerlove and Bessler (2001) state that (i) "information about the future can be acquired only at a cost" (p. 159) and that (ii) "central to almost all treatments of the subject since the work of Keynes and Hicks in the 1930s is the separation assumption, in which dynamic decision problems were modeled by separating expectation formation from optimizing behavior" (p. 157). With respect to (i), Nerlove and Bessler identify no studies in their review that explore the implications of this cost. With respect to (ii), Nerlove and Bessler not only adopt this specification throughout their review, but later argue that the separation assumption typically must be supplemented by a representative decision maker assumption as well as a certainty equivalent approach.

For the framework advanced in this chapter, both of these simplifying assumptions are relaxed. Our results suggest that relaxing these assumptions in the maintained hypotheses for analyzing producer behavior under risk comes at great expense. If what is rational for an individual agent depends on the cost of information relative to its associated benefits, then the optimal expectation mechanism must be determined endogenously. As a result, alternative regimes as simple as naïve expectations may in fact be rational for certain agents or at certain times once both the cost of information and the cost of information processing is recognized.

The analytical framework of this chapter, which endogenizes both information choices and expectation mechanisms, leads to some interesting insights and implications. Results suggest conditions under which only partial information will be collected and processed by rational agents. One surprising result is that increases in price variability can cause reduced information use. These and other conceptual results suggest that standard assumptions cloud our understanding of agricultural economic research, and as a result may have caused serious mischaracterization of risk behavior.

By incorporating the cost of acquiring information and the associated accuracy of expectations, we show that endogenizing the information decision leads to heterogeneity of expectations depending on the heterogeneity of risk preferences. Relaxing the standard assumption separating expectations formation from optimizing behavior appears to lead to a formulation where the anomalies that have given rise to the behavioral literature are potentially explained. As a result of broadening the concept of rationality to include the cost of information (possibly outcome-specific information), conventional risk models can be expanded to consider Knightian cases of uncertainty in a manner that is observationally equivalent to many observations otherwise used to support behavioral models.

Based on these results, we suggest two challenges for risk-modeling research. First, what is the incremental value for empirical models that endogenize the information decision relative to the literature that estimates risk behavior based on standard fixed-mechanism expectations formulations (such as summarized by Nerlove and Bessler 2001)? Second, in moving from individual decision problems as characterized by the cost-consistent rational expectations formulation of this chapter to aggregate market behavior, what are the appropriate learning process specifications, the associated convergence properties, and characteristics of any resulting equilibrium? As variance assessments as well as the underlying information choices are endogenized, a number of thorny conceptual issues arise. We have attempted to identify some of these issues based on a brief review of the current state of professional knowledge that exists for resolving such questions. The limited availability of applicable research suggests that generalizations of existing modeling specifications to the case of cost-consistent rational expectations is not only an important agenda item for future competitive market risk analysis of producer behavior, but also a largely untapped area for future research.

REFERENCES

Alpert, M., and H. Raiffa. 1982. "A Progress Report on the Training of Probability Assessors." In D.P. Kahneman, P. Slovic, and A. Tversky, eds., *Judgment Under Uncertainty: Biases and Hueristics*. New York: Cambridge University Press.

Antle, J.M. 1987. "Econometric Estimation of Producers' Risk Attitudes." *American Journal of Agricultural Economics* 69: 509-522.

___. 1989. "Nonstructural Risk Attitude Estimation." *American Journal of Agricultural Economics* 71: 774-784.

Antonovitz, F., and R. Green. 1990. "Alternative Estimates of Fed Beef Supply Response to Risk." *American Journal of Agricultural Economics* 72: 475-487.

Antonovitz, F., and T. Roe. 1986. "A Theoretical and Empirical Approach to the Value of the Information in Risky Markets." *Review of Economics and Statistics* 68: 105-114.

Aradhyula, S.V., and M.T. Holt. 1989. "Risk Behavior and Rational Expectations in the U.S. Broiler Market." *American Journal of Agricultural Economics* 71: 892-902.

Baak, S.J. 1999. "Tests for Bounded Rationality with a Linear Dynamic Model Distorted by Heterogeneous Expectations." *Journal of Economic Dynamics and Control* 23: 1517-1527.

Behrman, J.R. 1968. *Supply Response in Underdeveloped Agriculture: A Case Study of Four Major Annual Crops in Thailand, 1937-1963*. Amsterdam: North-Holland Publishing Co.

Binswanger, H. 1980. "Attitudes Toward Risk: Experimental Measurement in Rural India." *American Journal of Agricultural Economics* 62: 395-407.

Blume, L.E., and D. Easley. 1982. "Learning to be Rational." *Journal of Economic Theory* 26: 340-351.

Bobenrieth, E. 1999. "An Estimation of the 'Fat Tail' Phenomenon in Daily Commodity Price Changes." Working Paper. Departmento de Economia, Universidad de Concepción, Concepción, Chile.

Bray, M.M. 1982. "Learning, Estimation and the Stability of Rational Expectations." *Journal of Economic Theory*, 26 (1982):318-339.

Bray, M.M., and D.M. Kreps. 1988. "Learning, Estimation and the Stability of Rational Expectations." In G.R. Feiwel, ed., *Arrow and the Ascent of Modern Economic Theory*. Basingstoke, Hampshire: MacMillan.

Brock, W., and C. Hommes. 1997. "Rational Routes to Randomness." *Econometrica* 65: 1059-1095.

Buschena, D. 2001. "Non-Expected Utility: What Do the Anomalies Mean for Risk in Agriculture?" In R.E. Just and R.D. Pope, eds., *A Comprehensive Assessment of the Role of Risk in U.S. Agriculture*. Boston, MA: Kluwer Academic Publishers.

Camerer, C. 1995. "Individual Decision Making." In J.H. Kagel and A.E. Roth, eds., *Handbook of Experimental Economics*. Princeton, NJ: Princeton University Press.

Chavas, J.-P. 1999. "On the Economic Rationality of Market Participants: The Case of Expectations in the U.S. Pork Market." *Journal of Agricultural and Resource Economics* 24: 19-37.

___. 2000. "On Information and Market Dynamics: The Case of the U.S. Beef Market." *Journal of Economic Dynamics and Control* 24: 833-853.

Chavas, J.-P., and M.T. Holt. 1996. "Economic Behavior Under Uncertainty: A Joint Analysis of Risk Preferences and Technology." *Review of Economics and Statistics* 78: 329-335.

Cyert, R.M., and M.H. DeGroot. 1974. "Rational Expectations and Bayesian Analysis." *Journal of Political Economy* 82: 521-536.

Dhrymes, P.J. 1973. "Restricted and Unrestricted Reduced Forms: Asymptotic Distribution and Relative Efficiency." *Econometrica* 41: 119-134.

Eckstein, Z. 1984. "A Rational Expectations Model of Agricultural Supply." *Journal of Political Economy* 92: 1-19.

Frydman, R. 1982. "Towards An Understanding of Market Processes: Individual Expectations, Learning and Convergence to Rational Expectations Equilibrium." *American Economic Review* 72: 652-668.

Frydman, R., and E.S. Phelps. 1983. *Individual Forecasting and Aggregate Outcomes.* New York: Cambridge University Press.

Gardner, B.L. 1976. "Futures Prices in Supply Analysis." *American Journal of Agricultural Economics* 58: 81-84.

Goodhue, R.E., G.C. Rausser, and L.K. Simon. 2000. "Comparative Market Interactions and Rational Learning." Working Paper, Department of Agricultural and Resource Economics, University of California, Berkeley.

Goodhue, R.E., L.K. Simon, and G.C. Rausser. 1998. "Privatization, Market Liberalization and Learning in Transition Economies." *American Journal of Agricultural Economics* 80: 48-68.

Goodwin, R.M. 1947. "Dynamical Coupling with Especial Reference to Markets Having Production Lags." *Econometrica* 15: 181-204.

Goodwin, T.H., and S.M. Sheffrin. 1982. "Testing the Rational Expectations Hypothesis in an Agricultural Market." *Review of Economics and Statistics* 64: 658-667.

Greenspan, A. 2001. Federal Reserve Board Semiannual Monetary Policy Report to Congress before the Committee on Banking, Housing and Urban Affairs, U.S. Senate. February 13, 2001.

Grossman, S.J., and J.E. Stiglitz. 1976. "Information and Competitive Price Systems." *American Economic Review* 66: 246-253.

Hey, J.D., and C. Orme. 1994. "Investigating Generalizations of Expected Utility Theory Using Experimental Data." *Econometrica* 62: 1291-1326.

Hogarth R.M., and H.J. Einhorn. 1992. "Order Effects in Belief Updating: The Belief Adjustment Model." *Cognitive Psychology* 24: 1-55.

Holt, M.T., and G. Moschini. 1992. "Alternative Measures of Risk in Commodity Supply Models: An Analysis of Sow Farrowing Decisions in the United States." *Journal of Agricultural and Resource Economics* 17: 1-12.

Jordan, J.S., and R. Radner. 1982. "Rational Expectations in Microeconomic Models: An Overview." *Journal of Economic Theory* 26: 201-223.

Just, D.R. 2001. "Information, Processing Capacity, and Judgment Bias in Risk Assessment." In R.E. Just and R.D. Pope, eds., *A Comprehensive Assessment of the Role of Risk in U.S. Agriculture.* Boston, MA: Kluwer Academic Publishers.

___. Forthcoming. "Learning and Information." Unpublished Ph.D. dissertation, Department of Agricultural and Resource Economics, University of California, Berkeley.

Just, R.E. 1974. "An Investigation of the Importance of Risk in Farmers' Decisions." *American Journal of Agricultural Economics* 56: 14-25.

___. 1977. "Estimation of an Adaptive Expectations Model." *International Economic Review* 18: 629-644.

___. 2001. "Addressing the Displacement of Risk by Uncertainty in Agriculture." Working Paper, University of Maryland, College Park.

___. 2002. "Risk Research in Agricultural Economics: Opportunities and Challenges for the Next Twenty-Five Years." *Agricultural Systems* (special issue: "Advances in Risk Impacting Agriculture and the Environment") (forthcoming).

R.E. Just, and R.D. Pope. 1999. "Implications of Heterogeneity for Theory and Practice in Production Economics." *American Journal of Agricultural Economics* 81: 711-718.

Just, R.E., and G.C. Rausser. 1981. "Commodity Price Forecasting with Large-Scale Econometric Models and the Futures Market." *American Journal of Agricultural Economics* 63: 197-208

___1983. "Expectations and Intertemporal Pricing in Commodity Futures and Spot Markets." Invited Paper Presented at the NCR-134 Conference on Applied Commodity Price Analysis, Forecasting, and Market Risk Management, Des Moines, Iowa, April 28-29.

Just, R.E., and Q. Weninger. 1999. "Are Crop Yields Normally Distributed?" *American Journal of Agricultural Economics* 81: 287-304.

Kahneman, D., P. Slovic, and A. Tversky (eds.). 1982. *Judgment Under Uncertainty: Heuristics and Biases.* New York: Cambridge University Press.

Knight, F.H. 1921. *Risk, Uncertainty and Profit.* Boston: Houghton Mifflin Company. (Reprinted 1964, Silver Lake Publishing Company, New York.)

Lin, W. 1977. "Measuring Aggregate Supply Response Under Instability." *American Journal of Agricultural Economics* 59: 903-907.

Lin, W., G. Dean, and C. Moore. 1974. "An Empirical Test of Utility vs. Profit Maximization in Agricultural Production." *American Journal of Agricultural Economics* 56: 497-508.

Lewis, G. 1981. "The Philips Curve and Bayesian Learning." *Journal of Economic Theory* 24: 240-269.

Mills, E.S. 1955. "The Theory of Inventory Decisions." Unpublished Ph.D. dissertation, University of Birmingham, England.

___. 1962. *Price, Output, and Inventory Policy: A Study in the Economics of the Firm and Industry.* New York: John Wiley.

Nerlove, M. 1958. *The Dynamics of Supply: Estimation of Farmers' Response to Price.* Baltimore, MD: Johns Hopkins University Press.

___. 1983. "Expectations, Plans and Realizations in Theory and Practice." *Econometrica* 51: 1251-1279.

Nerlove, M., and D.A. Bessler. 2001. "Expectations, Information and Dynamics." In B.L. Gardner and G.C. Rausser, eds., *Handbook of Agricultural Economics* (Vol. 1). Amsterdam: Elsevier-North-Holland.

Oskamp, S. 1982. "Overconfidence in Case-Study Judgments." In D.P. Kahneman, P. Slovic, and A. Tversky, eds., *Judgment Under Uncertainty: Biases and Hueristics.* New York: Cambridge University Press.

Pagan, A.R., and A. Ullah. 1988. "The Econometric Analysis of Models with Risk Terms." *Journal of Applied Econometrics* 3: 87-105.

Pesaran, M.H. 1987. *The Limits to Rational Expectations.* Oxford: Basil Blackwell Ltd.

Pingali, P.L., and G.A. Carlson. 1985. "Human Capital, Adjustments in Subjective Probabilities, and the Demand for Pest Control." *American Journal of Agricultural Economics* 67: 853-861.

Rabin, M. 1998. "Psychology and Economics." *Journal of Economic Literature* 36: 11-46.

Rausser, G.C., J.A. Chalfant, H.A. Love, and K.G. Stamoulis. 1986. "Macroeconomic Linkages, Taxes, and Subsidies in the U.S. Agricultural Sector." *American Journal of Agricultural Economics* 68: 399-412.

Rausser, G.C., and E. Hochman. 1979. *Dynamics of Agricultural Systems: Economic Prediction and Control.* Amsterdam: North-Holland Publishing Co.

Roe, T., and F. Antonovitz. 1985. "A Producer's Willingness to Pay for Information under Price Uncertainty: Theory and Application." *Southern Journal of Economics* 52: 382-391.

Ryan, T.J. 1977. "Supply Response to Risk: The Case of U.S. Pinto Beans." *Western Journal of Agricultural Economics* 2: 35-43.

Schultz, T.W. 1975. "The Value of the Ability to Deal with Disequilibria." *Journal of Economic Literature* 13: 827-846.

Townsend, R.M. 1978. "Market Anticipations, Rational Expectations and Bayesian Analysis." *International Economic Review* 19: 481-494.

___. 1983. "Forecasting the Forecasts of Others." *Journal of Political Economy* 91: 546-588.

Traill, B. 1978. "Risk Variables in Econometric Supply Response." *Journal of Agricultural Economics* 24: 53-61.

Tversky, A., and D. Kahneman. 1971. "Belief in the Law of Small Numbers." *Psychology Bulletin* 76: 105-110.

von Neumann, J., and O. Morgenstern. 1944. *Theory of Games and Economic Behavior.* Princeton, NJ: Princeton University Press.

Chapter 5

INFORMATION, PROCESSING CAPACITY, AND JUDGMENT BIAS IN RISK ASSESSMENT

David R. Just
University of California, Berkeley

INTRODUCTION

Many agricultural economists cite risk as one of the most important elements of the farm problem (Gardner 1992). The focus of this literature has been to estimate the level of risk aversion across decision makers, and then estimate the effects of risk aversion on market movements (see for example Binswanger 1980, Just and Pope 1987, Just and Zilberman 1983). Arrow (1971) attributed risk aversion to the curvature of the individual's utility of wealth function. This measure of risk aversion is based on the idea that diminishing marginal utility of wealth leads one to value larger amounts of money less than one would otherwise, and hence engage in fewer risks involving possible large gains or losses.

While diminishing marginal utility of wealth is a plausible cause of risk aversion, there are other factors that could also affect risk attitudes, such as perception, and capacity to understand risks. Further, there is strong evidence now coming to light that curvature of a utility function cannot be the cause of some observed risk aversion (Rabin 2000a). In light of this evidence it seems reasonable to explore other explanations that might provide further insight into decisions under uncertainty.

One of the fundamental assumptions often maintained when estimating standard measures of risk aversion is that individuals have access to the same estimated distributions of outcomes that modelers do – not just access to the numerical representation, but mental access to the distribution in such a way as to make the best decision given this distribution for the decision maker's circumstances and preferences. Herbert Simon (1955, 1959, 1978) has challenged these assumptions as ignoring information processing costs, and psychological limitations on human ability to understand. Simon suggested a change in focus from models of *substantive* rationality (or global optimization) to models of *procedural* rationality (or optimizing subject to constraints on ability and cost of processing information). Schultz (1975) further suggested that education and training may lessen the information processing costs, or enhance the ability of a decision maker to make sense of his or her surroundings.

Psychologists have made a study of individual biases in information processing and forecasting. This judgment bias literature has recently received attention from economists (for example, Rabin 2000c, Camerer 1995, Grether 1980, 1990, Nelson and Bessler 1989). In this chapter, I review some of the major findings of the psychological literature on judgment bias (focusing mostly on Kahneman, Slovic, and Tversky's 1982 anthology) and then discuss some of the implications for the economic literature on decision making under uncertainty. The first section contains a short review of the failings of utility function measures of risk aversion, and some alternative causes of risk behavior. The second section is a review of the most cited biases in judgment, and some interpretation of the results. The third section proposes directions for modeling judgment bias phenomena in a way that is useful for applied economists. Using stylized facts based on the judgment bias literature, I show how traditional estimation of risk aversion may be inadequate in many circumstances.

UTILITY CURVATURE VERSUS PROBABILITY CURVATURE

Most economists are now aware of the growing evidence against expected utility theory. Among the more recent objections are those criticizing concavity of the utility function as the only measure of risk aversion. Matthew Rabin (2000a, 2000b) is chief among the critics. While not the first to notice a discrepancy (see Hansson 1988), Rabin points to the difference in estimated risk-aversion levels observed over small gambles and those over larger gambles. His argument stems from a calibration theorem, which he proves. A useful and illustrative corollary of this theorem is given below (Rabin 2000b).

Corollary 1. *Suppose that for all w, $U'(w) > 0$ and $U''(w) < 0$. Suppose there exists $g > l > 0$ such that for all w, $.5U(w-l) + .5U(w+g) < U(w)$. Then for all positive integers k, and all $m < m(k)$, $.5U(w-2kl) + .5U(w+mg) < U(w)$, where*

$$m(k) = \begin{cases} \dfrac{\ln[1-(1-l/g)2\sum_{i=1}^{k}(g/l)^{i}]}{\ln(l/g)} & \text{if } 1-(1-l/g)2\sum_{i=1}^{k}(g/l)^{i} > 0 \\ \\ \infty & \text{if } 1-(1-l/g)2\sum_{i=1}^{k}(g/l)^{i} \leq 0 \end{cases}.$$

For a statement of the theorem and a proof of both the theorem and the corollary presented above, see the appendix of Rabin (2000b). This corollary allows comparison of risk behavior over even-chance bets (commonly called 50-50 bets) under the assumption of risk aversion. For example, a person who will always turn down a lottery with a .5 probability of winning \$110 and .5 probability of losing \$100, henceforth (.5, 110, .5, -100), will also always turn down the lottery represented by (.5, 2090, .5, -800). Expected utility assumes that concavity is the only explanator of risk attitude, and that individuals are approximately risk neutral for small gambles, meaning that individuals should accept fair bets if they are small enough. The problem with this assumption is that when we observe someone turning down a small fair bet we must assume that this is due to concavity of the utility function. Unless the utility function changes from concave to convex as prizes get larger, outrageous behavior is implied.

It is easy to confirm from the corollary above that if a globally risk averse individual turns down (.5, 125, .5, -100) then he will always turn down any bet with a .5 chance of losing \$600, no matter how large a gain may be had with the remaining .5 probability. Bernoulli (1954 reprint) introduced expected utility theory as a way to explain why individuals might not be willing to pay infinite amounts of money to play gambles with infinite expected gains (e.g., the St. Petersburg Paradox). Variants on the St. Petersburg Paradox will sometimes yield an infinite certainty equivalent unless the utility function is bounded, implying eventual risk aversion for large enough gambles (Markowitz 1952). It is exactly this property, only in the small, that Rabin exploits in arguing the inconsistency of traditional expected utility theory. Even if there are non-convexities in the utility function, a bounded utility function will imply similarly ridiculous behavior if compared to large enough positive gains.

To see this last point, consider the lottery (.5, x, .5, -100). Suppose without loss of generality that $U(-100) = 0$, and $\lim_{x\to\infty} x = 1$. The certainty equivalent of this lottery is given by $U(CE) = .5U(x) \leq .5(0+1) = .5$. This means that an individual will turn down any bet with a 50 percent chance of losing \$100 for some fixed amount of money, no matter what the possible gains. This also means that a person with wealth greater than $U^{-1}(.5)$ will never choose to take any bet that involves a 50 percent chance of a loss of \$100.[1] If we limit ourselves to the notion that all risk aversion is due to concavity of utility, we must accept that either (1) more wealthy individuals are less willing to risk losses, or (2) individuals are willing to pay infinite amounts of money for some class of lotteries with infinite average payoffs, but only small probability of any large gains (as in the St. Petersburg Paradox).

---

[1]Note, given that the graph of utility is connected, and that utility is bounded and monotone increasing, there must be some level of wealth, w, for which $U(w-100) = 1/2 \lim_{x\to\infty} U(w+x)$.

Alternatively, some theorists have pointed out that risk-averse behavior may be due to misperceptions of probability, or transforming probabilities before expected utility optimization. Yaari (1987) proposed a model where all risk behavior was due to a warping of probabilities rather than of monetary values. More commonly, models are proposed that involve the optimization of some function,

$$V(p,x) = \sum_i \pi(p_i, x_i) U(x_i + w), \tag{1}$$

where p is a vector of probabilities associated with possible payoffs, x is the vector of possible payoffs, w is current wealth, U is a utility function over payoffs, and π represents a probability weighting function. This probability weighting function provides an alternate explanation for risk attitudes in the small, and can allow utility to be an unbounded function, thus eliminating the concavity problem of expected utility theory. One sufficient condition to eliminate the problems associated with bounded utility is that $\pi(p,x) = 0$ whenever $|x| > k$ for some k. In this case, infinite payoffs are given no weight and, hence, St. Petersburg paradoxes do not occur. This condition can be interpreted as an individual not believing prizes larger than k can occur. This particular weighting restriction does not seem plausible, but others show promise (see D.R. Just, forthcoming). Models of this sort try to incorporate psychological factors that bias the judgment of an individual. While the purpose is similar to the aims of the judgment bias literature, little has been done to incorporate the literature on judgment bias into decision models under uncertainty. Some exceptions are found in Eekhoudt et al. (1991) and Rabin (1998, 2000c) and among the literature on learning in games (see Camerer 1995 for a review).

DEVELOPMENTS IN THE JUDGMENT BIAS LITERATURE

There are two aspects of probability statements commonly measured to determine accuracy. The first of these is resolution (Camerer 1995), or the ability to distinguish between high- and low-probability events. Individuals with high resolution will make probability assessments that are near either 1 or 0. Resolution is a measure of how certain you are of your prediction: in other words, a measure of knowledge.

The other component of accuracy is calibration (Camerer 1995). Calibration measures the proportion of time an event evaluated to have probability x occurs. For example, a weather forecaster is well calibrated if about 10 percent of the time when he predicts a 10 percent chance of rain it actually rains. Calibration is a measure of how well an individual understands and processes probabilities, and has been the major focus of the judgment bias literature.

Overconfidence

Overconfidence was a problem discovered by psychologists when evaluating the accuracy of their own predictions (Oskamp 1982). Overconfidence is the phenomenon of individuals reporting distributions that are narrower than the true distribution. This effect seems to be pervasive among nearly all types of assessments. Alpert and Raiffa (1982) conducted experiments using over 1,000 MBA students from Harvard and the Massachusetts Institute of Technology. Their experiments elicited various quantiles for several numbers regarding which students had varying knowledge.

For example, one of the questions asked for the current number of students enrolled in the doctoral program at the Harvard Business School. The true number fell outside the 98 percent confidence interval of 61 percent of the students. On average, over all questions asked, one-third of true values fell within interquartile ranges on the first trial. This shows that individuals systematically had narrowed distributions from what might have given them true calibration.

Lichtenstein, Frischhoff, and Phillips (1982) found that experts were *less* confident than novices when it came to psychological assessment. They found that overconfidence seems to diminish with the difficulty of the question, with the easiest of questions eliciting underconfidence. Many studies have looked at overconfidence and calibration across fields and have found some differences. Weathermen and economists in short-run (one quarter) predictions seem to be well calibrated. Economists tend to be overconfident when making long-run (one year or more) forecasts. Physics journal articles tend to display overconfidence in the value of physical constants (see Camerer 1995 for a list of other professions that have been studied). Lichtenstein, Frischhoff, and Phillips (1982) and Alpert and Raiffa (1982) report that individuals show some improvement when trained. However, the training does not seem to transfer across tasks. In other words, an individual may become very well calibrated at commodity price forecasts and still display overconfidence in predicting yearly rainfall. Within some tasks (such as blackjack), training appears to have no effect.

Possible Explanation. Murphy and Winkler (1974) report that the spread of the distribution is correlated with the calibration. In other words, wide distributions are more likely to be narrowed by individuals. Due to this finding and the fact that overconfidence disappears in the easiest of questions, Lichtenstein, Frischhoff, and Phillips (1982) hypothesize that overconfidence is due to individuals failing to recognize and adjust to the difficulty of questions. Essentially, individuals will sacrifice calibration for resolution (which, of course, produces incorrect assessments). One popular explanation is that when assessing confidence intervals, individuals first come upon a point estimate and then find the interval by adjusting up and down (Tversky and Kahneman 1982b). Overconfidence is experienced as individuals simply underestimate the width of the distribution. This is called the "anchoring and adjustment" heuristic.

This bias differs from most of the other biases I will discuss in that it is not a comparison of before and after incorporation of information. Any updating has taken place over the course of the individual's life and overconfidence is observed to be prevalent at any point in time. The prevalence of overconfidence suggests that individuals systematically narrow the distributions of the information they observe.

Availability Bias

The availability bias can be thought of as people basing probability judgments on their ability to construct various scenarios mentally. For instance, Tversky and Kahneman (1982b) asked subjects which of these two alternatives is more frequent: words of three letters or more that start with 'R', or words that have 'R' for their third letter. Several other letters were tried also. A majority (about 66 percent) always said the letters occur more commonly in the first position. However, all letters used are more commonly found in the third position. Individuals had a harder time recalling words that had 'R' in the third position and, hence, biased their probability toward the more easily recalled words. The availability heuristic is not as well studied as overconfidence, but is nonetheless well documented. It is only a problem when availability contradicts true probabilities (as in the case of these letters).

Possible Explanation. As is intuitive, it is believed that individuals eliminate some information when storing for future use. In the case of the letters experiment, it is common when using dictionaries, etc., to look for words beginning with certain letters. It is convention to categorize words based on the first letter rather than the third. Again, this is a static problem rather than a before and after bias. Individuals have eliminated some parts of the distribution over their life experience. When certain events are salient (like extreme good or bad events), then these events might find larger probability weight simply because they have been given a prominent place in an individual's mind. If no events are salient, then individuals seem to exaggerate the mode of a distribution. This may have no impact on symmetric distributions, while asymmetric distributions will be drawn in toward the mode (away from the mean).

Law of Small Numbers

The law of small numbers is a facetious name given to the phenomenon of giving too much weight to new information. It is commonly observed among scientists (and particularly social scientists). Essentially individuals try to draw population properties from samples no matter how small that sample may be. Tversky and Kahneman (1982a) cite four effects of this bias affecting mostly scientists: (1) overestimated power of tests, (2) undue confidence in early trends and patterns, (3) undue expectations of replicability, and (4) finding of causal explanations for deviations that may be due to random variability. This bias is strongly related to the representative bias to be discussed next, and I believe related to overconfidence. Similar sorts of biases may have caused some of the problems with television network forecasts regarding the 2000 Florida presidential vote count.

Possible Explanation. Individuals largely ignore the size of sample when making inferences. This bias suggests that individuals have some sort of bias away from sampling variability. They form their beliefs assuming low degrees of variability (or, conversely, high degrees of representativeness). This may be produced by a process similar to that producing overconfidence. In other words, individuals narrow the distribution of the likelihood information transmitted. This means that individuals would have too narrow confidence intervals when beliefs were later elicited.

Representative Bias

Representativeness is similar to the law of small numbers. The representative bias occurs when individuals ignore base rates or prior information (Kahneman and Tversky 1982). This is what every Bayesian scientist accuses frequentists of doing (only frequentists do it on purpose!). This means they make predictions essentially using maximum likelihood estimation. Grether (1980) conducted experiments to test for the use of base rates using a modified Bayes' rule for estimation. Individuals were informed regarding the distribution of bingo balls in two bingo cages. One of the cages was chosen by the flip of a coin and a small sample of balls was drawn from that cage. Individuals were then asked to assess the probability that the sample came from each of the cages. Grether estimated the weights on prior and likelihood functions using the likelihood ratio form and consistently found that individuals underweighted prior information (although they did not ignore it altogether). He also noted that the weights were significantly different across scenarios. This phenomenon is called the "representative" heuristic because undue probability is given to the population with properties that are similar to (or represented by) the sample.

The representative bias is one of the most studied, particularly in financial markets. Several studies have conducted experiments to test for representativeness in experimental commodity markets. Camerer (1987) found significant representativeness in certain trading situations. De Bondt and Thaler found that representativeness may cause overreaction on the New York Stock

Exchange. Arrow (1982) suggests that the representative heuristic typifies securities and commodities markets. Individuals make decisions that are too large for the little information they have.

Possible Explanation. It appears that individuals will sometimes overreact to information. This is not always the case and can occur in varying degrees. In any case, the literature on overconfidence suggests that individuals will display this bias when likelihood functions are narrower than base rates. This means that individuals when faced with new information that is narrow (or less diffuse) will ignore base rates and make predictions based on the height of the likelihood function. It is not surprising that individuals display this sort of bias because it is essentially the method used by scientists in estimation to *avoid bias*. This makes sense if the scientists continue to add to their dataset when new information arrives. However, if each new set of information is used in place of previous information, learning is not consistent in a statistical sense. The fact that individuals do not always ignore base rates may be observed in the following discussion on conservatism.

Conservatism

Conservatism is when an individual weights prior information too heavily in probability assessment (Camerer 1995). Edwards (1982) cites a wide array of psychological studies confirming the existence of conservatism. Edwards ran some experiments nearly identical to Grether's (1980) bingo cage experiments and found conservatism. He also ran some other computer-based experiments asking subjects to assess the probability of four different events based on simulated radar screens. He found overwhelming support for conservatism in his data. Subjects seemed to take five periods to do the updating work that Bayes' rule would do in one.

When plotting the subjective likelihood ratios, Edwards (1982) found that they appeared to be linear (as they would be with perfect Bayesian updating) but that the slope was too shallow. This indicated an underweighting of the likelihood function. He then suggested use of a modified Bayes' rule that raises the likelihood to some power. This is equivalent to the Bayes-based model of Grether (1980).

Williams (1987) was also interested in forecasting ability. He cites strong evidence suggesting that forecasting errors in agricultural commodity markets and consumer price forecasts are autocorrelated. In most cases, forecasters were slow to update, displaying considerable conservatism (although not always). More interesting are the experiments Williams ran to see if markets with no change in the underlying distribution of market factors would lead eventually to rational expectations. He used a double-auction experiment over several periods where each individual had been assigned the same value for goods each period. Individuals were also asked to forecast prices. The data display significant positive autocorrelation in forecast errors. He was also able to reject the notion that individuals were converging to behavior consistent with rational expectations. In other words, no amount of stability seems to induce individuals to know their environment (at least for the size of rewards that Williams offered).

Possible Explanation. Edwards (1982) suggests that updating (or combining likelihood and prior information) requires significant mental exertion. Individuals may save on these costs by reducing the degree to which they incorporate new information into their beliefs. Edwards noted that subjects updated at different rates for differing situations, and hypothesized that updating depended on the difficulty of processing in that particular situation. He suggests that further research would need to be done to determine exactly what affects the difficulty before a model could be suggested. Haruvy, Erev, and Sonsino (2001) cite significant evidence that individuals fail to learn as quickly as Bayes' rule suggests when likelihood information is very diffuse. In other words a tight distribution may be easy to process using Edwards' (1982) hypothesis, and may allow individuals to update more quickly (maybe even too quickly). A wide distribution may be ignored because of the processing costs involved in gleaning information from a distribution with little information to be gleaned. This is consistent with the experimental evidence reported by Hogarth and Einhorn (1992), which suggests that the speed of learning (or the weight given to prior beliefs and new information) will depend on how unusual or complicated that information may be.

Alternate Models

Several models have been suggested to explain the biasing phenomenon above. Most of these fall into two categories: anchoring and adjustment, and Bayesian-based adjustment. The most prevalent models are the anchoring and adjustment models used by Kahneman and Tversky (1982), Hogarth and Einhorn (1992), and several others. Here I will explain the example given by Hogarth and Einhorn. The mathematical model can be written as

$$S_k = S_{k-1} + w_k[s(x_k) - R],$$

where S_k is the degree of belief in a hypothesis after updating k times and S_k is assumed to be in the unit interval. The initial belief is S_0, $s(x_k)$ is the subjective update in response to x_k (the kth piece of evidence), R is a reference point to which the new evidence is compared, and w_k is the weight (between 0 and 1) given to the new information. This model essentially allows a tradeoff between the weight, w, and strength, s, of information.

By using this model, it is possible to account for many of the biases above (most often representativeness is modeled this way). Hogarth and Einhorn (1992) suggest that weights and strengths are dependent upon the level of complication in interpreting information and the response mode. They conducted experiments by showing sets of data to individuals and then eliciting beliefs. From these elicited beliefs, they estimated a separate w and s for several sets of data with varying degrees of complication and length and with varying response modes. These estimates show significant differences in the weighting mechanisms in each of the situations, implying that complicated information may be weighted less heavily. However, because anchoring and adjustment models yield only a belief in a point estimate, the usefulness of this model is limited. For instance it would be hard to enter these beliefs in many of the expected utility models commonly used in economics. Moreover, in order to make this model useful where discrete values are possible, some model of how w and s vary across situations is desirable.

Ferrell and McGoey (1980) constructed a model of calibration. They assume that individuals assess the probability of X taking on a value by partitioning the probability space of X and assigning subjective weights to each partition. The novelty of the model is that partition resolution is not adjusted across problems. Interestingly this model implies underconfidence in easy problems and overconfidence in more difficult problems, where difficulty is measured by how often the correct value falls in the inner-quantile. While this is compelling as a model of static belief, it is hard to estimate in practice because parameters must be estimated for each set within the partition and for the number of sets within the partition. Nevertheless this model might have some interesting applications in economics. The costs of different partitions might be readily modeled, and then histogram data used to optimize subjective expected utility.

As mentioned above, the generalized Bayesian models due to Edwards (1982) and Grether (1980) use a weighting of likelihood and prior in the likelihood ratio form

$$P = \frac{P_1^r L_1^{1-r}}{P_2^r L_2^{1-r}},$$

where P is the posterior odds of event 1 over event 2. This model seems to describe the data very well when a single update is considered. The problem is that r must vary across situations (Grether 1980). Considering the huge differences in predictions that may result from differing weights, it is important to model the process that determines r.

I propose the following model of belief updating (D.R. Just, forthcoming):

$$P_{t+1}(x) = \frac{p_t(x)^{R(l,p,z)} l(\theta \mid x)^{L(l,p,z)}}{\int_{-\infty}^{\infty} p_t(x)^{R(l,p,z)} l(\theta \mid x)^{L(l,p,z)} dx},$$

where p_t is the density representing beliefs in period t, l is the likelihood function representing incoming information, z is a vector of environmental factors that might affect information processing costs, and R and L are weights representing recall and learning respectively. This model, which I call the limited learning model, lets the weights given to new information and previous beliefs depend upon the distributions representing these sets of information and other factors like education or presentation. In other work I have shown that in allowing R and L to depend only on the diffusion of likelihood and prior information, the model can account for the judgment biases explained above. By defining a reasonable prior, the limited learning model also explains many of the commonly observed violations of expected utility theory (D.R. Just, forthcoming).

Conclusions from the Judgment Bias Literature

From the empirical evidence it is clear that more difficult-to-process information may be given less weight than is optimal from a Bayesian perspective. When the information is more easily processed, it may be given more weight than Bayes' rule implies. For this reason it is important to model when each situation may occur. By improving models of beliefs under uncertainty to include availability of and psychological factors affecting use of information, models of decision making under uncertainty can begin to be evaluated. Without incorporating the learning and belief updating process, decisions under uncertainty become muddled in analyses of decision making because two separate processes are being modeled as one. I will further explore this theme in the section entitled "An Example."

DIRECTIONS FOR RESEARCH

One can think of several areas of the risk literature that could benefit from incorporating psychological limits and costs associated with beliefs. In this section, I identify three areas where systematically biased beliefs may be of particular importance: modeling the psychological costs of incorporating information, the impact of available information on beliefs, and the estimation and prediction of risk preferences. This list is by no means complete, but I believe these categories to have the greatest weight in applying economics to policy decisions.

Modeling the Psychological Costs of Incorporating Information

Information is not free. Even information that may be obtained without monetary cost will require time in communication, and time and exertion in order to make sense of the information once communicated. This means that some information that yields (or is perceived to yield) little benefit will not be fully incorporated into the belief system of an individual. Many of the biases found in the psychological literature may be due simply to actors realizing the psychological or physical costs of incorporating all information available. Which information will be ignored and which will be used is of some importance in modeling the type of market behavior that will result from these beliefs.

The first observation is due to the law of small numbers bias. Individuals have a hard time figuring out what weight to give small samples in their belief system. In fact, individuals may ignore sample size completely. This suggests some cost of considering sample size when incorporating new information. Because of this, individuals may make important decisions based on little information. Decisions to insure against rare events, or to take precautions to avoid disaster, may be biased away from what is best for the individual. The fact that an earthquake has never destroyed my house may lead me to believe earthquakes are more rare than they actually are. Similar thoughts may lead many to insure against earthquake damage the year after a major earthquake, even though the probability of another quake is more remote (Camerer 1995).

It is important when modeling beliefs at any particular point in time to remember that beliefs are an accumulation of the belief updating process over the entire lifespan of the individual. The predominance of overconfidence bias suggests that individuals tend to ignore information that is more diffuse in its predictions. Haruvy, Erev, and Sonsino (2001) cite this tendency in constructing a model of learning where speed of learning depends on variance of the distribution being learned. In fact, all of the experimental examples of representativeness and conservatism can be explained by individuals to some extent disregarding the more diffuse information (whether it was through the prior or likelihood). See D.R. Just (forthcoming).

Assuming that the cost of combining information depends on the diffusion of the information to be combined may go a long way in describing belief-updating behavior, and hence risk behavior. Thus, when information is poor (or does not yield precise predictions), actors in markets may tend to ignore changes that ought to alter behavior. For example, weak signals of a downturn in productivity may have little effect on stock market prices. Alternatively, farmers facing weak signals of slow demand may under-assess the possibility of bankruptcy. When information is particularly precise in its predictions, individuals may place too much weight on these predictions. One example of this may have been the television network forecasts of 2000 presidential ballot returns in the state of Florida. Twice the networks predicted an outcome when the return data was inconclusive. Similar examples might be found in apparent stock market and commodity market bubbles.

The costs of incorporating information must also include a cost of availability. Availability may have several dimensions. Some events are easier to remember (such as extremely bad or good weather years). Other events may be hard to remember (such as the number of years when weather was very close to average). Media may be able to bias our understanding of probabilistic events in this way (Tversky and Kahneman 1982b). For instance, despite a lower rate of farm failures than general business failures during the 1980s, many considered the rate of farm failures to be catastrophically high, and worth government attention. It could be that newspapers or other media over-reported the number of farms facing problems, causing the image of farm failure to be more available than business failure in general. Also, some information is simply not published in a format that decision makers can use (Wolf, Just, and Zilberman 2001). Information that is not widely published, or is published in formats that are hard to use, will likely be under-represented in beliefs. Conversely, information that is widely published may be given too much weight.

The degree to which these biases are important in agricultural economics will depend on the degree to which they affect not only farmers, but agribusiness firms and policymakers. National legislators originating from a certain district may have beliefs that are much more representative of their own district than of the nation as a whole. This may lead to conservation, adoption, and crop insurance programs ill-equipped to deal with the wide variety of risks within the farm sector. Further, policymakers, farmers, and input and output suppliers may conceivably (in fact, probably) have different beliefs as to the degree of risk present in a market. Research on the nature of risk is not likely to change these biases because it has been shown that individuals who are very knowledgeable about uncertainty still display significant bias in probability assessment (see the discussion above). However, by incorporating measures of belief updating costs in models of uncertainty, it may be possible to suggest policies designed to take into account the learning process. Extension and land grant university researchers may be in a prime position to educate farmers, agribusiness decision makers, and legislators on the effects of these biases in the markets once they are known. By proscribing policies that include information dissemination and training, researchers might be able to encourage a more efficient and equitable distribution of welfare through better-informed decisions.

How Does Information Formatting and Availability Affect Beliefs?

Schultz (1975) suggests that education and training (or human capital) may have significant effects on one's ability to make sense of information in the decision making process. Wolf, Just, and Zilberman (2001) draw attention to this relationship within agriculture by showing the differences in information purchasing patterns of those with varying levels of education. From their research, several conclusions may be drawn about the structure of the information service sector within agriculture. First, it appears that the government, through land grant universities, extension, and USDA, produces the majority of original information in the agricultural economy. This information, however, is not formatted in a way that farmers, exporters, input suppliers, or others find easy to use. This creates the need for a class of businesses and organizations designed to transform government and other information into information products more accessible to individual decision makers. In essence, less educated decision makers, or those firms with fewer resources, buy information that has been processed and customized to complement their own internal capacities.

Commodity associations tend to deliver information about regulatory changes, commercial consultants tend to deliver information about production problems, and agricultural media tend to deliver information about new technologies. Larger firms hire in-house analysts who coordinate the information-gathering effort, while small firms use less-customized information. Some information is distributed directly to end users by the government, and to a lesser extent, some is created outside the government. Many information services do nothing more than deliver government reports that pertain to a particular field of interest. The value added is in reduced search costs, which might otherwise inhibit availability. Even more interesting is the dependence of decision makers on informal information obtained through conversations with friends, business associates, and acquaintances.

This description, while incomplete, paints a picture of widely heterogeneous information needs within some industries where producers are thought to be very homogeneous (such as California fresh tomato growers). Incredible amounts of effort seem to be exerted in changing the format of information, and in making it easier to understand, apply, or find. While much of the information has the same origin, some meaning is transformed or lost in the process of altering formats, particularly in the realm of informal information. This means informational biases may not depend solely on the decision maker's ability to understand, but on the ability of the service sector to avoid biasing the information it compiles. Where there is much cross-fertilization and circulation among the various information intermediaries, including decision support within the realm of belief updating may be very complicated.

The fact that large and small firms may have differing capacity for analyzing and obtaining information is a particularly disturbing notion. Diminishing marginal utility of wealth has been used often to explain why some larger firms might display risk attitudes that differ from those of small firms. The fact that firms of varying size also have varying analytic capacity, and hence varying beliefs, provides a provocative alternative to the standard explanation. These types of differences may allow large firms to increase their own benefits of contracts with smaller firms. Intuitively, if a smaller firm believes any inherent risk is smaller than it is, it will be more willing to share this risk with the larger firm. The effects of informational bias need to be explored to determine whether there is a small firm effect. If these effects exist, they may lead to market distortions requiring government intervention.

Policies designed to reduce the risk of smaller firms may have a smaller than anticipated effect if firms are not trained or given the analytic capacity to understand the implications of such a policy. Government policies designed to make farmers (and policymakers) better decision makers become more important under this hypothesis. In particular, government provision is important for those types of information that may be systematically selected by information intermediaries. Furthermore, information provision in a format that is readily usable by the decision makers is important. This is already one of the goals of extension work. By training decision makers to better understand the risks they face, their decisions will better reflect their own preferences, leading to more efficient markets.

Estimation of Risk Preferences

The estimation of risk preferences has been the subject of many papers (see for example Binswanger 1980). However, if there are systematic distortions of probability due to the costs involved in understanding risk, these estimates may be misleading. Further, it may be that the measures of risk aversion proposed by Arrow (1971) are inadequate in determining risk attitudes. The purpose of determining risk attitudes is presumably to predict how individuals might act in situations that are not observed (such as under a mean-preserving spread). This type of prediction of behavior may depend on more than the curvature of a utility function. Under the assumptions of this chapter, one also must know the information environment that will exist in the new circumstances and whether it will affect individuals' abilities to understand their new environment.

Even if preferences may be written as in (1), current risk-aversion measures are misleading. This is because the ability to perceive truth is likely to depend on the distribution of payoffs and other factors. An aversion to or affinity for mean-preserving spreads will depend not only on diminishing marginal utility of wealth, but on how the mean-preserving spread is understood. The question is, then, whether we measure risk aversion with respect to the true uncertainty, or with respect to the uncertainty an individual has perceived. If we wish to do the latter, then current measures of risk aversion are adequate. However, in order to estimate these measures with accuracy, the process of perception must be understood and modeled. On the other hand, if we wish to measure risk aversion with respect to the true underlying uncertainty, then we need a measure that accounts for human factors in addition to wealth. In many cases, the reaction to decreased risk (through disaster relief, etc.) will not only change the underlying uncertainty, but change the distribution of information available. Modeling based purely on curvature of the utility function will not capture these effects. Some examples of important human factors are the information delivery method, individuals' education levels, and the degree of uncertainty itself.

Most theories explaining decision making under uncertainty, including expected utility theory and Machina's generalized expected utility theory, require that preferences over lotteries, $V(F)$ with F the distribution of payoffs, be Frechet differentiable with respect to the metric given by $d(F_1, F_2 = \int |F_1(x)\text{-}F_2(x)| dx$. Requiring that preferences change little when there are small changes in the distribution of outcomes seems reasonable. The absolute minimum requirement imposed by almost all theories is that preferences are a functional defined on the space of lotteries. Upon recognizing that beliefs may change despite no change in the underlying risk, it becomes clear that neither of these properties will hold. In this case, it may be more reasonable to represent preferences as $V(F, h)$, where h is a measure of the environmental situation leading to perception. In this case, Frechet differentiability of preferences with respect to F given h might be reasonable, and once again assuming that preferences are a true functional, mapping L x H, or the set of lottery and environment pairs, into the real line is reasonable. Individuals have often been observed to change preferences between lotteries when presentations differ (Kahneman and Tversky 1979). Dominated lotteries may be chosen when the dominant lotteries are convoluted and, hence, only detectable at a large mental cost (see Camerer 1995 for a list of examples). These are important effects to incorporate into any policy dealing with risk.

An Example

In this section I illustrate how ignoring human factors in decision making under risk may lead to incorrect conclusions. As a basis for the example, I take the model of land allocation found in Just and Zilberman (1983). They consider a farmer who allocates land between two different crops. I will call these crops new (n) and old (o). The farmer is assumed to have a limited amount of land L. Suppose each farmer is subject to an informational constraint h in assessing the newly available crop. The optimization condition may be written as

$$\max_{L_o, L_n} EU[\pi_o L_o + \pi_n L_n \mid h],$$

subject to

$$L_o + L_n \leq L \,,$$

$$L_o, L_n \geq 0 \,,$$

where L_i is the amount of land allocated to crop i, and π_i is the stochastic profit per acre planted of crop i. By assuming an interior solution, the optimal land allocation rule is approximated by

$$L_o = \frac{\mu_o - \mu_n(h)}{[\sigma_o^2 \sigma_o^2(h) - 2\sigma_{on}(h)]\phi(w)} + \frac{\sigma_o^2(h) - \sigma_{on}(h)}{\sigma_o^2 \sigma_o^2(h) - 2\sigma_{on}(h)} L \,, \tag{2}$$

where μ_i represents mean profit per acre for crop i, σ_i^2 represents the variance of profits per acre, σ_{ij} represents covariance of profits per acre between the two crops, and $\phi(W)$ is the Arrow-Pratt measure of absolute risk aversion.

Consider first that case where all individuals have the same informational constraint and the same mean and variance of returns per acre for the old crop. Just and Zilberman (1983) suggested using (2), ignoring human factors, to estimate absolute risk aversion as a function of wealth W. For example, consider three divisions of wealth, W_1, W_2, and W_3, each a dummy variable indicating that the individual falls within the indicated wealth level. We can then rewrite (2) as

$$L_o = A_1 W_1 + A_2 W_2 + A_3 W_3 + BL + \varepsilon \,, \tag{3}$$

to estimate parameters A_1, A_2, A_3, and B based on observed land allocation and income level. These estimates permit approximation of the change in absolute risk aversion in the following way:

$$A_3 = A_1 \frac{\phi(W_1)}{\phi(W_3)} = A_2 \frac{\phi(W_2)}{\phi(W_3)} \,,$$

yielding two equations in three unknowns. Because of sign restrictions, we can at least compare absolute risk aversion levels across wealth.

Considering human factors, this equation becomes more complicated. Suppose, for example, the distributions of profits with crops (n) and (o) are identical but have zero covariance. Correlation is evidently more difficult to identify than marginal distributions (see Kahneman, Slovic, and Tversky's 1982 section on correlation). Suppose individuals are able to understand the distributions of each crop in terms of mean and variance, but that covariance requires analytic capacity to learn. In other words, for all h, $\sigma_n^2\,(h) = \sigma_n^2$, $\mu_n(h) = \mu_n$, so

$$L_o \frac{\mu_o - \mu_n}{[\sigma_o^2 \sigma_o^2 - 2\sigma_{on}(h)]\phi(W)} + \frac{\sigma_o^2 - \sigma_{on}(h)}{\sigma_o^2 \sigma_o^2 - 2\sigma_{on}^{(h)}} L.$$

As hypothesized above, let analytic capacity be correlated with wealth. I use Monte Carlo estimation to provide some insight into the bias introduced by not considering human factors where absolute risk aversion is equal to .01, and $L \sim N(100, 10)$ for all observations. This means that individuals are slightly risk-loving. In a sample of 60,000, I let one-third fall into each income group. Profits were assumed to have the following uncorrelated distributions: $\pi_o \sim N(100,30)$, $\pi_n \sim N(110,60)$. In order to create a correlation between income level and perception of correlation, for each income level the covariance was assumed to be distributed uniformly over the interval $[0, t_i]$ where $t_1 = 25$, $t_2 = 12.5$, and $t_3 = 5$. This means that the maximum perceived correlation coefficient is .59, .29, and .11 for the low, medium, and high income groups, respectively. An individual with 100 units of land and correct perception will allocate 55.56 percent of his land to the old crop. An individual who believes σ_{on} will allocate 62.5 percent.

Estimating (3) yields the parameter estimates $A_1 = -12.35$, $A_2 = -13.95$, $A_3 = -14.57$, and $B = .70$. Using these estimates obtains

$$\phi_L = 1.13\phi_M,$$
$$\phi_L = 1.18\phi_H.$$

Hence, the estimates show decreasing absolute risk aversion despite the constant absolute risk aversion assumed in the underlying model.

Several studies already take into account many of the factors that might affect risk preferences (see, for example, Bar-Shira, Just, and Zilberman 1997, Dillon and Scandizzo 1978, Moscardi and de Janvry 1977, Binswanger 1980, 1981, 1982, and Shahabuddin, Mestelman, and Feeny 1986). However, these factors (education, experience, access to information) need to be modeled correctly in order to allow correct interpretation. Simply allowing dummy variables to affect the coefficient of absolute risk aversion is a misspecification in even the simplest of models (as in the model above). For instance, heterogeneous perception of correlation in the example above can also be interpreted to depend on education, which may be correlated with wealth. In this case, one might incorrectly conclude that the less educated are more risk averse. Recognizing that education may affect ability to understand, this interpretation becomes muddled. While the less educated avoid risk more often, their avoidance may be due to inability to understand the risks they face rather than a preference to avoid risk.

CONCLUSIONS

This chapter explores alternate explanations for risk behavior. In particular, some observed risk behavior seems to be due to limitations on ability to reason and make sense of information. The public policy debate in the agricultural arena has often centered on price, yield, or revenue risk. While eliminating some of this risk may be politically sound, it seems odd to subsidize in this manner if farmers correctly perceived the risks involved when choosing to farm. Alternatively it may be reasonable, if perceptions do not always resemble the real world, that some farmers and others might at times find themselves in completely unexpected circumstances requiring some sort of aid to maintain lifestyle. One aim of government risk policy should be oriented toward training farmers to interpret information and allowing them greater access to information that might not be publicly provided. In order to create a more reasonable policy agenda, researchers should recognize the difference between risk perception and risk attitude. The overriding goal of any government policy aimed at reducing the negative effects of risk should be to give farmers the tools they need to make the choices they most desire.

REFERENCES

Alpert, M., and H. Raiffa. 1982. "A Progress Report on the Training of Probability Assessors." In D. Kahneman, P. Slovic, and A. Tversky, eds., *Judgment under Uncertainty: Heuristics and Biases*. New York: Cambridge University Press.

Arrow, K.J. 1971. *Essays on the Theory of Risk Bearing*. Chicago, IL: Markham Publishing Co.

___. 1982. "Risk Perception in Psychology and Economics." *Economic Inquiry* 20: 1-9.

Bar-Shira, Z., R.E. Just, and D. Zilberman. 1997. "Estimation of Farmers' Risk Attitude: An Econometric Approach." *Agricultural Economics* 17: 211-222.

Bernoulli, D. 1954. "Exposition of a New Theory on the Measurement of Risk" (originally published in 1738). *Econometrica* 22: 23-63.

Binswanger, H.P. 1980. "Attitudes Toward Risk: Experimental Measurement in Rural India." *American Journal of Agricultural Economics* 62: 395-407.

___. 1981. "Attitude Toward Risk: Theoretical Implication of Experiment in Rural India." *Economic Journal* 91: 867-890.

___. 1982. "Empirical Estimation and Use of Risk Preference Discussion." *American Journal of Agricultural Economics* 64: 391-393.

Camerer, C. 1987. "Do Biases in Probability Judgment Matter in Markets? Experimental Evidence." *American Economic Review* 77: 981-997.

___. 1995. "Individual Decision Making." In J.H. Kagel and A.E. Roth, eds., *The Handbook of Experimental Economics*. Princeton, NJ: Princeton University Press.

De Bondt, W.F.M., and R.H. Thaler. 1985. "Does the Stock Market Overreact?" *Journal of Finance* 40: 793-805.

Dillon, J.L., and P.L. Scandizzo. 1978. "Risk Attitudes of Subsistence Farmers in Northeast Brazil: a Sampling Approach." *American Journal of Agricultural Economics* 60: 425-434.

Edwards, W. 1982. "Conservatism in Human Information Processing." In D. Kahneman, P. Slovic, and A. Tversky, eds., *Judgment under Uncertainty: Heuristics and Biases*. New York: Cambridge University Press.

Eekhoudt, L.R., L. Bauwens, E. Briys, and P. Scarmure. 1991. "The Law of Large (Small?) Numbers and the Demand for Insurance." *Journal of Risk and Insurance* 58: 438-451.
Ferrell, W.R., and P.J. McGoey. 1980. "A Model of Calibration for Subjective Probabilities." *Organizational Behavior and Human Performance* 26: 32-53.
Gardner, B.L. 1992. "Changing Economic Perspectives on the Farm Problem." *Journal of Economic Literature* 30: 62-101.
Grether, D.M. 1980. "Bayes Rule as a Descriptive Model." *Quarterly Journal of Economics* 95: 537-557.
___. 1990. "Testing Bayes Rule and the Representative Heuristic: Some Experimental Evidence." *Journal of Economic Behavior and Organization* 17: 31-57.
Hansson, B. 1988. "Risk Aversion as a Problem of Cojoint Measurement." In P. Gardenfors and N.-E. Sahlin, eds., *Decision, Probability, and Utility.* New York: Cambridge University Press.
Haruvy, E., I. Erev, and D. Sonsino. 2001. "The Medium Prizes Paradox: Evidence From a Simulated Casino." Presented at the Allied Social Science Association Meetings, New Orleans, LA.
Hogarth, R.M., and H.J. Einhorn. 1992. "Order Effects in Belief Updating: the Belief Adjustment Model." *Cognitive Psychology* 24: 1-55.
Just, D.R. Forthcoming. "Learning and Information." Ph.D. dissertation, Department of Agricultural and Resource Economics, University of California Berkeley.
Just, R.E., and R.D. Pope. 1987. "Stochastic Specification of Production Function and Economic Implications." *Journal of Econometrics* 7: 67-68.
Just, R.E., and D. Zilberman. 1983. "Stochastic Structure, Farm Size and Technology Adoption in Developing Agriculture." *Oxford Economic Papers* 35: 307-328.
Kahneman, D., P. Slovic, and A. Tversky (eds.). 1982. *Judgment under Uncertainty: Heuristics and Biases.* New York: Cambridge University Press.
Kahneman, D., and A. Tversky. 1979. "Prospect Theory: An Analysis of Decision under Risk." *Econometrica* 47: 263-292.
___. 1982. "Subjective Probability: A Judgment of Representativeness." In D. Kahneman, P. Slovic, and A. Tversky, eds., *Judgment under Uncertainty: Heuristics and Biases.* New York: Cambridge University Press.
Lichtenstein, S., B. Frischhoff, and L.D. Phillips. 1982. "Calibration of Probabilities: The State of the Art to 1980." In D. Kahneman, P. Slovic, and A. Tversky, eds., *Judgment under Uncertainty: Heuristics and Biases.* New York: Cambridge University Press.
Machina, M.J. 1982. "Expected Utility Analysis Without the Independence Axiom." *Econometrica* 50: 277-323.
Markowitz, H. 1952. "The Utility of Wealth." *Journal of Political Economy* 60: 151-158.
Moscardi, E., and A. de Janvry. 1977. "Attitudes Toward Risk Among Peasants: An Econometric Approach." *American Journal of Agricultural Economics* 59: 710-716.
Murphy, A.H., and R.L. Winkler. 1974. "Subjective Probability Forecasting Experiments in Meteorology: Some Preliminary Results." *Bulletin of the American Meteorological Society* 55: 1206-1216.
Nelson, R.G., and D.A. Bessler. 1989. "Subjective Probabilities and Scoring Rules: Experimental Evidence." *American Journal of Agricultural Economics* 71: 363-369.
Oskamp, S. 1982. "Overconfidence in Case-Study Judgments." In D. Kahneman, P. Slovic, and A. Tversky, eds., *Judgment under Uncertainty: Heuristics and Biases.* New York: Cambridge University Press.
Rabin, M. 1998. "Psychology and Economics." *Journal of Economic Literature* 36: 11-46.
___. 2000a. "Diminishing Marginal Utility of Wealth Cannot Explain Risk Aversion." Working Paper No. E00-287. Department of Economics, University of California, Berkeley.
___. 2000b. "Risk Aversion and Expected-Utility Theory: A Calibration Theorem." Working Paper No. E00-279. Department of Economics, University of California, Berkeley.

___. 2000c. "Inference by Believers in the Law of Small Numbers." In progress. Department of Economics, University of California, Berkeley.

Sandmo, A. 1971. "On the Theory of the Competitive Firm under Price Uncertainty." *American Economic Review* 61: 65-73.

Schultz, T. 1975. "The Value of the Ability to Deal with Disequilibria." *Journal of Economic Literature* 13: 827-837.

Shahabuddin, Q., S. Mestelman, and D. Feeny. 1986. "Peasant Behaviour Towards Risk and Socio-Economic and Structural Characteristics of Farm Households in Bangladesh." *Oxford Economic Papers* 38: 122-130.

Simon, H.A. 1955. "A Behavioral Model of Rational Choice." *Quarterly Journal of Economics* 69: 99-118.

___. 1959. "Theories of Decision-Making in Economics and Behavioral Science." *American Economic Review* 49: 253-283.

___. 1978. "Rationality as Process and as Product of Thought." *American Economic Review* 68: 1-16.

Tversky, A., and D. Kahneman. 1982a. "Belief in the Law of Small Numbers." In D. Kahneman, P. Slovic, and A. Tversky, eds., *Judgment under Uncertainty: Heuristics and Biases*. New York: Cambridge University Press.

___. 1982b. "Judgment under Uncertainty: Heuristics and Biases." In D. Kahneman, P. Slovic, and A. Tversky, eds., *Judgment under Uncertainty: Heuristics and Biases*. New York: Cambridge University Press.

Williams, A.W. 1987. "The Formation of Price Forecasts in Experimental Markets." *The Journal of Money, Credit and Banking* 19: 1-18.

Wolf, S., D.R. Just, and D. Zilberman. 2001. "Between Data and Decisions: The Organization of Agricultural Economic Information Systems." *Research Policy* 30: 121-141.

Yaari, M.E. 1987. "The Dual Theory of Choice under Risk." *Econometrica* 55: 95-115.

Part 2

CONCEPTUAL ADAPTATIONS OF RISK MODELS FOR AGRICULTURE

Chapter 6

DUAL APPROACHES TO STATE-CONTINGENT SUPPLY RESPONSE SYSTEMS UNDER PRICE AND PRODUCTION UNCERTAINTY

Robert G. Chambers and John Quiggin
University of Maryland and Australian National University

INTRODUCTION

Following a research direction originally set by Debreu (1959), Arrow (1953), Hirshleifer (1965), and Yaari (1969), Chambers and Quiggin (1992, 1996, 1997, 1998, 2000, 2001a, 2001b) and Quiggin and Chambers (1998a, 1998b, 2000) have studied the axiomatic foundations and theoretical applications of state-contingent production models. Among other results, they have shown that dual cost structures exist for state-contingent technologies and that these dual cost structures can be used to simplify the analysis of stochastic decision making. The guiding principle of their work was elucidated almost 50 years ago by Debreu. The state-contingent approach "allows one to obtain a theory of uncertainty free from any probability concept and formally identical with the theory of certainty."

Recently Chambers and Quiggin (2001b) have extended their results on dual cost structures and Coyle's (1992, 1999) analysis of duality for stochastic decision making to indirect certainty equivalents. The purpose of this chapter is to discuss some of the empirical implications of these recent developments.

In what follows, we first present a specification of a state-contingent technology for a single-product firm (the extension to multiple products can be found in Chambers and Quiggin 2001b) operating under conditions of both price and production uncertainty. We then state the basic result of Chambers and Quiggin (2001b) for this specification and use that result to discuss possible strategies for empirical examination of supply-response systems under conditions of both price and production uncertainty. We then turn to an examination of special cases including risk aversion, complete aversion to risk, and risk neutrality.

THE MODEL AND GENERAL RESULTS

Uncertainty is modeled by nature making a choice from a finite set of states $\Omega = \{1,2,...,S\}$. The state-contingent production technology, following Chambers and Quiggin (2000), is modeled by a continuous input correspondence, $X : \Re_+^S \rightarrow \Re_+^N$, which maps vectors of state-contingent outputs, z, into inputs capable of producing them,

$$X(z) = \{x \in \Re_+^N : x \text{ can produce } z\} \qquad z \in \Re_+^S.$$

The scalar $z_s \in \Re_+$ denotes the ex post or realized output in state s. In addition to continuity, the input correspondence satisfies[1]

X.1 $\quad X(0_S) = \Re_+^N$, and $0_N \in X(z)$ for $z \geq 0_S$ and $z \neq 0_S$

X.2 $\quad z' \geq z \Rightarrow X(z) \subseteq X(z')$

X.3 $\quad$ if $|z^k| \rightarrow \infty$ as $k \rightarrow \infty$, then $\cap_{k\rightarrow\infty} X(z^k) = 0$

X.4 $\quad \lambda X(z^0) + (1 - \lambda)X(z^1) \subseteq X[\lambda z^0 + (1 - \lambda)z^1] \quad 0 < \lambda < 1.$

Individual producers face stochastic output prices, $p \in \Re_{++}^S$, and nonstochastic input prices, $w \in \Re_{++}^N$. Their preferences are defined over ex post income, $y \in \Re^S$, which is the sum of their holding of a financial asset with state-contingent returns $q \in \Re^S$ and their flow profit from production. Hence, their returns in state s are

$$y_s = q_s + p_s z_s - wx.$$

Their evaluation of these ex post incomes are given by a continuous and nondecreasing certainty equivalent function, $e : \Re^S \rightarrow \Re$, with the property that

$$e(\mu\mathbf{1}) = \mu, \qquad \mu \in \Re.$$

By standard duality theorems (Färe 1988), there is a cost function dual to $X(z)$ and defined

$$c(w, z) = \min\{wx : x \in X(z)\}$$

[1] These properties are discussed in detail in Chambers and Quiggin (2000, Chapter 2). Note in particular that they correspond to standard properties placed on input correspondences associated with nonstochastic technologies (Färe 1988).

if $X(z)$ is nonempty and ∞ otherwise. The cost function satisfies

C.1 $c(\boldsymbol{w}, \boldsymbol{z})$ is positively linearly homogeneous, nondecreasing, concave, and continuous in $\boldsymbol{w}$

C.2 Shephard's lemma

C.3 $c(\boldsymbol{w}, \boldsymbol{z}) \geq 0$, $c(\boldsymbol{w}, 0_S) = 0$, and $c(\boldsymbol{w}, \boldsymbol{z}) > 0$ for $\boldsymbol{z} \geq 0_S$, $\boldsymbol{z} \neq 0_S$

C.4 $\|\boldsymbol{z}^k\| \to \infty$ as $k \to \infty \Rightarrow c(\boldsymbol{w}, \boldsymbol{z}^k) \to \infty$ as $k \to \infty$

C.5 $c(\boldsymbol{w}, \boldsymbol{z})$ is convex and continuous on R^S_{++}.

Moreover, by standard duality theorem (Färe 1988),

$$X(\boldsymbol{z}) = \cap_{w>0} \{\boldsymbol{x} : \boldsymbol{wx} \geq c(\mathbf{w}, \boldsymbol{z})\}.$$

The *indirect certainty equivalent* is defined

$$I(\boldsymbol{w}, \boldsymbol{p}, \boldsymbol{q}) = \sup_{z} \{e[\boldsymbol{pz} + \boldsymbol{q} - c(\boldsymbol{w}, \boldsymbol{z})\mathbf{1}]\}. \tag{1}$$

Chambers and Quiggin (2001b) prove

Lemma 1. *$I(\boldsymbol{w}, \boldsymbol{p}, \boldsymbol{q})$ is continuous in $(\boldsymbol{w}, \boldsymbol{p}, \boldsymbol{q})$, nondecreasing in $\boldsymbol{p}$ and $\boldsymbol{q}$, and nonincreasing and quasi-convex in $\boldsymbol{w}$.*

Denote

$$\boldsymbol{z}(\boldsymbol{w}, \boldsymbol{p}, \boldsymbol{q}) \in \arg\max_{z} \{e[\boldsymbol{pz} + \boldsymbol{q} - c(\boldsymbol{w}, \boldsymbol{z})\mathbf{1}]\}.$$

By the theorem of the maximum (Berge 1997, p. 116), the elements of $\boldsymbol{z}(\boldsymbol{w}, \boldsymbol{p}, \boldsymbol{q})$ are upper semi-continuous. Moreover, upon applying Shephard's lemma (Färe 1988) to $c[\boldsymbol{w}, \boldsymbol{z}(\boldsymbol{w}, \boldsymbol{p}, \boldsymbol{q})]$ in the case of a unique cost minimizing solution, we obtain

$$\boldsymbol{x}(\boldsymbol{w}, \boldsymbol{p}, \boldsymbol{q}) = \nabla_w c[\boldsymbol{w}, \boldsymbol{z}(\boldsymbol{w}, \boldsymbol{p}, \boldsymbol{q})],$$

where $\boldsymbol{x}(\boldsymbol{w}, \boldsymbol{p}, \boldsymbol{q})$ is the vector of optimal input demands and ∇ denotes the gradient with respect to the subscripted vector or element of the vector as appropriate. Hence, we obtain the following generalization of Hotelling's lemma for the generalized Sandmovian model. If c is differentiable in $\boldsymbol{w}$ at $\boldsymbol{z}(\boldsymbol{w}, \boldsymbol{p}, \boldsymbol{q})$ and I is differentiable with $\nabla_q I(\boldsymbol{w}, \boldsymbol{p}, \boldsymbol{q}) > 0^S$, then

$$z_s(\boldsymbol{w},\boldsymbol{p},\boldsymbol{q}) = \frac{\nabla_{p_s} I(\boldsymbol{w},\boldsymbol{p},\boldsymbol{q})}{\nabla_{q_s} I(\boldsymbol{w},\boldsymbol{p},\boldsymbol{q})}, \quad s \in \Omega$$

$$\boldsymbol{x}(\boldsymbol{w},\boldsymbol{p},\boldsymbol{q}) = -\frac{\nabla_w I(\boldsymbol{w},\boldsymbol{p},\boldsymbol{q})}{\nabla_q I(\boldsymbol{w},\boldsymbol{p},\boldsymbol{q})\boldsymbol{1}}. \tag{2}$$

Because

$$\boldsymbol{x}(\boldsymbol{w},\boldsymbol{p},\boldsymbol{q}) = \nabla_w c[\boldsymbol{w}, \boldsymbol{z}(\boldsymbol{w},\boldsymbol{p},\boldsymbol{q})],$$

a standard comparative-static decomposition of price effects exists for the optimal input demands. Hence, in the smooth case, the compensated input demands are downward-sloping and symmetric as a consequence of the concavity of $c(\boldsymbol{w}, \boldsymbol{z})$. The overall effect of a change in an input price, as in profit maximization, can be broken into two parts, the already mentioned compensated effect and an expansion effect associated with the induced change in the state-contingent output vector. So, for example, we have

$$\frac{\partial x_k(\boldsymbol{w},\boldsymbol{p},\boldsymbol{q})}{\partial w_j} = \frac{\partial^2 c(\boldsymbol{w},\boldsymbol{z})}{\partial w_k \partial w_j} + \sum_{s\in\Omega} \frac{\partial^2 c(\boldsymbol{w},\boldsymbol{z})}{\partial w_k \partial z_s} \frac{\partial z_s(\boldsymbol{w},\boldsymbol{p},\boldsymbol{q})}{\partial w_j}.$$

Unlike the case of profit maximization but similar to standard demand analysis, there is no general reason to expect either the overall effect to be symmetric or for input demands to be downward-sloping in their own price. These observations enable one to generalize the comparative-static results of Pope (1980) to the case of both price and production uncertainty.

Alternatively, if one is not interested specifically in the input demands themselves, then quasi-convexity places curvature restrictions on the indirect certainty equivalent, which in the differentiable case are reflected by restrictions on the principal minors of the bordered Hessian of the certainty equivalent in input prices

$$\begin{bmatrix} 0 & \nabla_w I(\boldsymbol{w},\boldsymbol{p},\boldsymbol{q})' \\ \nabla_w I(\boldsymbol{w},\boldsymbol{p},\boldsymbol{q}) & \nabla_{ww} I(\boldsymbol{w},\boldsymbol{p},\boldsymbol{q}) \end{bmatrix}.$$

TWO POLAR CASES

The preceding results are quite general, and, as advertised, are independent of any probability measure. However, given existing data collection practices, they likely will be very difficult to identify and efficiently estimate. Moreover, they apply regardless of the individual's risk preferences. Adding assumptions about the individual's preferences towards risk adds structure to the certainty equivalent. And, just as in non-stochastic dual theory, placing additional structure upon either preferences or the technology also places additional structure upon the indirect certainty equivalent. The most common restriction that will typically be placed upon preferences is that the producer has preferences that are quasi-concave over stochastic incomes. As pointed out by Debreu (1959) and Malinvaud (1972), this corresponds to risk aversion for a set of subjective probabilities defined by the supporting hyperplane to the least-as-good set in the neighborhood of the sure thing.

Chambers and Quiggin (2001b) show that quasi-concavity of the certainty equivalent in state-contingent incomes implies that the indirect certainty equivalent is quasi-concave in $\boldsymbol{q}$. In other words, the indirect certainty equivalent inherits the certainty equivalent's basic aversion to risk. A straightforward consequence of this inherited quasi-concavity is the fact that the level sets of $I(\boldsymbol{w}, \boldsymbol{p}, \boldsymbol{q})$ in $\boldsymbol{q}$ space will be supported by hyperplanes whose normals are given by the *virtual* or *risk-neutral* probabilities. In the smooth case, these virtual probabilities are given by[2]

$$\pi_s(\boldsymbol{w}, \boldsymbol{p}, \boldsymbol{q}) = \frac{\nabla_{q_s} I(\boldsymbol{w}, \boldsymbol{p}, \boldsymbol{q})}{\nabla_q I(\boldsymbol{w}, \boldsymbol{p}, \boldsymbol{q})\mathbf{1}}.$$

They can be thought of intuitively as the subjective probabilities that would lead a risk-neutral individual to make the same production choices as the risk averter.

For quasi-concave preferences, it is an interesting fact that the polar cases of aversion to risk – risk neutrality for a given set of subjective probabilities (linear preferences) and complete aversion to risk (Leontief preferences) – both exhibit constant risk aversion in the sense of Safra and Segal (1998). That is, they satisfy both constant absolute risk aversion and constant relative risk aversion. In fact, Chambers and Quiggin (2001c) show that the class of quasi-concave preferences consistent with constant risk aversion is the class of preferences whose least-as-good sets are linear everywhere except at the bisector where they can be kinked.

Thus, it follows immediately that both risk neutrality and complete aversion to risk are special cases of a more general result due to Chambers and Quiggin (2001b):

Lemma 2. *If preferences exhibit constant risk aversion*

$$I(\boldsymbol{w}, \boldsymbol{p}, \boldsymbol{q}) = I^A(\boldsymbol{w}, \boldsymbol{p}, \boldsymbol{q})$$

---

[2] Even if the indirect certainty equivalent is not smooth, when it is quasi-concave these probabilities are always well defined by the superdifferentials of the benefit function associated with the indirect certainty equivalent (Chambers and Quiggin 2001c).

where

$$I^A(\mathbf{w},\mathbf{p},\mathbf{q}) = \max\{e(\mathbf{p}\mathbf{z}+\mathbf{q}) - c(\mathbf{w},\mathbf{z})\},$$

with I^A continuous and positively linearly homogeneous in ($\mathbf{w}$, $\mathbf{p}$, $\mathbf{q}$), nondecreasing in $\mathbf{p}$ and $\mathbf{q}$, nonincreasing and convex in $\mathbf{w}$, and

$$I^A(\mathbf{w},\mathbf{p},\mathbf{q}+\delta 1) = I^A(\mathbf{w},\mathbf{p},\mathbf{q}) + \delta, \quad \delta \in \Re.$$

These properties are fairly self explanatory, apart from the final property. Constant absolute risk aversion implies that sure increases in income do not affect the individual's marginal evaluation of the riskiness of a risky prospect. The last property just guarantees that result. It generalizes the result well known from portfolio theory that an individual with constant absolute risk aversion picks an optimal portfolio that is independent of his wealth level.

Under expected utility preferences, it is well known that constant risk aversion is possible only if the individual is risk neutral. In that case, $I^A(\mathbf{w}, \mathbf{p}, \mathbf{q})$ would correspond to the expected profit function plus the expected value of the producer's portfolio. However, it is also well known that other risk-averse preference functionals can exhibit constant risk aversion. For example, both maximin preferences

$$e(\mathbf{y}) = \min\{y_1,\ldots,y_S\},$$

which exhibit complete aversion to risk and linear mean-standard deviation preferences satisfy constant risk aversion. We initiate our study of supply response systems by considering the case of complete aversion to risk and risk neutrality.

Maximin Preferences

Outside of studies using a programming approach and focusing on food safety and food security issues in a developing context, maximin preferences seem to have received relatively little attention in the applied literature. Because they represent a polar case, logically they seem no more implausible or less interesting than risk-neutral preferences. However, they have received less attention and, therefore, are presumably less well understood, so we address them first.

Maximin preferences are characterized by the fact that their least-as-good sets are supported at the equal-income vector by the class of hyperplanes whose normals contain no negative elements. Put another way, they can be recognized as risk averse in the sense of Yaari (1969) and Quiggin and

Chambers (1998a) for all possible probabilities. As noted, these preferences are not consistent with expected-utility theory, but they are consistent with generalizations of expected-utility theory including the rank-dependent model. Analytically, they are also interesting because the dual approach offers a clear method of deducing input demands and supplies even though preferences are not smoothly differentiable.

As demonstrated by Chambers and Quiggin (2000, 2001a), individuals with maximin preferences completely self-insure in the sense that they choose their state-contingent output vector so that

$$p_s z_s - p_t z_t = q_t - q_s \,.$$

If producers did not behave in this fashion then they could always lower at least one state-contingent output, thus realizing a cost saving, without affecting their welfare level. Hence, income is stabilized across states of nature. Although we do not pursue it here, a number of comparative-static results regarding changes in prices and changes in $\boldsymbol{q}$ follow immediately from this income stabilization result.

To proceed, let that stable level of income be denoted as r; the producer's problem now reduces to

$$V(\boldsymbol{w},\boldsymbol{q},\boldsymbol{p}) = \max_{r}\{r - c(\boldsymbol{w},\boldsymbol{z}) : z_s = r - q_s/p_s\}\,.$$

We refer to $V(\boldsymbol{w}, \boldsymbol{q}, \boldsymbol{p})$ as the *sure-profit function*, and we note that it is isomorphic to a normalized profit function of the form analyzed by Lau (1978).[3] Therefore, it is trivial that the properties of normalized profit functions apply here as well. In particular, in addition to the results we have already established, by the convexity of the cost function in state-contingent outputs and standard arguments in optimization theory (proof in Appendix),

Theorem 3.

$$V[\boldsymbol{w},\lambda\boldsymbol{q}^0 + (1-\lambda)\boldsymbol{q}',\boldsymbol{p}] \geq \lambda V(\boldsymbol{w},\boldsymbol{q}^0,\boldsymbol{p}) + (1-\lambda)V(\boldsymbol{w},\boldsymbol{q}',\boldsymbol{p})\,.$$

In the case of unique solutions, we then have the following analogue to Lau's (1978) restatement of Hotelling's lemma in terms of normalized profit functions. Input demands and the optimal sure income are obtainable as

[3] Note that Lau (1978) discusses duality in terms of the classic Legendre transformation. Although we do not pursue that course here, it obviously provides an alternative path to establishing dual relations here as well.

$$x(w, p, q) = -\nabla_w V(w, q, p),$$

$$r(w, p, q) = V(w, q, p) - w\nabla_w V(w, q, p), \qquad (3)$$

where $r(w, p, q)$ represents the optimal solution to the problem.

A few comments are in order here. First, upon appropriate specification of a parametric form for $V(w, q, p)$, equations (3) provide a basis for econometric estimation and evaluation of the supply-response system associated with this preference structure. Second, just as normalized profit functions completely characterize the underlying production structure, the sure-thing profit function completely characterizes the underlying production structure here as well. Third, by preceding arguments, it follows immediately that optimal state-contingent supplies can be captured directly from this estimated system of equations as

$$z_s(w, p, q) = \frac{r(w, p, q) - q_s}{p_s}.$$

Fourth, the optimal input demands and revenue here bear a very close resemblance to optimal input demands that would be obtained for a risk-neutral individual apart from the dependence of input demands upon the individual's portfolio. These optimal input demands and the optimal revenue, by the basic portfolio choice theorem, are independent of any sure increase in the portfolio. And finally, all of these results, which apply for risk-averse individuals with highly nonlinear preferences, have been obtained independently of any probability measure. One important cause of this result is the fact that individuals with Leontief preferences are, in fact, risk averse for all members of the probability simplex. As we shall see in the next section, probabilities figure importantly in the specification and estimation of optimal demands and supplies under risk neutrality.

Risk Neutrality

Risk-neutral preferences offer another important special case of the class of constant risk-averse preferences, and likely they are the best understood. In this instance, the indirect certainty equivalent is isomorphic to a multiple-output profit function familiar from non-stochastic producer theory, with the role of prices there now being played by the product of prices and probabilities. Here, for the first time in our development, we actually encounter a formal specification of a probability structure. We assume that these probabilities are known or, at least, knowable to the modeler. We have in this case

$$
\begin{aligned}
I(\boldsymbol{w},\boldsymbol{p},\boldsymbol{q}) &= \sup\{\Sigma_{s\in\Omega}\pi_s p_s z_s - c(\boldsymbol{w},\boldsymbol{z})\} + \Sigma_{s\in\Omega}\pi_s q_s \\
&= \overline{V}(\boldsymbol{p},\boldsymbol{w};\pi) + \Sigma_{s\in\Omega}\pi_s q_s\,,
\end{aligned}
$$

where $\pi_s \geq 0$ are the probabilities. The properties of profit functions are well known, so we will not discuss them in detail here. But we do note that this further restriction on the form of the functional structure of the certainty equivalent yields functional restrictions that reach beyond those contained in Lemma 2.

For unique solutions, we have

$$
\begin{aligned}
\nabla_{p_s}\overline{V}(\boldsymbol{p},\boldsymbol{w};\pi) &= \pi_s z_s(\boldsymbol{w},\boldsymbol{p},\boldsymbol{q})\,, \\
\nabla_{w}\overline{V}(\boldsymbol{p},\boldsymbol{w};\pi) &= -x(\boldsymbol{w},\boldsymbol{p},\boldsymbol{q})\,, \\
\nabla_{q} I(\boldsymbol{w},\boldsymbol{p},\boldsymbol{q}) &= \pi\,,
\end{aligned}
\tag{4}
$$

where we preserve the notational dependence upon $\boldsymbol{q}$ to conserve on notation, but it is immediate that both optimal supplies and demands are independent of $\boldsymbol{q}$. Earlier we saw that constant risk aversion ensures that optimal supplies and demands are independent of non-stochastic variations in the portfolio. Here, we see that risk neutrality strengthens that to complete independence. It is perhaps this characteristic more than anything else that distinguishes risk neutrality observationally from quasi-concave risk-averse preferences. Finally, in the risk-neutral case, the virtual probabilities just correspond to the producer's subjective probabilities.

Given current data collection practices, it is going to be rare to encounter a sufficiently rich data set that allows one to identify and estimate the state-contingent supplies themselves. Therefore, in this section we focus on the standard practice of estimating an expected supply function in conjunction with the input demands. It follows immediately from (4) that a supply response system built around an expected supply function and input demands can be constructed completely from the indirect certainty equivalent as

$$
\begin{aligned}
E_\pi z(\boldsymbol{w},\boldsymbol{p},\boldsymbol{q}) &= \sum_{s\in\Omega}\nabla_{p_s}\overline{V}(\boldsymbol{p},\boldsymbol{w};\pi) = \sum_{s\in\Omega}\pi_s\frac{\nabla_{p_s} I(\boldsymbol{w},\boldsymbol{p},\boldsymbol{q})}{\nabla_{q_s} I(\boldsymbol{w},\boldsymbol{p},\boldsymbol{q})} \\
x(\boldsymbol{w},\boldsymbol{p},\boldsymbol{q}) &= -\nabla_{w}\overline{V}(\boldsymbol{p},\boldsymbol{w};\pi).
\end{aligned}
\tag{5}
$$

These equations can provide an empirical platform from which to investigate risk-neutral supply response under conditions of both price and production uncertainty. In estimation, observed supply plus some stochastic error component will then be used in place of expected supply.[4] Once this system of equations is used to estimate a parametric representation of $\overline{V}$ (**p**, **w**; π), then (4) can be used to generate estimates of the state-contingent supplies as

$$z_s(\boldsymbol{w}, \boldsymbol{p}, \boldsymbol{q}) = \frac{\nabla_{p_s} \overline{V}(\boldsymbol{p}, \boldsymbol{w}; \pi)}{\pi_s}.$$

One special case that has been the focus of much recent interest is the estimation of the cost function when the producers are risk neutral and face no price uncertainty (Pope and Just 1996). In that case, dual to the cost function is the expected profit function

$$\overline{V}(\boldsymbol{p}, \boldsymbol{w}; \pi) = \max_{z} \{p \Sigma_{s \in \Omega} \pi_s z_s - c(\boldsymbol{w}, \boldsymbol{z})\}.$$

If so desired, this expected-profit maximization problem can be further decomposed as

$$\overline{V}(\boldsymbol{p}, \boldsymbol{w}; \pi) = \max\{p\bar{z} - \hat{c}(\boldsymbol{w}, \bar{z})\},$$

where $\bar{z} = \Sigma_{s \in \Omega} \pi_s z_s$ and

$$\hat{c}(\boldsymbol{w}, \bar{z}) = \min_{z} \{c(\boldsymbol{w}, \boldsymbol{z}) : \bar{z} = \Sigma_{s \in \Omega} \pi_s z_s\}.$$

It is easy to demonstrate that $\hat{c}$, which corresponds to the minimal cost of producing the expected output, satisfies the properties of a standard cost function.

By an application of Hotelling's lemma, the optimal level of expected supply is

$$\bar{z}(\boldsymbol{p}, \boldsymbol{w}) = \nabla_p \overline{V}(\boldsymbol{p}, \boldsymbol{w}; \pi), \tag{6}$$

while the optimal input demands are given by

$$\boldsymbol{x}(\boldsymbol{p}, \boldsymbol{w}) = -\nabla_w \overline{V}(\boldsymbol{p}, \boldsymbol{w}; \pi). \tag{7}$$

---

[4] Note that this stochastic error structure can be generated directly from the state-contingent specification given a probability measure, or it can represent a composite of the state-contingent error plus one introduced for the purposes of econometric investigation.

Expressions (6) and (7) can provide an empirical basis for econometric estimation of the parameters of the expected profit function by appending a suitable error structure to each of these equations.

Once the parameters of the expected-profit function have been estimated, the parameters of the underlying cost function for expected output can be obtained by solving the duality mapping

$$\hat{c}(\boldsymbol{w}, \bar{z}) = \max_{p} \left\{ p\bar{z} - \bar{V}(\boldsymbol{p}, \boldsymbol{w}; \pi) \right\}.$$

This method of estimating the parameters of the cost function for an expected-profit maximizer, which trivially extends to the case of multiple outputs, represents a fully dual alternative to the procedure suggested by Pope and Just (1996) that exploits all of the economic information available to the researcher.

GENERAL RISK-AVERSE SUPPLY-RESPONSE SYSTEMS

The specification of a general supply-response system follows the trail already blazed in the cases of complete aversion to risk and risk neutrality. In fact, intuition dictates that the supply-response systems that we isolate fall somewhere in between those two poles depending upon the producer's attitudes towards risk. When individuals are risk averse in the sense that their certainty equivalent is quasi-concave, $I(\boldsymbol{w}, \boldsymbol{p}, \boldsymbol{q})$ is continuous in $(\boldsymbol{w}, \boldsymbol{p}, \boldsymbol{q})$, nondecreasing in $\boldsymbol{p}$, nondecreasing and quasi-concave in $\boldsymbol{q}$, and nonincreasing and quasi-convex in $\boldsymbol{w}$. Moreover, given the assumption that the researcher can identify the producer's subjective probabilities, then we have

$$E_{\pi} z(\boldsymbol{w}, \boldsymbol{p}, \boldsymbol{q}) = \sum_{s \in \Omega} \pi_s \frac{\nabla_{p_s} I(\boldsymbol{w}, \boldsymbol{p}, \boldsymbol{q})}{\nabla_{q_s} I(\boldsymbol{w}, \boldsymbol{p}, \boldsymbol{q})}$$

$$\boldsymbol{x}(\boldsymbol{w}, \boldsymbol{p}, \boldsymbol{q}) = -\frac{\nabla_w I(\boldsymbol{w}, \boldsymbol{p}, \boldsymbol{q})}{\nabla_q I(\boldsymbol{w}, \boldsymbol{p}, \boldsymbol{q})\mathbf{1}}.$$

This system of equations can be used as the basis for specifying a system of estimable equations using observed or ex post supply much in the same fashion that we currently estimate demand systems. Many of the same problems encountered in traditional non-stochastic demand analysis, plus a host of new ones, will be encountered here as well. Once the parameters of the indirect certainty equivalent have been estimated, we can obtain by computation estimates of the state-contingent supplies as

$$z_s(\boldsymbol{w},\boldsymbol{p},\boldsymbol{q}) = \frac{\nabla_{p_s} I(\boldsymbol{w},\boldsymbol{p},\boldsymbol{q})}{\nabla_{q_s} I(\boldsymbol{w},\boldsymbol{p},\boldsymbol{q})}.$$

A primary reason for operating in terms of the certainty equivalent rather than in terms of the utility function is that the certainty equivalent is a cardinal measure while our general preference structure is ordinal. Because it is cardinal, it is, in principle, observable or elicitable using survey techniques. Personally, we have had little practice in obtaining such information, so we can offer few specifics here. But presuming that it can be elicited through appropriate survey techniques in conjunction with information on input utilization, then estimation can be based completely upon the elicited certainty equivalent and the observed input demands by specifying an appropriate statistical version of

$$e = I(\boldsymbol{w}, \boldsymbol{p}, \boldsymbol{q}),$$

$$\boldsymbol{x}(\boldsymbol{w},\boldsymbol{p},\boldsymbol{q}) = -\frac{\nabla_w I(\boldsymbol{p},\boldsymbol{w};\boldsymbol{q})}{\nabla_q I(\boldsymbol{p},\boldsymbol{w};\boldsymbol{q})\mathbf{1}}.$$

Estimation can then proceed without information on the producer's subjective probabilities.

Although we do not pursue it here, another alternative to estimation is to specify parametric representations of the virtual probabilities

$$\pi_s(\boldsymbol{w}, \boldsymbol{p}, \boldsymbol{q}), \quad s \in \Omega,$$

substitute these into a parametric representation of $\overline{V}(\boldsymbol{p},\boldsymbol{w};\pi)$, and then use (4) as a means of estimating the corresponding supply response system.

Both the maximin system and the risk-neutral system represent special cases of the approach being discussed here. Hence, given an appropriate parametric specification of the indirect certainty equivalent, parametric tests of various assumptions about the producer's attitudes towards risk can be carried out in a traditional manner. More generally, the entire battery of results that Chambers and Quiggin (2001b) isolate for concave preferences, constant absolute risk aversion, constant relative risk aversion, the convolution of constant absolute risk aversion and quasi-concave preferences, in addition to the results reported above on constant risk aversion, complete risk-averse preferences, and risk-neutral preferences, can be used to construct approaches for empirically testing the validity of these various restrictions using observed inputs and supplies.

DATA ISSUES

We have refrained from writing down actual functional representations of these supply-response systems because ultimately those forms will depend on the nature and type of data available for the predetermined variables $\boldsymbol{w}$, $\boldsymbol{p}$, and $\boldsymbol{q}$. In considering estimation of state-contingent supply-response systems, several issues arise. The first is how variables such as prices and weather conditions should be represented in a state-contingent setting. It is useful to distinguish between variables that are naturally discrete, such as the occurrence or absence of disease outbreaks or hailstorms, and variables that are naturally continuous, such as prices and rainfall. The most common approach to the problem of a finite-dimensional representation of continuous random variables has been based on the use of central moments (mean, variance, kurtosis, etc.). However, it is also possible to approximate a continuous distribution over a finite state space by the use of step functions. The larger the number of states, the smoother the approximation becomes. Most econometric studies of production under uncertainty have used the former approach.

Chambers and Quiggin (2001d), in the context of cost functions, examine and compare approaches based on the use of a finite set of moments (of the state-contingent outputs) with approaches based on the use of a finite set of states. One important result is that in neither case is there a resolution of the logical and technical difficulties associated with the use of a stochastic production function to represent a general state-contingent technology. A second, more fundamental result is that the two approaches yield equivalent results if the number of moments is equal to the dimension of the state space. There, however, the issues are different than in the current case because the random variable (output) that is being summarily represented by moments is endogenous to the actions of the producer. Moreover, that random variable is typically taken as predetermined in the estimation of the cost structure. Hence, using a moment-based approach inherently entangles the state-contingent technology with the underlying distribution of the states of nature in a typically unidentifiable fashion.

For the most part, the issues faced with supply response systems of the type presented here are simpler. In the systems discussed above, estimation requires creating approximations for the distributions of random variables ($\boldsymbol{p}$ and $\boldsymbol{q}$) that are taken as predetermined in estimation.[5] Thus, the problem is how to best represent random variables in a fashion that is flexible enough to approximate their true distribution but parsimonious enough to permit efficient econometric estimation. Traditionally, empirical studies have confined

[5] Of course, $\boldsymbol{q}$ is ultimately endogenous and some of the same issues that arise in the moment-based approach for cost functions must be faced here as well. This also presumes that one is satisfied with estimating an expected supply equation directly and then inferring the state-contingent supplies. However, to our knowledge all empirical supply response systems under production uncertainty are based on expected supply or some obvious transformation of it.

attention to the first two or three moments, primarily because of data limitations, but partly because of the difficulty of attaching theoretical significance to higher-order moments. In representing a variable such as prices, using a finite-state specification, a comparably parsimonious representation might allow for three states of nature corresponding to the top quintile (perhaps using a point estimate of the tenth percentile), the bottom quintile (using a point estimate of ninetieth percentile), and the middle three quintiles (perhaps using a point estimate of the median or a trimmed mean, excluding the top and bottom quintiles).

The second problem is the need to obtain appropriate data on the relevant state-contingent distributions. Given well-developed futures and options markets, it is possible to derive a complete characterization of the market's collective subjective price distribution. In particular, given a futures price (which determines the mean) and a single option market, the Black-Scholes formula can be used to infer the variance.

This approach raises the problem that, since farmers generally do not trade in futures markets, they need not share market expectations. Hence, it may be preferable to use historical data to derive price expectations. At the cost of assuming a strong form of rational expectations, it would alternatively be possible to derive price distributions from an econometric model of the relevant market.

If data on farmers' portfolio positions are available, it is necessary to estimate a joint distribution of prices and returns to financial assets. Each of the approaches above could be used for this purpose. Using standard tools of finance theory, it is possible to estimate a beta for different sets of assets and derive a joint probability distribution from the relevant marginal distributions. Alternative approaches based on historical data and econometric modeling may also be applied.

Financial assets are not the only possibilities for $\boldsymbol{q}$. In particular, there is a wealth of data on farm households with off-farm labor income. There are also well-developed ways to estimate (for a sample including full-time farmers) the off-farm wage rate. This would be a complicated problem, since we need to allow for allocation of labor between on-farm and off-farm work, but the very complexity should give us more traction in estimation.

In closing, we end with a caveat that is a direct consequence of the results of our analysis. If data on farmers' portfolio positions are not available, they should be obtained. As we have demonstrated, a necessary consequence of the net-returns assumptions is that optimal state-contingent supplies and input demands, even in the case of constant risk aversion, are dependent upon the portfolio for risk-averse individuals. Hence, supply-response systems that do not contain such information, or that are based only on a non-stochastic component of the portfolio (wealth), suffer from specification bias in the form of omitted variables. Of course, there are ways to avoid this bias, including the imposition of arbitrary functional structure on the certainty equivalent, but in principle even these represent testable hypotheses that should be confronted with the data.

CONCLUSIONS

We have briefly discussed the estimation of supply-response systems under price and production uncertainty using properties of the indirect certainty equivalent. Using results of Chambers and Quiggin (2001b), we have considered quasi-concave preferences, constant risk aversion (and the special cases of risk neutrality and complete aversion to risk), and more general aversion to risk.

APPENDIX: PROOF OF THEOREM

Let r^0 be optimal for $\boldsymbol{q}^0$ and r' for $\boldsymbol{q}'$, and $z_s^0 = \dfrac{r^0 - q_s}{p_s}$ and $z_s' = \dfrac{r' - q_s}{p_s}$.

Then

$$\begin{aligned} V[\boldsymbol{w}, \lambda \boldsymbol{q}^0 + (1-\lambda)\boldsymbol{q}', \boldsymbol{p}] &\geq \lambda r^0 + (1-\lambda) r' - c[\boldsymbol{w}, \lambda \boldsymbol{z}^0 + (1-\lambda)\boldsymbol{z}'] \\ &\geq \lambda[r^0 - c(\boldsymbol{w}, \boldsymbol{z}^0)] + (1-\lambda)[r' - c(\boldsymbol{w}, \boldsymbol{z}')], \end{aligned}$$

where the second equality follows by C.5.

REFERENCES

Arrow, K. 1953. "Le Role des Valeurs Boursiers pour la Repartition la Meilleur des Risques." *Cahiers du Seminaire d'Economie.* Paris: CNRS.

Berge, C. 1997. *Topological Spaces.* Mineola, MN: Dover Publications.

Chambers, R.G., and J. Quiggin. 1992. "A State-Contingent Approach to Production Under Uncertainty." Working Paper No. 92-03, Department of Agricultural and Resource Economics, University of Maryland.

___. 1996. "Nonpoint-Source Pollution Control As a Multi-Task Principal-Agent Problem." *Journal of Public Economics* 59: 95-116.

___. 1997. "Separation and Hedging Results with State-Contingent Production." *Economica* 64: 187-209.

___. 1998. "Cost Functions and Duality for Stochastic Technologies." *American Journal of Agricultural Economics* 80: 288-295.

___. 2000. *Uncertainty, Production, Choice, and Agency: The State-Contingent Approach.* New York: Cambridge University Press.

___. 2001a. "Decomposing Input Adjustments Under Price and Production Uncertainty." *American Journal of Agricultural Economics* 83: 20-34.

___. 2001b. "A Note on Indirect Certainty Equivalents for the Firm Facing Price and Production Uncertainty." Working Paper No. 01-04, Department of Agricultural and Resource Economics, University of Maryland.

___. 2001c. "The State-Contingent Properties of Stochastic Production Functions." *American Journal of Agricultural Economics* (forthcoming).

___. 2001d. "Primal and Dual Approaches to the Analysis of Risk Aversion." Working Paper No. 01-08, Department of Agricultural and Resource Economics, University of Maryland.

Coyle, B. 1992. "Risk Aversion and Price Risk in Duality Models of Production: A Linear Mean Variance Approach." *American Journal of Agricultural Economics* 74: 849-859.

___. 1999. "Risk Aversion and Yield Uncertainty in Duality Models of Production." *American Journal of Agricultural Economics* 81: 553-567.

Debreu, G. 1959. *The Theory of Value*. New Haven, CT: Yale University Press.

Färe, R. 1988. *Fundamentals of Production Theory*. Berlin: Springer-Verlag.

Hirshleifer, J. 1965. "Investment Decision Under Uncertainty: Choice-Theoretic Approaches." *Quarterly Journal of Economics* 79: 509-536.

Lau, L.J. 1978. "Application of Profit Functions." In M. Fuss and D. McFadden, eds., *Production Economics: A Dual Approach to Theory and Applications*. Amsterdam: North-Holland Elsevier Publishing Co.

Malinvaud, E. 1972. *Lectures on Microeconomic Theory*. North Holland: Amsterdam.

Pope, R.D. 1980. "The Generalized Envelope Theorem and Price Uncertainty." *International Economic Review* 27: 75-86.

Pope, R.D., and R.E. Just. 1996. "Empirical Implementation of Ex Ante Cost Functions." *Journal of Econometrics* 72: 231-249.

Quiggin, J., and R.G. Chambers. 1998a. "Risk Premiums and Benefit Measures for Generalized Expected Utility Theories." *Journal of Risk and Uncertainty* 17: 121-138.

___. 1998b. "A State-Contingent Production Approach to Principal-Agent Problems With an Application to Point-Source Pollution Control." *Journal of Public Economics* 70: 441-472.

___. 2000. "Increasing and Decreasing Risk Aversion for Generalized Preferences." Working Paper, Department of Agricultural and Resource Economics, University of Maryland.

Safra, Z., and U. Segal. 1998. "Constant Risk Aversion." *Journal of Economic Theory* 83: 19-42.

Yaari, M. 1969. "Some Remarks on Measures of Risk Aversion and on Their Uses." *Journal of Economic Theory* 1: 315-329.

Chapter 7

CAN INDIRECT APPROACHES REPRESENT RISK BEHAVIOR ADEQUATELY?

Rulon D. Pope and Atanu Saha
Brigham Young University, Utah, and Analysis Group/Economics, New York

INTRODUCTION

In order to answer the question posed as the title to this chapter, one must have a clear picture of the relevant risk issues and the questions that need to be asked and hopefully answered. This set of economic issues involving risk in agriculture is quite broad, but we take as a typical and no doubt an easier one the supply and income response to policy changes such as the 1996 Farm Bill or policies that mitigate production uncertainty (Lin et al. 2000, Holt 1999, Goodwin and Vandeveer 2000, Ramaswami 1993). Inherent in any formal econometric indirect approach is considerable structure. Conventionally, agricultural economists have focused on static expected utility of wealth models of choice (Anderson, Dillon, and Hardaker 1977). This is not to say that behavioral anomalies (relative to expected utility maximization) are not considered important, but at this juncture, alternatives to expected utility seem to be viewed as having insufficient net benefit in most agricultural applications. Expected utility models are relatively easy to interpret and convenient to apply, yet still present many significant challenges, as will be apparent below. Further, intertemporal models of choice also are central to many if not most agricultural questions (e.g., Saha, Innes, and Pope 1993, Taylor 1984). Yet these considerations magnify the difficulties of coherently applying economic theory, and are not discussed here. In this chapter, we focus on the application of static indirect expected utility including mean-variance and conditional moment methods to agricultural risk problems. We use simple calculus tools and production uncertainty to illustrate the benefits and costs of applying risk responsive models to behavior.[1]

---

[1] See Chambers and Quiggin (2000) for a rigorous treatment of the theory of state-dependent choice under risk.

SOME PRELIMINARIES

To discuss the issues here, we assume the smooth traditional utility over wealth static model so commonly used in production economic problems. Let wealth be W. Wealth is a function of random variables, X, and decisions, z, and θ represents known or readily measured parameters of the problem, and background wealth is w_0. Thus, expected utility of wealth is

$$E\{U[w_0 + \pi(z, X, \theta)]\} = U\{w_0 + E[\pi(z, X, \theta)] - RP[w_0, z, dF(X, \theta)]\}, \quad (1)$$

where RP is the risk premium that depends on initial wealth, w_0, decisions, z, and the distribution of profit, written $dF(.)$. There are two distinct ways to model expected utility in a non-state preference approach. The first derives from the presumption that all that is known about X is contained in the empirical probabilities $dF_e(x)$. This approach is particularly consistent with a frequentist interpretation of probabilities. Thus, in the typical expected utility problem in agriculture, all one "observes" is, for example, weather and prices for discrete random draws that are characterized by probabilities $dF_e(X)$. Hence, expected utility is

$$E\{U[w_0 + \pi(z, X, \theta)]\} = \int_a^b U[w_0 + \pi(z, x, \theta)] dF_e(x), \quad (2)$$

on support $[a, b]$, and the integral is the Stieltjes integral handling both the discrete and continuous cases. It should be noted that the empirical distribution function is the unbiased minimum-variance nonparametric estimator of any distribution function $F(x)$. And any biased estimator of $F(x)$ will yield a biased estimate of EU for some U (Meyer and Pope 1980) and hence possibly biased choice functions.[2] Optimal choices, z^*, are characterized by

$$\int_a^b U_W[w_0 + \pi(z, x, \theta)] \pi_z(z, x, \theta) dF_e(x) = 0, \quad (3)$$

where $U_W = \partial U / \partial W$ is the marginal utility of wealth and $\pi_z = \partial \pi / \partial z$ is marginal effect of decisions on profit. Indirect utility, G, is characterized by

$$\begin{aligned} \max_z E[w_0 + \pi(z, X, \theta)] &= \int_a^b U\{w_0 + \pi[z^*(w_0, dF_e(x), x, \theta)]\} dF_e(x) \\ &= G[dF_e(s), w_0, s, \theta]. \end{aligned} \quad (4)$$

[2] We note that the minimum variance unbiased estimator of a normal is a Beta (see Meyer and Pope 1980 for a discussion).

This is consistent with the approach taken in Lambert and McCarl (1985).

To illustrate, suppose that wealth takes on T discrete states $x_1,...,x_T$, with corresponding probabilities $\tau_1,...,\tau_T$. Indirect expected utility is

$$G(w_0,x,\tau,\theta)=\sum_{t=1}^{T}\tau_t U\{w_0+\pi[z^*(w_0,x,\tau,\theta),x_t,\theta]\}\,. \tag{5}$$

Applying the envelope theorem using initial wealth yields

$$G_{w_0}=E\{U_w[W(z^*,.)]\}\,. \tag{6}$$

Adding more structure, let profit be $\pi=\varpi z$,where z represents netput decisions with associated known unit prices ϖ. Applying the envelope theorem yields

$$G_{\varpi}=z^*\sum_{t=1}^{y=T}\tau_t U_W[w_o+\pi(z^*,x_t,\theta)]\,. \tag{7}$$

Applying a primal approach to (3) is relatively straightforward econometrically, particularly in a Generalized Method of Moments (GMM) framework where a mean zero shock is appended to the first-order condition (Chavas and Holt 1996, Saha 1997, Saha, Shumway, and Talpaz 1994). However, econometrically attempting to apply equations (5)-(7) using the envelope theorem can be quite difficult due to the complexity of including the $2T$ components of τ and x in a coherent way directly and through z^*. Likely more burdensome than the direct approach, one must work out curvature and nullity type restrictions for each of these parameters.

A second approach imposes some form of smoothness on the cumulative distribution function. Often in the latter case, an additional assumption is invoked that expected utility can be written in terms of moments or parameters that define moments. In such case we write $dF(x,m_x)$ as the density function, where m_x represents moments of the distribution of X. In such case, indirect expected utility is written as

$$\begin{aligned}\max_z E\{U[w_0+\pi(z,X,\theta)]\} &= \int U\{w_0+\pi[z^*(w_0,m_x,\theta),m_x,\theta]\}dF(x,m_x)\\ &= G(w_0,m_x,\theta),\end{aligned} \tag{8}$$

where the support is often presumed to be the positive orthant. To illustrate (8), suppose that

$$\pi=R(z,X)-\omega z\,, \tag{9}$$

where R represents revenue, z represents input choices, and ω represents competitive input prices. If X is normally distributed, the distribution function is $F_x(x, \mu_x, \sigma_x)$ and indirect utility is of the form

$$G(w_0, \omega, \mu_x, \sigma_x) \tag{10}$$

and depends on initial wealth (assumed fixed), input prices, and the first two moments of $X : \mu_x$ and σ_x. This implies that the data space needed to implement indirect expected utility is reduced to a few sufficient statistics. In most situations, this provides substantial econometric simplicity if indirect utility methods are to be used.

An approach similar to (10) is to assume that direct expected utility can be written in terms of a finite number of moments of wealth or profit:

$$\max_z E[U(W)] = \max_x \widetilde{U}[w_0, m_\pi(z, \theta), \theta] = G\{w_0, m_\pi[z^*(w_0, \theta), \theta], \theta\}. \tag{11}$$

That is, direct expected utility can be written in terms of w_0 and moments of π such that G is obtained under optimization. Under constant absolute risk aversion (CARA), a moment-generating or characteristic function can be used to generate $\widetilde{U}$ (Collender and Zilberman 1985). In the case of normality of W, expected utility is $-\exp\{-\lambda[\mu_w - (\widetilde{\lambda}/2)\sigma_w^2]\}$, and the resulting maximand is

$$\mu_w - (\widetilde{\lambda}/2)\sigma_w^2, \tag{12}$$

where μ_w is the mean of wealth and σ_w^2 is the variance of wealth (Freund 1956). Expected utility is increasing and concave $\widetilde{\lambda} \geq 0$ (risk aversion).

Another way to rationalize using moments is to assume that utility functions have polynomial form. Taylor's series approximations of utility or marginal utility will also generate a form as in (10). For example, quadratic utility generates an expected utility function that is linear in mean, mean squared, and the variance of wealth:

$$U(w) = a + bW + cW^2; EU(W) = a + b\mu_w + c(\sigma_w^2 + \mu_w^2).$$

Because of the simplicity of working with moments, much of what follows reduces the problem to preference over moments. Undoubtedly some risk problems involve highly asymmetric distributions and thus at least third moments should be included. However, because our discussion is only illustrative, we generally simplify by considering only two moments, mean and standard-deviation (MS) or mean-variance (EV). As we have seen, the MS approach can be rationalized based upon restricting the distribution to a particular class (e.g., normality or lognormality) or restricting utility to a particular class (e.g., quadratic). These rationales may be conceptually or empirically suspect. For example, wealth may not be normal for all (any) decisions and quadratic utility exhibits increasing absolute risk aversion. It is important to place MS/EV analysis in context.

To our knowledge, the mean-standard deviation (MS) decision criterion was first proposed by Fisher, in 1906. However, this decision framework has gained prominence since the studies by Markowitz (1952) and Tobin (1958) and has been widely used in the analytical as well as the empirical risk literature. Rosenzweig and Binswanger (1993) provide a recent empirical application. Sinn (1983) and Meyer (1987) have shown that neither normally distributed random payoff nor quadratic utility for agent's preference is necessary for the preference ordering under the expected utility and MS approaches to be consistent. The consistency condition is met when the choice set is composed of random variables that belong to a "linear class" within which all distributions can be "transformed into one another merely by a shift and a proportional extension" (Sinn 1983, p. 56). Equivalently, the consistency condition is satisfied when the "choice set [is]... composed of random variables which differ from one another only by location and scale parameters" (Meyer 1987, p. 422).

MEAN-STANDARD DEVIATION (MS) DECISION MAKING

One of the most cogent of rationales for the MS criteria is based on Meyer's (1987) location/scale analysis. Therefore, we briefly review it. These results will be essential in order to interpret empirical results.

Consistency condition: Meyer's location and scale condition is sufficient for a ranking of a set of random variables by the expected utility (EU) and mean-standard deviation (MS) criteria to be consistent with one another.

Definition. *Two cumulative distribution functions* $G_1(\cdot)$ *and* $G_2(\cdot)$ *differ only by location and scale* (LS) *parameters* α *and* β *if* $G_1(x) = G_2(\alpha + \beta x)$, *with* $\beta > 0$.

The consistency condition requires that the choice set be composed of random variables that differ from one another only by LS parameters. More formally, assume a choice set in which all random variables Y_i differ from one another only by location and scale parameters. Let X be the random variable obtained from one of the Y_i using $X = (Y_i - \mu_i)/\sigma_i$, where μ_i and σ_i are the mean and the standard deviation of Y_i. Thus, *all* Y_i are equal in distribution to $\mu_i + \sigma_i X$. Hence the expected utility from Y_i for any agent with utility function $U()$ can be written as

$$Eu(Y_i) = \int_a^b U(\mu_i + \sigma_i x) dF(x) \equiv V(\sigma_i, \mu_i),$$

where the support of X is $[a,b]$.

Properties of the MS preference function $V(\sigma, \mu)$:

Property 1. $V_\mu(\sigma,\mu) \Leftrightarrow U' \geq 0$

Property 2. $V_\sigma(\sigma,\mu) \Leftrightarrow U'' \leq 0$.

Define $S(\sigma,\mu) = -V_\sigma(\cdot)/V_\mu(\cdot)$; then

Property 3. $U' \geq 0$ and $U'' \leq 0 \Rightarrow S(\sigma,\mu) \geq 0$

Property 4. $V(\sigma,\mu)$ is a concave function of μ and $\sigma \Leftrightarrow U'' \leq 0$.

Properties 1-4 are depicted in Figure 1. Property 1 implies that movements in the vertical direction in (σ, μ) space are movements to higher indifference curves. Property 2 implies that, under risk aversion, as one moves from left to right in the (σ, μ) space, lower indifference curves are encountered. Property 3 implies that, under risk aversion, the indifference curves are positively sloped in (σ, μ) space. Finally, Property 4 implies that, under risk aversion, the preference set is convex, that is, all points preferred or indifferent to a given point form a convex set.

Property 5. $\dfrac{\partial S(\sigma,\mu)}{\partial \mu} \leq (=,\geq) 0 \Leftrightarrow U()$ displays DARA (CARA, IARA).

This property is explained in Figure 2. As one moves in a vertical direction in (σ,μ) space, the slope of the indifference curves decreases (stays constant, increases) under DARA (CARA, IARA). For example, as shown in Figure 2, under CARA, the indifference curves are vertically parallel.

Property 6. $\dfrac{\partial S(t\sigma,t\mu)}{\partial t} \geq (=,\leq) 0 \Leftrightarrow U()$ displays IRRA (CARA, DRRA).

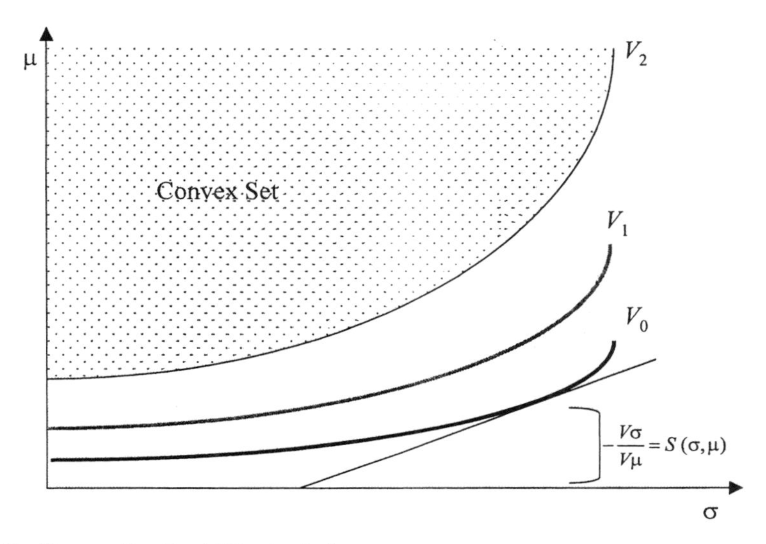

Figure 1. Properties 1–4 Illustrated

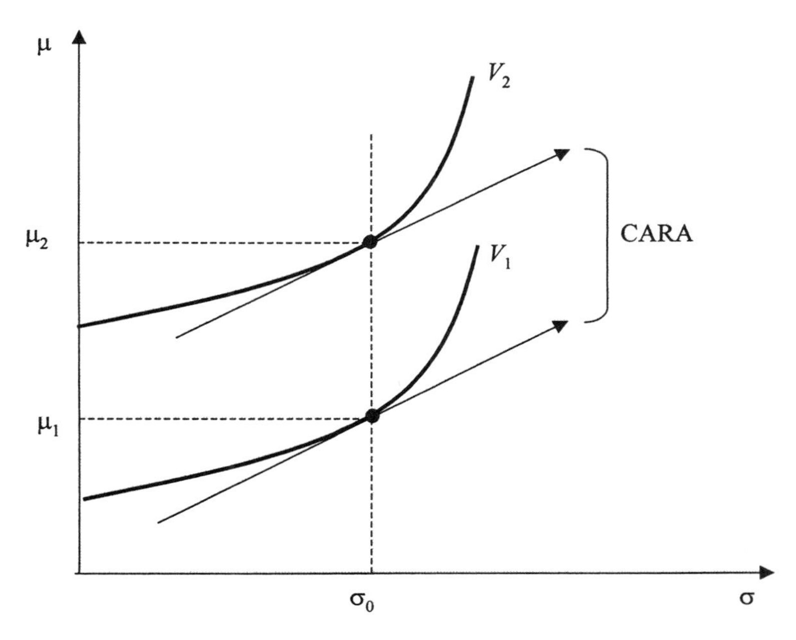

Figure 2. Property 5 Illustrated

This property implies that, as one moves out along a ray, the slope of the indifference curves decreases (stays constant, increases) under IRRA (CARA, DRRA). This property is illustrated in Figure 3.

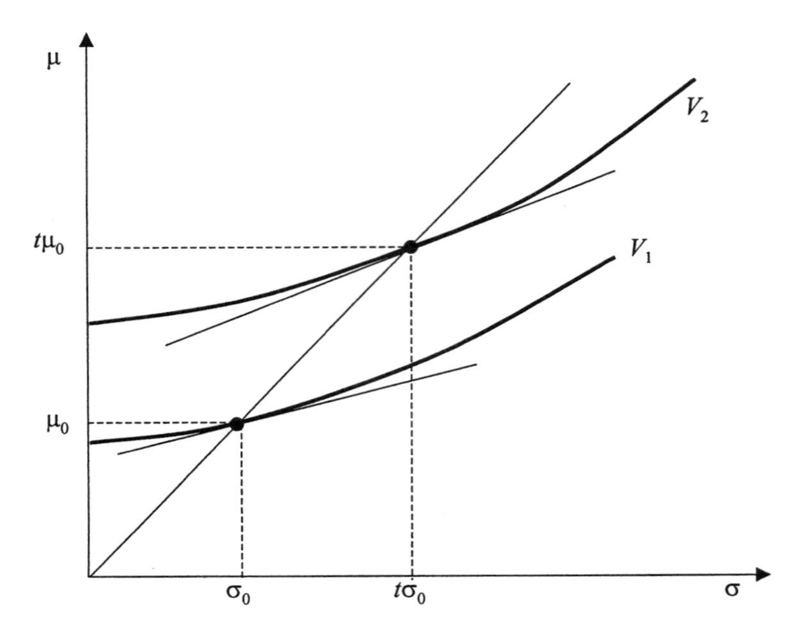

Figure 3. Property 6 Illustrated

Therefore, when random variables are in the location-scale family, it is possible to reduce expected utility to preference over two moments. This will greatly simplify attempts to apply indirect expected utility and facilitate interpretation.

Like stochastic dominance, there are production approaches that do not explicitly use or even try to measure risk preferences explicitly but attempt to place a partial ordering on choices. Suppose that $\widetilde{U}$ in (11) depends on only two moments: the mean and the standard deviation of wealth. In such case, $\widetilde{U} = \widetilde{U}(\mu, \sigma)$, where μ and σ are the mean and standard deviation of wealth.

CONDITIONAL EXPECTED PROFIT

From Properties 1 and 2, a firm that is strictly risk averse can solve the following sequential pair of problems:[3]

Stage 1: $$\max_{z} \{\mu(z,\theta) : \sigma = h(z,\theta)\} = \mu[z^{*}(\sigma,\theta),\theta] = \widetilde{G}(\sigma,\theta) \tag{13}$$

[3] We assume that all constraints are binding so that classical optimization problems obtain. If one wanted to have a consistent procedure that enveloped both risk-averse and risk-neutral cases, then the variance or standard deviation constraint would need to be an inequality and Kuhn-Tucker theory would apply.

and

$$\text{Stage 2:} \qquad \max_{\sigma}\{\widetilde{U}[w_0 + \widetilde{G}(\sigma,\theta),\sigma]\} = G(w_0,\theta), \tag{14}$$

where $\widetilde{G}$ is the maximal mean for any arbitrary value of the standard deviation or variance of wealth, and G is the indirect expected utility function.[4] In this section, we focus on the former on a standard deviation constraint. We shall assume that technology is binding in the sense that firms are on the mean and variance (or MS) frontier of production itself.

The function $\widetilde{G}$ in (13) is not a conventional profit or expected profit function. It is completely analogous to the mean-variance or mean-standard deviation portfolio frontier where the efficient mean-standard deviation or variance is given. In many respects, it is analogous to a conventional cost function, which is a conditional profit function. Second, initial wealth only enters the maximization in (13) trivially if initial wealth is certain. In this case, w_0 in (13) can be eliminated. Because initial wealth is seldom observed or accurately reported, this can be a large convenience for some problems (see Meyer, this volume). Aside from the usual portfolio or investment problems, the approach appears to have been applied only a single time assuming production uncertainty (Pope and Just 1997). Yet it seems to have merit and so it is briefly discussed here. Because our purpose here is largely exploratory and pedagogical, we focus on the simplest case: output or price uncertainty for a single product competitive firm with policy or marketing issues undeveloped.

Let us assume that w_0 is certain and omit it from μ. Assuming an interior solution, the efficient conditional choices are characterized by

$$\begin{aligned} L_z &= \mu_z(z,\theta) - \lambda h_z(z,\theta) = 0 \\ L_\lambda &= \sigma - h(z,\theta) = 0, \end{aligned} \tag{15}$$

where λ is the Lagrange multiplier and where it is assumed that the Lagrangean, L, has a solution and subscripts denote differentiation. Solving $z^*(\sigma,\theta)$ and $\lambda^*(\sigma,\theta)$ and inserting into μ yields the indirect function, $\widetilde{G}(\sigma,\theta)$. In order to make further headway, it is necessary to put more structure on $\mu(z,\theta)$ and $h(z,\theta)$. Suppose that revenue in (9) is $Pf(z)$, where P is unknown random price (incomplete risk markets), z are inputs, ω are known input prices, and $f(z)$ denotes known technically efficient production. In this case, expected revenue $E[R(z,\theta)] = \mu_p f(z)$ and $h(z,\theta) = \sigma_p f(z)$. Hence, maximizing expected

---

[4] This presumes that the problem has a solution. Customary second-order conditions and constraint qualifications apply: letting L_{BB} denote the Hessian in parameters B of the Lagrangean, $zL_{BB}z < 0$ for all $h_B z = 0, z \neq 0$, given a nontrivial constraint.

profit subject to the standard deviation constraint is just equivalent to minimizing the cost of producing $y = f(z)$. Note that the problem can be reformulated as

$$\max_z \{\mu_p \sigma / \sigma_p - \omega z : \sigma / \sigma_p = f(z)\} \rightarrow c(\sigma / \sigma_p, \varpi). \tag{16}$$

Therefore, when there is only price uncertainty, conditional expected profit maximization trivially reduces to cost minimization.[5] In what follows, production uncertainty is assumed. To keep notation parsimonious, we often eliminate from consideration price risk. Assume that production is uncertain of the general form

$$y = f(z) + h(z)\varepsilon, E(\varepsilon) = 0, E(\varepsilon^2) = \sigma_\varepsilon^2 \tag{17}$$

and maximizing expected profit subject to the σ constraint yields

$$\max_z \{Pf(z) - \omega z : \sigma = Ph(z)\sigma_\varepsilon\} = \tilde{G}(\omega, \sigma, \sigma_\varepsilon). \tag{18}$$

Applying Silberberg's (1974) dual results assuming a binding constraint on the interior,

$$\Im = Pf(z) - \omega z + \lambda[\sigma - P\sigma_\varepsilon h(z)] - \tilde{G}(\omega, \sigma_\varepsilon, \sigma), \tag{19}$$

it follows from the envelope theorem:

$$\begin{aligned}
&\Im_P = 0 \Rightarrow y^* - \lambda^* h^* \sigma_\varepsilon = \partial\tilde{G} / \partial P, \; or, \\
&y^* = \partial\tilde{G} / \partial P - (\partial\tilde{G} / \partial\sigma_\varepsilon)(\sigma_\varepsilon / P) \; presumably \; positive \\
&\Im_\omega = 0 \Rightarrow -z^* = \partial\tilde{G} / \partial\omega \le 0 \\
&\Im_{\sigma_\varepsilon} = 0 \Rightarrow -\lambda^* Ph^* = \partial\tilde{G} / \partial\sigma_\varepsilon \le 0 \\
&\Im_\sigma = \lambda^* = \partial\tilde{G} / \partial\sigma \ge 0.
\end{aligned} \tag{20}$$

Because ω does not appear in the constraint it follows from the negative definiteness of $\Im$ subject to constraint that $\tilde{G}$ is convex in ω. More generally, letting $\alpha = \{P, \omega, \sigma_\varepsilon\}$, the Hessian matrix of $\Im$ is negative-semi-definite in α subject to constraint:

[5] When production is also uncertain and price and production are dependent, minimizing cost subject to expected revenue is the appropriate cost function. As will be seen below, it is appropriate in this case to maximize expected profit subject to a variance of profit constraint.

$$z\Im_{\alpha\alpha}z \le 0 \text{ subject to } zh_{\alpha} = 0, z \ne 0. \tag{21}$$

Because expected profit is linear in parameters, this reduces to the quasi-convexity of $\widetilde{G}$ in α. That is, $z'\widetilde{G}_{\alpha\alpha}z \ge 0$ given $z\partial(Ph\sigma_{\varepsilon})/\partial\alpha = 0, z \ne 0$. A sufficient condition that might be useful in empirical checks for quasi-convexity is merely convexity of $\widetilde{G}$ in α, implying that

$$\begin{aligned} &\partial^2\widetilde{G}/\partial P^2 = \partial(y^* - \widetilde{G}_{\sigma}Ph^*\sigma_{\varepsilon})/\partial P \ge 0 \\ &-\partial^2\widetilde{G}/\partial\omega\partial\omega = \partial z^*/\partial\omega \le 0 \\ &\partial^2\widetilde{G}/\partial(\sigma_{\varepsilon})^2 = \partial(PG_{\sigma}h^*)/\partial\sigma_{\varepsilon} \le 0, \end{aligned} \tag{22}$$

where $y^* = f(z^*)$ is optimal expected output. It is always apparent that $\widetilde{G}$ cannot be strictly convex in all dimensions because it is homogeneous of degree one in (P,ω), i.e., $\widetilde{G}(tP,t\omega,.) = t\widetilde{G}(P,\omega,.)$, $t > 0$. Equations (20)-(22) give all of the results necessary to apply duality to real data.

Although the approach above is very doable and can be easily modified to handle both production and price uncertainty, the envelope results in (20) are not as straightforward as one might like. For example, one cannot simply recover expected output by using variations in output price. A simple approach under production uncertainty is to merely hold the standard deviation of output fixed by constraint in the first stage. Thus, consider the two-stage problem:

Stage 1: $$\max_z \{Pf(z) - \omega z : \sigma/\sigma_{\varepsilon} = h(z)\} = \mu(P,\omega,\sigma/\sigma_{\varepsilon}), \tag{23}$$

Stage 2: $$\max_{\sigma/\sigma_{\varepsilon}} E\{U[W_0 + \mu(P,\omega,\sigma/\sigma_{\varepsilon}) + P(\sigma/\sigma_{\varepsilon})\varepsilon]\} = G(W_0,P,\omega,\sigma_{\varepsilon}). \tag{24}$$

Under the usual regularity conditions, these are consistent with direct expected utility maximization. Stage 1 is the most interesting here. Unlike (20), note that the more conventional envelope properties apply in the first stage:

$$\begin{aligned} &\Im = Pf(z) - \omega z + \lambda[\sigma/\sigma_{\varepsilon} - h(z)] - \mu(P,\omega,\sigma/\sigma_{\varepsilon}) \\ &\partial\Im/\partial P = 0 \to \partial\mu/\partial P = E(y^*) \ge 0 \\ &\partial\Im/\partial\omega = 0 \to -\partial\mu/\partial\omega = z^* \ge 0 \\ &\partial\Im/\partial(\sigma/\sigma_{\varepsilon}) = 0 \to \partial\mu/\partial(\sigma/\sigma_{\varepsilon}) = \lambda^*. \end{aligned} \tag{25}$$

This yields a conditional expected profit function, and the negative definiteness (or semi-definiteness) of the Hessian of $\Im$ subject to constraint implies that $\mu(.)$ is simply convex in P and ω, as under certainty.

There are many production economic issues that are altered by the presence of risk aversion and the σ constraint. Among those that might be considered are the following: homotheticity and returns to scale, returns to risk, risk constant substitution, efficiency, welfare, and economies of scope (see Pope and Just 1997). To illustrate just a few of these issues for the production uncertainty example, consider efficient economic choice. One cannot reasonably conclude that those with larger mean or average profits are more economically efficient when risk-taking motives or the risk environments are different any more than one could say that a person who has a higher return on his or her investment portfolio is more efficient. A reasonable index of economic efficiency for constant P, ω, and σ_ε is

$$I = (P\bar{y} - \omega z) / \widetilde{G}(P, \omega, \sigma / \sigma_\varepsilon) \ , \tag{26}$$

where $\bar{y}$ is actual average output and firm 1 is more efficient than firm 2 if $\mu_1 > \mu_2$ for a given σ. For discrete data, one may think of $\widetilde{G}$ as being the convex hull of MS data. When subjective means and variances of prices differ across firms, one could decompose total inefficiency into its constituent parts.

In conjunction with (19) and (20), there are a host of elasticities that could be examined: expected profit with respect to variance, supply and input demand own-price, cross-price, and elasticities with respect to variance or standard deviation. Consider the elasticity of mean profit with respect to the standard deviation of profit, $\widetilde{G}_\sigma(\sigma / \widetilde{G})$, which measures the percentage change in expected profit with respect to a one percent change in the standard deviation of profit. If the elasticity is small, then there is a very inelastic "return-risk" trade-off, and marginal differences in risk preferences will lead to small differences in average profits but larger differences in risk taking.

To apply the approach in equations (19)-(20), within a mathematical programming framework, σ or σ^2 could be parametrically varied, given knowledge of θ, and thus an efficient set of conditional choices could be developed. However, in econometric applications one would presumably need to estimate σ. This type of problem is quite common in all risk models and it is what presents the largest challenge. Ignoring any potential problem from generated regressors (Pagan and Ullah 1988), one can estimate σ_ε and y^* as in Just and Pope (1978), Antle (1983), or time series improvements such as GARCH (or its variants) when necessary. These will estimate σ. Then, a quasi-convex $\widetilde{G}$ can be specified in ω, P, and σ_ε. This can be inserted into $\widetilde{G}$, and applying equation (14) yields the estimating equations.

Note that when $h(z) = f(z)$, maximizing expected profit while constraining $f(z) = \sigma / \sigma_\varepsilon$ implies only cost minimization subject to expected output, just as under the risk-neutral case. A special form of a conditional expected profit function is the cost function. For the most part, these concepts and applications are developed for a revenue uncertain firm where input prices are known. The focus is generally on factor substitution and economies of scale and scope.

EX ANTE COST FUNCTIONS

The advantages and use of a cost function are well known. One does not need to specify risk preferences to estimate and use it, just as in the previous section. Even when response to risk is an essential part of the question at hand, there is some value to decomposing a problem into one that is relatively straightforward (the cost function) and one that might require extensive examination of the form and structure of the response. For the most part, these concepts and applications are developed for a revenue-uncertain firm where input prices are known. The focus is generally on factor substitution and economies of scale and scope. Clearly, because expected utility is monotonic in wealth, cost minimization is a consistent first-stage objective. When output and input prices are certain, there is really no significant conceptual or empirical issue (Batra and Ullah 1974). Firms solve

Stage 1: $$\max_{z}\{-\omega z : y = f(z)\} = c(y,\omega);$$

Stage 2: $$\max_{y}\{\mu_p y - c(y,\omega)\} = \pi(\mu_p,\omega).$$

When production is uncertain, difficulties arise as to substance and how to operationalize costs. Early economic effort focused on production variability and plant design (reference and findings). Subsequent efforts focused on cost functions appropriate under risk neutrality, implying that the two-stage problem becomes

Stage 1: $$\max_{z}\{-\omega z : E(y) = f(z)\} = c[E(y),\omega)];$$

Stage 2: $$\max_{y}\{PE(y) - c(y,\omega)\} = \pi(p,\omega).$$

Conceptually, these two problems are identical with the exception of the interpretation of the first argument of cost. That is, in applications, expected output conditions costs and factor demand curves. It is easy to show that if one wishes to analyze $c[E(y),\omega]$ but instead the researcher estimates $c(y,\omega)$, then a typical error in variable problem arises, leading to inconsistent estimates of parameters (Pope 1979, Pope and Chavas 1994, Pope and Just 1998a), implying that one must use appropriate instruments or indicator functions unless additional assumptions are made. That is, if the conditional mean of $c[E(y),\omega]$ is desired, then one cannot estimate it by the usual approaches (e.g., least squares) by estimating $c = c(y,\omega)$, even under constant returns to scale. If one knew for sure that the firm was risk neutral and that a profit-maximizing $E(y)$ exists and is known, it would be the appropriate instrument. However, this seems to imply that one might as well estimate a profit function to begin with.

When risk aversion is present, a simple way to state the cost function problem is

$$\text{Stage 1:} \quad \min_{z}\{wz : \overline{R} = E[R(z,P,\theta)] - RP(w_0,w,\theta)\} = c(\omega,w_0,\theta,\overline{R}), \tag{27}$$

where RP is the risk premium and R is revenue. Stage 2 solves

$$\text{Stage 2:} \quad \max_{\overline{R}}\{\overline{R} - c(w_0,\omega,\theta,\overline{R})\} = G(w_0,\omega,\theta), \tag{28}$$

where G is the indirect expected utility (certainty equivalent) and $\overline{R}$ is expected revenue. In some cases, constraining $\overline{R}$ in the first stage reduces to simple constrained cost. For example, if production errors are multiplicative (e.g., $y = f(z)X,\ E(X) = 1$), then $\overline{R}$ is constrained by constraining $E(y) = f(z)$, as in the risk-neutral case. Hence, a cost function useful for the case of risk aversion but devoid of explicit risk preference parameters in the multiplicative error case is

$$\min_{z}\{\omega z : E(y) = f(z)\} = c[\omega, E(y)]. \tag{29}$$

Thus, in this case, one needs to consistently estimate the parameters of c. Pope and Just (1998a), assuming that econometric errors are optimization errors along an expected output isoquant, propose a consistent approach to the estimation of parameters of c. When applied to U.S. agriculture, they found that an estimated ex ante cost function gave more reasonable estimates of parameters. In general, they found that elasticities were smaller with the ex ante than the ex post estimates.[6]

When production does not have the multiplicative error, matters become more complicated. For example, when $y = f(z) + h(z)\varepsilon$, $E(\varepsilon) = 0$, $E(\varepsilon^2) = \sigma_\varepsilon^2$. Properties of cost functions under risk aversion in this case can easily be established similar to the profit function case in Pope and Just (1997).

---

[6] This approach was extended in Pope and Just (1998b) to include average technical errors.

REVENUE FUNCTIONS

Though revenue functions have received less attention, it makes sense to maximize expected revenue subject to a variance constraint and a cost constraint. The latter is constrained by holding fixed in the aggregate each level of input. Hence, for example, the firm can solve under input non-jointness, independence, and price certainty:[7]

$$\max_{x_{ij}}\left\{\sum_{i=1}^{N} P_i E f_i(x_i) : \bar{x}_j = \sum_{i=1}^{N} x_{ij}, j = 1,\ldots,M; \sigma^2 = \sum_{i=1}^{N} P_i^2 h_i^2(x_i)\sigma_{\varepsilon i}\right\} \tag{30}$$
$$= R(P, \bar{x}, \sigma^2),$$

where $x_i = (x_{i1}, x_{i2}, \ldots, x_{iM})$ and $\bar{x}_j$ is the aggregate input use of input j, and associated vector $\bar{x}$.

MIXED SYSTEMS

If one of the conditional methods discussed above is relevant, then one can create a mixed system of primal and dual methods. For example, if one has a conditional expected profit function, $\mu(P, \omega, \sigma, \sigma_\varepsilon)$, then application to a direct utility function $U = \widetilde{U}(W_0 + \mu, \sigma)$ yields first-order conditions

$$\frac{\partial \widetilde{U}}{\partial \mu}\frac{\partial \mu}{\partial \sigma} + \frac{\partial \widetilde{U}}{\partial \sigma} = 0\,. \tag{31}$$

Placing the second term to the right of the equals sign yields the familiar marginal benefit = marginal cost calculations appropriate to risk aversion (see Property 2).

When moments are not appropriate, then stage 2 in (24) can be solved yielding again the interpretation that the marginal benefit of risk equals the marginal cost, which is minus the marginal risk premium:

$$\frac{\partial \mu}{\partial(\sigma/\sigma_\varepsilon)} = \frac{-PE[U_W(W)\varepsilon]}{E[U_W(W)]}\,. \tag{32}$$

---

[7] Adding price uncertainty will imply that the firm will maximize expected revenue subject to the constraint.

Similar mixed systems could use revenue or cost functions and involve maximization of direct utility in the second stage using the indirect functions derived in the first stage.

INDIRECT EXPECTED UTILITY

Letting α be exogenous or passive variables, indirect expected utility is defined simply by

$$\max_{z} E[U(W_0, z, \alpha, \theta)] = E[U(W_0, z^*, \alpha, \theta)] = G(W_0, \alpha, \theta). \tag{33}$$

A small change in α implies that

$$G_\alpha = E[U_W(W)(W_\alpha)]. \tag{34}$$

Thus, it is apparent that the wealth effect through the marginal utility of wealth is important. It was recognized at least as early as Sandmo (1971) that the wealth effect played a role in expected utility of wealth similar to the one of income in the theory of consumer under certainty. All of the properties of behavior implied by expected utility under output price uncertainty were obtained in Chavas and Pope (1985) and econometrically applied in Appelbaum and Ullah (1997). Compensated demand curves (keeping expected utility constant) are downward-sloping and compensated supply curves are upward-sloping regardless of risk preferences. Treating y as output and z as inputs, the matrix of compensated slopes,

$$\begin{bmatrix} \partial y^{*c} / \partial \overline{P} & \partial y^{*c} / \partial \omega \\ -\partial z^{*c} / \partial \overline{P} & -\partial z^{*c} / \partial \omega \end{bmatrix}, \tag{35}$$

is symmetric and positive semi-definite, where the c superscript stands for compensated responses,[8] e.g., $\partial z^{*c} / \partial \omega = \partial z^* / \partial \omega - z^{*\prime}(\partial z^* / W_0)$. Thus, indirect expected utility is quasi-convex in ω and P similar to consumer theory. Considering production uncertainty to illustrate a few possible results, let W be of the

[8] Write the compensated response in terms of an endogenous value of initial wealth W_0. Then perturb a parameter, α. It then follows that the amount of wealth change required to keep expected utility constant is given by $\partial EU(W)/\partial\alpha = 0 \rightarrow \partial W_0 / \partial \alpha = -E[U_W(W)W_\alpha]/E[U_W(W)]$. When only output price is uncertain, this reduces to $\partial W_0 / \partial \overline{P} = y^*$, and $\partial W_0 / \partial \omega = -z^*$. Then, for example, using $z^{*c}(\omega, \overline{P}) = z^*[W_0(\omega, \overline{P}), \omega, \overline{P}]$ and differentiating one obtains $\partial z^{*c} / \partial \omega = \partial z^* / \partial \omega + (\partial z^* / \partial W_0)(\partial W_0 / \partial \omega)$ and compensated slopes in terms of observed behavior. Similar calculations apply for the compensated supply results.

form $W = W_0 + Pf(z) + Ph(z)X - \omega z$, where X is random and conventionally $E(X) = 0$.

From the maximization hypothesis, it follows that

$$\Im_{\alpha\alpha} = \partial^2 E[U(W)] / \partial\alpha^2 - \partial^2 G / \partial\alpha^2 \tag{36}$$

is negative semi-definite. That is,

$$\tilde{z}'[EU_W(W)W_{\alpha\alpha} + EU_{WW}(W)W_\alpha W'_\alpha]\tilde{z} - \tilde{z}' G_{\alpha\alpha}\tilde{z} \le 0 . \tag{37}$$

Hence, if W is convex in α, then the first term in (39) is nonnegative and it must follow that

$$\tilde{z}'[EU_{WW}(W)W_\alpha W'_\alpha - G_{\alpha\alpha}]\tilde{z} \le 0 . \tag{38}$$

If W_α is certain under any structure of risk preferences, the first term in (39) will be negative or positive as the individual is risk averse or risk loving. However, (38) does imply that $\tilde{z}' G_{\alpha\alpha} \tilde{z} \ge 0$, for all $\tilde{z} G_\alpha = 0$, $\tilde{z} \ne 0$. That is, given convexity of W in α with W_α certain, then G is quasi-convex in α.[9] Prominent price-taking cases where W is convex in α with a certain gradient include initial wealth, mean prices under multiple output certain production, and input prices.

This type of reasoning can also be applied to cases that include production uncertainty. It is particularly convenient to work with mean-variance utility functions to illustrate maximizing behavior under production uncertainty. Consider the linear mean-variance case in equation (12). Letting $z^*(P, \omega, \sigma^2)$ be argmax of equation (12), then the indirect certainty equivalent (a monotonic transformation of $E[U(W)]$) is

$$G(P, \omega, \sigma_\varepsilon^2, \tilde{\lambda}) = Pf(z^*) - \omega z^* - (\tilde{\lambda}/2)P^2 h^2(z^*)\sigma_\varepsilon^2 , \tag{39}$$

where $\tilde{\lambda}$ is the Arrow-Pratt measure of absolute risk aversion. The primal-dual is

$$\Im = Pf(z) - \omega z - (\tilde{\lambda}/2)P^2 h^2(z)\sigma_\varepsilon^2 - G(P, \omega, \sigma_\varepsilon^2, \tilde{\lambda}) \tag{40}$$

with envelope results

---

[9] This basic set of results is obtained differently but found in Chavas and Pope (1985) and Saha and Just (1996).

$$\Im_P = 0 \rightarrow \partial G / \partial P = E(y^*) - 2(\partial G / \partial \sigma_\varepsilon^2)(\sigma_\varepsilon^2 / P)$$
$$\Im_\omega = 0 \rightarrow \partial G / \partial \omega = -z^*$$
$$\Im_{\sigma_\varepsilon^2} = 0 \rightarrow \partial G / \partial \sigma_\varepsilon^2 = -(\tilde{\lambda} / 2) P^2 h^2 (z^*).$$

Thus, indirect utility is decreasing in ω, and decreasing in σ_ε^2 under risk aversion, and increasing in P under risk aversion or risk neutrality.

Further, because $\Im_{\alpha\alpha}$ is negative semi-definite in parameters, α, it follows that G is convex in $\alpha = (\omega, \sigma_\varepsilon^2)$. The only nonlinear parameter in the direct objective function is P. The maximization hypothesis yields the econometric restriction that

$$\begin{bmatrix} G_{PP} - 2G_{\sigma_\varepsilon^2}(\sigma_\varepsilon^2 / P^2) & G_{P\omega} & G_{P\sigma_\varepsilon^2} \\ G_{\omega P} & G_{\omega\omega} & G_{\omega\sigma_\varepsilon^2} \\ G_{\sigma_\varepsilon^2 P} & G_{P\omega} & G_{\omega_\omega} \end{bmatrix} \tag{42}$$

is positive semi-definite. This together with the envelope conditions in (41) give a workable system.

This approach can be extended to the addition of both production and price uncertainty and to more general mean-variance settings using the certainty equivalent. Envelope and curvature conditions are found in Coyle (1999, Proposition 2). The clear promise of working with the indirect utility function is that often convenient forms can be used to generate the items of interest: presumably the mean output, the variance of output, and input demands. This is apparent from equations (41)-(42). The disadvantage is that if one is interested in testing or imposing curvature conditions, it is more difficult than in the standard certainty theory of profit maximization. In addition, there remain econometric issues that often surface when the variance is estimated and inserted into an econometric system (Pagan and Ullah 1988). However, the relevant comparison is not with certainty but with the direct expected utility maximization and primal first-order conditions. In this comparison, for many purposes, using indirect functions is very useful for welfare analysis and to formulate a coherent demand-supply system incorporating risk.

VERY INDIRECT MEANS

There is a body of literature that tries to infer the value of risk-reduction through indirect means. One normative starting place is to find the costs of a risk-reduction policy or choice and infer that this places a lower bound on the value than a risk averter would place on it. For example, one can attempt to infer the value of life from the cost of a smoke detector (e.g., Dardis 1980).

In a more general approach, one has initial distribution $F^1(\pi)$ and subsequent (post-policy) distribution of $F^2(\pi)$. A natural measure of the value of the policy is compensating variation, CV, defined as

$$G[w_0 - CV, F^2(\pi)], \theta = G[w_0, F^1(\pi)], \theta . \tag{43}$$

That is, the demand price for the policy change yielding F^2 is CV. When use of moments is appropriate, CV can be defined as

$$V(w_0 + m_{1\pi}^2 - CV, m_{2\pi}^2) = V(w_0 = m_{1\pi}^1, m_{2\pi}^1) . \tag{44}$$

This is the value that one would try to elicit for example in questionnaires attempting to get at a willingness to pay for the policy (such as a crop insurance premium or reduction in food risk). Note that when decisions other than the policy affect profits, then the moments on the left side of (44) incorporate post-policy behavior (e.g., perhaps involving moral hazard), including the wealth effect induced by CV. Presumably if the economic agent expended K to get the second distribution, then we can imply that $CV \geq K$. Using K to bound CV may be crude, but may be of some value. For example, the value to farmers of the crop insurance program $(\Sigma_i CV_i)$ is at least as large as the unsubsidized premiums farmers pay $(\Sigma_i K_i)$.

Even at this level of abstraction, there are things that can be said qualitatively about the magnitude of CV for an individual's risk preferences. If F^2 is derived from F^1 as merely a shift in the mean, then CV is positive for all increasing utility functions (see Property 1). If F^2 is derived from F^1 as an increase in the spread while preserving the mean, then CV is positive for all risk averters (see Property 2). If both the mean and the standard deviation are scaled in a location/scale family, the willingness to pay will be independent of the initial risk-return endowment under CARA (see Property 6).

CONCLUSIONS

Applications of indirect (including dual) approaches to modeling risk response are few in number but they do illustrate the viability of the approach. The work of Coyle (1999) exemplifies the challenges and strengths of such methods. Envelope and curvature conditions are obtainable just as they are under certainty. Although we have not pursued a formal theory of duality here, the essence of duality is that appropriately formulated, indirect or dual problems contain the same information as a primal problem. Thus, duality is not at issue conceptually so much as the essence of the tradeoffs between method used, light shed on the problem, and the cost of using each approach. One should not, however, maintain that the problems are identical in econometric practice. Different econometric error structures are assumed, with different issues of exogeneity, to name just two substantive differences that are likely to occur between primal and dual applications.

There remain many challenges in applying either of these methods to general production systems. One has to decide functional forms, find ways to measure perceptions, and model whether, and how, perceptions change. For example, would it be reasonable to expect that the variance of output (conditional on inputs) changes within a farm as well as across farms? How are these variance perceptions linked? Similar questions hold for price and other moments when they are uncertain. If one is studying individual- or farm-level data, does the researcher get the rational expectations from the market or does one use a naïve or extrapolative technique for modeling farm-level expectations? In the former case, the market itself must be modeled even if the study is at the farm level. At this time, measuring unobservables (see Holt and Chavas, this volume) seemingly dwarfs the decision as to whether one uses an indirect or direct objective function. This leads us to suggest that a parsimonious method of specification that pays careful attention to the decision problem and measurement of beliefs and preferences may have the greatest empirical return given the current state of applications of production economics to agriculture. Either approach might accomplish these results. However, perhaps at this stage in the development of risk modeling, more effort should be devoted to specifying the economic problem complete with expectations formulations and adjustment. When this has a sound footing, then the effort to decide whether direct or indirect, primal or dual methods are the appropriate technique for agricultural economists to use to study production behavior, is appropriately spent.

REFERENCES

Anderson, J.R., J.L. Dillon, and J.B. Hardaker. 1977. *Agricultural Decision Analysis*. Ames: Iowa State University Press.

Antle, J.M. 1983. "Testing the Stochastic Structure of Production: A Flexible Moment-Based Approach." *Journal of Business and Economic Statistics* 1: 192-201.

Appelbaum, E., and A. Ullah. 1997. "Estimation of Moments and Production Decisions Under Uncertainty." *Review of Economics and Statistics* 79: 631-637.

Batra, R., and A. Ullah. 1974. "Competitive Firm and the Theory of Input Demand Under Price Uncertainty." *Journal of Political Economy* 82: 537-548.

Chambers, R.G., and J. Quiggin. 2000. *Uncertainty, Production, Choice, and Agency: The State-Contingent Approach*. New York: Cambridge University Press.

Chavas, J.-P., and M.T. Holt. 1996. "Economic Behavior Under Uncertainty: A Joint Analysis of Risk Preferences and Technology." *Review of Economics and Statistics* 78: 329-335.

Chavas, J.-P., and R. Pope. 1985. "Price Uncertainty and Competitive Firm Behavior: Testable Hypotheses from Expected Utility Maximization." *Journal of Economics and Business* 37: 1-13.

Collender, R., and D. Zilberman. 1985. "Land Allocation Under Uncertainty for Alternative Specifications of Return Distributions." *American Journal of Agricultural Economics* 67: 779-786.

Coyle, B. 1992. "Risk Aversion and Price Risk in Duality Models of Production: A Linear Mean-Variance Approach." *American Journal of Agricultural Economics* 74: 849-859.

___. 1999. "Risk Aversion and Yield Uncertainty in Duality Models of Production: A Mean-Variance Approach." *American Journal of Agricultural Economics* 81: 553-567.

Dardis, R. 1980. "The Value of a Life: New Evidence from the Marketplace." *American Economic Review* 70: 1077-1082.

Fisher, I. 1906. *The Nature of Capital and Income.* London: Macmillan.

Freund, R. 1956. "Introduction of Risk into a Programming Model." *Econometrica* 24: 253-263.

Goodwin, B., and M. Vandeveer. 2000. "An Empirical Analysis of Acreage Distortions and Participation in the Federal Crop Insurance Program." Paper presented at the USDA Economic Research Service conference, "Crop Insurance, Land Use and the Environment," September 20-21.

Holt, M. 1999. "A Linear Approximate Acreage Allocation Model." *Journal of Agricultural and Resource Economics* 24: 383-397.

Holt, M.T., and J.-P. Chavas. 2001. "The Econometrics of Risk." In R.E. Just and R.D. Pope, eds., *A Comprehensive Assessment of the Role of Risk in U.S. Agriculture.* Boston, MA: Kluwer Academic Publishers.

Just, R., and R. Pope. 1978. "Stochastic Specification of Production Functions and Economic Implications." *Journal of Econometrics* 7: 67-86.

___. 1979. "On the Relationship of Input Decisions and Risk." In J. Roumasset, J.M. Boussard, and I. Singh, eds., *Risk Uncertainty, and Agricultural Development.* New York: ADC.

Lambert, D.A., and B.A. McCarl. 1985. "Risk Modeling Using Direct Solution of Nonlinear Approximations of the Utility Function." *American Journal of Agricultural Economics* 67: 846-852.

Lin, W., Westcott, P., Skinner, R., Sanford, S., and D. Ugarte. 2000. "Supply Response Under the 1996 Farm Act and Implications for the U.S. Field Crops Sector." Market and Trade Economics Division, ERS, USDA, Technical Bulletin No. 1888 (July).

Markowitz, H. 1952. "Portfolio Selection." *Journal of Finance* (March): 77-91.

Meyer, J. 1987. "Two-Moment Decision Models and Expected Utility Maximization. *American Economic Review* 77: 421-430.

___. 2001. "Expected Utility as a Paradigm for Decision Making in Agriculture." In R.E. Just and R.D. Pope, eds., *A Comprehensive Assessment of the Role of Risk in U.S. Agriculture.* Boston, MA: Kluwer Academic Publishers.

Meyer, J., and R. Pope. 1980. "Unbiased Estimation of Expected Utility and Agricultural Risk Analysis." Unpublished paper, Department of Economics, Texas A&M University.

Pagan, A., and A. Ullah. 1988. "The Econometric Analysis of Models with Risk Terms." *Journal of Applied Econometrics* 3: 87-105.

Pope, R. 1979. "The Effects of Production Uncertainty on Input Demands." In D. Yaron and C.S. Tapiero, eds., *Operations Research in Agriculture and Water Resources.* Amsterdam: North Holland Publishing, Co.

Pope, R., and J.-P. Chavas. 1994. "Cost Functions Under Production Uncertainty." *American Journal of Agricultural Economics* 76: 196-204.

Pope, R., and R. Just. 1996. "Empirical Implementation of Ex Ante Cost Functions. *Journal of Econometrics* 72: 231-249.

___. 1997. "Expected Profit Maximization Under Risk Aversion." Unpublished paper, Department of Economics, Brigham Young University, Utah.

___. 1998a. "Cost Function Estimation Under Risk Aversion." *American Journal of Agricultural Economics* 80: 296-302.

___. 1998b. "Taking Duality Semi-Seriously." Unpublished paper presented to SRIEF 170, Gulf-Shores, Alabama, and University of Arizona.

Ramaswami, B. 1993. "Supply Response to Agricultural Insurance: Risk Reduction and Moral Hazard Effects." *American Journal of Agricultural Economics* 75: 914-925.

Rosenzweig, M.R., and H.P. Binswanger. 1993. "Wealth, Weather Risk and the Composition and Profitability of Agricultural Investments." *The Economic Journal* 103: 56-78.

Saha, A. 1997. "Risk Preference Estimation in the Nonlinear Mean Standard Deviation Approach." *Economic Inquiry* 35: 770-782.

Saha, A., R. Innes, and R. Pope. 1993. "Production and Savings Under Uncertainty." *International Review of Economics and Finance* 2: 365-375.

Saha, A., and R. Just. 1996. "Duality Under Uncertainty." Unpublished paper.

Saha, A., C.R. Shumway, and H. Talpaz. 1994. "Joint Estimation of Risk Preference Structure and Technology Using Expo-Power Utility." *American Journal of Agricultural Economics* 76: 173-184.

Sandmo, A. 1971. "On the Theory of the Competitive Firm under Price Uncertainty," *American Economic Review* 61: 65-73.

Silberberg, E. 1974. "A Revision of Comparative Statics Methodology in Economics, or, How to Do Economics on the Back of an Envelope." *Journal of Economic Theory* 7: 159-172.

Sinn, H.W. 1983. *Economic Decisions under Uncertainty* (2nd edition). Amsterdam, New York, and Oxford: North Holland Publishing Company.

Taylor, C.R. 1984. "Stochastic Dynamic Duality: Theory and Empirical Applicability." *American Journal of Agricultural Economics* 66: 351-357.

Tobin, J. 1958. "Liquidity Preference as Behavior Towards Risk." *Review of Economic Studies* 67: 1-26. (Reprinted in D.D. Hester and J. Tobin, eds., *Risk Aversion and Portfolio Choice*, Cowles Foundation for Research in Economics, New Haven, CT, 1967.)

Quiggin, J. 1982. "A Theory of Anticipated Utility." *Journal of Economic Behavior and Organization* 3: 323-343.

Chapter 8

THE SIGNIFICANCE OF RISK UNDER INCOMPLETE MARKETS

Jean-Paul Chavas and Zohra Bouamra-Mechemache
University of Wisconsin, Madison
Institut National de la Recherche Agronomique, Toulouse, France

INTRODUCTION

The efficiency of complete competitive markets is well known (e.g., Allais 1943, 1981, Arrow and Debreu 1954, Debreu 1959, Mas-Colell, Whinston, and Green 1995, Luenberger 1995). Efficiency results apply as well to the allocation of risk (e.g., Debreu 1959). Yet markets (and especially risk markets) are typically incomplete. This is particularly true in the agricultural sector where weather uncertainty and unstable commodity markets can generate significant income risk, often borne by farmers. This has stimulated a policy debate on the efficiency of risk allocation and the relative role of market mechanisms versus government interventions. This is a complex issue. The limitations of alternative approaches have been pointed out. Market failures have been contrasted with government failures. In addition, the prevalence of both incomplete markets and incomplete contracts has made the economic analysis of risk allocation somewhat difficult.

Our objective is to develop a general conceptual model of resource allocation under risk and to investigate its implications for the analysis of efficiency under transaction costs. Following Debreu, we rely on a state-contingent representation of economic decisions among a set of economic agents involved in production, consumption, and exchange. The analysis is presented under general ordinal risk preferences (which include as a special case the standard expected utility model). To reflect possible market imperfections, we introduce transaction costs in a general equilibrium context. The transaction costs involve resources used as part of the exchange process among economic agents. While transaction costs may be low in some markets (e.g., financial markets), they can be large and significant in many commodity and insurance markets (due to transportation costs, information costs, administrative costs, etc.). In this context, transaction costs represent "frictions" in the functioning of markets (e.g., Foley 1970, Hahn 1971, Starr 1989). They can contribute to explaining the existence of incomplete markets and incomplete contracts. They also affect the efficiency of risk allocation and the functioning of risk markets. We focus on a monetary economy, where money is used as a (costless) means of exchange.

In the first section that follows we present our general model of production, consumption, and exchange activities under risk. In the section following that, we define different economic concepts useful in the analysis of economic efficiency. They rely on the benefit function proposed by Luenberger (1992a, 1992b). The benefit function provides a general and convenient way to characterize risk aversion and to analyze its implications for the efficiency of resource and risk allocation. Following Allais (1943, 1981), we rely on the concept of zero-maximal equilibrium (corresponding to the maximization of aggregate benefits followed by their complete redistribution to consumers). Zero-maximality being intuitive and quite powerful, it provides a convenient basis for investigating the efficiency of resource allocation under risk. Extending the work of Ostroy (1980), Luenberger (1992a, 1994), and Chavas and Bouamra (2001), we examine four concepts under risk and transaction costs: Pareto optimality, zero maximality, Lagrange equilibrium, and competitive equilibrium. We evaluate the relationships between Pareto efficiency, zero maximality, and competitive equilibrium.

In the section entitled "The Effects of Transaction Costs," we analyze the negative influence of transaction costs on efficiency, and on exchange and market activities. Then in the section following that we investigate the role of information. We show how imperfect information contributes to lowering efficiency by generating less refined decision rules and less effective risk redistribution schemes. In addition, asymmetric information has adverse effects on trade and contracts. In this context, we argue that information costs and bounded rationality help explain the prevalence of incomplete contracts and incomplete risk markets. We discuss the implications for information management and the efficiency of risk and resource allocation. The section on special cases relates our analysis to previous literature. For example, we show that our model includes as a special case the standard microeconomic model of household production under price risk and/or production risk. The final section of our chapter offers some concluding remarks.

THE MODEL

Consider the case of n agents involved in the allocation of m commodities under uncertainty. Let $N = \{1,2,...,n\}$ denote the set of agents. Each agent can get involved in the production and/or consumption of the m commodities. Let $x_i = (x_{1i},...,x_{mi})' \in R^m$ be the $(m \times 1)$ vector of netputs (where outputs are positive and inputs are negative) involved in the production activities of the ith agent, $i \in N$. And let $y_i = (y_{1i},...,y_{mi})' \in R^m$ be the $(m \times 1)$ vector of quantities of the m commodities consumed by the ith agent, $i \in N$. In addition to producing and consuming, the n agents can trade with each other. Let $t_{ij} \in R_+^m$ be the quantities of the m commodities exchanged from the ith agent to the jth agent, $i, j \in N$. We want to investigate the allocation of production, consumption, and exchange activities $\{x_i, y_i, t_{ij} : i, j \in N\}$.

The n agents make decisions under uncertainty. The uncertainty is represented by mutually exclusive states of nature. The sth state of nature is given by $e_s, s \in S$, where S is the set of all possible states. The issue is to analyze the efficiency of the decision rules $x^e = \{x_i(e_s) : i \in N; s \in S\}$, $y_i^e = \{y_i(e_s) : s \in S\}$, $i \in N$, and $t^e = \{t_{ij}(e_s) : i, j \in N; s \in S\}$, where $x_i(e_s)$, $y_i(e_s)$, and $t_{ij}(e_s)$ are the choices made under the sth state of nature.

Let the set X represent the production technology for x^e, where $x^e \in X$ means technical feasibility. This allows for private goods and public goods, as well as externalities in production activities across agents. Also, let Y_i be the consumption set for y_i^e, where $y_i^e \in Y_i$ means consumption feasibility for the ith agent, $i \in N$. We let $y^e = \{y_i^e : i \in N\}$, where $y^e \in Y = Y_1 \times Y_2 \times ... \times Y_n$. Also, consider that exchange t^e is costly and requires the use of resources (e.g., transportation activities). Denote by $z_i(e_s)$ the $(m \times 1)$ vector of commodities used by the ith agent in its exchange activities under state s. And let $z^e = \{z_i(e_s) : i \in N; s \in S\}$ represent the amount of resources used to support transactions among the n agents. The transaction technology is denoted by the set T, where $(t^e, z^e) \in T$ denotes feasibility. We assume that the sets X, Y, and T are closed with non-empty interiors, that the consumption set Y is convex and has a lower bound, and that $(\underline{0}, \underline{0}) \in T$, i.e., that the absence of exchange requires no resources z. Also, we assume that each agent's exchange with itself (as represented by t_{ii}) is costless.

This formulation is very general. However, it does involve two simplifying assumptions. First, it assumes that every consumer is also a producer. Note that this could be easily relaxed. For example, the case where the number of producers differs from the number of consumers can be obtained by simply restricting the feasible sets X and Y such that non-producing agents produce nothing, and non-consuming agents consume nothing. Second, while the production set X allows for private goods and public goods, as well as externalities across agents, the consumption set $Y = Y_1 \times ... \times Y_n$ considers only private goods. This will simplify the analysis presented below. Note that this does not appear overly restrictive to the extent that externalities and public goods for households can be taken into consideration in a household production framework (e.g., Deaton and Muellbauer 1980, Chapter 10) through the production set X.

Each agent has preferences represented by the ex ante utility function $u_i(y_i^e) : Y_i \to R$, $i \in N$. Throughout, we assume that the utility functions $u_i(y_i^e)$ are continuous and quasi-concave. When considering consumption across states of nature, the quasi-concavity of $u_i(y_i^e)$ implies risk aversion (see below). In this general form, the ex ante utility function $u_i(y_i^e)$ does not rely on the existence of probabilities describing the uncertain states of nature. However, it is often empirically convenient to make explicit use of probability assessments. Assume that the ith agent assesses the probability of the sth state to be $\text{Prob}_i(e_s)$. This probability may be estimated from sample information. Alternatively, in the case of non-repeatable events, $\text{Prob}_i(e_s)$ may be the agent's subjective evaluation of the relative likelihood of the sth state (e.g., DeGroot 1970, Chapter 6). More generally, in a Bayesian framework, the probabilities may be obtained from combining subjective assessments with sample information (through Bayes' theorem).[1]

When probability assessments are used, the ex ante utility function is $u_i[y_i^e; \text{Prob}_i(e_1), \text{Prob}_i(e_2),\ldots]$, $i \in N$. Note that, in this general form, the utility function can be non-linear in the probabilities. However, this includes as a special case the expected utility model $u_i(\cdot) = \sum_{s\in S}\{\text{Prob}_i(e_s) \cdot v_i[y_i(e_s)]\}$, $i \in N$, where $v_i(\cdot)$ is a von Neumann-Morgenstern utility function representing the ith agent's risk preferences. In this case, the utility function $u_i(\cdot)$ is strongly separable across states and linear in the probabilities. Although the expected utility model is sometimes seen as being overly restrictive (especially because of its linearity in probabilities), it is commonly used in conceptual as well as empirical analyses of economic behavior under risk. First, it is convenient for empirical work. Second, expected utility can provide a good local approximation to more general utility functions in the neighborhood of a given probability distribution (Machina 1987).

Following Debreu (1959), let $w = (x^e, y^e, z^e, t^e) = \{w_k(e_s)\colon k \in K; s \in S\}$, where $w_k(e_s)$ denotes the kth decision made under the sth state of nature, and K is the set of all decisions made by all agents. Thus, w represents all decisions as *state-contingent choices*. The state space S can be partitioned into various mutually exclusive subsets. Assume that the kth decision variable is based on an information partition $P_k = \{P_{kp}\colon p = 1, 2,\ldots\}$ of the state S. For each k, the P_{kp}'s are mutually exclusive subsets of S and satisfy $\{\cup P_{kp}\colon p = 1, 2,\ldots\} = S$. Each set P_{kp} contains the states that are not distinguishable for the purpose of making the kth decision. It means that the kth decision rule must satisfy the following information constraint:

---

[1] In this context, learning can take place by observing that the true state is some information subset Ω of S: $\Omega \subset S$. Then, for the ith agent, the probability of facing state s is updated from $\text{Prob}_i(e_s)$ to

$$\text{Prob}_i'(e_s) = \text{Prob}_i(e_s)/[\textstyle\sum_{s'\in\Omega} \text{Prob}_i(e_{s'})] \text{ for } s \in \Omega$$

$$= 0 \text{ for } s \notin \Omega.$$

$$w_k(e_s) = w_k(e_{s'}) \text{ for any two } s \text{ and } s' \in P_{kp} \text{ for each } p, k \in K. \quad (1)$$

Equation (1) reflects how information affects the decision w_k. It includes as special cases two extreme situations. At one extreme, perfect information corresponds to $P_k = \{e_s : s \in S\}$. Then, w_k is chosen ex post and can vary across each state of nature as (1) imposes no restriction. At the other extreme, no information corresponds to $P_k = \{S\}$. Then, (1) implies that the decision w_k is chosen ex ante and constrained to be the same across all states. In intermediate situations, equation (1) restricts the kth decision to be the same across states that are not distinguishable, but allows it to differ across states otherwise. This formulation allows for information asymmetry as different agents can face different information. And it allows for learning to the extent that different decisions made by a given agent may be made at a different time and based on different information. We denote by $P = \{P_k : k \in K\}$ the information structure supporting the decision making process for all decisions across all agents.

In the analysis presented below, we proceed in two steps. In the first step, we consider that the information available for each decision w_k is fixed. This allows for information to vary across decisions for a particular agent (e.g., if the agent learns over time) as well as across agents (representing asymmetric information). As such, we first treat the information structure $P = \{P_k: k \in K\}$ as exogenous. This allows us to investigate the effects of exogenous changes in information (as represented by P) on resource allocation. In a second step, we will treat the information structure P as an object of choice, thus endogenizing the amount of information available in the decision making process. This will provide additional insights into the economics of information.

Definition 1. *An allocation* $w = (x^e, y^e, z^e, t^e) = \{w(e_s): s \in S\}$ *is* *feasible* *if it satisfies (1), and*

$$\Sigma_{j\in N}\, t_{ij}(e_s) \leq x_i(e_s) - z_i(e_s),\ i \in N,\ s \in S, \quad (2a)$$

$$y_i(e_s) \leq \Sigma_{j\in N}\, t_{ji}(e_s),\ i \in N,\ s \in S, \quad (2b)$$

where $x^e \in X$, $y^e \in Y$, *and* $(z^e, t^e) \in T$.

Equation (2a) states that the ith agent cannot export more than its production (x_i), net of resources used in exchange (z_i). And equation (2b) states that the ith agent cannot consume more than it obtains from itself (t_{ii}) or from others ($\Sigma_{j\neq i} t_{ji}$). These two restrictions also guarantee that aggregate consumption cannot exceed aggregate production, net of aggregate resources used in exchange.

EQUILIBRIUM CONCEPTS

We are interested in investigating the efficiency of resource allocation under risk. For that purpose, we rely on some key economic concepts.

Definition 2. *An allocation w^* is Pareto efficient if it is feasible and there is no other feasible allocation w that can make one agent better off without making any other worse off.*

We assume that money is used as means of exchange. For that purpose, we identify the mth commodity as "money." We denote one unit of money by $g = (0, 0,\ldots, 0, 1) \in R^m$. And we denote one unit of *sure money* (where g is obtained in every state) by $g^e = \{g(e_s)\colon g(e_s) = g;\ s \in S\}$. Although we allow for transaction costs for the first $(m-1)$ commodities (as represented by the feasible set T), we will assume throughout that money can be exchanged costlessly among the n agents.

Definition 3. *Define the benefit function of the ith agent as*

$$b_i(y_i^e, u_i) = \max_\beta \{\beta : (y_i^e - \beta g^e) \in Y_i;\ u_i(y_i^e - \beta g^e) \geq u_i\} \text{ if a feasible } \beta \text{ exists},$$

$$= -\infty \text{ otherwise},$$

for $i \in N$. And the aggregate benefit function is

$$B(y^e, u) = \textstyle\sum_{i \in N} b_i(y_i^e, u_i),$$

where $u = (u_1,\ldots,u_n)$.

The function b_i has an intuitive interpretation. It measures the sure amount of money the ith agent is willing to give up facing the consumption vector y_i^e to reach the utility level u_i. And the benefit function B is the corresponding aggregate measure. As shown by Luenberger (1992b), under the assumptions that the set Y_i is convex and the function $u_i(y_i^e)$ quasi-concave, the function $b_i(y_i^e, u_i)$ is concave in y_i^e and non-increasing in u_i. Thus, the aggregate benefit function $B(y^e, u)$ is concave in y^e and non-increasing in u.

The benefit function b_i also provides a convenient characterization of risk aversion for the ith agent. To see that, let y_i^c be the vector of consumption goods set equal to the expected value of $y_i(e_s)$: $y_i^c = \sum_{s \in S} \text{Prob}_i(e_s) \cdot y_i(e_s)$. Denoting by $(y_i^c,\ldots,y_i^c)$ the *sure* consumption of y_i^c (i.e., the consumption of y_i^c obtained in every state), consider the benefit function $b_i[y_i^c,\ldots,y_i^c, u_i(y_i^e)]$. The function $b_i[y_i^c,\ldots,y_i^c, u_i(y_i^e)]$ defines the *risk premium*: it is the amount of money the ith agent is willing to pay to replace the state-contingent consumption bundle y_i^e by the sure consumption bundle $(y_i^c,\ldots,y_i^c)$. By definition, the

i^{th} agent is *risk averse* if the risk premium (measuring the implicit cost of private risk bearing) is non-negative: $b_i[y_i^c, \ldots, y_i^c, u_i(y_i^e)] \geq 0$ for all y_i^e. Note that the quasi-concavity of $u_i(y_i^e)$ implies that $b_i[y_i^c, \ldots, y_i^c, u_i(y_i^e)] \geq 0$ for all y_i^e. This is another way to show that the quasi-concavity of $u_i(y_i^e)$ identifies risk-averse preferences. It follows that the concavity of $b_i(y_i^e, \cdot)$ is also associated with risk aversion. For example, under the differentiability of the benefit function, risk aversion means that $\partial^2 b_i/\partial y_i^{e2}$ is a negative semi-definite matrix. This provides some useful information in the neighborhood of the riskless case. To see that, let $y_i^{ec} = \{y_i^c + \alpha\, e_s : s \in S\}$, where e_s is the sth realization of a random variable with mean zero and unit variance, and α is a $(m \times 1)$ vector of mean-preserving spread parameters. Denote the benefit function evaluated at y^{ec} by $\beta_i(\alpha) = b_i[y_i^{ec}, u_i(y_i^e)]$. Then, in the neighborhood of the riskless case (where $\alpha = 0$), the risk premium can be approximated as

$$b_i[y^c, u_i(y_i^e)] \approx \tfrac{1}{2}\, \text{trace}\{-\partial^2\beta_i/\partial\alpha^2(0) \cdot \text{Var}[y(e)]\},$$

where $-\partial^2\beta_i/\partial\alpha^2(0)$ is a positive semi-definite matrix under risk aversion (Luenberger 1995, p. 394). In the context of the expected utility model with $m = 1$, this identifies $-\partial^2\beta_i/\partial\alpha^2(0)$ as the Arrow-Pratt absolute risk aversion coefficient (Arrow 1970, Pratt 1964).

Following Luenberger (1992a), we define the concept of zero maximality. Let f be a function defined over a set H. Then, h is said to be maximal if h maximizes f over H. And h is said to be zero maximal if, in addition, $f(h) = 0$. Applying this concept to the aggregate benefit function $B(y^e, u)$ under feasibility, we obtain:

Definition 4. *An allocation w is* <u>*maximal*</u> *if it satisfies*

$$V(u, P) = \max_w \{B(y^e, u): \text{equ. (1), (2a), (2b)}; x^e \in X, y^e \in Y; (z^e, t^e) \in T\}, \quad (3)$$

where $P = \{P_k: k \in K\}$ represents the information structure associated with all decisions. And w is <u>*zero maximal*</u> *if, in addition, u is chosen in (3) such that $V(u, P) = 0$.*

Building on the work of Luenberger (1992a), the relationships between zero maximality and Pareto efficiency are presented next.

Proposition 1. *Assume that at least one agent is non-satiated in money (i.e., with $u_i(y_i)$ being strictly increasing in the mth commodity for some i). If the allocation w^* is Pareto efficient, then it is zero maximal*

Proof. The allocation w^* is feasible. It follows from definitions 3 and 4 that $b_i[y_i^{e*}, u_i(y_i^{e*})] \geq 0$ for all $i \in N$. Assume that $b_j > 0$ for some agent j. This implies $B > 0$. Since money can be exchanged costlessly, the positive monetary benefit B can be feasibly redistributed to the agent who is non-satiated in money. This would make this agent better off without making any one else worse off, thus contradicting Pareto efficiency. It follows that Pareto efficiency implies zero maximality.

Proposition 2. *If w^* is zero maximal, then it is Pareto efficient compared to all feasible allocations in the interior of Y.*

Proof. Assume that there is a feasible allocation w in the interior of Y where $u_i(y_i^e) \geq u_i(y_i^{e*})$ for all $i \in N$, but with $u_j(y_j^e) > u_j(y_j^{e*})$ for some agent j. This means that w^* is not Pareto efficient. From definition 3, $b_i[y_i^e, u_i(y_i^{e*})] \geq 0$ for all $i \in N$, and $b_j[y_j^e, u_i(y_j^{e*})] > 0$. It follows that w cannot be zero maximal. Thus, zero maximality implies Pareto efficiency.

Propositions 1 and 2 establish general relationships between Pareto efficiency and zero maximality. In particular, they do not require the convexity of the sets X and T. In addition, when Pareto efficiency and zero maximality are equivalent (as stated in Propositions 1 and 2), the function $V(u,P)$ defined in (3) provides a nice and intuitive characterization of Pareto efficiency. Indeed, $V(u,P)$ can be interpreted as the maximum distributable monetary surplus. From (3), a maximal allocation makes the distributable surplus as large as possible. And zero maximality means that this surplus must be entirely redistributed to the n agents. Then, Pareto efficiency can be interpreted as a situation where the distributable surplus is first maximized and then entirely redistributed.

In addition, the zero maximality condition $V(u,P)=0$ is an implicit equation for $u=(u_1,...,u_n)$ that characterizes the Pareto utility frontier. To see that, note that $V(u,P)$ is non-increasing in u. It follows that a non-negative distributable surplus, $V(u,P) \geq 0$, corresponds to a feasible allocation. In other words, the set of u that satisfies $V(u,P) \geq 0$ defines the space of reachable utility levels. And $\{u : V(u,P)=0\}$ identifies the upper bound of this space as the Pareto utility frontier. Since monetary transfers across agents are assumed costless, note that one can always move along the Pareto utility frontier by implementing some income transfers across agents. However, in the presence of income effects, we can expect such movements to affect the Pareto optimal allocation z^*.

As mentioned above, these results do not require the convexity of the sets X and T. They allow the production decisions x to involve private goods and public goods, as well as externalities. They involve possible transaction costs in the first $(m-1)$ commodities. They represent economic situations under risk and risk aversion, allowing for learning and asymmetric information. They allow for state-dependent transfers that can redistribute risk across agents. As such, they identify a role for insurance contracts in the Pareto efficiency of risk allocation: such contracts can improve consumer welfare by efficiently redistributing consumption risk away from the more risk-averse individuals. Our analysis allows for possible income effects. Finally, at this point, the efficiency of allocation is evaluated without relying on market prices. As such, our results can be interpreted as characterizing efficient contracts, thus presenting a generalization of the Coase theorem under risk, transaction costs, and income effects.

Next, we would like to know whether the Pareto efficient allocations just identified could be supported by a market economy. To answer this question, the following equilibrium concept will prove useful.

Definition 5. *Let* $q_{ri}^e = \{q_{ri}(e_s): s \in S\}$ *and* $q_{ci}^e = \{q_{si}(e_s): s \in S\}$, *where* $q_{ri} = (q_{ri1},\ldots,q_{rim}) \in R_+^m$ *and* $q_{ci} = (q_{ci1},\ldots,q_{cim}) \in R_+^m$, $i \in N$. *Letting* $q_r^e = \{q_{ri}^e: i \in N\}$ *and* $q_c^e = \{q_{ci}^e: i \in N\}$, *an allocation* w^* *is a* *Lagrange equilibrium* *if it is a saddle point of the Lagrangean*

$$L(w, q_r^e, q_c^e, u, P) = B(y^e, u) + \textstyle\sum_{i\in N} q_{ri}^e \cdot [y_i^e - z_i^e - \sum_{j\in N} t_{ij}^e] + \sum_{i\in N} q_{ci}^e \cdot [\sum_{j\in N} t_{ji}^e - x_i^e], \tag{4a}$$

i.e., if the feasible point $(w^*, q_r^{e*}, q_c^{e*})$ *satisfies*

$$L(w, q_r^{e*}, q_c^{e*}, u, P) \le L(w^*, q_r^{e*}, q_c^{e*}, u, P) \le L(w^*, q_r^e, q_c^e, u, P), \tag{4b}$$

for all feasible (w, q_r^e, q_c^e), *where u is chosen such that* $L(w^*, q_r^{e*}, q_c^{e*}, u, P) = 0$.

Note that the $L(\cdot)$ in (4a) is the Lagrangean of the constrained optimization problem (3) defining a maximal equilibrium, where q_r^e and q_c^e are the Lagrange multipliers associated with the feasibility constraints (2a) and (2b), respectively. These Lagrange multipliers have the standard interpretation of measuring the shadow prices of these constraints. In other words, q_r^e measures the shadow prices of resource scarcity on the supply side, while q_c^e measures the shadow prices of resource scarcity on the demand side.

Proposition 3. *If an allocation* w^* *is a Lagrange equilibrium, then it is zero maximal.*

Proof. From the saddle-point theorem (e.g., Takayama 1985, p. 74), the saddle-point problem (4a)–(4b) implies that w^* solves the constrained optimization problem (3). It also implies the complementary slackness condition:

$$\textstyle\sum_{i\in N} q_{ri}^{e*} \cdot [y_i^{e*} - z_i^{e*} - \sum_{j\in N} t_{ij}^{e*}] + \sum_{i\in N} q_{ci}^{e*} \cdot [\sum_{j\in N} t_{ji}^{e*} - x_i^{e*}] = 0. \tag{5}$$

Thus, $L(w^*, q_r^{e*}, q_c^{e*}, u, P) = 0$ yields $B(y^{e*}, u) = 0$. It follows that w^* is a zero maximal equilibrium.

Proposition 4. *Assume that the sets X and T are convex and that there exists a feasible point w such that equations (2a)–(2b) are non-binding. Then, if an allocation w^* is zero maximal, it is a Lagrange equilibrium.*

Proof. Under the stated assumptions, the feasible set in (3) is convex and Slater's condition holds with respect to the constraints (2a)–(2b). Then, $B(y^e,\cdot)$ being concave and the constraints (2a)–(2b) being linear, the maximal allocation given in (3) implies that the Lagrange multipliers $(q_r^{e*}, q_c^{e*}) \geq 0$ exist and that $(w^*, q_r^{e*}, q_c^{e*})$ is a saddle-point of the Lagrangean in (4a)–(4b) (see Takayama 1985, p. 75). In turn, the complementary slackness condition (5) holds. Then, $B(y^{e*}, u) = 0$ implies that $L(w^*, q_r^{e*}, q_c^{e*}, u, P) = 0$. It follows that w^* is a Lagrange equilibrium.

Propositions 3 and 4 present general relationships between zero maximality and Lagrange equilibrium. Note that, while Proposition 3 holds in general, Proposition 4 requires convexity assumptions on the sets X and T. These convexity assumptions are needed to guarantee the existence of the shadow prices (q_r^{e*}, q_c^{e*}) in Lagrange equilibrium. As expected, these shadow prices (q_r^{e*}, q_c^{e*}) can play the role of market prices in market economies. These arguments are presented next.

Definition 6. *Denote commodity prices facing the ith agent by $q_i^* = (q_{i1}^*,\ldots,q_{im}^*) \in R_+^m$ and $q_i^{e*} = \{q_i^*(e_s): s \in S\}$, where $q_i^{e*} \cdot g^e = 1$, $i \in N$. Letting $q^{e*} = \{q_i^{e*}: i \in N\}$, an allocation w^* along with market prices q^{e*} is a competitive equilibrium if*

- *w^* is a feasible allocation,*

- $y_i^{e*} \in \text{argmin}_y \{q_i^{e*} \cdot y_i^e: u_i(y_i^e) \geq u_i, y_i^e \in Y_i\}, i \in N,$ (6a)

- $x^{e*} \in \text{argmax}_x \{\Sigma_{i\in N}\, q_i^{e*} \cdot x_i^e: x^e \in X\},$ (6b)

- $(z^{e*}, t^{e*}) \in \text{argmax}_{z,t} \{\Sigma_{i\in N} \Sigma_{j\in N} (q_j^{e*} - q_i^{e*}) \cdot t_{ij}^e - \Sigma_{i\in N}\, q_i^{e*} \cdot z_i^e: (z^e, t^e) \in T\}$ (6c)

and

- $q_i^{e*} \cdot [x_i^{e*} - z_i^{e*} - \Sigma_{j\in N}\, t_{ij}^{e*}] + q_i^{e*} \cdot [\Sigma_{j\in N}\, t_{ji}^{e*} - y_i^{e*}] = 0, i \in N.$ (6d)

First, a competitive equilibrium requires feasibility. Condition (6a) represents economic rationality for consumption units. Conditions (6b) and (6c) are profit maximization behavior for production and exchange activities, respectively. Finally, condition (6d) states the budget constraint for each agent. Note that solving for the term $(q_i^{e*} \cdot t_{ij}^{e*})$ in (6d) gives

$$q_i^{e*} \cdot t_{ij}^{e*} = q_i^{e*} \cdot y_i^{e*} - q_i^{e*} \cdot z_i^{e*} - q_i^{e*} \cdot (\Sigma_{j\neq i}\, t_{ij}^{e*}) = q_i^{e*} \cdot x_i^{e*} - q_i^{e*} \cdot (\Sigma_{j\neq i}\, t_{ji}^{e*}),\ i \in N,$$

or

$$q_i^{e*} \cdot (\Sigma_{j\neq i}\, t_{ij}^{e*}) - q_i^{e*} \cdot (\Sigma_{j\neq i}\, t_{ji}^{e*}) = q_i^{e*} \cdot y_i^{e*} - q_i^{e*} \cdot z_i^{e*} - q_i^{e*} \cdot x_i^{e*},\ i \in N.$$

This can be interpreted as a "balance of payment" constraint which states that, for any agent $i \in N$, the value of net exports must equal profit, minus the cost of trade, minus consumer expenditures.

The relationships between Lagrange equilibrium and competitive market equilibrium under transaction costs are presented next.

Proposition 5. *If the feasible allocation w^* is a Lagrange equilibrium, then it is a competitive equilibrium.*

Proof. Since exchange t_{ii} is assumed to be costless, the first inequality in (4b) implies that $q_{ri}^{e*} = q_{ci}^{e*}$ for all $i \in N$ (otherwise, t_{ii}^{e*} and thus $L(\cdot)$ would be unbounded, a contradiction). Let $q_i^{e*} = q_{ri}^{e*} = q_{ci}^{e*}, i \in N$.

Since exchanging money is assumed to be costless, the first inequality in (4b) implies that $q_{rim}^{e*} = q_{cim}^{e*}$ for all $i \in N$ (otherwise, t_{ijm}^{e*} or t_{jim}^{e*} and thus $L(\cdot)$ would be unbounded, a contradiction). Let $q_{im}^{e*} = q_i^{e*} \cdot g^e = 1$ for all $i \in N$.

The allocation w^* is feasible. Then, the first inequality in (4b) implies

$$y_i^{e*} \in \text{argmax}_y\ \{b_i(y_i^e, u_i) - q_i^{e*} \cdot y_i^e : y_i^e \in Y_i\},\ i \in N.$$

Given $q_i^{e*} \cdot g^e = 1$, Luenberger (1992b, pp. 472-473) has shown that this implies (6a). The first inequality in (4b) also implies (6b) and (6c). Finally, (6d) follows from the complementary slackness condition (5), with $q_{ri}^e = q_{ci}^e = q_i^e$.

This identifies the key role played by two of our assumptions. First, assuming that exchange with itself (t_{ii}) is costless implies that the prices faced by each agent are unique in the sense that $q_i^{e*} = q_{ri}^{e*} = q_{ci}^{e*},\ i \in N$. And assuming that money (the mth commodity) can be exchanged costlessly, the price of sure money g^e is the same for all agents: $q_{im}^{e*} = q_i^{e*} \cdot g^e$ for all $i \in N$. Without a loss of generality, it is normalized to be equal to 1. This means that money is used as a basis for evaluating all welfare measures.

Proposition 6. *If the feasible allocation w^* is a competitive equilibrium and $B(y^{e*}, u^*) = 0$ (where $u^* = \{u_i(y_i^{e*}) : i \in N\}$), then it is a Lagrange equilibrium.*

Proof. Consider the conditions stated in the definitions of competitive equilibrium. Given $q_i^{e*} \cdot g^e = 1$ and $q_i^{e*} = q_{ic}^{e*} = q_{is}^{e*}$ for all $i \in N$, retracing back the steps presented in the proof of Proposition 5 yields equations (6a), (6b), and (6c). And (6d) implies equation (5). Combining these results generates the first inequality in (4b).

Under (5), $L(w^*, u, q_s^{e*}, q_c^{e*}) = B(y^{e*}, u)$, which is equal to zero as desired when evaluated at $u = \{u_i(y_i^{e*}) : i \in N\}$.

Note that, for any $q_{ri}^e \geq 0$ and $q_{ci}^e \geq 0$, and since w^* satisfies (2a) and (2b) by feasibility, we have

$$\Sigma_{i \in N}\, q_{si}^{\ e} \cdot [y_i^{e*} - z_i^{e*} - \Sigma_{j \in N}\, t_{ij}^{\ e*}] + \Sigma_{i \in N}\, q_{ci}^{\ e} \cdot [\Sigma_{j \in N}\, t_{ji}^{\ e*} - x_i^{e*}] \geq 0.$$

This implies that

$$0 \leq B(y^{e*}, u^*) + \Sigma_{i \in N}\, q_{si}^{\ e} \cdot [y_i^{e*} - z_i^{e*} - \Sigma_{j \in N}\, t_{ij}^{\ e*}] + \Sigma_{i \in N}\, q_{ci}^{\ e} \cdot [\Sigma_{j \in N}\, t_{ji}^{\ e*} - x_i^{e*}],$$

which is the second inequality in (4b). We conclude that w^* is a Lagrange equilibrium.

Propositions 1–6 present formal relationships between four concepts: Pareto efficiency, zero-maximality, Lagrange equilibrium, and competitive equilibrium. As derived, such relationships hold under risk and transaction costs. They are summarized in the following chart.

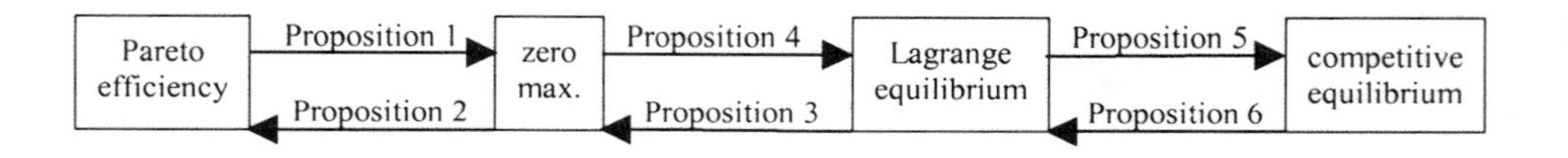

Note that Propositions 1 and 2 do not require convexity assumptions on the set X or T. As a result the linkages between Pareto efficiency and zero maximality hold without such assumptions. Similar comments apply to Propositions 3, 5, and 6. However, Proposition 4 requires convexity assumptions for the sets X and T. The reason is that, without convexity, the Lagrange multipliers q_r^{e*} and q_c^{e*} may not exist. From Propositions 5 and 6, these Lagrange multipliers become the market prices q^{e*} in competitive market equilibrium. This corresponds to the well-known result that, in general, convexity assumptions are needed to guarantee the existence of market prices q^{e*} in competitive market equilibrium.[2]

What is the nature of the state-contingent prices q^e? To gain insights on their properties, consider the following functions. From (6a), define the ith agent's expenditure function $E_i(q_i^e, u_i) = \min_y \{q_i^e \cdot y_i^e : u_i(y_i^e) \geq u_i,\, y_i^e \in Y_i\}$, $i \in N$, with corresponding aggregate expenditure function $E(q^e, u) = \Sigma_{i \in N}$

[2] Note that the convexity assumptions made in Proposition 4 can be relaxed somewhat. For example, this can be done along the lines presented by Heller (1976) and Starr (1989).

$E_i = (q_i^e, u_i)$. And let the production profit function be $\pi_x(q^e) = \max_x \{\Sigma_{i \in N} q_i^e \cdot x_i^e : x^e \in X\}$, and the exchange profit function be $\pi_T(q^e) = \max_{z,t} \{\Sigma_{i \in N} \Sigma_{j \in N} (q_j^e - q_i^e) \cdot t_{ij}^e - \Sigma_{i \in N} q_i^e \cdot z_i^e : (z^e, t^e) \in T\}$. This yields the aggregate profit function $\pi(q^e) = \pi_x(q^e) + \pi_T(q^e)$. Note that the profit functions $\pi(q^e)$, $\pi_x(q^e)$, and $\pi_T(q^e)$ are each convex in q^e. And the expenditure functions $E(q^e, u)$ and $e_i(q^e, u_i)$ are each concave in q^e (Diewert 1974, Berge 1963). From equation (4b), the Lagrange equilibrium can then be expressed as

$$q^{*e} \in \text{argmin}_q \{\pi(q^e) - E(q^e, u) : q^e \geq 0\}, \tag{7}$$

where u is chosen such that $\pi(q^{e*}) - E(q^{e*}, u) = 0$.

The solution to expression (7) then gives the competitive market prices q^{e*} that support a competitive equilibrium under risk. It corresponds to the concept of zero minimality discussed in Luenberger (1992a, 1994). Expression (7) thus provides a convenient representation of the determinants of competitive prices under uncertainty.

THE EFFECTS OF TRANSACTION COSTS

Our analysis has incorporated transaction costs in economic analysis. In this section, we consider the case where transaction costs may change and investigate their effects on welfare and resource allocation. This is modeled through a change in the transformation set T representing the feasibility of transactions among all agents. We want to investigate the case where the transformation technology shifts from T to T' and its effects on the economy.

We start from the zero maximal equilibrium given in (3), where the monetary surplus under T is denoted by $V(u, P, T)$. Then, two choices for u are relevant. First, consider the case where u is chosen such that aggregate net benefit is zero under technology T: $V(u, P, T) = 0$. Then, $[V(u, P, T') - V(u, P, T)] = V(u, P, T')$ measures the aggregate net income gain (or loss if negative) associated with a move from T to T'. In other words, $[V(u, P, T') - V(u, P, T)]$ is a simple measure of aggregate efficiency gains (compensating variation) generated by a change in the transaction technology in the economy. Second, consider the case where u is chosen such that aggregate net benefit is zero under technology T': $V(u, P, T') = 0$. Then, $[V(u, P, T') - V(u, P, T)] = -V(u, P, T)$ measures the aggregate net income loss (or gain if negative) associated with replacing T' by T. It follows that $[V(u, P, T') - V(u, P, T)] = -V(u, P, T)$ is a simple aggregate efficiency measure (equivalent variations) generated by giving up the transaction technology T'.

Consider the case where technological change from T to T' reduces transaction costs, with $T \subset T'$. Since this expands the feasibility set in the maximization problem (3), it follows that

$$[V(u, P, T') - V(u, P, T)] \geq 0.$$

This is an important result. It shows that reducing transaction costs tends to generate efficiency gains. In this context, any technological or institutional change that reduces transaction costs (e.g., improvements in infrastructure and information technology) would enhance economic efficiency. This suggests that private management and/or public policy that reduces transaction costs can contribute to significant efficiency gains.

To obtain additional insights in the role of transaction costs, define the transaction cost function $C(q^e, t^e) = \min_z \{\Sigma_{i\in N}\, q_i^e \cdot z_i^e\text{: } (z^e, t^e) \in T\}$. Assume that the function $C(q^e, t^e)$ is differentiable in t^e on the feasible set. Then, the saddle-point problem (4) implies the familiar Kuhn-Tucker conditions with respect to $t_{ijk}(e_s)$, the quantity of the kth commodity exchanged from agent i to agent j under state s:

$$\partial L/\partial t_{ijk}(e_s) = q_{jk}^*(e_s) - q_{ik}^*(e_s) - \partial C/\partial t_{ijk}(e_s) \leq 0 \text{ for } t_{ijk}(e_s) \geq 0, \qquad (8a)$$

and

$$[q_{jk}^*(e_s) - q_{ik}^*(e_s) - \partial C/\partial t_{ijk}(e_s)] \cdot t_{ijk}(e_s) = 0. \qquad (8b)$$

In the context of a competitive market equilibrium, equation (8a) implies that $[q_{jk}^*(e_s) - q_{ik}^*(e_s)] \leq \partial C/\partial t_{ijk}(e_s)$, i.e., that the price difference for commodity k between agents i and j $[q_{jk}^*(e_s) - q_{ik}^*(e_s)]$ cannot exceed the marginal transaction cost $\partial C/\partial t_{ijk}(e_s)$ under state s. And when trade takes place from agent i to agent j for the kth commodity $[t_{ijk}(e_s) > 0]$, then (8a) and (8b) imply that $[q_{jk}^*(e_s) - q_{ik}^*(e_s)] = \partial C/\partial t_{ijk}(e_s)$. In this case, the price difference $[q_{jk}^*(e_s) - q_{ik}^*(e_s)]$ must equal the marginal transaction cost $\partial C/\partial t_{ijk}(e_s)$ under state s. This can be interpreted as the first-order condition for profit-maximizing trade. It follows that, in the absence of transaction costs where $\partial C/\partial t_{ijk}(e_s) = 0$, the *law of one price* applies since $q_{jk}^*(e_s) = q_{ik}^*(e_s)$. Alternatively, when $\partial C/\partial t_{ijk} > 0$, transaction costs create a price wedge between $q_{jk}^*(e_s)$ and $q_{ik}^*(e_s)$. In such a situation, the law of one price clearly fails to apply. This allows the development of "local markets" where participants in each market are endogenously determined (depending on transaction technology and price differences). This is relevant in economic geography, where spatial prices typically vary across agents. This is also relevant in risk markets (e.g., the case of market segmentation for insurance contracts). And when transaction costs are "high enough" so that $\partial C/\partial t_{ijk}(e_s) > [q_{jk}^*(e_s) - q_{ik}^*(e_s)]$ for some i and j satisfying $[q_{jk}^*(e_s) - q_{ik}^*(e_s)] \geq 0$, then the incentive to trade disappears as (8b) implies $t_{ijk}(e_s) = 0$. Then, under state s, the kth commodity becomes non-traded between agents i and j. If this happens for all states, this implies the *absence of state-dependent exchange* for the kth commodity. In addition, if it happens for all agents, this implies the *absence of market* for the kth commodity. This illustrates well the adverse effects of transaction costs on market activities. Alternatively, as suggested by equations (8a)–(8b), reducing transaction costs can stimulate exchange activities. This would increase the gains from trade and improve opportunities for agents to specialize in the activities where they have a comparative advantage. These translate into better access to resources, and enhanced consumer welfare. It means that low transaction costs are critical in the creation and proper functioning of competitive markets in general, and of risk markets in particular.

THE ROLE OF INFORMATION

So far, we have considered the information structure $P = \{P_k: k \in K\}$ as fixed. In this section, we evaluate the effects of information on the efficiency of resource allocation. First, we consider the effects of an exogenous change in the information structure from P to P'. We assume that P' is at least as fine as P as defined next.

Definition 7. *Information structure* $P' = \{P_k': k \in K\}$ *is at least as fine as* $P = \{P_k': k \in K\}$ *if, for every* $\gamma \in P_k$ *and* $\gamma' \in P_k'$, *either* $\gamma' \subset \gamma$ or $\gamma \cap \gamma' = \varnothing$, *for each* $k \in K$.

This means that each decision k made under P' is based on information that is at least as precise as under P, $k \in K$. From equation (1), replacing P' by P allows more opportunities for decisions to become state-dependent. In other words, compared to P, the information structure P' imposes fewer restrictions on the decision rules $w(e_s)$. This implies that the feasible set in (3) is larger under P' than under P, and thus that

$$V(u, P') \geq V(u, P).$$

This means that better information tends to generate an outward shift in the Pareto utility frontier and yield efficiency gains. Such gains can be measured by the increase in the monetary surplus: $V(u, P') - V(u, P) \geq 0$. As discussed above, two attractive candidates for u are: u satisfying $V(u, P) = 0$ (corresponding to compensating variations); and u satisfying $V(u, P') = 0$ (corresponding to equivalent variations). In either case, $[V(u,P') - V(u,P)] \geq 0$ can be interpreted as measuring the (gross) social value of information.

If information were free, this result would be clear: more information would always be desirable in the sense that it contributes to more refined state-dependent decision rules that can improve the efficiency of resource allocation under risk. In this context, it may be useful to consider the two extreme information structures: $P^- = \{P_k^-: P_k^- = \{S\}, k \in K\}$, where all decisions are made ex ante without any information; and $P^+ = \{P_k^+ : P_k^+ = \{e_s : s \in S\}\}$, where all decisions are made ex post under perfect information. Then, for any information structure P associated with the state space S, using (1) and (3), the following result obtains:

$$V(u, P^-) \leq V(u, P) \leq V(u, P^+).$$

This provides a lower bound and an upper bound on how the distributable surplus varies with P. Since the upper bound $V(u, P^+)$ is associated with perfect information, this suggests that, if information were free, then any information structure short of perfect information would typically be inefficient. While interesting, this intuitive result does not appear particularly realistic. It suggests a need to explore in more depth the role of information cost.

Choice of Information

We now consider the case where information is costly. For that purpose, assume that obtaining and processing information requires the use of some of the m commodities. This is represented by the sets $X(P, \alpha)$ and $T(P, \alpha)$, where α is a parameter reflecting the information technology. Again, consider two information structures P and P', where P' is at least as fine as P. We assume that the information technology satisfies $X(P', \alpha) \subset X(P, \alpha)$ and $T(P', \alpha) \subset T(P, \alpha)$. This simply means that obtaining information involves information-gathering activities that use resources. Such resources are no longer available for other production, exchange, or consumption activities. In the case where $X(P', \alpha) = X(P, \alpha)$ and $T(P', \alpha) = T(P, \alpha)$ for all feasible P and P', this would correspond to the situation of "free information" discussed above. More realistically, we consider the case where information is costly.

One interesting possibility is the case where the information technology satisfies $X(P^+, \alpha) = \varnothing$ and $T(P^+, \alpha) = \varnothing$. This is a situation where obtaining perfect information is not feasible. This identifies *bounded rationality*, where there are severe limitations to obtaining and processing information. This could happen when the economic environment of the decision makers is particularly complex (as represented by a large number of states in S). Then, in the quest for perfect information, information-gathering activities could increase up to a point where there is no resource left for other activities. In addition, even if enough resources could be found to generate new information, the decision makers' ability to process this information may be constrained by the capacity of their brain to retain it and use it in an effective manner. Under such scenarios, bounded rationality implies that perfect information P^+ is not feasible. In this context, economic institutions and decision making processes must function under imperfect information.

This raises the question: How much information should be used in resource allocation? After introducing the feasible sets $X(P, \alpha)$ and $T(P, \alpha)$ in our earlier analysis, the answer can be obtained from the surplus $V(u, P, \alpha)$ in equation (3) (where $X = X(P, \alpha)$ and $T = T(P, \alpha)$). To attain Pareto efficiency, zero maximality suggests choosing the information structure P^* that satisfies

$$P^* \in \text{argmax}_P\{V(u, P, \alpha)\}, \tag{9}$$

where u is chosen such that $V(u, P^*, \alpha) = 0$. This corresponds to using an information structure P that maximizes the monetary surplus, and then redistributing this surplus entirely to the n agents. The optimization problem (9) involves trading off the benefits of better information (generating more refined state-dependent decision rules) with its cost (as measured by the opportunity cost of the information-gathering activities). In situations where the benefits of new information are greater than its cost, then the information is worth getting. Alternatively, if the cost of new information were greater than its benefits, neglecting this information would be efficient. This means that efficient decision rules would not depend on such information, corresponding to a world of incomplete contracts and/or incomplete risk markets.

It is of interest to analyze the role of the information technology. For that purpose, represent an improvement in information technology by a change from α to α' such that $X(P, \alpha) \subset X(P, \alpha')$ and $T(P, \alpha) \subset T(P, \alpha')$ for all P. This means that a move from α to α' tends to reduce the amount of resources used to obtain the information structure P. This would reduce the social cost of information. Since a change from α to α' tends to enlarge the feasible set, it follows from (3) that

$$V(u, P, \alpha') \geq V(u, P, \alpha).$$

This is an intuitive result: reducing information cost increases the distributable surplus and generates an outward shift in the Pareto utility frontier. The efficiency gains come from two sources: (1) lower information cost frees resources that can be used in other activities to improve consumer welfare; and (2) if lower cost contributes to better information,[3] this can improve efficiency through the use of more refined state-dependent decision rules. Note that this latter effect would apply to production and consumption as well as exchange activities. Then, reducing information cost would contribute to more refined contracts as well and stimulate the development of state-contingent exchange. Alternatively, increasing information costs would generate less refined state-dependent decision rules, i.e., less refined contracts and fewer state-dependent exchanges. This suggests that information cost can help explain why both contracts and risk markets are typically incomplete.

Note that our formulation is fairly general. First, from Propositions 1 and 2, the feasible sets $X(P, \alpha)$ and $T(P, \alpha)$ do not have to be convex for (3) or (9)

[3] This is the case if information behaves like a "non-inferior" good.

to identify a Pareto efficient allocation. This permits possible non-convexities in technology (e.g., due to increasing returns to scale) that may be inconsistent with the existence of competitive markets. In other words, our analysis does not require competitive markets to exist: it allows for non-market mechanisms used in resource allocation (e.g., contracts, government policy). Second, the sets $X(P, \alpha)$ and $T(P, \alpha)$ allow for interactions between learning and production/exchange activities. This includes the case of learning by doing, where information acquisition is joint with other economic activities. Third, the sets X and T allow for private goods and public goods, as well as externalities. This means that they can represent situations where information is a public good and/or it is associated with external effects across agents. Note that these external effects can be positive (e.g., new inventions) as well as negative (e.g., information about pollution). In such cases, P^* in equation (9) still identifies a Pareto efficient information structure.

The linkages between information and exchange are worth stressing. From equation (1), it is clear that refined state-dependent decisions are possible only under refined information. When applied to exchange, equation (1) implies that no agent can implement a trade that depends on information not available to him or her. There can be no markets for contracts that depend on information that is not available to someone in the economy.[4] More generally, net trade between two groups of agents can at most depend on the information that is common to both groups. Since risk markets/contracts require state-dependent decisions, their development requires the interested parties to be well informed. In addition, because common information is needed to trade state-dependent contracts, heterogeneity of information across agents has additional adverse effects on such markets/contracts. Radner (1968, pp. 46-49) illustrated this point through an example showing that imperfect and *asymmetric information* can yield the disappearance of all risk markets (with all efficient exchange becoming state-independent). This indicates the importance of both the amount and distribution of information in the development of risk markets as well as contracts. Insurance markets provide a well-known example where poor and asymmetric information can contribute to market failure.

If imperfect and asymmetric information adversely affects efficiency and the functioning of risk markets, this suggests a need to develop schemes that save on costly information. The issue is how to reduce the reliance on state-contingent markets while avoiding adverse effects on efficiency. Several approaches have been explored in the literature. One approach is to rely on security markets. Arrow (1964) has shown that, under weak separability conditions, sequential security markets can be substitutes to complete contingent claim markets. Another approach involves trading only on a subset of security markets with prices spanning the contingent claim prices q^{e^*}. Again, under weak separability, the resulting spanning equilibrium can be shown to

[4] For example, contracts and prices for delivery at a given time cannot depend on events that occur at a later date.

be Pareto efficient (Duffie 1988). More generally, the challenge of risk markets and contract design is to save on costly information while attempting to capture most of the benefits of state-contingent contracts. Yet, because of information cost and/or bounded rationality, incomplete risk markets and incomplete contracts prevail. In this context, the role of the courts can be seen as an institutional response to incomplete contracts: one of their objectives is to deal with contract disputes associated with contingencies not anticipated in incomplete contracts.

Finally, it is of interest to examine briefly the nature of information management. There are scenarios under which a centralized management of information can be efficient. They include situations where there are economies of scale in obtaining information, and where information is a public good or involves significant externalities. In this context, a decentralized decision making process would typically fail to be efficient without appropriate policy intervention. This suggests a role for government to generate the associated public goods, or to intervene in the management of the information externalities (whether they are positive or negative). It could involve public institutions, regulations, and/or Pigouvian taxes inducing each agent to choose efficient bundles. Examples include national defense (where information involves strong economies of scale), basic research (generating public goods), and pollution information (associated with significant externalities).

But there are also many scenarios suggesting a decentralized management of information. They can be motivated in part by the bounded rationality of centralized decision makers. In situations where the number of agents is large, the economic environment is complex (with a large number of states), and there is significant heterogeneity across agents, one can expect that centralized decision makers face severe limitations in obtaining information. This would provide an incentive to decentralize the decision making process. Again, the optimal form of economic organization would involve tradeoffs between information cost and the benefits of using more refined decision rules. When the benefits of information tend to be "local" and the costs of information are relatively low, decentralized decision making can be efficient. This applies to production and consumption, as well as exchange activities. In the context of exchange, both transaction costs and information costs need to be relatively low to motivate any transaction between agents (see above). If such costs are low enough to motivate a transaction, two possible mechanisms are relevant: a market mechanism, and contracts. Market mechanisms tend to arise when the number of potential market participants interested in exchanging standard commodities is relatively large. Alternatively, contracts develop when the number of parties involved is small and/or the object of exchange is nonstandard (as defined by quality, timing, etc.). In either case, good ability to obtain and process information seems crucial to support active risk markets and generate efficient risk allocation. Then, more refined state-contingent decision rules as well as risk redistribution away from the more risk-averse individuals can generate significant efficiency gains.

SOME SPECIAL CASES

Our analysis has presented a model of risk allocation under fairly general conditions. For example, the production sets X and T allow for private goods and public goods, as well as externalities. While Propositions 1–6 indicate when a Pareto efficient allocation can be supported by a competitive equilibrium, the nature of a decentralized decision making process remains to be investigated in more detail. As noted above, equation (6b) involves a joint choice of $x^e = \{x_i^e: i \in N\}$ for all agents. While this may be relevant in some cases (e.g., in the presence of economies of scope across agents), we want to investigate here the situation where the decision making process is decentralized.

For that purpose, consider the case where all production goods in X are private (i.e., when there is no public good and no externalities). In such a case, the feasible set can be written as $X = \{X_i: i \in N\}$, where X_i is the production set for the ith agent. This corresponds to a situation where technology is non-joint across agents. Then, equation (6b) becomes

$$x_i^{e*} \in \text{argmax}_x \{q_i^{e*} \cdot x_i^e: x_i^e \in X_i\}\ i \in N. \tag{6b$'$}$$

Assuming that it has a solution, equation (6b′) is a standard profit-maximizing condition for each agent i, $i \in N$. Several points are worth stressing. First, profit maximization applies for each agent even under risk and risk aversion. Second, the prices $q_i^{e*} = \{q_i^*(e_s): s \in S\}$ are in general state-contingent. Third, the solution x_i^{e*} of (6b′) generates efficient production decisions. Thus, in the absence of externalities and public goods, given state-dependent prices q_i^{e*}, a decentralized choice of production decisions made according to (6b′) is efficient. In this case, the price signals q^{e*} reflect the social scarcity of resources and efficiently guide the production decisions of each agent without a need for additional information.

In the case where (6b′) applies, our approach generates some well-known results as special cases. To see that, consider the situation where the ith agent receives no income transfers from others and makes no profit from trade. Then, combining (6a), (6b′), and (6d) gives

$$(x_i^{e*}, y_i^{e*}) \in \text{argmax}_{x,y} \{u_i(y_i^e): q_i^{e*} \cdot x_i^e - q_i^{e*} \cdot y_i^e = 0, x_i^e \in X_i, y_i^e \in Y_i\}. \tag{10}$$

Equation (10) is a general form of the household production model under risk. It is consistent with profit maximization (6b′) (or the associated cost-minimizing choice of inputs). It involves maximizing ex ante utility subject to a household budget constraint. The budget constraint states that consumption expenditures $(q_i^{e*} \cdot y_i^e)$ must equal the income generated by the agent production activities $(q_i^{e*} \cdot x_i^e)$. Equation (10) allows for price uncertainty (when the prices $q_i^e = \{q_i(e_s): s \in S\}$ vary across states), as well as production uncertainty (when the netputs $x_i^e = \{x_i(e_s): s \in S\}$ vary across states). It also allows for learning if different input decisions are based on different information sets. It includes as a special case the expected utility model (when $u_i(y_i^e)$ is a linear function of probabilities). For example, under the expected utility model and production certainty, (10) reduces to the classical model of firm behavior under price uncertainty discussed by Sandmo (1971). And in the absence of uncertainty, (10) becomes the standard household production model (see Deaton and Muellbauer 1980, Chapter 10).

CONCLUDING REMARKS

The linkages between Pareto efficiency and competitive equilibrium have been the subject of much research (e.g., Debreu 1959, Arrow and Debreu 1954). We investigated the adverse effects of transaction costs, risk, and information costs on the efficiency of resource allocation. We showed how trade and markets are more likely to develop under low transaction and information costs. Alternatively, high transaction costs can generate the absence of market for some commodities. And poor and asymmetric information contributes to incomplete risk markets and incomplete contracts. In this context, we developed general propositions establishing relationships between Pareto efficiency and competitive equilibrium under risk, information cost, and transaction costs.

Our analysis stresses several points. First, it relies on the benefit function and the concepts of zero-maximal equilibrium. As first proposed by Allais (1943, 1981), the concept of zero-maximality is intuitive and quite powerful in welfare and efficiency analysis. It provides a convenient basis for investigating the efficiency of resource allocation under risk. Second, since information and transaction costs can imply that some markets are inactive, this means that complete markets are not necessary for Pareto efficiency under risk and transaction costs. Third, in the presence of high transaction and information costs, the incentive for exchange is expected to be low. This suggests that competitive market structures (with a large number of traders) are unlikely to arise under high transaction and information costs. In such situations, non-market mechanisms (e.g., contracts, government regulations) are needed to obtain an efficient resource allocation. Fourth, such comments apply to risk markets as well. For example, insurance contracts are adversely affected by high information costs and significant information asymmetry. The challenge is to design contracts (e.g., insurance), markets (e.g., security and derivative markets), government programs (e.g., price support program), and social safety nets (e.g., bankruptcy protection, disaster payments) that implement state-contingent income transfers that can reduce the social cost of private risk-bearing and improve the efficiency of risk allocation. Finally, lowering transaction and information costs is expected to enhance the efficiency of contracts and markets, to stimulate the demand for exchange, and to improve risk allocation. These improvements follow from improved decision rules and more effective risk transfer mechanisms (through markets and contracts, as well as government policy). This suggests that policies targeted to lower transaction and information costs would play a crucial role in enhancing the efficiency of risk allocation.

We presented a conceptual framework to analyze efficiency and market equilibrium under externalities, transaction costs, risk, and risk aversion. The challenge is now to implement this framework to applied studies. It is hoped that this chapter will help generate useful insights in the investigation of agricultural and environmental economic issues, as well as in the analysis of incomplete markets and of incomplete contracts.

REFERENCES

Allais, M. 1943. *Traité d'Economie Pure* (Vol. 3). Paris: Imprimerie Nationale.

___. 1981. "La Théorie Générale des Surplus." *Economies et Sociétés*. Institut des Sciences Mathématiques et Economiques Appliquées.

Arrow, K.J. 1964. "The Role of Securities in the Optimal Allocation of Risk-Bearing" *Review of Economic Studies*. 31(April 1964): 91-96.

___. 1970. *Essays in the Theory of Risk-Bearing*. Amsterdam: North Holland Publishing Co.

Arrow, K.J., and G. Debreu. 1954. "Existence of an Equilibrium for a Competitive Economy." *Econometrica* 22: 265-290.

Berge, C. 1963. *Topological Spaces*. New York: Macmillan.

Chavas, J.-P., and Z. Bouamra-Mechemache. 2001. "Economic Efficiency and Market Equilibrium under Transaction Costs." Working paper, Department of Agricultural and Applied Economics, University of Wisconsin, Madison.

Deaton, A., and J. Muellbauer. 1980. *Economics and Consumer Behavior*. New York: Cambridge University Press.

Debreu, G. 1959. *Theory of Value*. New York: Wiley.

DeGroot, M.H. 1970. *Optimal Statistical Decisions*. New York: McGraw-Hill Book Co.

Diewert, W.E. 1974. "Applications of Duality Theory." In M.D. Intriligator and D.A. Kendrick, eds., *Frontiers of Quantitative Economics* (Vol. II). Amsterdam: North Holland.

Duffie, J.D. 1988. *Security Markets: Stochastic Models*. New York: Academic Press.

Foley, D.K. 1970. "Economic Equilibrium with Costly Marketing." *Journal of Economic Theory* 2: 276-291.

Hahn, F.H. 1971. "Equilibrium With Transaction Costs." *Econometrica* 39: 417-439.

Heller, W.P., and R.M. Starr. 1976. "Equilibrium with Non-Convex Transaction Costs: Monetary and Non-Monetary Economies." *Review of Economic Studies* 43: 195-215.

Luenberger, D.G. 1992a. "New Optimality Principles for Economic Efficiency and Equilibrium." *Journal of Optimization Theory and Applications* 75: 221-264.

___. 1992b. "Benefit Functions and Duality." *Journal of Mathematical Economics* 21: 461-481.

___. 1994. Luenberger, D.G. 1994. "Dual Pareto Efficiency." *Journal of Economic Theory* 62: 70-85.

___. 1995. *Microeconomic Theory*. New York: McGraw-Hill, Inc.

Machina, M.J. 1987. "Choice under Uncertainty: Problems Solved and Unsolved." *Journal of Economic Perspectives* 1: 121-154.

Mas-Colell, A., M.D. Whinston, and J. Green. 1995. *Microeconomic Theory.* New York: Oxford University Press.

Ostroy, J.M. 1980. "The No-Surplus Condition as a Characterization of Perfectly Competitive Equilibrium." *Journal of Economic Theory* 22: 183-207.

Pratt, J.W. 1964. "Risk Aversion in the Small and in the Large." *Econometrica* 32: 122-136.

Radner, R. 1968. "Competitive Equilibrium under Uncertainty." *Econometrica* 36: 31-58.

Sandmo, A. 1971. "On the Theory of the Competitive Firm under Price Uncertainty." *American Economic Review* 61: 65-73.

Starr, R.M. (ed.) 1989. *General Equilibrium Models of Monetary Economies.* Boston: Academic Press.

Takayama, A. 1985. *Mathematical Economics* (2nd edition). Cambridge: Cambridge University Press.

Chapter 9

CONTRACTS AND RISK IN AGRICULTURE: CONCEPTUAL AND EMPIRICAL FOUNDATIONS*

Brent Hueth and David A. Hennessy
Iowa State University

INTRODUCTION

Though farm-level risks, risk attitudes, and associated behavioral consequences have been extensively investigated by agricultural economists, comparatively little research has focused on why farmers face risk. It is evident that markets for agricultural production and price risk are generally incomplete. Even in commodities with well-developed markets for price futures, farmers still face considerable production risk; for example, attempts to establish private (unsubsidized) insurance markets for multiple-peril yield risks have universally failed (Knight and Coble 1997). The natural conclusion to draw from this and other supporting evidence is that farmers face risk because it is organizationally efficient. The costs of farm-level risk are well documented. It follows that there must be significant benefits associated with farmers' exposure to risk. Contract theory, and principal-agent theory in particular, provides a decision environment where these benefits are explicitly modeled, and hence where farmers' exposure to risk can be endogenized.[1]

Of course, endogenizing farmer exposure to risk has merit only if doing so extends our understanding of the policy and social welfare implications of risk. We believe it does, particularly for commodities where contract production plays a prominent role. As motivation for the remainder of this chapter, we offer three examples of where contract theory can deepen our understanding of behavioral and institutional responses to risk:

---

* The authors thank, without implicating, Marvin Hayenga, John Lawrence, and Ethan Ligon for many informative discussions on the topic of this chapter, and Robert Chambers for helpful comments on a near-final version of the text.

[1] The approach we discuss below emphasizes the role various types of intermediaries play in sharing risk with farmers, not farmers' ability to alter the risk they face. This is a rather limiting feature of the production technologies typically assumed in studies of contract design.

- Empirical research assessing risk attitudes and the effect of risk on farm-level decision making often uses market-level prices. Thus, contracts (and other risk-sharing instruments) that may mitigate farmers' exposure to risk are typically not explicitly recognized. Since the objects of study for contract theory include, among other things, equilibrium risk-sharing mechanisms, it seems likely that interest in such objects will spawn efforts that complement analyses based on market-level price data.

- Agricultural contracts have been around for some time, and yet there are few instances of agricultural economists providing normative input to private sector intermediaries and farmers on contract design.[2] Two specific design features where theory can serve as a useful guide include relative performance schemes (e.g., tournaments) and quality incentives. In both cases, quantifying risk and farm-level behavioral responses to risk – two topics that have received considerable attention from agricultural economists – can play a prominent role in designing optimal contracts.

- One suggested motivation for agricultural policy is price and income stabilization. But the government's ability to stabilize farm income is severely restricted by its limited capacity to contract directly on income. Contract theory provides a framework for directly incorporating this type of informational constraint. Doing so can perhaps lead to more efficient policy design, or at least to a better understanding of the implications of existing policies.

These three examples are certainly not exhaustive. One can envision other useful applications of contract theory to the study of risk in agricultural markets. For example, the theory has been employed extensively in a developing country context to study informal risk-sharing in agrarian economies (e.g., Bardhan 1989). Although many interesting informal institutions exist for risk sharing in developing countries, informational constraints and "interlinkages" are likely no less important in the many formal risk-sharing institutions observed in developed countries. In any case, we hope to have conveyed to the reader a sense of the potential payoff from developing a better understanding of the contract theory toolbox. In the following section, we present the theory and review its application to the study of risk in agriculture.

Before doing this, however, we provide an overview and analysis of agricultural contracting activities for two specific commodities. We do not attempt a comprehensive summary of contracting in agriculture, because doing so is well beyond the scope of this chapter.[3] Instead, we briefly review output contracting in

[2] Agricultural insurance contracts are a notable exception.

[3] Explicit ex ante contracts have long been prominent in U.S. land rental, crop insurance, and price insurance markets. They are also used extensively in crop seed and breeding livestock production, fruit and vegetable markets, and credit markets.

the U.S. pork and poultry sectors and use these two commodities to draw out relevant contract and organizational design issues that we study further in subsequent sections. There seems to be a sense among many agricultural economists, and among other industry observers, that trends in these two livestock sectors may presage explicit forms of ex ante contracting that will increasingly arise in other agricultural sectors. This evidence should therefore serve as further motivation for deepening our understanding of the factors affecting contract design.

CONTRACTING IN AGRICULTURAL MARKETS: TWO EXAMPLES

Poultry Contracting. Sometime near 1930, egg and poultry meat production ceased to be a fragmented sideline enterprise of many farmers and assumed a far more formal, task-specialized structure. During the Great Depression and through the 1950s, poultry production became increasingly concentrated, especially in southern U.S. states such as Georgia, Alabama, Arkansas, and Mississippi. Processors (or "integrators") in these areas expanded output, obtaining substantial scale economies, and for the most part adopted a contract-based procurement structure with farm operators.

Modern broiler production contracts are typically between a grower, who provides labor and housing, and a processor. The processor hatches the chicks and retains ownership of the birds through to slaughter. The processor also owns the feed, and determines the amount and composition of feed available to a flock. Growers are expected to follow a set of operating guidelines, while the processor also provides transportation to and from the production unit. Remuneration to the grower is determined by production performance relative to a peer group, and the typical remuneration scheme is not conditioned on any market price. Thus grower exposure to risk is highly delineated, being confined to on-farm production risk. Further protection is often provided by clauses to support a minimum guaranteed payment and/or a disaster payment scheme.

Pork Contracting. Commencing circa 1985, the U.S. hog sector began experiencing a transformation similar to that observed much earlier in poultry. Though initially slow to take hold, in recent years evidence of this transformation has strengthened. For example, as late as 1988 operations producing fewer than 1,000 head per year still accounted for 32 percent of production, while 50,000 or more head per year operations accounted for about 7 percent. By 1997, the respective numbers were 5 percent and 37 percent. The new larger units are overwhelmingly engaged in contract production. Between January 1999 and January 2001, U.S.-sold hogs that were priced through spot market purchases fell from 36 percent to 17 percent (Smith 2001). It is estimated that 27 percent of U.S. hog production was owned in 2001 by packers or companies with packing plants, with approximately 56 percent procured either through contracts with integrators or directly with growers. This new prominence in packer ownership is despite limitations, or even prohibitions, on such activities in many of the traditional Midwestern and Great Plains hog-producing states such as Iowa and Minnesota.

Hog contracts take two general forms: marketing contracts and production contracts. In the former the grower owns the hogs, whereas in the latter the integrator does. The latter is the most rapidly growing form of contract. A wide variety of production contract formats have been agreed upon, and some are quite similar to the typical poultry production contract. The grower usually owns the buildings and provides the labor, while it is often the contractor's responsibility to provide the feed. In contrast with poultry, hog production contracts typically provide incentives relative to absolute benchmarks such as target feed conversion efficiency, a target rate of daily gain, and a target mortality statistic.

Marketing contracts also take a variety of forms. In some, the contract seems to function as a guarantee of placement, and the price paid is tied to the cash or futures price at time of delivery. However, the grower may have to make commitments on the quantity and quality delivered, and on the time of delivery. In other contracts, the contractor assumes some risk by guaranteeing to share price losses (and gains) if a reference price falls outside a certain window. Effectively, the contractor writes over-the-counter puts at a low reference price and buys over-the-counter calls at a higher reference price. Cash-flow assistance (or "ledger") contracts are also popular. Here the contractor agrees to provide cash assistance to partially bridge the difference between a reference market price and a contract price. An account is maintained on the cumulation of these flows, and the grower must pay off any outstanding balance (generated when the reference market price is below the contract price) when the reference market price is above the contract price.

Implications. Although brief, our discussion of contracting activities in the pork and poultry sectors highlights a few important issues with respect to agricultural contracting and risk in general.[4] First, agricultural contracts do not lead to full insurance or even near complete protection from risk for producers. Both pork and poultry producers face considerable production and quality risk (e.g., Knoeber and Thurman 1995, Goodhue 2000, Martin 1997), while some pork producers (those using marketing contracts) also face price risk. Neither type of producer receives anything close to a "wage" contract, affirming the assertions made in our introduction.[5] Second, it is natural to make a connection between the

---

[4] More detailed discussion of the contracting and organizational attributes of these two sectors can be found in Martinez (1999), and Perry, Banker, and Green (1999).

[5] The contract design model we develop below does not incorporate dynamic incentives. For example, contractees can be dismissed after performing poorly or contract terms may be conditioned on past performance in some other way. Considering dynamic incentives makes apparent that virtually no contractees face a "wage" contract, i.e., compensation completely independent of performance.

existence of spot markets for live animals (in the case of pork) and contract growers' exposure to price risk. The formula-type marketing contracts observed in the pork industry, which condition growers' compensation on current spot market prices, apparently serve some important function. And such contracts are only feasible when spot markets exist. Spot markets clearly enhance the price-discovery process. Less is known about how spot markets improve contract efficiency, and about why contractors choose to use formula contracts instead of simply offering a contract price (thus protecting growers from price risk). Third, the evidence presented above suggests that asset ownership and the allocation of decision rights between growers and integrators have important consequences for economic efficiency. There has been very little empirically oriented research to address such issues in the context of contract farming.

These points represent just a few examples of the types of questions one can ask about contract design. Contract theories are not as mature as theories of market behavior, and this presents a challenge to those who are interested in answering practical kinds of questions. Although challenging, attempts to square contract theories with observed contract activity in agriculture clearly improve our understanding of agricultural production systems.

Markets vs. Contracts

Somewhat apart from contract design, which we consider in the next section, is the question of why many agricultural firms are moving from spot markets to ex ante contracts. A commonly adopted loose taxonomy of motives emphasizes risk-sharing and coordination benefits (Milgrom and Roberts 1992). In the case of pork production contracts, the risk-sharing benefits are likely small in comparison with other (coordination) benefits. Hog forward and futures markets have performed adequately as risk management instruments in the past, and pure marketing contracts (where integrators have little or no involvement in on-farm decisions) can accommodate longer-term risks. But what are these coordination benefits from superseding the market, and why are they emerging now rather than in the past? Streeter, Sonka, and Hudson (1991) suggest that changing consumer demands are interacting with information management technologies to alter the bounds of the agricultural firm; Hennessy (1996) points to difficulties in ensuring quality with market transactions. In any case, the market vs. contract question is a classic one that has been the subject of an enormous amount of research (e.g., Williamson 1985, Coase 1952). However, unless risk-sharing considerations are giving rise to contract production in agriculture, which seems not to be the case, it is a question that lies mostly outside the scope of this chapter. We therefore set this question aside and move on to a discussion of contract design.

MODELING CONTRACT DESIGN

In its simplest form, a contract represents an exchange of promises between two parties. Most contracts pertain to events that will occur at a non-trivial period in advance, and are typically motivated by expected gains from specialization. Contract design is straightforward when specialization occurs in a setting where both parties are perfectly informed about each others' abilities (and all other economically relevant "characteristics") and there is no uncertainty between the time that promises are made and outcomes are realized. Having agreed to some mutually acceptable division of post-contract surplus, the parties need only articulate a surplus-maximizing course of action. Uncertainty can be added without complicating matters so long as there are no post-contractual information asymmetries and parties can fully commit to their ex ante promises.

Of course, most economic environments where people use contracts are subject to violations of one or more of these ideal contracting conditions, and contract theory attempts a systematic treatment of efficient contract design when such violations are present. Adverse selection and moral hazard models consider pre- and post-contractual information asymmetries, respectively, and moral hazard models are further divided into models with hidden information (one agent observes post-contractual information that the other does not) and hidden actions (one party chooses a post-contractual action that the other party does not observe). A variety of modeling strategies have been employed to address contracting parties' ability to commit. One extreme is to suppose that contracts are always renegotiated to maximize ex post surplus. This is the approach employed in models where one or more economically relevant (and observable) variables may be "nonverifiable" (making contract enforcement impossible). The other extreme is to assume full commitment: even when there are interim opportunities to benefit from contract renegotiation, it is assumed that parties are bound to their initial agreement. One can also consider intermediate forms of commitment where, for example, the parties only renegotiate when it is in both parties' interest to do so.[6]

Our task in this chapter is to study agricultural contracting and risk. Thus, it is most natural to consider moral hazard models (where post-contractual uncertainty plays an important role).[7] We therefore limit ourselves to formal treatment of only this class of models.[8] We focus on the simple analytics of the moral hazard model, examine its predictions, consider various extensions of the static

[6] Though somewhat dated, Hart and Holmström (1987) still represents an excellent overview of contract theory. Recent texts include Salanié (1997) as well as Macho-Stadler and Perez-Castrillo (1997). Hart (1995) provides an overview of incomplete contract theory.

[7] Adverse selection models have been used by a number of authors to study farm policy design, but only in settings where post-contractual risk is ignored and farmers are assumed risk neutral. See Chambers (forthcoming) for an overview.

[8] Though possibly relevant, we choose to leave out formal discussion of incomplete contracting. Modeling of contract incompleteness has focused mostly on settings with risk-neutral parties, limiting application of the models to the study of risk.

single-agent model, and appraise applications of the model to the study of agricultural contracts.

The Principal-Agent Framework

The economic framework for analysis of moral hazard problems was pioneered by Mirrlees (1974), Ross (1973), Harris and Raviv (1979), Spence and Zeckhauser (1971), and Holmström (1979). In it, one party, the principal, contracts out a task to an agent. After the contract is signed, the agent receives a signal $\theta \in [\theta_0, \theta_1]$ and chooses an action a. Common ex ante beliefs on the distribution of θ are given by the probability density $p(\theta) > 0$. Ex post surplus is stochastic and is governed by a cumulative distribution function $F(\pi|a,\theta)$, conditioned on the agent's action and signal.[9] Among other things, the contract divides the surplus π between the principal and the agent. The agent's (reduced-form) production technology $F(\cdot)$ is assumed twice continuously differentiable in π and a with $F_\pi(\pi|a,\theta) = f(\pi|a,\theta) > 0$ for all $\pi \in [\pi_0,\pi_1]$. A contract commits $w(\pi,\theta)$ to the risk-averse agent, and the residual π-$w(\pi,\theta)$ accrues to the (possibly) risk-averse principal. The principal maximizes her expected utility of residual surplus by choosing the structure of $w(\pi,\theta)$.

The principal may be constrained in several ways. First, the agent may elect not to accept the contract. An alternative best occupation is available to the agent, yielding expected utility level $\bar{u}$. Second, the principal may be constrained in contracting on or even observing the signal θ. And third, the principal may not be able to contract on the agent's action a.

This setup describes the economic environment in which agricultural contracting occurs reasonably well: farm production is stochastic; farmers engage in a wide variety of activities (e.g., forward contracting, product diversification, "excess" nitrogen use) consistent with risk aversion; and, given the organizational structure of most farm operations (relatively small, autonomous, and owner-operated production units), farmers almost certainly have better information about their production environment than do their intermediaries. Also, we could include additional production uncertainty that is always observed by both parties

[9] In what follows, we treat the agent's production environment as a "black box" and focus on contract design under various informational asymmetries. Chambers and Quiggin (2000) refer to this black box (a discrete version of $F[\pi|a,\theta]$) as the "outcome-state" representation of an agent's production technology and develop an alternative (state-contingent) framework for modeling principal-agent problems where the relationship between states of nature, the agent's action, and production outcomes is explicitly modeled. This framework makes clear that what is often referred to as a "production technology" in contract design models only partially characterizes the agent's underlying production environment. A challenging empirical question is whether one can ever recover information about state-contingent production possibilities and behavior when states of nature and agent actions are unobserved (the case most often considered in principal-agent models).

without changing any of the discussion that follows. Each party would simply "integrate out" this uncertainty when computing their respective expected utilities.

In what follows, we study each of the three aforementioned constraints in sequence. Imposing the first constraint allows us to characterize the full-information Pareto optimum for the parties. This leads to the familiar result that a risk-neutral intermediary will fully insure a risk-averse agent, which of course is empirically inconsistent with the structure of most agricultural contracts. Adding informational constraints leads to surplus-contingent (and hence risky for the agent) equilibrium contracts.

Full-Information Contract Design

Whenever it is observed by both parties, θ will condition the sharing arrangement $w(\pi,\theta)$ and the agent's action $a(\theta)$. Realization of surplus π results in utility $V[\pi\text{-}w(\pi,\theta)]$ for the principal and $U\{w(\pi,\theta)\text{-}c[a(\theta)]\}$ for the agent, where $c[a(\theta)]$ represents the strictly increasing and convex monetary cost of the agent's action. Under full information, the arguments and schedule available to the principal are denoted by the set $S_p = [a(\theta),w(\pi,\theta)]$ (S_p represents the full-information "contract"). The agent's action is fully dictated by the principal, with an optimal contract solving

$$\max_{S_p} \int V[\pi - w(\pi,\theta)]\, dF[\pi \,|\, a(\theta),\theta] \tag{1}$$

subject to

$$\int U\{w(\pi,\theta) - c[a(\theta)]\}\, f[\pi \,|\, a(\theta),\theta]\, p(\theta) d\pi d\theta \geq \bar{u} \ .$$

Note that the contract is offered ex ante (before θ is realized), but the principal optimizes ex post. The agent's action is chosen to equate expected marginal benefits and costs at each θ:

$$\int \big[V[\pi - w(\pi,\theta)] + \lambda U\{w(\pi,\theta) - c[a(\theta)]\}\big] dF_a[\pi \,|\, a(\theta),\theta] =$$

$$\lambda \int U'\{w(\pi,\theta) - c[a(\theta)]\} c_a[a(\theta)]\, dF[\pi \,|\, a(\theta),\theta] \ ,$$

where λ is the Lagrange multiplier for the agent's participation constraint in problem (1), and where subscripts represent partial derivatives with respect to the indicated argument.

For each π and θ, choice of an optimal surplus-sharing arrangement solves

$$\lambda = \frac{V'[\pi - w(\pi,\theta)]}{U'\{w(\pi,\theta) - c[a(\theta)]\}}. \qquad (2)$$

Because λ is constant (for given $\bar{u}$), it turns out that an optimal payment schedule provides the agent with full insurance if risk-neutral intermediation is available. This is easily verified by observing that we may write $V'(\cdot) = 1$ in such a case, and that $w(\pi,\theta) - c[a(\theta)]$ is therefore constant. Furthermore, it is not difficult to verify that, in this full information environment, and for an interior solution, the optimal action taken by the agent will maximize expected net revenue.

Though there is considerable heterogeneity in the structure of agricultural markets, one constant is farmer exposure to risk. Predictions of the full information model with a risk-neutral principal therefore do not accord well with empirical observation. It seems reasonable to suppose that risk-neutral intermediation is available in at least *some* agricultural markets (and that farmers are risk averse), and yet farmers face considerable price and production risk in their operations. Of course, risk aversion on the part of principals can rationalize growers' exposure to risk, though the type of risk sharing that is observed seems inconsistent with this explanation (Hueth and Ligon 1999). However, irrespective of whether intermediaries are risk neutral or risk averse, information asymmetries may lead to contracts that are contingent on realized surplus. In the next two sections we consider two such asymmetries. We begin with hidden information.

Hidden Information

Suppose that only the agent observes θ, and rule out any other sources of uncertainty (i.e., assume $F(\cdot)$ is degenerate). To simplify matters slightly, we also assume that profits are given by $\pi(a,\theta)$, strictly increasing in a and θ and with $\pi_{a\theta}(a\ \theta) > 0$. For any θ, define $\tilde{a}(\pi,\theta)$ as the state-conditioned action that induces π, and let $\tilde{c}(\pi,\theta) \equiv c[\tilde{a}(\pi,\theta)]$ represent the cost of producing π, given θ. Because a can be imputed from observation of π for given θ, there is no loss of generality in supposing that the agent chooses π directly.

Since an efficient production plan requires information about θ, the principal may wish to ask that the agent report this information once it is learned. Unfortunately, the agent will not always want to report truthfully. Imagine, for example, that the principal offers the agent complete risk protection, as in the full-information contract of the previous section, and that the agent observes $\theta > \theta_0$. The agent, knowing that a truthful report will lead the principal to ask for an action greater than $a(\theta_0)$, will prefer to report θ_0. This example suggests that the principal, if she wishes to obtain accurate information about θ, must offer the agent a contract in which truthful reporting is an equilibrium strategy. Moreover, revelation principle arguments (e.g., Myerson 1989 or Harris and Townsend

1985) can be used to verify that, whatever the interests of the principal, a feasible allocation (state-contingent production plan and payment) can always be implemented with a truthful direct revelation mechanism where the agent is asked to report θ in return for a contract $[\pi(\theta), w(\theta)]$ that induces truthful reporting.

Thus, we can restrict attention to mechanisms where the principal chooses the agent's state-contingent action $\pi(\theta)$ and payment $w(\theta)$ to solve an appropriately modified version (to reflect ex ante contract choice) of (1) with the added constraint

$$\theta \in \arg\max_{\hat{\theta}} U\{w(\hat{\theta}) - \tilde{c}[\pi(\hat{\theta}), \theta]\}. \tag{3}$$

It turns out that with a risk-averse agent, it is generally quite difficult to completely characterize the solution to this problem.[10] Nevertheless, from (3) we observe that any incentive-compatible (and differentiable) contract must satisfy

$$w'(\theta) = \tilde{c}_{\pi}[\pi(\theta), \theta]\, \pi'(\theta).$$

Thus, if the principal's optimal mechanism satisfies $\pi'(\theta) \geq 0$ (the agent should work harder when his action is more productive), then the agent's wage schedule must also be increasing. An indirect mechanism for implementing such an outcome is some (possibly nonlinear) reward schedule $\tilde{w}(\pi)$ that offers the agent a share of the surplus π. In this sense, the agent faces risk. He enters a contract not knowing the value of some stochastic variable that will affect his production environment. Once this information is observed, joint surplus maximization requires that it be used appropriately; and this requires rewarding the agent with a higher payment in states where surplus maximization demands a more costly action.

Although useful as a benchmark, this information structure does not fit an agricultural context particularly well. Hidden information is certainly relevant in a typical contract relationship. For example, a farmer might know something (that is ex ante unobserved) about the opportunity cost of his time that is difficult for an intermediary to observe; and efficient use of the farmer's time should exploit this information. However, it seems extreme to assume that the farmer's action can be imputed from observation of π (and truthful reporting of θ). A typical farm production cycle for most any commodity is characterized by a variety of stochastic events. In the model developed so far, introducing additional sources of uncertainty (or multiple unobserved actions) makes it impossible for the principal to contract directly on the agent's action. Having observed the outcome π, she may no longer be able to infer a – the agent's action becomes "hidden." In the next section, we analyze this type of constraint formally. For simplicity, we first consider the extreme case in which there is no hidden information. In the

[10] Guesnerie and Laffont (1984) provide a comprehensive treatment.

subsequent section we briefly consider the hybrid case with both hidden information and a hidden action.

Hidden Actions

Relative to the full-information environment, suppose that the principal can no longer include a in her contract. She can recommend some a, but the recommendation will be followed only if, given the sharing arrangement $w(\pi)$, the agent maximizes his expected utility by doing so. Mathematically, this can be expressed in the form of an incentive compatibility constraint:

$$a \in \arg\max \int U\,[w(\pi) - c(a)]\, dF(\pi \mid a) \ . \tag{4}$$

Assuming the principal is risk neutral, point-wise optimization of problem (1) with constraint (4) appended yields an optimal agent remuneration schedule satisfying[11]

$$\frac{1}{U'[w(\pi) - c(a)]} = \lambda + \mu \{ r(\pi \mid a) + \eta [w(\pi) - c(a)]\, c_a(a) \}, \tag{5}$$

where μ is the non-negative multiplier associated with the agent's incentive compatibility constraint (4), $\eta(\cdot)$ is the agent's Arrow-Pratt measure of absolute risk aversion, and $r(\pi|a) = f_a(\pi|a)/f(\pi|a)$. Because the right-hand side of (5) includes a term containing π, the principal's optimal sharing rule will depend on this outcome, and the agent will face some degree of risk. Again, the farmer is endogenously exposed to risk.

Under constant absolute risk aversion, it is apparent that when the function $r(\pi|a)$ is nondecreasing in π, $w(\pi)$ is also nondecreasing. That is, monotonicity characterizes the optimal agent remuneration schedule. Monotonicity of $r(\pi|a)$ has a rather intuitive interpretation. Choose any two supported observations on surplus adhering to $\pi^l \leq \pi^h$, and choose any two admissible actions adhering to $a^l \leq a^h$. Then the likelihood of the high observation on surplus relative to the low observation is monotone increasing in the level of the action: $f(\pi^h|a^l)/f(\pi^l|a^l) \leq f(\pi^h|a^h)/f(\pi^l|a^h)$. It would be of some interest to test the validity of this condition for specific noncontractible actions and surplus measures. Test statistics on the appropriateness of a monotone likelihood ratio have been developed. For exam-

[11] This is the so-called "first-order approach" for incorporating the incentive constraint (4) into the principal's optimization. Although analytically convenient, it is well recognized that this approach is valid only for a restricted class of problems; simply put, second-order conditions on constraint (4) may not adhere (see Jewitt 1988 for restrictions on $U(\cdot)$ and $F(\cdot)$ that ensure the appropriate concavity). More generally, computational methods can be used to ensure global optimality in (4) (e.g., Phelan and Townsend 1991).

ple, Dardanoni and Forcina (1998) establish a nonparametric test with tight, and easily calculated, bounds on the asymptotic distribution of the test statistic.

Two extensions of the hidden action model have important implications for the study of risk in agricultural contracts. First, Kim (1995) develops a criteria for ranking the informational efficiency of performance measures based on the distribution of the function $r(\pi|a)$. In particular, any modification of the technology governing π (or inclusion of additional performance measure(s), i.e., signals, jointly distributed with π) that generates a mean-preserving spread in the signal-conditioned distributions of $r(\pi|a)$ results in a more efficient information system. Various quality attributes are often included as performance measures in agricultural contracts (examples were provided in our discussion of poultry and pork contracts). Kim's result can perhaps serve as a guide for empirically evaluating the welfare consequences of alternative contract structures.

Second, Holmström's (1982) analysis extends the basic hidden action model to a setting with multiple agents. It is supposed that a vector of signals concerning performance can be contracted upon that may include independent measures of each agent's output. When the multivariate distribution of these signals is possessed of a strong form of separability in actions and signals, this separability provides a vector of sufficient statistics, one for each agent, and contracts should be written on these statistics alone. In the case of linear stochastic production relations and multivariate normal observable outputs, one for each agent, Holmström (1982) shows that a sufficient statistic for an agent is the difference between a weighted average output and the agent's output. As we indicated previously (and as we will see in more detail below), something resembling such a sufficient statistic shows up in poultry contracts.

Hidden Information and Hidden Actions

As one might imagine, analyzing the hybrid case with both hidden information and hidden actions complicates analysis somewhat relative to either of the extreme cases presented thus far. As earlier, contracts cannot be conditioned on the state θ (which, recall, for simplicity we suppressed in the previous section). Thus, for any wage schedule $w(\pi)$ offered by the principal, the agent will choose his unobserved action such that

$$a(\theta) \in \arg\max \int U\left[w(\pi) - c(a)\right] dF(\pi \mid a, \theta),$$

resulting in a state-conditioned equilibrium action $a(\theta)$ and Lagrangean multiplier $\mu(\theta)$. It is straightforward to verify (see Holmström 1979) that the addition of this constraint to problem (1) (again appropriately adjusted to reflect ex ante contract choice) results in a modified version of $r(\pi|a)$ in (5) of the form

$$\hat{r}(\pi|a) = \frac{\int \mu(\theta) f_a(\pi|a,\theta) p(\theta)\, d\theta}{\int f(\pi|a,\theta)\, p(\theta)\, d\theta} .$$

With this modification, the intuition for optimal contract design is much like in the previous section, replacing $r(\pi|a)$ with $\hat{r}(\pi|a)$.

The theory thus far presented has been applied by researchers to study agricultural contracts in a variety of contexts. In the following section, we briefly review some of these applications, focusing in particular on land and output contracts in developed country agriculture. The model has also been used to study agricultural insurance contracts, but this work has been summarized elsewhere (e.g., Chambers, forthcoming; Hueth and Furtan 1994).

APPLICATIONS OF THE THEORY AND CONSIDERATION OF RISK

Several applications can demonstrate how risks are addressed by contracting.

Sharecropping

Textbook treatments of the principal-agent model often use sharecropping between a landlord and tenant as a motivating example. Sharecropping contracts have also been studied empirically in a variety of settings. Otsuka, Chuma, and Hayami (1992) provide a comprehensive survey of theory and evidence for share contracts in agrarian economies. Allen and Lueck (e.g., 1999) have examined whether risk sharing is an important part of land rental contracts in U.S. agriculture. Somewhat surprisingly, they find that an increase in exogenous variability across commodities (measured as an increase in yield variability) reduces the likelihood that landlords and tenants will choose crop share contracts over cash rent contracts. They interpret this result as an indication that risk-sharing motives are relatively unimportant in the design of crop share arrangements.

Given the model presented in the previous section, two comments can be made on this interpretation. First, it is important to note that the model *does not* provide an unambiguous prediction regarding the effect of an increase in exogenous risk on contract choice. To see this, suppose we take as a maintained hypothesis that farmers and landlords are equally risk averse.[12] It follows that, for a particular landlord and farmer, a cash rental agreement will be observed only if there is some kind of exogenous (to the model) cost associated with a share

[12] For sharecropping in U.S. agriculture, this seems reasonable since landowners are often retired farmers with socioeconomic attributes similar to their tenants'.

contract. Suppose there is, and that a cash rental contract is optimal for this pair.[13] Now consider an increase in exogenous risk associated with the crop that will be produced on this land. Does this make use of the share contract more attractive? The answer depends on how expected total surplus under each arrangement changes in response to the increase. Expected surplus for the farmer operating under a cash rental agreement goes down. And presumably he will pay the landlord a lower rent as a result. But expected surplus under the share arrangement also goes down. Because an exogenous increase in risk reduces expected surplus under both contract structures, its effect on contract choice is in general ambiguous.

The second point we can make is to note that studying contract choice amounts to taking existing contract structures as exogenous. Thus, even if risk-sharing benefits do not seem to drive contract *choice*, they may nevertheless help explain the structure of observed contracts. To illustrate, let us refine the maintained hypothesis of our example. In particular, suppose that in addition to both parties being risk averse, we assume that the landlord and farmer operate in a full-information setting. For example, suppose that the landlord lives on his land and fully observes the farmer's actions. Assume further that both parties have log preferences (and that $w(\pi) - c(a) > 0$ for all a and π). From equation (2), it is then straightforward to verify that an equilibrium sharing arrangement satisfies

$$w(\pi, \theta) = \frac{\lambda}{1+\lambda}\pi + \frac{1}{1+\lambda}c[a(\theta)].$$

Thus, if we assume full information and identical (logarithmic) preferences for each party, an equilibrium contract awards fixed shares of total surplus and production costs. It is noteworthy that some actual crop share contracts are similarly structured. In any case, this example demonstrates how risk aversion might explain a particular contract design, even though risk sharing by itself may be dominated by other forces (e.g., various sorts of "transaction costs" like the one mentioned in footnote 13), in the explanation of contract choice.

Price and Quality Risk

Grower contracts for fresh produce typically make a grower's payment contingent on prices that are downstream from the grower's marketing agent. For example, among other things, a grower might be promised a fraction of the revenues received from sale of his produce in some downstream market. In such an environment, intermediaries have an incentive to under-report actual prices

[13] For example, U.S. labor law counts participation in a share contract as a form of employment. Thus, a landlord who is in retirement may be unable to collect Social Security benefits if he uses a share contract. Though not particularly interesting from an economic point of view, as a practical matter such a factor may prove decisive.

received, and there is abundant evidence that in fact they sometimes do.[14] Such activities point to significant costs of using price-contingent contracts, in addition to the costs associated with grower exposure to price risk. What are the benefits? Agency theory suggests contracts should be contingent only on variables that allow a principal to more precisely control an agent's actions. If idiosyncratic variation in quality shows up in downstream prices – variation that is not observed by a grower's marketing agent – then market prices potentially convey information that can be used to achieve this end.

Hueth and Ligon (1999) investigate this issue and find evidence that downstream prices do provide information about idiosyncratic variation in quality. This primarily occurs via deductions that are assessed by downstream buyers on negotiated prices between these buyers and farmers' marketing agents (i.e., contractors). A marketing agent negotiates prices with buyers based on her assessment of produce quality. If, upon delivery and inspection, some buyer believes the marketer overestimated quality, a price deduction (relative to the negotiated price) is assessed. When growers face the risk of such deductions, they have greater incentive to ensure quality. Note, however, that market prices also respond to aggregate supply and demand conditions, and there seems little reason to expose growers to these risks. This issue is taken up in the next section when discussing the role of markets in providing relative performance incentives.

Relative Performance Evaluation

Various aspects of relative performance incentives have been studied in the context of agricultural contracts. We briefly review some of these studies.

Tournaments in Poultry Production. As reported earlier, contracts are central to the U.S. poultry production system. Further, an important feature of broiler contracts is relative performance evaluation of grower outcomes. Specifically, for a typical broiler contract let the jth grower's reward per live pound delivered to the principal be p^j, while x^j and q^j are the respective amounts of feed supplied and live output delivered. Then, for the set of growers $N = \{1,2,...,n\}$ the typical unit reward function is

$$p^j = b_0 + b_1\left[\frac{1}{n}\sum_{k\in N}\frac{x^k}{q^k} - \frac{x^j}{q^j}\right],\ b_0 > 0,\ b_1 > 0. \qquad (6)$$

[14] In California, for example, agricultural contracting and marketing activities are regulated by the Market Enforcement Branch of the California Department of Food and Agriculture. An important task of the Market Enforcement Branch is to adjudicate on alleged cases of this type of fraud.

That is, the reward structure is linear in feed conversion efficiency relative to cohort feed conversion efficiency. Among the more important features of (6) is that it substantially reduces the effects of common shocks to feed conversion efficiencies on grower compensation, because the grower's performance is compared with the mean performance of group N. To facilitate this mitigation of common shocks, the group is generally chosen so that flocks are grown under relatively homogeneous conditions (e.g., genetics, feed, location, time period). Knoeber and Thurman (1995) measure the extent to which the reward structure in (6) protects growers from common production risk using firm-level data for a single poultry integrator. They find that relative performance evaluation reduces total production risk by half.

Relative vs. Absolute Performance. In contrast to poultry, contracts for pork producers normally use linear *absolute* performance; i.e., replace the mean in equation (6) with a fixed number (the "benchmark"). Given the broad technical similarities between the enterprises, this distinction is somewhat of a puzzle. Tsoulouhas and Vukina (1999) suggest that the reason is due to the risk of integrator bankruptcy. Hog integrators are less well established than poultry integrators, and are highly leveraged because of rapid expansion during the 1990s. Thus, they may be more likely to go bankrupt. Linear relative performance schemes shield growers from a systemic fall in productivity. An absolute linear performance scheme, as is often in place in hog contracts, reduces the risk of bankruptcy because the systemic production risk is assumed by the grower. Growers with large relationship-specific investments in the form of buildings may be willing to assume systemic production risk, rather than increase the risk of bankruptcy. As the hog industry matures, and bankruptcy concerns diminish, this model predicts that relative performance schemes will become more common.

Grower Heterogeneity and Relative Performance Evaluation. Holmström's (1982) results on the design of relative performance schemes are derived in a setting with homogeneous agents. Goodhue (2000) introduces grower heterogeneity and notes that, even when production is linear in effort with additive, normally distributed production shocks, optimal schemes generally do not depend on average output as in (6). Tsoulouhas and Vukina (2000) consider a related issue. Poultry growers often complain that tournament schemes are unfair because the set of growers in a group continually changes. For a given set of production outcomes, group composition can substantially affect payment outcomes. Thus, in addition to traditional price and production risk, tournament schemes subject growers to "group composition risk." Holmström's 1982 analysis clearly identifies the benefits of relative performance evaluation. The analyses by Goodhue (2000) and Tsoulouhas and Vukina (2000) suggest that agent heterogeneity introduces two important costs: Optimal relative performance schemes become more complex and therefore more costly to implement, and unobserved agent heterogeneity introduces a new source of risk that can offset risk reductions associated with relative performance evaluation.

Markets and Relative Performance Evaluation. An often observed feature of agricultural contracts is the use of some market price as a reference point from which compensation is determined. For example, as noted earlier, some hog contracts pay on a formula that uses a spot price reference. Similarly, production contracts for specialty grains (e.g., high oil corn) pay growers a premium on current market prices. Why use market prices in this manner? If anything, it would seem that doing so *increases* risk relative to a contract based on some mutually agreed upon ex ante price. Again, going back to the model presented earlier, a market price should be used if it helps control growers' actions. Hueth and Ligon (2001) point out that market prices provide relative performance incentives, where the comparison group for a single grower is the set of all growers for a market. The equilibrium marginal incentive to exert effort (i.e., equilibrium output price) is highest for individuals who perform well in bad (high price) states. This amounts to a bonus for good performance relative to the average. Incentives of this sort are likely to be most useful when market price uncertainty is due primarily to common production uncertainty, as tends to be the case for agricultural outputs.

CONTRACT INCOMPLETENESS, UNFORESEEN CONTINGENCIES, AND RISK

Agricultural contracts are controversial. Popular press (and some academic) accounts of contract activities in agriculture lament the lack of autonomy growers have in such contracts, and claim that growers are coerced into making specific investments (that presumably later lead to some form of holdup). The model presented above seems to have nothing whatsoever to say about such issues. In a sense, the hidden action model affords the agent greater autonomy, relative to the full information model, because his unobserved action is not specified in the contract. This is a stretch, however, since autonomy is most naturally associated with authority and with one's right to make future choices (that are not specified ex ante) as one pleases. Similarly, there are no "specific investments" (physical assets, or otherwise) in the principal-agent model, and contract renegotiation is ruled out; by construction there are no opportunities for holdup. Thus, in attempting to evaluate concerns that many people seem to have with agricultural contracts, one needs to enrich the principal-agent model to include these issues, or turn to another class of models altogether. Theoretical research is active on both fronts (e.g., Segal and Whinston 2000, Hart 1995, Tirole 1999), though still somewhat incomplete.

We briefly describe an example of an application of notions from incomplete contract theory to give a sense for how the model can be applied to an agricultural context. Consider the harvest timing decision of a grower under contract. For a given set of performance incentives, and for given decisions made during the course of the growing season, a farmer chooses the harvest date that maximizes his return. An intermediary, who possibly faces a capacity constraint (e.g., because the crops of many contract growers happen to have matured on more or less the same date), may wish to alter the "optimal" harvest dates of one or more of its growers. If the farmer is explicitly granted authority over harvest timing, then any harvest adjustments from the farmer's optimal date will require that the farmer be compensated for associated losses. In practice, compensating for losses requires a fairly precise method of determining the magnitude of yield and quality reductions associated with suboptimal harvest timing. In most contexts, such a method does not exist.[15] Although both parties to a contract may be able to provide reasonable *estimates* of losses, it is unlikely that such estimates will be verifiable. As a result, whenever the parties' assessments do not coincide and an adjustment is desired, some form of bargaining will determine the actual compensation paid.

This example represents a specific instance of contract incompleteness, and provides motivation for why the allocation of decision rights can be important in a contract relationship. Even though the optimal harvest timing decision may be noncontractible (due to unforeseen contingencies), authority over this decision can be, and often is, specified in an ex ante contract. In this example, awarding the farmer the right to choose his preferred harvest date has both benefits and costs. The farmer is motivated to work hard, but ignores the consequences of his decisions on the intermediary. Granting authority over harvest timing to the intermediary reverses these incentives to some extent. In our discussion of contract theory we chose to ignore incomplete contracting models. Nevertheless, examples such as the one just described can perhaps be fruitfully investigated with this theory.

POLICY AND AGRICULTURAL CONTRACTS

As noted earlier, the move from open spot market trading toward contract-based hog production is almost complete in the United States. There are indications that similar trends may emerge in other commodity sectors. Whatever the relative weightings on the motives for these changes, it is clear that the commoditized open market paradigm is no longer the only relevant stylization of market behavior in agriculture.

[15] The authors are aware of at least one agricultural contract in which harvest timing *is* contractible, and where a specialized instrument is used to calibrate and measure losses associated with a "suboptimal" harvest date. This contract is an exception, however; more often, ex post bargaining is used to adjust harvest dates.

We note two likely consequences of this fact. First, traditional mechanisms for delivering farm income support may become increasingly difficult to implement. For example, existing loan deficiency payment rates for commodity corn and soybeans are determined using local spot prices ("posted county prices"), and these are established using regional market prices. Also, federally subsidized multiple peril crop insurance relies on farmers' actual yield histories for commodity production to establish individual rates. Verifiable yield histories can substantially reduce the cost of providing insurance, and may be difficult to obtain as farmers switch to production of specialty grain products. Given these problems, income support policies targeted to production practices (e.g., cost sharing for best management practices and green payments) may become comparatively more viable.

Additionally, contract activities create new regulatory concerns. Federal lawmakers have responded, and it looks increasingly likely that many state legislatures will respond too. At the federal level, price reporting by large cattle, hog, and lamb packers has become mandatory commencing April 2001. The intent is to increase transparency in quality and volume premia. At the state level, sixteen State Attorneys General have drafted a template "Producer Protection Act" that focuses on grower-integrator relations (Iowa Attorney General 2001). Among the clauses that are suggested for inclusion are requirements that contracts be "in plain language and contain disclosure of material risks," that farmers be protected against strategic threats of contract termination, and that confidentiality clauses and tournament schemes be prohibited.

The implications of enacting a confidentiality clause are unclear. Many of those who advocate passage of the Producer Protection Act are also concerned about downstream collusion. A ban on nondisclosure clauses or, in the case of the federal legislation, mandated public disclosure of prices, could strengthen the potential for tacit collusion. Albæk, Mølgaard, and Overgaard (1997) have studied the effects of the 1993 decision by Denmark's antitrust agency to collect and disclose prices firm-by-firm for ready-mixed concrete in three regional markets. The decision was motivated by the Competition Act of 1990, which required the government to promote transparency in competition. Over time, conditions evolved so that prices should have declined. And yet they increased. Some consensus emerged that the practice of publishing prices was not pro-competitive. By late 1996, the agency discontinued publishing these data.

DATA NEEDS AND EMPIRICAL APPROACHES TO STUDYING CONTRACTS

Empirical study of agricultural contracts presents researchers with new methodological and data collection challenges, and also many opportunities. In this section, we briefly highlight some of these issues and suggest possible directions for future research.[16]

Contract theory focuses on the design of incentive systems. The key behaviors that drive outcomes are unobserved, and can only be inferred from observation of contracts and contract outcomes. Empirically, both ex ante contracts *and* ex post contract outcomes carry information about the economic environment under study. Moreover, in some sense, a contract carries more information than a single agent's production or consumption decision. For example, although a single production decision (the object of study for producer theory) is generally not sufficient for any kind of meaningful inference, a single contract (the object of study for contract theory) can be. These observations suggest that empirical study of contracts can, by focusing attention on a different set of observables, complement previous empirical approaches to the study of risk.

More specifically, farmer decision making has been extensively studied at the market level. Unfortunately, contracts do not easily aggregate and any study of producer decision making in the presence of contract incentives will likely require contract-level data. Of course, such data is normally collected by the private sector intermediaries who contract with farmers.[17] Such data represents a potentially rich source of information for empirical study of contracts and farm-level behavior. However, given the proprietary nature of contract-related data, gaining access to such data will likely require developing empirical methodologies for evaluating design issues that these intermediaries find useful.

Additionally, insights can be gained from observation of cross sectional variation on contract structures. For example, it would be useful to identify and test hypotheses about how exogenous variation in contracting environments affect equilibrium contract attributes (e.g., incentives structures, asset ownership, authority allocations). Indeed, agricultural markets seem ideally suited to this type of exercise. They are sufficiently alike that interesting comparisons can be made across commodities and production regions, and yet there are sufficient differences to allow for useful "natural experiments."

In summary, this brief discussion suggests specific data needs for further study of agricultural contracts: detailed and comprehensive description of contract structures for specific commodities; contract outcomes under individual contracts; and descriptive evidence on variation in the terms and structure of agricultural contracts (and appropriate covariates) across commodities and contracting environments. Unfortunately, none of these data are presently available in the way that many other kinds of data on agricultural markets are. Policy concerns such as those expressed in the previous section, and an interest in addressing them intelligently, should motivate collection of such data.

---

[16] Also, see Goodhue (1999), who provides a discussion of similar issues.

[17] In some cases, data collection occurs at an industry level through a third party. The services provided by eTomato (www.etomato.agris.com) for processing tomato producers, by e-markets (www.e-markets.com) for U.S. pork producers, and by Vantagepoint (www.vantagepoint.com) for U.S. grain producers are examples.

CONCLUSIONS

In the introduction to this chapter, we claimed that contract theory had much to add to discussions of economic behavior in the presence of risk. In presenting the principal-agent model, and in discussing applications of the model to the study of agricultural contracts, we hope to have validated this claim. Contract theory allows for analyses of risk that include institutional responses, in addition to the behavioral responses of individuals. Continuing developments in contract theory, and the growing importance of contract relationships in agriculture, should lead to an increasing number of efforts to apply the theory to the study of agricultural contracts. Because uncertainty is a key ingredient of any contract relationship, such research should also improve our collective understanding of the policy and social welfare implications of risk.

REFERENCES

Albæk, S., P. Mølgaard, and P.B. Overgaard. 1997. "Government-Assisted Oligopoly Coordination? A Concrete Case." *Journal of Industrial Economics* 45: 429-443.

Allen, D.W., and D. Lueck. 1999. "The Role of Risk in Contract Choice." *Journal of Law, Economics, and Organization* 15: 704-736.

Bardhan, P. (ed.). 1989. *The Economic Theory of Agrarian Institutions*. Oxford: Clarendon Press.

Chambers, R.G. Forthcoming. "Information, Incentives, and the Design of Agricultural Policies." In B. Gardner and G. Rausser, eds., *Handbook of Agricultural Economics*. Amsterdam: North Holland.

Chambers, R.G., and J. Quiggin. 2000. *Uncertainty, Production, Choice, and Agency: The State-Contingent Approach*. Cambridge, UK: Cambridge University Press.

Coase, R.H. 1952. "The Nature of the Firm." Reprinted in G.J. Stigler and K.E. Boulding, eds., *Readings in Price Theory*. Homewood, IL: Richard D. Irwin.

Dardanoni, V., and A. Forcina. 1998. "A Unified Approach to Likelihood Inference on Stochastic Orderings in a Nonparametric Context." *Journal of the American Statistical Association* 93: 1112-1123.

Goodhue, R.E. 1999. "Input Control in Agricultural Production Contracts." *American Journal of Agricultural Economics* 81: 616-620.

___. 2000. "Broiler Production Contracts as a Multi-Agent Problem: Common Risk, Incentives and Heterogeneity." *American Journal of Agricultural Economics* 82: 606-622.

Guesnerie, R., and J.-J. Laffont. 1984. "A Complete Solution to a Class of Principal-Agent Problems with an Application to the Control of a Self-Managed Firm." *Journal of Public Economics* 25: 329-369.

Harris, A., and A. Raviv. 1979. "Optimal Incentive Contracts with Imperfect Information." *Journal of Economic Theory* 20: 231-259.

Harris, M., and R.M. Townsend. "Allocation Mechanisms, Asymmetric Information and the 'Revelation Principle'." In George R. Feiwel, ed., *Issues in Contemporary Microeconomics of Welfare*. New York: State University of New York Press.

Hart, O. 1995. *Firms Contracts and Financial Structure*. New York: Oxford University Press.

Hart, O., and B. Holmström. 1987. "The Theory of Contracts." In T. Beweley, ed., *Advances in Economic Theory: Fifth World Congress*. Cambridge: Cambridge University Press.

Hennessy, D. 1996. "Information Asymmetry as a Reason for Food Industry Vertical Integration." *American Journal of Agricultural Economics* 78: 1034-1043.

Holmström, B. 1979. "Moral Hazard and Observability." *Bell Journal of Economics* 10: 74-91.

___. 1982. "Moral Hazard in Teams." *Bell Journal of Economics* 13: 324-340.

Hueth, D., and W. Furtan. 1994. *Economics of Agricultural Crop Insurance: Theory and Evidence*. Norwell, MA: Kluwer Academic Publishers.

Hueth, B., and E. Ligon. 1999. "Producer Price Risk and Quality Measurement." *American Journal of Agricultural Economics* 81: 512-524.

___. 2001. "Agricultural Markets as Relative Performance Evaluation." *American Journal of Agricultural Economics* 83: 318-328.

Iowa Attorney General. 2001. "Section by Section Explanation of the Producer Protection Act," http://www.state.ia.us/government/ag/agcontractingexplanation.htm (accessed on April 16, 2001).

Jewitt, I. 1988. "Justifying the First Order Approach to Principal-Agent Problems." *Econometrica* 56: 1177-1190.

Kim, S.K. 1995. "Efficiency of an Information System in an Agency Model." *Econometrica* 63: 89-102.

Knight, T.O., and K. Coble. 1997. "Survey of U.S. Multiple Peril Crop Insurance Literature Since 1980." *Review of Agricultural Economics* 19: 128-156.

Knoeber, C.R., and W.N. Thurman. 1995. "'Don't Count your Chickens . . .': Risk and Risk Shifting in the Broiler Industry." *American Journal of Agricultural Economics* 77: 486-496.

Macho-Stadler, I., and J.D. Perez-Castrillo. 1997. *An Introduction to the Economics of Information: Incentives and Contracts*. New York: Oxford University Press.

Martin, L. 1997. "Production Contracts, Risk Shifting, and Relative Performance Payments in the Pork Industry." *Journal of Agricultural and Applied Economics* 29: 267-278.

Martinez, S.W. 1999. "Vertical Coordination in the Pork and Broiler Industries: Implications for Pork and Chicken Products." Agricultural Economic Report No. 777, Economic Research Service, U.S. Dept of Agriculture.

Milgrom, P., and J. Roberts. 1992. *Economics, Organization and Management.* Englewood Cliffs, NJ: Prentice Hall.

Mirrlees, J. 1974. "Notes on Welfare Economics, Information and Uncertainty." In M. Balch, D. McFadden, and S. Wu, eds., *Essays in Economic Behavior under Uncertainty*. Amsterdam: North-Holland.

Myerson, R.B. 1989. "Mechanism Design." In J. Eatwell, M. Milgate, and P. Newman, eds., *Allocation, Information, and Markets.* New York: Macmillan.

Otsuka, K., H. Chuma, and Y. Hayami. 1992. "Land and Labor Contracts in Agrarian Economies: Theories and Facts." *Journal of Economic Literature* 30: 1965-2018.

Perry, J., D. Banker, and R. Green. 1999. "Broiler Farms' Organization, Management, and Performance." Agriculture Information Bulletin No. 748, Economic Research Service, U.S. Dept of Agriculture.

Phelan, C., and R. Townsend. 1991. "Computing Multiperiod, Information Constrained Optima." *Review of Economic Studies* 58: 853-882.

Rogerson, W.P. 1985. "Repeated Moral Hazard." *Econometrica* 53: 69-76.

Ross, S. 1973. "The Economic Theory of Agency: The Principal's Problem." *American Economic Review* 63: 134-139.

Salanié, B. 1997. The Economics of Contracts: A Primer. Cambridge, MA: The MIT Press.

Segal, I., and M. Whinston. 2000. "The Mirlees Approach to Mechanism Design with Renegotiation (with Applications to Hold-Up and Risk-Sharing)." Stanford Economics Working Paper.

Smith, R. 2001. "Cash Hog Market Prices Likely to Be Gone in Two Years." *Feedstuffs* 73 (March 12): pp. 1 and 5.

Spence, M., and R. Zeckhauser. 1971. "Insurance, Information, and Individual Action." *American Economic Review* 61: 380-387.

Streeter, D.H., S. Sonka, and M.A. Hudson. 1991. "Information Technology, Coordination, and Competitiveness in the Food and Agribusiness Sector." *American Journal of Agricultural Economics* 73: 1465-1471.

Tirole, J. 1999. "Incomplete Contracts: Where Do We Stand?" *Econometrica* 67: 741-781.

Tsoulouhas, T., and T. Vukina. 1999. "Integrator Contracts with Many Agents and Bankruptcy." *American Journal of Agricultural Economics* 81: 61-74.

___. 2000. "Regulating Broiler Contracts: Tournaments Versus Fixed Performance Standards." Stanford Economics Working Paper.

Williamson, O.E. 1985. "The Economic Institutions of Capitalism." New York: The Free Press.

Part 3

ADEQUACY OF GENERAL METHODOLOGICAL APPROACHES FOR RISK ANALYSIS

Chapter 10

PROGRAMMING METHODS FOR RISK-EFFICIENT CHOICE

C. Robert Taylor and Thomas P. Zacharias
Auburn University and National Crop Insurance Service

INTRODUCTION

> "...I [am] still puzzled at the insistence of many writers on treating the uncertainty of result in choice as if it were a gamble on a known mathematical chance..." (Frank Knight, in Preface to the 1933 reissue of *Risk, Uncertainty and Profit*).

Programming models were prominent in early theoretical and empirical research on risk-efficient choices, beginning primarily with Freund's (1956) seminal incorporation of risk into a quadratic programming (QP) model. Building on the QP formulation, subsequent model developments in agricultural economics generally dealt with introducing risk into a computationally feasible programming format, or dealt with introducing different types of risk-aversion assumptions such as safety-first, or mean-variance (EV), into a programming format. Models that incorporate risk have pertained primarily to an individual's or a firm's decision, although a few programming models have been proposed to apply in the aggregate.

The rich heritage of programming models for risk-efficient choices is well documented in the professional literature. Consequently, after a cursory review of some of these models, we turn in this chapter to unresolved issues and possible future directions for theoretical and applied research for risk-efficient choice.

PAST PROGRAMMING MODELS

Programming models incorporating risk are known by a veritable alphabet soup of acronyms and shorthand names. Stylized programming formulations included quadratic programming (QP), mean absolute deviation (MAD), absolute negative total deviation (ANTD), minimization of total absolute deviations (MOTAD) in a linear programming framework, target MOTAD, direct expected utility maximization nonlinear programming (DEMP), direct expected utility maximization nonlinear programming with numerical quadrature (DEMPQ), Gaussian quadrature for risk programming (GQRP), lattice programming methods (LAT), safety-first linear programming with discrete joint probability distributions (LP-DPD), semivariance (SV), chance-constrained linear programming (CCLP), focus-loss constrained programming (FLCP), convex programming (CP), separable programming (SP), marginal risk constrained (MRC), marginal risk constrained linear programming (MRCLP), linear-quadratic Gaussian (LQG) stochastic control under uncertainty, Ito's stochastic control (ISC), stochastic dynamic programming (SDP), and discrete stochastic programming (DSP).[1]

Programming formulations for a sector include models to theoretically examine the effects of crop yield risk on market equilibrium (Hazell and Scandizzo 1977), to analyze price stabilization policy (Hazell and Pomareda 1981), and to analyze stochastic technological coefficients and uncertain input supplies (Paris and Easter 1985).

ISSUES FOR THE FUTURE

After a burst of interest in incorporating risk into programming models in the 1970s and 1980s, the programming approach practically disappeared from the journals in recent years (Just 2000). Dramatic advances in numerical computational power combined with new risk and uncertainty paradigms leading to programming models suggest a possible rebirth of this approach to decision problems in agricultural and resource economics. In the remainder of this chapter, we discuss several issues pertinent to the future of programming models of risk-efficient choice.

Before proceeding, it might be worthwhile to informally offer some thoughts on the perceived disappearance or demise of risk-efficiency programming models in the 1980s. Without question, Meyer's (1977) seminal paper had a major impact on the profession, particularly those working in the area of farm management and production economics. After Meyer, Kramer and Pope (1981) followed with one of the first applications of stochastic dominance (SD). In the same year, King and Robison (1981) provided the basic software infrastructure that led to a widespread adoption of stochastic dominance applications. During the next several years stochastic dominance applications in farm-level analysis became very popular and the source of a great many journal publication. It would be interesting to perform a count of programming model applications published in our journals before and after the early 1980s, along with a count of stochastic dominance papers.

---

[1] Discussion of the various programming models as well as references to these models can be found in the books by Kennedy (1986) and Hazell and Norton (1986).

Stochastic dominance analysis had and may still have several key advantages over programming applications. A key advantage of SD was its ease of use in combination with simulation models, particularly of the FLIPSIM variety (Richardson and Nixon 1986). Moreover, for agricultural economists with joint extension and applied research programs, SD analyses were easier to explain than other risk models. The concepts of first- and second-degree dominance can be demonstrated without much difficulty. There was also a "joint product" effect, that is, those working in risk management and applied marketing were already building empirical distributions anyway. It is probably fair to say that SD analyses became a "dominant" or preferred tool of the profession during the 1980s.

In the remainder of this chapter, we discuss several issues pertinent to the future of programming models of risk-efficient choice. Issues include (1) numerical computational power, (2) the importance of risk vs. risk aversion vs. the appearance of risk aversion, (3) the distinction between risk and uncertainty, (4) propagation of model uncertainty, (5) philosophical statistics – frequentists, Bayesians, and Fuzzies, (6) temporal coherence and temporal resolution of uncertainty, (7) dual control formulations, (8) aggregate risk, (9) game theory, and (10) post-optimality analysis. We conclude with discussion of future directions for modeling research.

Numerical Computational Power

As noted previously, many of the early programming formulations of risk-efficient choice were manifestations of the attempt to introduce risk into a computationally feasible framework. Recent advances in computational power have made this computational curse fade to the point where it should not generally be of prime consideration in model formulation or solution.

As noted by Taylor (1993b), the numerical curse of dimensionality in solving large, complex stochastic optimization models has faded dramatically, but has been replaced by other curses. Other curses include (a) the formulation curse, which refers to conceptualization difficulties with stochastic dynamic models of empirical problems; (b) the curse of a complex decision rule, which is difficult to visualize, rationalize, and understand; and (c) the acceptance curse, which applies not only to professionals not well trained in complex models, but also to potential clients that might use results from such models.[2]

We have tremendous excess capacity with desktop PCs, almost all of which go unused nights and weekends. Moreover, we now have relatively inexpensive dual processor PCs. One processor could be used for complex optimization models, which take days to numerically solve, while the other

[2] Ciriani and Gliozzi (1995) provide a fairly up-to-date discussion of optimization methods for large-scale models, while Banks (1995) presents some of the simulation software (which could also be used for optimization).

processor could be used for routine chores (e.g., e-mail, word processing) while the model is being solved. Once the model is solved, array visualizers available with some of the programming language packages can be used to better understand and to gain more insight into the solutions of complex models.

To summarize, our numerical computational power for solving models often exceeds our capability to correctly formulate and parameterize decision problems.

The Importance of Risk vs. Risk Aversion vs. The Appearance of Risk Aversion

A careful distinction between the importance of risk, risk aversion, and the *appearance* of risk aversion is not always apparent in agricultural and resource economics. Furthermore, it is not always apparent that results from a particular theoretical or empirical model of risk aversion are due to incorporation of risk into the model, to the inclusion of risk aversion, or to the interaction of both.

Cases of risk-influenced optimal solutions to static decision models under risk neutrality were considered by Just (1975a), Pope (1982a, 1982b), and Just and Pope (1978, 1979). As shown by Antle (1983), dynamic models often show that "...input and output price risk and production risk generally affect productivity and optimal resource allocation, whether or not the decision maker is risk averse." The stochastic, dynamic duality relationships derived by Taylor (1984) also implicitly show the importance of risk in optimal solutions under risk neutrality. Chavas (1994) shows that temporal uncertainty in the presence of sunk costs also influences risk-neutral behavior.

The appearance of risk aversion under risk neutrality but with a progressive income tax was considered by Taylor (1986). Using a stochastic dynamic portfolio management model, Zimmerman and Carter (1996) recently showed that "...poorer households behave in ways that *appear* to be more risk-averse than for wealthier households even though the underlying utility parameters are the same for everyone" (p. 8).

Formulation and solution of more realistic models of agricultural decisions are likely needed for the profession to gain more insight into the distinction between and the relative importance of risk, risk aversion, and the appearance of risk aversion. The case can certainly be made (Antle 1983) that the profession has been preoccupied with risk aversion, perhaps at the expense of overlooking risk and apparent risk aversion.

Although we typically consider risk attitudes of the decision maker, practical experience suggests that risk attitudes of individuals or organizations on the periphery of the firm may also influence decisions. In particular, risk preferences of bankers may be fuzzily superimposed on agricultural producers' risk preferences and may even dominate input, marketing, and contracting decisions of some agricultural firms. Where such complexities exist, it may erroneously be inferred that the producer is risk averse, when in fact it is the banker, for example, who is risk averse.

The Distinction Between Risk and Uncertainty

Eighty years ago, Frank Knight attempted to make a careful distinction between risk and uncertainty – a distinction that continues to be blurred in economists' writings to this day. It may be useful in this regard to think of two extremes. At one extreme is pure risk where probabilities can be numerically assigned exactly from objective, physical data. At the other extreme, however, is pure uncertainty. Knight claimed that the case of pure uncertainty is unanalyzable because probabilities of purely uncertain events are unmeasurable.

As we move from pure risk toward pure uncertainty, it seems more difficult to conceptualize and define random variables. Furthermore, as we move toward pure uncertainty, probabilities are increasingly difficult to specify, subjectively or objectively.

The topological region between these two precisely defined extremes is blurred at best and contains epistemological gaps at worst. In the terminology of fuzzy set theory (Zadeh 1978), there is a fuzzy continuum between the crisp cases of pure risk and pure uncertainty. Most applied problems in agricultural and resource economics appear to fit on this fuzzy continuum and not at either extreme.

As we enter the twenty-first century, there is tremendous uncertainty about the future of U.S. agriculture, uncertainty about whether an individual producer can get a contract, especially a preferential contract, uncertainty about farm programs, uncertainty about the structure of agriculture, uncertainty about intimidation and retaliation by integrators, uncertainty about exchange rates, uncertainty about environmental regulations, uncertainty about the future existence of commodity markets, and so forth. Yet as a profession we have devoted incredible resources to analyzing crop insurance, for example. And although crop insurance is at the forefront of the current farm policy debate, uncertainty not related to traditional crop yield and price variability may play a more dominant role in farmers' decisions in the future.

Models and applications published in the agricultural economics literature have tended to force risk or uncertainty problems to fit in a classical probability mold developed for the case of pure risk. A common view of uncertainty is that probabilities can be subjectively assigned with no problem and that the analyst can proceed with models based on classical statistical axiomatic theory. Yet it seems to us that this approach may ignore some of the most important applied problems and the most intellectually challenging problems – those that tend toward the uncertainty end of the spectrum.

Propagation of Model Uncertainty

Another form of uncertainty pertains to uncertainty about the structure of the problem being modeled. Draper (1995) concluded that it was common in statistical theory and practice to acknowledge parametric uncertainty given a particular assumed structure, *S*, but less common to acknowledge structural uncertainty about *S* itself. Although Draper's conclusion was in the context of statistical inference and prediction, it is also a valid criticism of programming models for risk-efficient choice. Factors for which there are no observations and which tend toward the uncertainty end of the continuum are not often acknowledged in the programming model literature.

In our haste to quantify and publish – just give me the mean and variance and let me assume the risk-aversion parameter[3] – we overlook some of the most important and most intellectually challenging dimensions of the problem of decision making, namely, aspects of the problem tending toward pure uncertainty as well as uncertainty about the system itself.

Philosophical Statistics – Frequentists, Bayesians, and the Fuzzies

Literature in philosophical statistics, much of which emerged after the wave of risk programming models in agricultural economics, suggests that different philosophical approaches to statistics may be needed to address issues that are dominated by uncertainty. Many of the newer approaches also lead to programming models, but to gain some perspective on the new models we must first briefly discuss the competing philosophical paradigms.

The literature reveals considerable diversity of thought by statisticians and philosophers on how we should approach decision making in situations characterized by uncertainty.[4] Although an oversimplification, it is useful here to consider three philosophical camps: Frequentists, Bayesians, and what we will call the "Fuzzies."

Early statisticians adopted a "Frequentist" view, meaning that probability makes empirical sense only if it is defined as frequency (Shafer 1990). The Frequentist statistical methodology, developed primarily by Karl Pearson and R.A. Fisher, remains at the core of our statistical and econometric theory (Shafer 1990), and also remains at the core of much of the literature dealing with risk. The Frequentist camp evolved to the belief that probabilities could be subjectively or objectively specified (Savage 1954, Freund 1956, Machina

---

[3] In the context of an EV model, for example, it has always been curious to the authors, to say the least, that the one parameter about which we know little – the risk-aversion parameter – is treated as though it had a known value.

[4] Readers interested in a concise treatment of the diversity of philosophical views on statistics are referred to the 1988 Wald Memorial Lecture, "The Present Position in Bayesian Statistics," by D.V. Lindley, with comments by J.O. Berger, G.A. Barnard, J.M. Bernardo, D.R. Cox, S. French, J.B. Kadane, E.I. Lehmann, and M. Mouchart, and a rejoinder by Lindley.

and Schmeidler 1992), a belief carried over to early agricultural economics literature on decision making under risk and continuing today.

It may seem natural to some analysts to view uncertainty as a situation where probability can be modeled by another probability distribution. However, as cautioned by Nau (1992), "... the quality of being unsure about a probability cannot itself be satisfactorily modeled by another probability distribution" (p. 1737). Thus, the Frequentists' views of probability cannot be logically extended by taking a second-order view of probability.

A Bayesian approach, which retains much of the axiomatic philosophy of the Frequentists, emerged as an alternative to the Frequentist view. In a Bayesian framework, however, probability is defined in terms of a subjective degree of belief. A subjective probability distribution is used to summarize a decision maker's beliefs about a parameter of interest. A probability distribution, called the *prior*, can be specified before any sample information is available. Using Bayes' Theorem, sample information can be combined with the *prior* distribution to derive a *posterior* distribution which can then be used in a conventional decision framework.

Although the basic idea behind the Bayesian approach is fairly straightforward, it should be recognized that there are a wide variety of Bayesian models. (Good [1972] claimed that there are 46,656 varieties of Bayesians!)

Problems in implementing a Bayesian approach begin with problems in eliciting *prior* probabilities, and often continue with numerical integration difficulties. Despite these difficulties, the Bayesian approach has been applied in a few instances in agricultural economics (e.g., Carlson 1970).

Implementation of a Bayesian approach in a dynamic setting introduces considerably more difficulties, computationally and conceptually. Dynamic Bayesian models are troublesome because of unresolved questions about temporal coherence and about how to introduce sample information into specification of *priors* as time progresses. Numerical problems can also arise because of the problems of extensive numerical integration combined with the usual curses of numerically solving dynamic, stochastic optimization models.

An emerging alternative to classical Frequentist or Bayesian models can be traced to the recent development of fuzzy set theory and applications of that theory to decision making under uncertainty (Zadeh 1978, Shackle 1961, Cohen 1970, Shafer 1976). A whole new vocabulary is emerging with development of fuzzy set theory and its applications: fuzzy logic, a fuzzy continuum, fuzzy measures, fuzzy integrals, intuitionistic modal logic,[5] belief functions, possibilistic inference, anxiety measures,[6] a scenario approach,[7]

---

[5] See, for example, Lano (1991).

[6] With the anxiety measure approach, indecision and anxiety are assumed to be caused by the lack of clear information and well-defined criteria to make the decision (Kikuchi and Perincherry 1997).

[7] Scenario groups are used to characterize uncertainty. Solutions are obtained for each scenario group, then individual solutions aggregated to yield a non-anticipative policy that minimizes the regret of wrong decisions (Ahn, Escudero, and Guinard-Spielberg 1995).

linguistically expressed uncertainty,[8] linguistically fuzzy numbers, and linguistically fuzzy labels are all relatively new terms used in the context of decision making under risk and uncertainty. As Lindley (1987) notes, this new approach "... contains great complexities of language and ideas" (p. 22). Not only are the language and ideas complex, they are downright – should we say it – fuzzy!

The fuzzy set theory approach to decisions under uncertainty often leads to nonlinear programming (Lindley 1987, Gaivoronski 1995). Some model formulations, however, implement a complex graphical interface (Gebhardt and Kruse 1997).

Most agricultural economists appear to have been firmly grounded in Frequentist statistics, with limited exposure to Bayesian models, and virtually no exposure to fuzzy set theory. However, applications of the fuzzy models are beginning to show up in agricultural economics literature. In particular, fuzzy models recently developed include the MADM approach to alternative farming systems (Marks and Dunn 1999), fuzzy logic and compromise programming applied to portfolio management (Duval and Featherstone 1999), and fuzzy access rights and common property considerations (Goodhue and McCarthy 1998).

Frequentists and Bayesians have had an ongoing debate about the validity of each approach for decades. As Shafer (1990) states, this debate has essentially evolved to "... the worn out dogmas of either group" (p. 441).

Although the debate between the Frequentists and Bayesians continues, they appear to have loosely united against the emerging Fuzzies. As seen by both the Frequentists and Bayesians, "the only satisfactory description of uncertainty is probability," and alternative descriptions of uncertainty are unnecessary (Lindley 1987, Singpurwalla 1988). Stated another way, even if a problem is characterized by tremendous uncertainty, one should nevertheless assign subjective probabilities, no matter how uncertain, and proceed using a classical Bayesian or non-Bayesian approach. This view certainly predominates the past literature in agricultural economics.

From the nontraditional camp comes criticism of the crisp approach of both the Frequentists and Bayesians. Zadeh (1983), a pioneer of fuzzy set theory, asserts that "A serious shortcoming of (probability based) methods is that they are not capable of coming to grips with the pervasive fuzziness of information in the knowledge base, and, as a result, are mostly ad hoc in nature" (p. 199). He also stated that "The validity of (Bayes rule) is open to question since most of the information in the knowledge base of a typical expert system consists of a collection of fuzzy rather than non fuzzy propositions" (p. 199).

---

[8] Linguistically expressed uncertainty uses a terminology set such as {Impossible, Almost-Impossible, Slightly-Possible, Moderately-Possible, Possible, Quite-Possible, Very-Possible, Almost-Sure, and Sure}.

Fuzzy critics of the traditional approaches also assert (a) that decision makers are not capable of arbitrarily fine distinctions[9] (Berger 1990, Leamer 1986); (b) an intuitive conviction that personal probabilities cannot always be quantified exactly (Nau 1992, Berger 1990, Leamer 1986, Lindley 1990a, Good 1988); (c) that individuals' beliefs are seldom perfectly coherent (French 1990, Einhorn and Hogarth 1986); (d) that utility and probability are not necessarily independent as is implicitly assumed in the classical decision framework, which is the Allais paradox (Mouchart 1990); and (e) that changes in context can strongly affect the evaluation of uncertainty (Einhorn and Hogarth 1986). Barnard (1990) provides us food for thought in his argument that "inference from ignorance to a joint distribution (the Fuzzy approach) seems more direct than assuming ignorance priors (the Bayesian approach)" (pp. 68-69).

The traditional approach may force the analyst to be unrealistically precise, while the fuzzy set theory approach may lack coherence and internal consistency. And, although the fuzzy set theory approach may use probability in a very loose way, it nevertheless contains great complexities (Lindley 1987). But the key consideration, given the focus of this chapter, is that all three philosophical approaches to risk and uncertainty – the Frequentist, the Bayesian, and the Fuzzy – often lead to programming models. But in implementing any of the approaches, we must better recognize the inherent limitations of each approach.

Temporal Coherence and Temporal Resolution of Uncertainty

Just as the proper philosophical view of the statistical structure of decision making under risk and uncertainty has not been fully resolved, neither has the incorporation of risk-averse or uncertainty-averse behavior into intertemporal models (Zacharias 1993). From a practical standpoint, for example, we rarely know whether to impose a safety-first constraint on single-period returns in each period in a dynamic model, to impose a safety-first constraint on terminal wealth, or both (Krautkraemer, van Kooten, and Young 1992). Based on the literature review provided in Zacharias (1993), there would seem to be a fair range of farm management applications with which to experiment with the notion of intertemporal risk aversion or the presence of risk in intertemporal models. To date, it is not obvious that any such applications have been pursued to any great extent.

---

[9] Berger (1990) states that "...the unique-probability Bayesian paradigms all contain the unrealistic axiom that we are capable of arbitrarily fine distinctions in judgement; that, if one thought long enough, one could decide whether one's subjective probability of rain tomorrow was .38792567 or .38792566 (to paraphrase I.J. Good) ... The opposing position – that the ideal is, in practice, not approachable (i.e., that the measurement problems are insurmountable) – is, logically, a viable escape for non-Bayesians" (pp. 72-73).

Temporal coherence of probabilities and how information gained over time is incorporated into estimating probabilities has not been fully resolved either. Good's (1988) unique way of acknowledging the importance of temporal coherence of probability is that "Thinking will often cause you to change your mind; that is why dynamic probabilities are relevant" (p. 388). Thinking about dynamic probabilities may naturally lead to a Bayesian philosophical view. However, as emphasized by Kulhavy (1993), "The problem of real-time implementation of the Bayesian paradigm raises difficult theoretical and conceptual questions" (p. 480).

Dual Control Formulations

Dual control models have been proposed for decision making under uncertainty (Tse, Bar-Shalom, and Meier 1973, Rausser 1978, Bar-Shalom and Tse 1976, Chow 1975, 1976a, 1976b, Taylor and Chavas 1980, Feldstein 1971, Klein et al. 1978, and McRae 1972, 1975). Balvers and Cosimano (1990) present a model with adaptive learning about demand for a firm's product. These models, which are based essentially on a Frequentist view of decision under uncertainty, purport to combine estimation of an uncertain process with control (optimization) of that process, leading to temporal resolution of uncertainty and learning.

Both passively adaptive and actively adaptive formulations have been proposed. These models are intuitively appealing because they explicitly recognize the need for learning. Current decisions consider both control and estimation of the model. Current and future uncertainty are introduced in the decision rules, thus increasing information for improved performance in the future (Taylor and Chavas 1980). However, the closed-loop nature of actively adaptive strategies leads to stochastic dynamic computational problems. Thus, workable actively adaptive schemes are only approximations.

Applications of the actively adaptive approximations have met with mixed success. Unpublished work by C. Robert Taylor and Oscar Burt suggests that curvature of the objective function in parameter (not decision variable) space is a key to the performance of actively adaptive approximations compared to updated certainty equivalence (passively adaptive). If the objective function has some desirable (but uncertain!) degree of concavity in parameter space, the approximation works well. If, however, the objective function is convex or highly concave in parameter space, the approximation does not work well. Thus, performance of actively adaptive approximations seems dependent on choice of functional form. This suggests that something is wrong with the problem formulation; unfortunately, we cannot suggest what is wrong. Perhaps some of the criticisms made of traditional decision models made by proponents of a fuzzy set theory apply to previous formulations of dual control models.

Aside from strictly mathematical difficulties encountered in dual control applications or control theory applications in general, there was at least one conceptual impediment to expanded application. During the 1970s and 1980s, general economists (e.g., Kendrick) were heavily involved in macroeconomic applications of control theory. Several agricultural economists were also developing similar control models, the most common applications being buffer stock applications with implications for commodity price stability (Taylor and Talpaz 1979, Burt, Koo, and Dudley 1980, and Gardner 1979). The conceptual stumbling block came in the form of what Kydland and Prescott (1977) called "rules rather than discretion." Kydland and Prescott introduced the notion of game-theoretic behavior with respect to the state variable equations and called into question the validity and usefulness of macro-level control theory applications. Without the macro agricultural policy dimension, the research incentives to pursue advanced control theory applications in agricultural economics were greatly diminished.

Although there appear to be conceptual problems with dual control models, their intuitive appeal suggests that this is an important area for future research. As a final historic aside (or note), the adoption of control theory applications was probably hindered by the profession itself in the early 1980s, as evidenced by at least one formal exchange found in the 1982 *Western Journal of Agricultural Economics* (Burt 1982, Talpaz 1982). This discussion featured several of the prominent quantitatively oriented agricultural economists at the time. In essence, the discussion centered on both technical and philosophical issues involved with intertemporal solution models and methodologies. The extent to which this debate resulted in a segmentation of the profession is not clear, but it is probably safe to conclude that the result was somewhat divisive within the profession and complementarity across solution paradigms was not fostered.

Aggregate Risk

As noted in the introductory section of this chapter, there have been a few attempts at developing programming models to represent risk aversion in the aggregate. Such models, while contributing insight into the aggregate effects of risk aversion, suffer from uncertainties about aggregation biases. In models of agricultural markets, not only are there concerns with aggregating over different point estimates of price or differing distribution function estimates of price, but there are also concerns about aggregating over different levels of risk aversion and aggregating over different types of risk aversion.

Although aggregation problems must be recognized as having great potential import, aggregate models may nevertheless continue to offer insight into the effects of risk aversion and help identify aggregate paradoxical effects. For example, provision of subsidized crop insurance, with or without risk aversion by producers, will shift the aggregate supply curve downward and to the right, thereby lowering average price. This shifts the aggregate net farm income distribution to the left; in a safety-first sense, provision of subsidized crop insurance increases aggregate risk. Thus, there is an aggregate risk paradox.

To the extent that such an aggregate risk paradox exists and is widely recognized by market participants, we may need to develop an extension of rational expectations to include second- and higher-order terms. Obviously this would add great theoretical and empirical complexity to an already complex subject.

Game Theory

The resurgence of game theory in general economics has been quite remarkable. In the late 1950s and early 1960s, agricultural economists "dabbled" with game theory as a set of decision criteria to deal with the problems of risk and uncertainty (Swanson 1959). Somewhat later, Hazzell (1970) provided an illustration of various game theory criteria in farm management planning. However, with the concept of subjective probability and widespread adoption of the application of expected utility theory, game theoretic applications went by the wayside (Anderson, Dillon, and Hardaker 1977). Most recently, however, Johnson in his 1998 American Agricultural Economics Association Fellows Address utilized a game theory framework to discuss strategic behavior in land grant university funding problems. As the discussion here attempts to bridge or at least more formally acknowledge the spectrum between pure risk and pure uncertainty, use of non-probabilistic and probabilistic game theory models might deserve a much more careful look by our profession.

In terms of intertemporal decision problems, use of extensive form game theoretic exposition might prove useful in at least two regards. Quoting Kreps (1990), "In an extensive form game, attention is given to the timing of actions that players may take and the information they will have when they must take those actions" (p. 13).

First, diagrammatic exposition could prove useful in quasi-extension or applied research problem identification. Second, diagrammatic exposition should prove highly valuable in probabilistic dynamic decision problems in which risk is "resolved" at various stages in the process. The farm-level applications here are numerous, such as input timing. To "sweeten" the publication incentive pie, one need only reflect on the environmental implications associated with these types of applications.

Several recent developments in game theory would appear to have applicability in agricultural and resource economics. Behavioral game theory (Camerer 1997) aims to describe actual behavior and is driven by empirical observation. This theory attempts to chart a course between over-rational equilibrium analyses and under-rational adaptive analyses. Tversky and

Kahneman (1992) present prospect theory in which people value gains and losses from a reference point, and in which they dislike losses much more than they like gains.

Useful discussion and documentation of systematic violations of common game theoretic principles can be found in Camerer (1997). A useful discussion of static versus dynamic games, with complete or incomplete information, can be found in Gibbons (1992).

With the easing of computational restrictions and the emphasis on strategic behavior (and its non-probabilistic nature), agricultural economists should seriously consider a resurrection of the profession's early game theoretic formulations, and consider extension of these formulations along the lines suggested by recent adaptive game theory developments.

Post-Optimality Analysis

There is not a great deal to say on the issue of post-optimality analysis other than that analysts should probably do more of it. In the "early" days of widespread linear programming (LP) applications, sensitivity analysis seemed to receive more attention. One very obvious reason may have been the microeconomic insights provided in the solution of the typical farm management problem. As an instructor of agricultural production economics, one could simply "camp out" with the final solution tableau and use it to "teach" for a while. In addition, certain LP software packages provided a great deal of post-optimality support. It may be that once model formulation became more complex, the opportunity cost of examining the solution became extremely high relative to other pursuits, or to put it bluntly, it was time to move on if the manuscript had been accepted!

Three observations are offered here. First, it does not appear that agricultural economists are as well trained on "non-LP" post-optimality technology. Second, it does not appear that "non-LP" post-optimality technology is as well developed. Given our profession's rich LP heritage, it might be valuable to reconsider the role of post-optimality analysis in the future. Issues such as the stability of the solution, shadow prices, and dual properties probably deserve greater attention, particularly if concepts such as risk aversion or presence of risk cannot be readily distinguished. Lastly, post-optimality analysis in dynamic models is not trivial. One suggestion would be to consider the formulation of a Markov model using LP and exploit the capabilities of LP technology to gain some insight for dynamic problems.[10]

---

[10] It should be recognized, however, that the LP formulation of the traditional Markovian dynamic programming (DP) model is computationally much more expensive than solving DP using Bellman's recursive equation with a computer program written specifically for the problem of interest.

THE FUTURE

Recent extensions of fuzzy set theory into decision making under risk and uncertainty offer intriguing and intellectually challenging alternatives to the more traditional models developed by agricultural economists. These developments have a great complexity of ideas and language and, like traditional approaches, often lead to programming models (Lindley 1987).

Stylized fuzzy models are providing new insights into decision making. Perhaps more insightful, however, is the ongoing philosophical debate among the Frequentist, Bayesian, and Fuzzy approaches to decision making under risk and uncertainty. At the very least, the debate highlights inherent limitations of conventional decision models – limitations not often mentioned in the training of agricultural economists.

Realistic formulations of practical agricultural decision problems with a Frequentist, Bayesian, or Fuzzy model approach may often be too complex for traditional analytical analysis. With the near disappearance of the computational curse, however, numerical methods can now be used for theoretical analysis of problem formulations too complex for analytical evaluation. Numerical analysis may thus be useful in gaining insight into complex models of decision making under risk and uncertainty.

In the context of the ongoing debate between the Frequentists, Bayesians, and Fuzzies, Good (1988) argues that we should allow for the "cost of thinking and calculation" when applying the principle of maximization of expected utility. The question to be posed here is whether additional "thinking" about the issues raised in this chapter will increase expected utility; that is, will additional thinking about risk and uncertainty, particularly in the context of programming models, increase our understanding of decision making? We believe so, but we also believe that there is considerable *uncertainty* about the professional outcome. Worded another way, our semi-informed *priors* suggest that there is tremendous *uncertainty* about the expected professional payoff (aside from gaining tenure and promotion!) of additional research in the areas discussed in this chapter. Being not only academics, but academic economists, we will *assume* that this is not the unanalyzable pure uncertainty identified by Professor Knight!

While there are many intriguing and intellectually exciting possibilities with further development of complex conceptual and empirical models of risk and uncertainty, there is – unfortunately, one could say – the problem of the practical significance of programming (and other) models of risk-efficient choice. In our experience, only the simplest models such as stochastic dominance and results from simple stochastic simulation models have met with any degree of acceptance in the farm community. In the words of Frank Knight (1957), there is "...the unpleasant question of where to compromise between expounding more truth and being more useful" (p. liii).

At times it seems that agricultural economists are like Hiawatha, the brilliant (but fictitious) statistician, in not being able to "hit the target."[11] Most often we seem to aim at, but not hit, the target of clearly expounding more truth. But even when we aim at the target of being more useful, we seem to have too many arrows that do not even fall within our experimental plot!

As an ending, we would like to say that it has not escaped our attention that our discussion may be nothing more than cheap talk. To the extent that this is true, however, we hope that "Simply by making noises with our mouths, we can reliably cause precise new combinations of ideas to arise in each other's minds" (Pinker 1994, p. 15).

REFERENCES

Ahn, S., L.F. Escudero, and M. Guinard-Spielberg. 1995. "On Modeling Financial Trading Under Interest Rate Uncertainty." In A. Sciomachen, ed., *Optimization in Industry: Mathematical Programming and Modeling Techniques in Practice.* New York: John Wiley and Sons Ltd.

Anderson, J.R., J.L. Dillon, and B. Hardaker. 1977. *Agricultural Decision Analysis.* Ames: Iowa Sate University Press.

Antle, J. 1983. "Incorporating Risk in Production Analysis." *American Journal of Agricultural Economics* 65: 1099-1106.

Atwood, J.A., M.J. Watts, and G.A. Helmers. 1988. "Chance-Constrained Financing as a Response to Financial Risk." *American Journal of Agricultural Economics* 70: 79-89.

Balvers, R.J., and T.F. Cosimano. 1990. "Active Learning About Demand and the Dynamics of Price Adjustment." *The Economic Journal* 100: 882-898.

Banks, J. 1995. 1995. "Simulation and Related Software." In A. Sciomachen, ed., *Optimization in Industry: Mathematical Programming and Modeling Techniques in Practice.* New York: John Wiley and Sons Ltd.

Barnard, G.A. 1990. "The 1988 Wald Memorial Lectures: The Present Position in Bayesian Statistics: Comment." *Statistical Science* 5: 65-71.

Bar-Shalom, Y., and E. Tse. 1976. "Caution, Probing and the Value of Information in the Control of Uncertain Systems." *Annals of Economic and Social Measurement* 5: 323-336.

Berger, J.O. 1990. "The 1988 Wald Memorial Lectures: The Present Position in Bayesian Statistics: Comment." *Statistical Science* 5: 71-74.

Bernardo, J.M. 1990. "The 1988 Wald Memorial Lectures: The Present Position in Bayesian Statistics: Comment." *Statistical Science* 5: 75-76.

Burt, O.R. 1982. "Dynamic Programming: Has Its Day Arrived?" *Western Journal of Agricultural Economics* 7: 381-393.

Burt, O.R., W.W. Koo, and N.J. Dudley. 1980. "Optimal Stochastic Control of Wheat Stocks and Exports." *American Journal of Agricultural Economics* 62: 172-187.

Camerer, C.F. 1997. "Progress in Behavioral Game Theory." *The Journal of Economic Perspectives* 11: 167-188.

---

[11] This metaphor comes from a poem published by Maurice Kendall (1959). In the poem, Hiawatha, a brilliant statistician, organized a shooting contest to impress tribal marksmen with his knowledge of experimental design. The other contestants, ignorant benighted creatures, simply put in practice shooting at the target. Perhaps needless to add, Hiawatha never hit the target.

Carlson, G.A. 1970. "A Decision Theoretic Approach to Crop Disease Prediction and Control." *American Journal of Agricultural Economics* 52: 216-223.

Chavas, J.-P. 1994. Production and Investment Decisions Under Sunk Cost and Temporal Uncertainty." *American Journal of Agricultural Economics* 76: 114-127.

Chen, Y.Y. 1995. "Statistical Inference Based on the Possibility and Belief Measures." *Transactions of the American Mathematical Society* 347: 1855-1863.

Chow, G.C. 1975. "A Solution to Optimal Control of Linear Systems with Unknown Parameters." *Review of Economics and Statistics* 57: 338-345.

___. 1976a. "An Approach to the Feedback Control of Nonlinear Econometric Systems." *Annals of Economic and Social Measurement* 5: 297-310.

___. 1976b. "The Control of Nonlinear Econometric Systems with Unknown Parameters." *Econometrica* 44: 685-695.

Ciriani, T.A., and S. Gliozzi. 1995. "Optimization Technology for Large Scale Models." In A. Sciomachen, ed., *Optimization in Industry: Mathematical Programming and Modeling Techniques in Practice.* New York: John Wiley and Sons Ltd.

Cohen, L.J. 1970. *The Implications of Induction.* London: Methuen.

Collins, R.A., and P.J. Barry. 1986. "Risk Analysis with Single-Index Portfolio Models: An Application to Farm Planning." *American Journal of Agricultural Economics* 68: 152-161.

Cox, D.R. 1990. "The 1988 Wald Memorial Lectures: The Present Position in Bayesian Statistics: Comment." *Statistical Science* 5: 76-78.

Draper, D. 1995. "Assessment and Propagation of Model Uncertainty." *Journal of the Royal Statistical Society* 57: 45-97.

Dubois, D., H. Prade, and R.R. Yager (eds.). 1997. *Fuzzy Information Engineering.* New York: John Wiley & Sons, Inc.

Duval, Y., and A.M. Featherstone. 1999. "Fuzzy Logic and Compromise Programming in Portfolio Management." Paper presented at the Western Agricultural Economics Association Annual Meeting, Fargo, ND, July 11-14, 1999.

Einhorn, H.J., and R.M. Hogarth. 1986. "Decision Making under Ambiguity." *Journal of Business* 59: S225-S250.

Feldstein, M.S. 1971. "Production with Uncertain Technology: Some Economic and Econometric Implications." *International Economic Review* 12: 27-38.

French, S. 1990. "The 1988 Wald Memorial Lectures: The Present Position in Bayesian Statistics: Comment." *Statistical Science* 5: 78-80.

Freund, R.J. 1956. "Introduction of Risk into a Programming Model." *Econometrica* 24: 253-263.

Gaivoronski, A.A. 1995. "Stochastic Programming Approach to the Network Planning under Uncertainty." In A. Sciomachen, ed., *Optimization in Industry: Mathematical Programming and Modeling Techniques in Practice.* New York: John Wiley and Sons Ltd.

Galbraith, J.K. 1992. *The Culture of Contentment.* New York: Houghton Mifflin Company.

Gardner, B.L. 1979. *Optimal Stockpiling of Grain.* Lexington, MA: D.C. Heath.

Gebhardt, J., and R. Kruse. 1997. "POSSINFER: A Software Tool for Possibilistic Inference." In D. Dubois, H. Prade, and R.R. Yager, eds., *Fuzzy Information Engineering.* New York: John Wiley & Sons, Inc.

Gibbons, R. 1992. *Game Theory for Applied Economists.* Princeton, NJ: Princeton University Press.

Good, I.J. 1972. "Some Probability Paradoxes in Choice from Among Random Alternatives: Comment." *Journal of the American Statistical Association* 67: 374-375.

___. 1988. "The Interface between Statistics and Philosophy of Science." *Statistical Science* 3: 386-412.

Goodhue, R.E., and N.A. McCarthy. 1998. "Beyond Mobility: The Role of Fuzzy Access Rights and Common Property Considerations in Semi-Arid African Pastoralist Systems." Paper presented at the American Agricultural Economics Association Annual Meeting, Salt Lake City, UT, 1998.

Hazell, P.B.R. 1970. "Game Theory – An Extension of Its Application to Farm Planning under Uncertainty." *Journal of Agricultural Economics* 21: 47-60.

___. 1971. "A Linear Alternative to Quadratic and Semivariance Programming for Farm Planning Under Uncertainty." *American Journal of Agricultural Economics* 53: 53-62.

___. 1971. "Game Theory – An Extension of Its Application to Farm Planning Under Uncertainty." *Journal of Agricultural Economics* 21: 239-252.

Hazell, P.B.R., and R.P. Norton. 1986. *Mathematical Programming for Economic Analysis in Agriculture*. New York: Macmillan.

Hazell, P.B.R., and C. Pomareda. 1981. "Evaluating Price Stabilization Schemes with Mathematical Programming." *American Journal of Agricultural Economics* 63: 550-556.

Hazell, P.B.R., and P.L. Scandizzo. 1977. "Farmers' Expectations, Risk Aversion, and Market Equilibrium under Risk." *American Journal of Agricultural Economics* 59: 204-209.

Hewitt, R.E. 1982. "Multiperiod Optimization: Dynamic Programming vs. Optimal Control: Discussion." *Western Journal of Agricultural Economics* 7: 407-411.

Johnson, S.R. 1998. "Strategic Behavior, Institutional Change and the Future of Agriculture." *American Journal of Agricultural Economics* 80: 898-915.

Just, R.E. 1975a. "Risk Aversion Under Profit Maximization." *American Journal of Agricultural Economics* 57: 347-352.

___. 1975b. "Risk Response Models and Their Use in Agricultural Policy Evaluation." *American Journal of Agricultural Economics* 57: 836-843.

___. 2000. "Risk Research in Agricultural Economics: Opportunities and Challenges for the Next Twenty-Five Years." Paper presented at the annual meeting of Regional Project SERA-IEG-31, Economics and Management of Risk in Agriculture and Natural Resources, Gulf Shores, AL, March 23-25, 2000.

Just, R.E., and R.D. Pope. 1978. "Stochastic Specification of Production Functions and Economic Implications." *Journal of Econometrics* 7: 67-86.

___. 1979. "Production Function Estimation and Related Risk Considerations." *American Journal of Agricultural Economics* 61: 276-284.

Kadane, J.B. 1990. "The 1988 Wald Memorial Lectures: The Present Position in Bayesian Statistics: Comment." *Statistical Science* 5: 80-82.

Kendall, M.G. 1959. "Hiawatha Designs an Experiment." *The American Statistician* 13: 23-24.

Kendrick, P. 1978. "Control Theory with Applications to Economics." In K.J. Arrow and M.P. Intriligator, eds., *Handbook of Mathematical Economics*. Amsterdam: North-Holland.

Kennedy, J.O.S. 1986. *Dynamic Programming: Applications to Agriculture and Natural Resources*. Amsterdam: Elsevier Publishing Co.

Kikuchi, S., and V. Perincherry. 1997. "Use of Possibility Theory to Measure Driver Anxiety During Signal Change Intervals." In D. Dubois, H. Prade, and R.R. Yager, eds., *Fuzzy Information Engineering: A Guided Tour of Applications*. New York: John Wiley and Sons, Inc.

King, R.P., and L.J. Robison. 1981. "Implementation of the Interval Approach to the Measurement of Decision Maker Preferences." Michigan State University Agricultural Experiment Station Research Report 418.

Klein, R.W., L.C. Rafsky, D.S. Sibley, and R.D. Williz. 1978. "Decisions with Estimation Uncertainty." *Econometrica* 46: 1363-87.

Knight, F.V. 1957. *Risk, Uncertainty and Profit*. New York: Kelley and Millman, Inc.

Kramer, R.A., and R.D. Pope. 1981. "Participation in Farm Commodity Programs: A Stochastic Dominance Analysis." *American Journal of Agricultural Economics* 63: 119-128.

Krautkraemer, J.A., G.C. van Kooten, and D.L. Young. 1992. "Incorporating Risk Aversion into Dynamic Programming Models." *American Journal of Agricultural Economics* 74: 870-878.

Kreps, D.M. 1990. *Game Theory and Economic Modeling*. Oxford: Clarendon Press.

Kreps, D.M., and E.L. Porteus. 1979. "Temporal von Neumann-Morgenstern and Induced Preferences." *Journal of Economic Theory* 20: 81-109.

___. 1978. "Temporal Resolution of Uncertainty and Dynamic Choice Theory." *Econometrica* 46: 185-200.

Kulhavy, R. 1993. "Implementation of Bayesian Parameter Estimation in Adaptive Control and Signal Processing." *Statistician* 42: 471-482.

Kydland, F.E., and E.C. Prescott. 1977. "Rules Rather than Discretion: The Inconsistency of Optimal Plans." *Journal of Political Economy* 85: 473-492.

Lano, K. 1991. "Intuitionistic Modal Logic and Set Theory." *The Journal of Symbolic Logic* 56: 497-516.

Leamer, E.A. 1986. "Bid-Ask Spreads for Subjective Probabilities." In P. Goel and A. Zellner, eds., *Bayesian Inference and Decision Techniques*. New York: Elsevier.

Lehmann, E.L. 1990. "The 1988 Wald Memorial Lectures: The Present Position in Bayesian Statistics: Comment." *Statistical Science* 5: 82-83.

Lin, W., G. Dean, and C. Moore. 1974. "An Empirical Test of Utility vs. Profit Maximization in Agricultural Production." *American Journal of Agricultural Economics* 56: 497-508.

Lindley, D.V. 1987. "The Probability Approach to the Treatment of Uncertainty in Artificial Intelligence and Expert Systems." *Statistical Science* 2: 17-24.

___. 1990a. "The 1988 Wald Memorial Lectures: The Present Position in Bayesian Statistics." *Statistical Science* 5: 44-65.

___. 1990b. The 1988 Wald Memorial Lectures: The Present Position in Bayesian Statistics: Rejoinder." *Statistical Science* 5: 85-89.

Machina, M.J., and D. Schmeidler. 1992. "A More Robust Definition of Subjective Probability." *Econometrica* 60: 745-780.

Marks, L.A., and E.G. Dunn. 1999. "Evaluating Alternative Farming Systems: A Fuzzy MADM Approach." Paper presented at the American Agricultural Economics Association Annual Meeting, Nashville, TN, 1999.

MacRae, E.C. 1972. "Linear Decision with Experimentation." *Annals of Economic and Social Measurement* 1: 437-447.

McRae, E.C. 1975. "An Adaptive Learning Rule for Multiperiod Decision Problems." *Econometrica* 43: 893-906.

McRae, E.C., L.C. Rafsky, D.S. Sibley, and R.D. Williz. 1978. "Decisions with Estimation Uncertainty." *Econometrica* 46: 1363-1387.

Mehr, R.I. 1983. *Fundamentals of Insurance*. Homewood, IL: Richard D. Irwin, Inc.

Meyer, J. 1977. "Choice Among Distributions." *Journal of Economic Theory* 14: 326-336.

Mouchart, M. 1990. "The 1988 Wald Memorial Lectures: The Present Position in Bayesian Statistics: Comment." *Statistical Science* 5: 84-85.

Nau, R.F. 1992. "Indeterminate Probabilities on Finite Sets." *The Annals of Statistics* 20 (1992): 1737-1767.

Paris, Q., and C.D. Easter. 1985. "A Programming Model with Stochastic Technology and Price: The Case of Australian Agriculture." *American Journal of Agricultural Economics* 67: 120-129.

Pinker, S. 1994. *The Language Instinct*. New York: Morrow and Co.

Pope, R.D. 1982a. "Empirical Estimation and Use of Risk Preferences: An Appraisal of Estimation Methods That Use Actual Economic Decisions." *American Journal of Agricultural Economics* 64: 376-383.

___. 1982b. "Expected Profit, Price Change and Risk Aversion." *American Journal of Agricultural Economics* 64: 581-584.

Rausser, G.C. 1978. "Active Learning, Control Theory and Agricultural Policy." *American Journal of Agricultural Economics* 60: 476-490.

Richardson, J.W., and C.J. Nixon. 1986. "Description of FLIPSIM 5: A General Firm Level Policy Simulation Model." Texas Agricultural Experiment Station Bulletin No. 1528 (July).

Savage, L.J. 1954. *The Foundations of Statistics*. New York: Wiley and Sons, Inc.

Shafer, G. 1987. "Probability Judgement in Artificial Intelligence and Expert Systems." *Statistical Science* 2: 3-16.

___. 1990. "The Unity and Diversity of Probability." *Statistical Science* 5: 435-444.

Sciomachen, A. (ed.). 1995. *Optimization in Industry: Mathematical Programming and Modeling Techniques in Practice.* New York: John Wiley and Sons Ltd.

Shackle, G.L.S. 1961. *Decision, Order and Time in Human Affairs.* Cambridge: Cambridge University Press.

Shafer, G. 1976. *A Mathematical Theory of Evidence.* Princeton, NJ: Princeton University Press.

___. 1990. "The Unity and Diversity of Probability." *Statistical Science* 5: 435-444.

Singpurwalla, N.D. 1988. "Foundational Issues in Reliability and Risk Analysis." *SIAM Review* 30: 264-282.

Swanson, E.R. 1959. "Selection of Crop Varieties: An Illustration of Game Theoretic Technique." *Revista Internazionale di Scienze Economiche e Commercialli* 6: 3-14.

Talpaz, H. 1982. "Multiperiod Optimization: Dynamic Programming vs. Optimal Control: Discussion." *Western Journal of Agricultural Economics* 7: 407-411.

Taylor, C.R. 1984. "Stochastic Dynamic Duality: Theory and Empirical Applicability." *American Journal of Agricultural Economics* 66: 351-357.

___. 1986. "Risk Aversion versus Expected Profit Maximization with a Progressive Income Tax." *American Journal of Agricultural Economics* 68: 137-143.

___. 1991. "The Fading Curse of Dimensionality." In M.E. Wetzstein and C.R. Taylor, eds., *Analyzing Dynamic and Stochastic Agricultural Systems.* Proceedings of a technical symposium by the Consortium for Research on Crop Production Systems, March 1991.

___ (ed.). 1993a. *Applications of Dynamic Programming to Agricultural Decision Problems.* Boulder, CO: Westview Press.

___. 1993b. "Dynamic Programming and the Curses of Dimensionality." In C.R. Taylor, ed., *Applications of Dynamic Programming to Agricultural Decision Problems.* Boulder, CO: Westview Press.

___. 1994. Deterministic vs. Stochastic Evaluation of the Aggregate Effects of Price Support Programs." *Agricultural Systems* 44: 461-474.

Taylor, C.R., and J.-P. Chavas. 1980. "Estimation and Optimal Control of an Uncertain Production Process." *American Journal of Agricultural Economics* 62: 675-680.

Taylor, C.R. and H. Talpaz. 1979. "Approximately Optimal Levels of Wheat Stocks in the United States." *American Journal of Agricultural Economics* 61: 32-40.

Tintner, G. 1941. "The Pure Theory of Production under Technological Risk and Uncertainty." *Econometrica* 9: 305-312.

Tse, E., Y. Bar-Shalom, and L. Meier. 1973. "Wide-Sense Adaptive Dual Control of Stochastic Nonlinear Systems." *IEEE Transactions of Automatic Control* AC-18: 98-108.

Tversky, A., and D. Kahneman. 1992. "Advances in Prospect Theory: Cumulative Representations of Uncertainty." *Journal of Risk and Uncertainty* 5: 297-323.

Zacharias, T.P. 1993. "Representation of Preferences in Dynamic Optimization Models Under Uncertainty." In C.R. Taylor, ed., *Applications of Dynamic Programming to Agricultural Decision Problems.* Boulder, CO: Westview Press.

Zacharias, T.P., and E.R. Swanson. 1982. "A Method for Selecting From Among Soybean Varieties and Blends." Illinois Agricultural Economics Staff Paper No. 82 E-219 (April).

Zadeh, L.A. 1978. "Fuzzy Sets as a Basis for a Theory of Possibility." *Fuzzy Sets and Systems* 1: 3-28.

___. 1983. "The Role of Fuzzy Logic in the Management of Uncertainty in Expert Systems." *Fuzzy Sets and Systems* 11: 199-227

Zimmerman, F., and M.R. Carter. 1996. "Dynamic Portfolio Management Under Risk and Subsistence Constraints in Developing Countries." Staff Paper No. 402, Department of Agricultural Economics, University of Wisconsin (November).

Chapter 11

THE ECONOMETRICS OF RISK

Matthew T. Holt and Jean-Paul Chavas
North Carolina State University and University of Wisconsin, Madison

INTRODUCTION

The purpose of this chapter is to review some of the recent relevant literature on the empirical implementation and testing of risk in agricultural production decisions. As well, it suggests several potentially fruitful areas for future investigation. In many respects agriculture and risk are synonymous. Agricultural economists have long recognized risk as a primary distinguishing feature of the agriculture sector relative to other sectors of the economy. Two overriding factors have contributed significantly to the observed variability in price, income, and production realizations for agricultural producers over time: the fact that production – and especially field crop production – is stochastic, and the received empirical truth that the demands, both domestic and foreign, for many agricultural products are highly price inelastic. The inherent biological lags that describe most physical agricultural production processes and, as well, the reliance of the agricultural sector on highly volatile international markets have also added to the instability in agriculture markets. Given these fundamental characterizing features of agricultural production and marketing environments, a primary question is: Do agricultural decision makers account for these inherent risks in any meaningful and systematic way when making production and investment decisions? As well, to what extent does the intrinsic underlying stochastic nature of agricultural production make it necessary to modify extant economic models, both theoretical and empirical, of production and producer decision making? Attempts to address these two fundamental questions have been the underlying motivating force for the research programs of many professional agricultural economists during the past twenty-five years (Pope 1982).

During this period of time, several approaches to modeling the effects of risk in agriculture have been adopted. First, a central issue has been the investigation of risk preferences. This has been done through direct preference elicitation (e.g., Lin, Dean, and Moore 1974, Dillon and Scandizzo 1978, Binswanger 1981) or through the analysis of observed behavior (e.g.,

Moscardi and de Janvry 1977, Antle 1987, Saha, Shumway, and Talpaz 1994, Chavas and Holt 1996). The analysis has focused on the expected utility (EU) model. Although there is some concern that the EU model does not always give an accurate representation of behavior under risk (e.g., Machina 1987), it remains the cornerstone of empirical economic analysis. In this context, aversion to risk (where risk exposure makes individuals worse off) is associated with a concave von Neumann-Morgenstern utility function (Pratt 1964). The empirical evidence strongly indicates that most farmers are risk averse (e.g., Lin, Dean, and Moore 1974, Dillon and Scandizzo 1978, Binswanger 1981, Antle 1987, Chavas and Holt 1996).

Second, a large amount of research on agricultural risk has been normative in nature, focusing on making recommendations about risk management strategies. Typically, under risk aversion, firms (agents) are assumed to maximize expected utility subject to production and other technical constraints. The most often used approach is to assume that firms attempt to maximize a mean-variance utility function, the result being that quadratic programming methods may be used to obtain a model solution once the first and second moments of the relevant price or revenue distributions have been determined (Meyer 1987). The ultimate goal of these studies is typically to trace out mean-variance efficiency frontiers under risk aversion (e.g., Hardaker, Huirne, and Anderson 1997).

Third, empirical analyses of risk have attempted to econometrically estimate the effects of risk on actual production data. Various assumptions are typically made about the nature of underlying risk preferences, the precise process by which expectations are formed, and the exact nature of the underlying technical production relationship. The goal of such studies, at least in agricultural economics, has typically been to quantify the role of risk in production decisions, although several studies have attempted to test directly the structural implications of behavioral risk. The primary focus of this chapter is on the role of the latter, the econometric approach, in agricultural economics research. This said, several observations about the potential for merging the programming approach within an econometric framework will also be provided.

Finally, it should be stressed that any empirical analysis of risk requires risk measurements. This is a challenging issue. There is a strong consensus that risk is a prevalent characteristic of agricultural decision making, and that most decision makers are risk averse. Both risk preferences and risk exposure may, however, vary across individuals. This makes empirical risk analysis difficult. Typically, risk (e.g., weather risk, price risk, etc.) is represented by random variables with given probability distributions. In situations where risk involves repeatable events, probability distributions can be estimated from sample information. Empirically, it is often convenient to summarize probability distributions using sufficient statistics, and/or to focus on estimation of moments (e.g., the mean as a measure of central tendency, and the variance as a measure of volatility or spread). For non-repeatable events, economists often rely on a subjective interpretation of probabilities. In this instance probabilities are personal and can and do vary across individuals depending on their access to information and their ability to process it. This approach has the advantage of providing a rationale for why different people hold different beliefs. However, heterogeneity of expectations makes it challenging to assess individual risk exposure empirically. This seems particularly important when there is asymmetric information about the nature, source, and extent of risk.

In many respects this chapter is the logical successor to a 1982 article authored by Rulon Pope entitled "Empirical Estimation and Use of Risk Preferences: An Appraisal of Estimation Methods That Use Actual Economic Decisions," published in the *American Journal of Agricultural Economics*. In this article Pope details what were at the time current and emerging issues for investigating empirically the role of risk in agricultural production decisions. Where have we gone and where have we been in the ensuing nineteen years since this article was published? For which areas has substantial headway been made and which areas remain as likely targets for future investigation? What we argue in this chapter is that while progress has certainly been made – mostly due to improved estimation methods and enhanced computer capacity available at increasingly lower costs, neither of which could of course have been fully anticipated by Pope and others working in this field twenty years ago – many potentially productive areas of research have gone largely unexplored. In this chapter we attempt to highlight several of these areas.

The remainder of the chapter is organized as follows. To motivate the discussion, the next section reviews some of the basic, extant theory on producer behavior under uncertainty. Prior work on modeling and estimating the effects of risk on production decisions is reviewed in the section, "Empirical Work: Where Have We Been?" A discussion of possible future areas of empirical investigation is provided in the penultimate section, "Where Might We Go?" The final section concludes.

THE BASIC THEORY

Serious empirical investigations of the role of risk in agricultural supply and production decisions were subsequent to the theoretical contributions of Pratt, Arrow, and Sandmo. The basic expected utility (EU) paradigm that evolved from the efforts of these individuals (and others) provided agricultural economists with a framework from which to build models that embodied testable hypotheses with respect to risk response. The fundamental tenants of the EU model include price-taking firms operating in competitive markets, firms making production decisions before output prices are revealed, and a concave von Neumann Morgenstern utility function defined over wealth, profit, or income. At least some of these features appear to be reasonably descriptive of the agricultural production environment.

In the EU framework producers are assumed to maximize expected utility of wealth (income) when making production decisions. Under fairly loose regularity conditions the result is a system of optimal supply and input demand equations. In the simple case where there is no production risk, the basic expected utility model may be specified as

$$\max_{\boldsymbol{x}}[EU(w_0+\pi;\boldsymbol{\alpha})|\, y=f(\boldsymbol{x};\boldsymbol{\tau}),\, \boldsymbol{x}\in\Re^n_+], \tag{1}$$

where

$$\pi = py-\boldsymbol{r}^T\boldsymbol{x}. \tag{2}$$

In (1) and (2), we define $\boldsymbol{x}=(x_1,\ldots,x_n)^T$ as an n-vector of inputs, where a superscripted T denotes vector (matrix) transposition; $\boldsymbol{r}=(r_1,\ldots,r_n)^T$ is a corresponding n-vector of input prices; w_0 represents initial (non-random) wealth and π denotes profit; $U(w_0+\pi;\boldsymbol{\alpha})$ is the von Neumann Morgenstern utility function, where $\boldsymbol{\alpha}$ denotes a vector of preference parameters; $y=f(\boldsymbol{x};\boldsymbol{\tau})$ is a strictly concave and deterministic production function that defines scalar output, y, and where $\boldsymbol{\tau}$ is a scalar output, y, and where $\boldsymbol{\tau}$ is a vector of technology parameters; $p=\overline{p}+e$ denotes stochastic output price, where $E(p)=\overline{p}$, $E(e)=0$; and $E(e^k)=\sigma^k_p$, $k=2,\ldots,m$, where σ^k_p denotes the kth central moment of the price distribution. Also, E of course denotes the conditional expectation operator such that $E(z\,|\,\Omega)=\overline{z}$ for any random z, where Ω is the information set (sigma field) relevant at the time decisions are made. More formally,

$$E(\boldsymbol{z}\,|\,\Omega)=\int_0^{t_1}\cdots\int_0^{t_n} \boldsymbol{z}\,dF(\boldsymbol{z}\,|\,\Omega), \tag{3}$$

for any random vector $\boldsymbol{z}=(z_1,\ldots,z_n)^T$, where $F(\cdot)$ is the multivariate distribution function associated with $\boldsymbol{z}$.

Assuming a unique interior solution to the optimization problem in (1)-(2), such a solution may therefore be defined by

$$\boldsymbol{v}^* = \arg\max_{\{y,\boldsymbol{x}\}}[EU(w_0+py-\boldsymbol{r}^T\boldsymbol{x};\boldsymbol{\alpha})\,|\,y=f(\boldsymbol{x};\boldsymbol{\tau}),\,\boldsymbol{x}\in\Re^n_+], \tag{4}$$

where $\boldsymbol{v}^T=(y,\boldsymbol{x}^T)$. To further characterize (4), consider the Lagrangian,

$$L(y,\boldsymbol{x},\lambda;w,\boldsymbol{r},\boldsymbol{\alpha},\boldsymbol{\tau})=EU(w_0+py-\boldsymbol{r}^T\boldsymbol{x};\boldsymbol{\alpha})+\lambda[y-f(\boldsymbol{x};\boldsymbol{\tau})],$$

where $\lambda \in R_+$ is a Lagrange multiplier. Assuming full differentiability, the first-order necessary conditions associated with the optimization problem in (1)-(2) are

$$L_y(y^*,x^*,\lambda^*;w,\boldsymbol{r},\alpha,\tau) = E[U_{\widetilde{w}}(w_0 + py^* - \boldsymbol{r}^T\boldsymbol{x}^*;\alpha)p] + \lambda^* = 0, \tag{5a}$$

$$\begin{aligned} &L_{\boldsymbol{x}}(y^*,\boldsymbol{x}^*,\lambda^*;w,\boldsymbol{r},\alpha,\tau) = -EU_{\widetilde{w}}(w_0 + py - \boldsymbol{r}^T\boldsymbol{x};\alpha)\boldsymbol{r}^T \\ &-\lambda^* f_{\underline{x}}(\boldsymbol{x};\tau) = \boldsymbol{0}^T, \end{aligned} \tag{5b}$$

$$L_\lambda(y^*,\boldsymbol{x}^*,\lambda^*;w,\boldsymbol{r},\alpha,\tau) = y^* - f(\boldsymbol{x}^*;\tau) = 0, \tag{5c}$$

where $\widetilde{w} = w_0 + \pi$ and where subscripted letters denote derivatives. By combining (5a) and (5b), the Lagrange multiplier may be eliminated and the first-order conditions expressed more compactly as

$$\begin{aligned} &L_{\boldsymbol{x}}(y^*,\boldsymbol{x}^*,\lambda^*;w,\boldsymbol{r},\alpha,\tau) = -E\{U_{\widetilde{w}}[w_0 + py - \boldsymbol{r}^T\boldsymbol{x};\alpha] \\ &[pf_{\boldsymbol{x}}(\boldsymbol{x};\tau)\boldsymbol{r}^T]\} = \boldsymbol{0}^T \end{aligned} \tag{6a}$$

$$L_\lambda(y^*,\boldsymbol{x}^*,\lambda^*;w,\boldsymbol{r},\alpha,\tau) = y^* - f(\boldsymbol{x}^*;\tau) = 0. \tag{6b}$$

The solution of the $n + 1$ equation system in (6a) and (6b) defines optimal input demand and output supply decisions under risk aversion where output price is random. Assuming that certain regularity conditions associated with (5) are satisfied, the model provides a complete characterization of the production problem.[1] That is, the solutions to (6a) and (6b) may be characterized in general form as

$$x_i = x_i^*(\overline{p},\boldsymbol{\sigma}_p,w_0,\boldsymbol{r},\alpha,\tau), \quad i = 1,\ldots,n, \tag{7a}$$

$$y = y^*(\overline{p},\boldsymbol{\sigma}_p,w_0,\boldsymbol{r},\alpha,\tau), \tag{7b}$$

where $\boldsymbol{\sigma}_p = (\sigma_p^2,\sigma_p^3,\ldots,\sigma_p^k)^T$ is a vector of relevant higher-order moments of the underlying price distribution. Therefore, by using comparative static

[1] These regularity conditions are (1) the functions $U(\cdot)$ and $f(\cdot)$ are continuous and twice continuously differentiable; (2) the $(n + 1) \times 1$ vector (or conformable matrix, in the case of multiple outputs) $g_{\boldsymbol{v}} = \partial g(\boldsymbol{v};\tau)/\partial\boldsymbol{v}$ has rank one, where $\boldsymbol{v}^T = (y,\boldsymbol{x}^T)$ and $g(\cdot)$ is defined such that $g(\boldsymbol{v};\tau) = y - f(\boldsymbol{v};\tau)$; and (3) the second-order sufficient conditions hold: $\boldsymbol{z}^T L_{\boldsymbol{vv}}\boldsymbol{z} < 0$ for all $\boldsymbol{z} \in \Re^{n+1}$ such that $\boldsymbol{z} \neq \boldsymbol{0}$ and $g_{\boldsymbol{v}}\boldsymbol{z} = 0$.

analysis, Chavas, Pope, and Leathers (1988), Batra and Ullah (1974), Ishii (1977), Just and Zilberman (1983), and Sandmo (1971), among others, have attempted to characterize the properties of the solutions in (7) under general conditions. That is, the theoretical properties of risk aversion for production decisions under price uncertainty have been investigated by using a framework similar to that outlined above. The analysis depends on the nature of risk behavior. This can be characterized by the coefficient of absolute risk aversion, $\varphi = -U_{\tilde{w}\tilde{w}}/U_{\tilde{w}}$, where $U_{\tilde{w}} > 0$ by assumption. In general, risk aversion corresponds to situations where the risk premium (measuring the shadow cost of private risk bearing) is positive. As shown by Pratt (1964), this corresponds to $\varphi = -U_{\tilde{w}\tilde{w}}/U_{\tilde{w}} > 0$, implying a concave utility function. In addition, there is some consensus that private wealth accumulation tends to reduce the shadow cost of risk exposure. This corresponds to the decreasing absolute risk aversion (DARA) hypothesis where the risk premium decreases with initial wealth, or equivalently where $\varphi = -U_{\tilde{w}\tilde{w}}/U_{\tilde{w}}$ is a decreasing function of $\tilde{w}$ (Pratt 1964). The empirical evidence indicates that most farmers exhibit risk aversion as well as DARA (e.g., Binswanger 1981, Chavas and Holt 1996). In this context, the following results apply:

- Supply curves are upward-sloping in expected price under non-increasing absolute risk aversion (where $\varphi = -U_{\tilde{w}\tilde{w}}/U_{\tilde{w}}$ is non-increasing in $\tilde{w}$).
- A risk increase measured by a mean-preserving spread (e.g., generated by an increase in σ_p) will, under the same assumption, result in lower optimal output.
- If the sole source of uncertainty is associated with output price, standard cost minimization is consistent with risk aversion.
- For non-inferior inputs (where cost-minimizing inputs tend to increase with output), input demands are downward-sloping in their own price under non-increasing absolute risk aversion.
- For non-inferior inputs, a risk increase (e.g., generated by an increase in σ_p) has a negative impact on input demand under non-increasing absolute risk aversion.
- Under decreasing absolute risk aversion (DARA), increasing initial wealth (or decreasing fixed cost) tends to stimulate output supply as well as the demand for non-inferior inputs. In addition, under constant absolute risk aversion (CARA), where the risk premium or, equivalently, the absolute risk aversion coefficient $\varphi = -U_{\tilde{w}\tilde{w}}/U_{\tilde{w}}$ is independent of initial wealth w_0), changing initial wealth has no impact on output supply and input demand.

Of course these and other propositions can be verified in a meaningful way only by empirical investigation. As well, to be relevant to agriculture, the model outlined above needs to be extended to include (1) multiple outputs and (2) production uncertainty. Either or both of these extensions make it all the more difficult to obtain meaningful comparative statics results for the risk-averse firm except in the most restrictive of circumstances. As well, this later extension – allowing the basic model to include stochastic production – has been a major research thrust in the agricultural economics profession (Chambers and Quiggin 1998, 2000, Moschini 2001, Pope and Chavas 1994, Just and Pope 1978, Pope and Just 1996).

EMPIRICAL WORK: WHERE HAVE WE BEEN?

The basic EU model under price uncertainty outlined in the previous section has served as the backdrop for much empirical research in agricultural economics. By examining the solutions for the optimal input demand/output supply decisions in (7), it is apparent that any serious empirical investigation of risk must be based on a number of simplifying assumptions. Specifically, depending on the approach taken, it may be necessary to specify the nature of the underlying utility function, $U(\cdot)$; the underlying production technology, $f(\cdot)$; the nature of the underlying price (revenue) distribution; and, in nearly all instances, the precise manner in which conditional expectations about relevant moments of the underlying price (revenue) distribution are formed and updated. This latter point is important because it underscores the fact that any discussion of methods used to estimate risk effects must be inextricably linked to the framework used to specify and estimate the empirical distribution of risk (e.g., as measured by conditional moments).

Single-Equation Reduced-Form Methods

Many early and continuing attempts to infer risk effects in agricultural production focused on acreage supply decisions, both in developed and developing country contexts. In all instances acreage planting models that include risk have been estimated with time series data, and in linear form. With regards to the theory, recall that the optimal supply equation in (7b) is $y = y^*(\bar{p}, \sigma_p, w_0, \boldsymbol{r}, \alpha, \tau)$. If only first and second moments of price are appropriate to the decision, then a first-order Taylor series approximation of this supply equation yields

$$y = b_0 + (\partial y^* / \partial \bar{p})\bar{p} + (\partial y^* / \partial \sigma_p^2)\sigma_p^2 + (\partial y^* / \partial w_0) w_0 + (\partial y^* / \partial \boldsymbol{r})\boldsymbol{r}, \qquad (8)$$

or, alternatively,

$$y = b_0 + b_1\overline{p} + b_2\sigma_p^2 + b_3 w_0 + \boldsymbol{c}^T \boldsymbol{r}, \tag{9}$$

where in both (8) and (9) all remaining technology and risk preference terms are subsumed into the intercept term b_0, where $\boldsymbol{c} = (c_1, \ldots, c_n)^T$, and where the definitions of the b_i and c_i terms in (9) are apparent from the definitions in (8). In many instances either CARA preferences are assumed (implying that $\partial y^*/\partial w_0 = b_3 = 0$) or data on initial wealth are not available, so that the terms $(\partial y^*/\partial w_0) w_0$ in (8) (respectively, the $b_3 w_0$ term in (9)) are dropped from the empirical analysis.[2]

To introduce production risk, consider the case where production y is defined as

$$y = aq, \tag{10}$$

where a is acreage planted and q is random yield satisfying $q = \overline{q} + \eta$, where $E(q) = \overline{q}$ is expected yield, and $E(e) = 0$. In general, the distribution of yield is expected to depend on input decisions (e.g., fertilizer, pesticides, etc.). We anticipate the marginal effects of inputs on expected productivity to be non-negative and decreasing. But inputs can also affect the farmer's risk exposure (e.g., as measured by the variance of yield or revenue). While some inputs may be risk increasing (e.g., fertilizer; see Just and Pope 1978), others appear to be risk decreasing (e.g., pesticides). To the extent that farmers are risk averse, this would provide some incentive for farmers to stimulate (decrease) demand for inputs that are risk reducing (risk increasing). In that case, both production risk and price risk would influence production decisions.

In general, a key production decision is the acreage decision. Analyzing acreage decisions under risk has been the subject of much empirical research for two reasons. First, the acreage allocation has a very large impact on agricultural production. Second, acreage decisions are typically made at the beginning of the growing season, at a time when there is still much uncertainty about future market and weather conditions. The implication is that optimal acreage decisions typically depend on both price risk and production risk (e.g., weather risk). From equation (10), note that revenue can be expressed as $py = pqa$. Regarding equation (4), this suggests that the acreage decision has properties similar to output y after replacing output price p by the "revenue per acre" (pq) – under risk aversion, both price risk and production risk influence the acreage decision. With this interpretation in mind, corresponding to (9), the acreage decision may be specified as

---

[2] An exception is the study by Chavas and Holt (1990), where a proxy variable for initial wealth was used to estimate corn and soybean acreage supply decisions in the U.S.

$$a = \beta_0 + \beta_1 \bar{p} + \beta_2 \sigma_p^2 + \gamma^T \boldsymbol{r} + \varepsilon, \tag{11}$$

where $\bar{p}$ now denotes expected revenue per acre, and σ_p^2 is the variance of revenue per acre. Once a mechanism for generating the conditional first and second moments of revenue per acre is specified, the model in (11) is easily estimated by using linear least squares procedures. Models similar to that in (11) have been used to investigate the role of price or revenue risk in acreage planting decisions by Barten and Vanloot (1996), Behrman (1968), Chavas and Holt (1990), Holt (1994, 1999), Duffy, Shalishali, and Kinnucan (1994), Krause and Koo (1996), Krause, Lee, and Koo (1995), Lin (1977), Pope and Just (1991), and Traill (1978), among others. As well, Holt (1993), Holt and Aradhyula (1990, 1998), Holt and Moschini (1992), and Tronstad and McNeill (1989) have estimated models similar to (11) for risk-responsive livestock supply equations.

Single-Equation Reduced-Form Methods – Issues

Several observations are in order regarding (11). First, the equation (or system of equations, as the case may be), while generally estimable by least squares methods, is at best a reduced-form specification. It is therefore rather difficult to obtain insights into the nature of underlying risk preferences. Employing parametric tests of the restrictions developed by Pope (1988) has, at least to a certain extent, circumvented this problem. Specifically, Pope develops a set of linear relationships designed to test for CARA, constant relative risk aversion (CRRA), and constant partial relative risk aversion (CPRRA) in the structure of risk preferences. Tests for these specific forms of risk aversion may then be conducted by using either likelihood ratio (LR) or Wald-type tests. An important caveat is that for these tests to work, (11) must include a measure of initial *wealth*. Chavas and Holt (1990), Pope and Just (1991), and Holt (1994) have all reported results of (local) tests of CARA, CRRA, and CPRRA preferences in what are otherwise reduced-form supply models.

A second observation regarding (11) is, as already noted, that some method must be used to generate the conditional expectations of the first and second moments of revenue or price. Of course the role of price expectations and, by extension, price risk expectations has been a dominant feature of agricultural supply response analysis for many years (see, e.g., Chavas 2000 for a recent review). Following Ezekiel (1938), agricultural supply response was for many years thought to respond only to the previous period's price. Slightly more sophisticated is the adaptive expectations hypothesis, which assumes that agents adjust their expectations about future prices in a manner consistent with an MA(1) process (Muth 1960). The adaptive expectations hypothesis was introduced into the agricultural supply response analysis literature by Nerlove (1958), and remains as one of the more popular methods for modeling price and price risk expectations. Other methods used for modeling price expectations in agriculture include rational expectations, introduced originally by Muth (1961), rolling moving average methods, and quasi-rational expectations (see, e.g., Nerlove and Fornari 1998).

The simplest method for constructing empirical measures of expected price and price risk (or, for that matter, revenue and revenue risk expectations) is a simple rolling weighted average. Expectation models of this sort typically take the form

$$\bar{p}_t = \sum_{i=1}^{k} (m_i / m) p_{t-i}, \sum_{i=1}^{k} m_i = m, \tag{12a}$$

and

$$\sigma_{pt}^2 = \sum_{i=1}^{k} (m_i / m)(p_{t-i} - \bar{p}_{t-i})^2 , \tag{12b}$$

where the subscript t is a time index, $t = 1,\ldots,T$, and the m_i weights in (12a) and (12b) are typically restricted to be the same. In this setting both the m_i weights and the lag length, k, are generally determined a priori by the investigator. The above approach has been used by, among others, Brorsen et al. (1985), Chavas and Holt (1990, 1996), Holt (1999), and von-Massow and Weersink (1993). While the weighted moving average approach is relatively easy to implement empirically in that the moments in (12) may be constructed outside of the framework used to estimate (11) – that is, combining (12) with (11) truly results in a linear estimation problem – it leaves much to be desired. Pagan and Ullah (1988) have in fact provided a thorough criticism of this approach. At a minimum, a potentially more satisfactory approach would be to determine the weights and the lag length empirically. In part this is what is done when (12a) and (12b) are specified as finite polynomial lags as in, for example, Lin (1977).

One of the more popular and enduring methods for deriving expectations in risk models such as (11) is Just's generalization to the adaptive expectations model. The idea is that if price expectations evolve over time in a manner consistent with geometrically declining lags, then the mean price expectation may be determined from

$$\bar{p}_t = \theta \sum_{i=0}^{\infty} (1-\theta)^i p_{t-i-1} = \bar{p}_{t-1} + \theta(p_{t-1} - \bar{p}_{t-1}),\ \theta \in [0,1]. \tag{13a}$$

As previously mentioned, this expectations model was popularized in agricultural supply analysis by Nerlove (1958). But as Muth (1960) showed,

(13a) is an optimal predictor of p_t in the sense of providing minimum mean square error forecasts if and only if the process generating p_t is ARIMA(0,1,1). That is,

$$p_t - p_{t-1} = e_t - (1-\theta)e_{t-1}$$

implies

$$p_t = \theta\sum_{i=0}^{\infty}(1-\theta)^i p_{t-i-1} + e_t,$$

which is then consistent with (13a).

The intuitively logical extension of (13a) to the conditional second moment of price is

$$\sigma_{pt}^2 = \theta\sum_{i=0}^{\infty}(1-\theta)^i (p_{t-i-1} - \bar{p}_{t-i-1})^2, \qquad (13b)$$

which is the conditional variance model put forth originally by Just (1974). Of course the geometric weighting parameter θ in (13b) may or may not be constrained to be equal to that in (13a). One drawback of this specification is that in comparison with (13a) the specification in (13b) is consistent with an integrated generalized autoregressive conditional heteroscedastic (IGARCH) model (i.e., an IGARCH(1,0) model) in the squared innovations $\{e_t^2\}$, where the unconditional variance of the $\{e_t\}$ process is zero. See Engle and Bollerslev (1986) or Bollerslev and Mikkelsen (1996) for additional details. Such a specification is, of course, potentially restrictive, and there is apparently little if any empirical support for such a model reported in the existing literature.

Nevertheless, (13a) and (13b) provide a logical framework for estimating risk effects in the context of supply equation (11). In estimation the lag length is typically truncated at a finite point k, say, in which case a grid search may be done over the admissible values of θ, in which case conditional maximum likelihood (ML) estimates of the parameters in (11) are obtained. Estimates of models similar to (11) that utilize the expectations mechanisms in (13) are reported by Bar-Shira, Just, and Zilberman (1997), Barten and Vanloot (1996), Estes, Blakeslee, and Mittelhammer (1981), Holt and Aradhyula (1990), Just (1974), Pope and Just (1991), and White and Ziemer (1982).

A third option for modeling risk that has been utilized to some effect in recent years is the autoregressive conditional heteroscedastic (ARCH) and generalized ARCH (GARCH) class of models introduced respectively by Engle (1982) and Bollerslev (1986), as well as some of the semiparametric and nonparametric methods recently introduced. Turning first to the ARCH/GARCH class of models, assume that a finite order autoregressive process generates the price series $\{p_t\}$. That is,

$$[1-\Phi(L)]p_t = \delta_0 + e_t,\ e_t \mid \Omega_{t-1} \sim N(0,\sigma_p^2), \tag{14}$$

where $\Phi(L) = \Phi_1 L + \cdots + \Phi_k L^k$ is a polynomial in the lag operator L of order k, such that $Lx_t = x_{t-1}$, and where sufficient lags have been added to the autoregression so as to render e_t white noise.[3] That is, additional properties typically attributed to $\{e_t\}$ include $E(e_t e_s) = 0\ \forall t \neq s$. As may be readily verified, the conditional expectation (mean) of p_t based on (14a) is

$$\overline{p}_t = \delta_0 + \Phi_1 p_{t-1} + \cdots + \Phi_k p_{t-k}. \tag{15}$$

Indeed, in the case where (11) contains no risk terms, direct substitution of (15) for $\overline{p}_t$ yields

$$a_t = \beta_0 + \beta_1(\delta_0 + \Phi_1 p_{t-1} + \cdots + \Phi_k p_{t-k}) + \varsigma^T \boldsymbol{r}_t + \varepsilon_t, \tag{16}$$

which, when estimated jointly with (14), results in a form of the quasi-rational expectations model explored extensively by Nerlove, Grether, and Carvalho (1979), Nerlove and Fornari (1998), and Chavas (2000).

The issue of how to generate a time-varying conditional expectation of the variance of p_t from (14) remains, of course, an open question. Bollerslev's (1986) GARCH approach basically assumes that σ_p^2 is heteroscedastic and, moreover, that the specific form of heteroscedasticity generating σ_p^2 is essentially equivalent to an ARMA specification. That is, associated with the autoregressive model in (14) is

$$\sigma_{pt}^2 = \omega_0 + \sum_{i=1}^{q} \alpha_i e_{t-i}^2 + \sum_{i=1}^{r} \gamma_i \sigma_{pt-i}^2. \tag{17}$$

Equation (17) is commonly referred to as a GARCH(r, q) model, and the additional restrictions are that all of the parameters in (17) be non-negative. As well, if $r = 0$, then Engle's (1982) ARCH(q) model results.

---

[3] As a practical matter the polynomial $(1-\Phi(L))$ may contain a root on the unit circle in which case first differencing of the price series is required in order to contain asymptotically efficient estimates. Any number of unit root tests are now available for testing just such a hypothesis, including the augmented Dickey-Fuller (ADF) test, although the power of the ADF and related tests has generally been found to be low in small to medium sample sizes.

In a wide variety of empirical applications, the most popular specification reported for (17) has been the GARCH(1,1) model. Therefore, an adequate definition of the conditional first and second moments of price (revenue) in the GARCH-type approach is typically represented by

$$[1-\Phi(L)]p_t = \delta_0 + e_t,\ e_t \mid \Omega_{t-1} \sim N(0, \sigma^2_{pt}), \tag{18a}$$

$$\sigma^2_{pt} = \omega_0 + \alpha_1 e^2_{t-1} + \gamma_1 \sigma^2_{pt-1}. \tag{18b}$$

Combining (18) with (11) gives the dynamic supply equation

$$\begin{aligned} a_t = {} & \beta_0 + \beta_1(\delta_0 + \Phi_1 p_{t-1} + \cdots + \Phi_k p_{t-k}) \\ & + \beta_2(\omega_0 + \alpha_1 e^2_{t-1} + \gamma_1 \sigma^2_{pt-1}) + \varsigma^T r_t + \varepsilon_t, \end{aligned} \tag{19}$$

which, when estimated jointly with (18), is the counterpart to the quasi-rational expectations model in (17) that includes a risk term. Of course in this instance the bivariate structure of the error process for e_t and ε_t must be specified.

In the context of risk-responsive supply, models similar to (18) and (19) have been estimated and reported by Appelbaum and Kohli (1997), Aradhyula and Holt (1989), Holt (1993), Holt and Aradhyula (1990, 1998), and Holt and Moschini (1992). Overall the GARCH approach has been found to be superior to the rolling weighted average approach or to the adaptive expectations approach as a method for modeling conditional price variances. Several caveats are, however, in order. With the exception of Appelbaum and Kohli, each of the aforementioned studies examined use ARCH/GARCH processes to infer risk effects in the context of livestock supply.[4] The underlying reason is that ARCH-type models are most useful for revealing underlying conditional variance dynamics when there is volatility clustering; that is, when large shocks tend to follow large shocks and when small shocks tend to follow small shocks. Moreover, volatility clustering is typically a feature of data generated at a frequency higher than annual. This helps explain why GARCH-type models have not been adopted for modeling price (revenue) risk in annual acreage supply decisions.

One potential limitation of the ARCH (GARCH) approach is that if the underlying conditional variance dynamics are not adequately characterized by ARCH (GARCH), then in a full information maximum likelihood (FIML) context, biased estimates of the parameters in (11) will obtain. This is due to the in-mean effects associated with the conditional variance in equation (19).

---

[4] Appelbaum and Kohli (1997) estimate the role of import price risk for oil and non-oil imports in the U.S. manufacturing sector by using annual data for the 1957-87 period.

Clearly, in this circumstance it is imperative that tests of the validity of the ARCH/GARCH parameterization be performed (Pagan and Ullah 1988).

An alternative to the ARCH-GARCH approach is to estimate the conditional moments of the price (revenue) distribution by using, for example, a nonparametric kernel density estimator. Suppose that the possibly nonlinear autoregressive (i.e., reduced-form) process that explains price is given by

$$p_t = r(z_t) + e, \tag{20}$$

where $z_t = (p_{t-1}, \ldots, p_{t-k})^T$ are elements contained in Ω_{t-1}. Let the rth-order conditional moment of p be given by $m_r(z_t) = E(p_t^r \mid z_t)$ for $r = 1, \ldots$. For $r =$ 1, $m_1(z_t) = E(p_t \mid z_t)$ is the conditional mean of p_t; and for r = 2, $m_2(z_t) = E(p_t^r \mid z_t)$ is the conditional second moment of p_t given z_t. The problem is to develop a nonparametric kernel estimator of $m_r(z_t)$ based on the data $\{p_t, z_t\}$, $t = 1, \ldots, T$.

One approach to estimating these moments is to use the kernel estimator

$$\hat{m}_1(z_t) = \sum_{i=1}^{T} p_i w_i(z_t), \tag{21}$$

where

$$w_i(z_t) = K\left[\frac{z_i - z_t}{h}\right] \Bigg/ \sum_{j=1}^{T} K\left[\frac{z_i - z_t}{h}\right], \tag{22}$$

and where $K(\cdot)$ is the kernel function and h is the smoothing parameter or window width. The kernel $K(\cdot)$ is assumed to be non-zero, and to possess the properties of symmetry and integration to unity. One feasible candidate for $K(.)$ is $K(.) = N(\boldsymbol{0}, I)$, a multivariate normal density with mean vector zero and identity covariance matrix (Ullah 1988). In (22), h is the window width, otherwise known as the smoothing parameter, and it helps determine the "size" of the window around z_t for which the observations are averaged.[5] See Ullah (1988) for details on the choice of h, as well as the choice of the kernel $K(.)$.

Once (21) has been estimated for a given data set, it is a straightforward matter to obtain estimates for the remaining central moments. Specifically, by using (22), the nonparametric estimators for $m_r(z_t)$ are [6]

---

[5] In general, a "larger" h results in lower variance and a smoother curve, but a larger bias.

[6] A slight variation on the nonparametric estimator in (21) is the "leave one out estimator," defined as $\hat{m}_1(z_t) = \sum_{i=1, i \neq t}^{T} p_i w_i(z_t)$, where the w_i weights in (22) are also correspondingly adjusted by leaving out the tth observation.

$$\hat{m}_r(z_t) \quad = \sum_{i=1}^{T} p_i^r w_i(z_t) . \tag{23}$$

It therefore follows that the second conditional moment of p_t around the mean is

$$\hat{\mu}_2(p_t \mid z_t) = \hat{m}_2(z_t) - \hat{m}_1^2(z_t) . \tag{24}$$

See Appelbaum and Ullah (1997) for details on constructing higher-order conditional moments by using a nonparametric approach.

Estimation of risk effects via the nonparametric framework may be conducted by using (21) and (24) as proxy variables for $\overline{p}_t$ and σ_{pt}^2 in (11). Of course this approach will lead to unbiased estimates of the parameters in (11), but due to the generated repressors problem, may result in biased and inefficient estimates of precision (see, e.g., Pagan and Ullah 1988 or Hoffman 1991).[7] This potential limitation notwithstanding, the nonparametric approach has seen some, albeit limited, application in risk modeling. To date, aside from various applications in the empirical finance literature, Holt and Moschini (1992) and Appelbaum and Ullah (1997) report the only known applications of nonparametric estimators for select moments of the underlying price distribution that are in turn used to estimate risk effects in a production/supply context. Overall, the nonparametric approach appears to be a viable alternative to the parametric ARCH/GARCH approach. Additionally, unlike the ARCH/GARCH setup, the nonparametric methodology has the advantage that it can be employed in instances where only panel or cross-sectional data are available.

Multi-Equation Reduced-Form Methods

Many of the reduced-form methods for estimating risk effects in single-equation models of output supply have been extended to a multiple-equation context, where there are now n acreage (output) categories. To accomplish this, some authors have simply modified the specification in (11) to be vector-valued, and then attempted to include measures of all relevant variance and covariance terms. The resulting model might then assume the form

$$\boldsymbol{a} = \beta_0 + \beta_1 \overline{\boldsymbol{p}} + \beta_2 \, vec(\Sigma_{pt}) + \gamma \boldsymbol{r} + \varepsilon , \tag{25}$$

[7] Of course the non-parametric approach is not the only one for which the generated regressors problem is a potential issue in econometric models of risk.

where Σ_{pt} is the covariance matrix of the relevant stochastic variables (prices, revenues, etc.), $vec(.)$ is an operator that vectorizes the unique elements of Σ_{pt}, and β_1 and β_2 are now conformable matrices. As in the single-equation case, wealth terms may or may not be included in the multi-equation specification. A typical example of the specification and estimation of models of this sort may be found in von-Massow and Weersink (1993).

While direct estimation of the system in (25) by using, say, seemingly unrelated regression (SUR) techniques facilitates the inference of cross-equation risk effects, the potentially more interesting applications of (25) involve theoretically derived cross-equation restrictions. For example, Chavas (1985), Chavas and Pope (1985), Dalal (1990), Pope (1980, 1988), and Pope and Chavas (1985) have shown that a set of cross-equation reciprocity and homogeneity conditions (restrictions) exists for the risk-averse firm that maximizes expected utility while facing output price risk. Specifically, it may be shown that

$$(\partial \boldsymbol{a}^c / \partial \overline{\boldsymbol{p}}) = (\partial \boldsymbol{a} / \partial \overline{\boldsymbol{p}}) - (\partial \boldsymbol{a} / \partial w)\boldsymbol{a}^T = \beta_1 - (\partial \boldsymbol{a} / \partial w)\boldsymbol{a}^T , \tag{26}$$

where a^c is a wealth-compensated acreage (supply) decision, utility held constant, and $(\partial \boldsymbol{a} / \partial w)$ is the $n \times 1$ vector of wealth effects (slopes) corresponding to $(\partial y / \partial w_0)$ in (8). Under fairly general conditions it may be shown that the $n \times n$ matrix of wealth-compensated slopes is symmetric, positive semi-definite. As well, (26) shows that the uncompensated effects may be decomposed into two parts: the compensated expected price slopes (pure substitution effects) and a wealth effect.[8] Under CARA, of course, wealth effects vanish, and compensated and uncompensated slopes are identical.

To date the only known empirical application and test of symmetry restrictions similar to (26) with agricultural data has been by Chavas and Holt (1990). These authors concluded in an application involving corn and soybean acreage planting decisions that the symmetry restriction(s) could not be rejected. As well, they rejected the CARA specification for underlying risk preferences. One potential shortcoming of the Chavas-Holt study is that they did not include an exhaustive set of acreage supply decisions (i.e., no attempt was made to use an acreage adding-up identity).

A second systems approach to generating cross-equation restrictions, at least in the case of acreage planting decisions, has been explored by Barten and Vanloot (1996) and Holt (1999). The model is restrictive in that CARA preferences are assumed; the farmer is assumed to maximize certainty equivalent (CE) profit subject to a total land constraint and to otherwise

[8] As summarized by Pope (1982), in the case where input decisions are being modeled as well the approach may be extended to include additional reciprocity relationships across both the output supply and input demand relationships.

behave like a portfolio manager with respect to acreage allocation decisions. The basic setup is

$$\max_{\boldsymbol{a}} CE(\pi) = \left[\boldsymbol{a}^T \boldsymbol{r}^e - (\varphi / 2)\boldsymbol{a}^T \Sigma \boldsymbol{a}^T \middle| a_{tot} - \sum_{i=1}^{n} a_i \right], \tag{27}$$

where the n-vector $\boldsymbol{r}^e$ denotes expected per-acre returns, Σ is the covariance matrix of returns, φ is the coefficient of absolute risk aversion, and a_{tot} denotes total land available for planting. Under these conditions it is possible to derive a linear system of acreage share equations of the form

$$\boldsymbol{w} = \boldsymbol{b} + S\boldsymbol{r}^e + \varepsilon, \tag{28}$$

where $w_i = a_i/a_{tot}$ is the ith acreage share and S is a symmetric, positive semi-definite matrix of (at most) rank $n - 1$. Although direct information on risk preferences may not be obtained from (28) – the coefficient φ and the unique elements of Σ are not identifiable from S – it is the case that symmetry and a form of homogeneity are implied in (28). These restrictions may be imposed during estimation or otherwise tested for. As per usual with systems of share equations, maximum likelihood estimates of (28) may be obtained by deleting an equation and by estimating the remaining $n - 1$ share equations using iterative SUR techniques. In an empirical application to state-level data for the 1991-95 period, Holt (1999) finds that both sets of restrictions are consistent with the data. Lastly, it should also be noted that the model in (27)-(28) is similar in many respects to a specification considered previously by Pope (1978, 1982).

Structural Approaches

To this point the discussion has focused on how single- or multi-equation reduced form models may be used to estimate risk effects, and therefore presumably to infer something about behavioral attitudes toward risk. As has already been seen, inferences about the nature of underlying preferences are typically difficult to make in the context of a reduced-form model. For these and related reasons, several attempts have been made to estimate models that include risk by using a primal or structural approach. The basic idea is to specify a utility function, a technical production relationship, and the form(s) of the distribution functions generating the stochastic variables in the model. In the simplest of cases, closed-form expressions for the first-order conditions may be obtained (i.e., the solutions in (7) may be derived analytically). Alternatively, the first-order conditions in (6) may be viewed as a set of implicit equations, the parameters of which are to be estimated econometrically by using some variant of an instrumental variable (IV) or FIML estimator. That is, if an n-vector of error terms, ε, is appended to (6a), and an error term, e, is appended to (6b), the parameters in the resulting $n+1$ nonlinear simultaneous equations system,

$$E\{U_{\tilde{w}}(w_0 + py - \boldsymbol{r}^T \boldsymbol{x}; \alpha)[pf_{\boldsymbol{x}}(\boldsymbol{x}; \tau) - \boldsymbol{r}^T]\}^T = \varepsilon, \tag{29a}$$

$$y^* - f(\boldsymbol{x}^*; \tau) = e, \tag{29b}$$

could, in principle, be estimated. Studies in the literature that employ this or a related approach include Antle (1987), Chavas and Holt (1996), Love and Buccola (1991), and Saha, Shumway, and Talpaz (1994). Of these, only Chavas and Holt, and Saha, Shumway, and Talpaz have truly undertaken a direct, simultaneous estimation of the parameters in the utility function, the production function, and the distribution functions for random variables.

The direct estimation approach poses several serious numerical issues, not the least being that numerical integration called for in (6a) must be performed repeatedly, as required in any FIML estimation approach. Chavas and Holt (1996) deal with this issue by using Monte Carlo integration, while Saha, Shumway, and Talpaz (1994) use numerical integration techniques. In both instances, parameters associated with price/revenue distributions were estimated outside of the overall modeling framework. Another issue is the specification of the production function, and the precise form of the stochastic component of production. As Just and Pope (1978) have illustrated, multiplicative risk components in a stochastic production function imply severe restrictions on the relationship between inputs and production risk. More flexible specifications of the error structure are called for in order to avoid imposing unwarranted restrictions on the structure of the model. Finally, it is desirable to have a fairly general specification of the utility function so that various maintained hypotheses such as CARA, DARA, CRRA, IRRA, etc., may be directly tested for. Saha, Shumway, and Talpaz (1994) propose an Expo-Power utility function, while Chavas and Holt (1996) present a related alternative.[9] While the direct or primal approach may be computationally demanding, it has the added benefit that multiple sources of risk (i.e., price and production, for example) may be accounted for within the model's structure.

---

[9] Chavas and Holt (1996) describe an additional problem in estimating systems such as (29). Specifically, if there is a linear-additive (intercept) parameter in the $EU_{\tilde{w}}(.)$ component of (29a), then it will not be possible to uniquely identify all the diagonal elements of $\text{var}(\varepsilon)$. This is because an arbitrary positive re-scaling of both $EU_{\tilde{w}}(.)$ and the square roots of the diagonal elements of $var(\varepsilon)$ (i.e., the standard deviations) will leave the equation unchanged. Chavas and Holt (1996) achieve identification by placing suitable restrictions on the terms in the model's full covariance matrix.

An alternative but equally viable approach, at least in the case of price risk, is to use the duality results under uncertainty developed by Appelbaum and Kohli (1997), Chavas (1985), Chavas and Pope (1985), Coyle (1999), Dalal (1990), and Pope (1980). Following Appelbaum and Kohli (1997), the basic idea is that – assuming there is a fixed cost κ, so that $\pi = py - \boldsymbol{r}^T\boldsymbol{x} - \kappa$, and ignoring wealth effects –the problem in (1) may be restated as

$$\max_{\boldsymbol{x}}[EU(y - \boldsymbol{r}^T\boldsymbol{x} - \kappa; \boldsymbol{\alpha}) \mid y = f(\boldsymbol{x}; \boldsymbol{\tau}), \boldsymbol{x} \in \Re_+^n] = V(\boldsymbol{r}, \bar{p}, \kappa, \sigma_p^2, \boldsymbol{\sigma}_p^k). \quad (30)$$

In (30), $V(.)$ denotes the indirect or dual expected utility function and $\boldsymbol{\sigma}_p^k$ is a vector of higher-order moments of the price distribution. As well, $V(.)$ is continuous and convex in the moments.

The firm's output supply and input demand equations may be readily derived from (30) by applying the envelope theorem. Doing so obtains

$$\frac{\partial V}{\partial r_i} = -E[u'(\pi)]x_i, \quad (31)$$

$$\frac{\partial V}{\partial \bar{p}} = E[u'(\pi)]y, \quad (32)$$

$$\frac{\partial V}{\partial \kappa} = -E[u'(\pi)]. \quad (33)$$

Applying the equivalent of Roy's identity to (31)-(33) then yields[10]

$$x_i(\boldsymbol{r}, \bar{p}, \kappa, \sigma_p^2, \boldsymbol{\sigma}^k) = (\partial V / \partial r_i)/(\partial V / \partial \kappa), \quad (34)$$

$$y(\boldsymbol{r}, \bar{p}, \kappa, \sigma_p^2, \boldsymbol{\sigma}^k) = -(\partial V / \partial \bar{p})/(\partial V / \partial \kappa). \quad (35)$$

A corresponding set of symmetry or reciprocity conditions may also be immediately obtained from (30)-(33). Just as in standard dual cost or profit function models, a suitably flexible specification for $V(\cdot)$ may be defined, and empirically estimable sets of input demand and output supply equations may be derived. Of course some method must still be employed to generate estimates for the conditional moments of price. But having done so, the symmetry results from theory may then be used in estimation to reduce the

[10] Similar results can be derived if fixed costs are zero, or if (nonstochastic) initial wealth replaces fixed costs. See Appelbaum and Kohli (1997) for additional details.

number of free parameters. To date, the only empirical application of these procedures reported in the literature is by Appelbaum and Kohli (1997) and Coyle (1999).

Where We Have Been – Summary

In the nearly twenty years following Pope's first appraisal of the empirical estimation of risk effects in agricultural production models, considerable headway has been made. Methods have been refined and improved for actually estimating relevant conditional moments of price (revenue). Modern time series techniques such as ARCH/GARCH models have been fruitfully employed, as well as semiparametric and nonparametric techniques. Importantly, the rational expectations paradigm has now been successfully extended to models that include behavioral risk terms (Holt 1993, 1994, and Holt and Aradhyula 1990, 1998). Useful results have also been developed for testing the structure of risk preferences in otherwise reduced-form supply models. Due to improved numerical simulation and integration procedures, progress has also been made in defining empirically tractable primal approaches to estimating risk preferences. Some limited application of duality under uncertainty has also been witnessed, although arguably more work in this important area remains. Given this overview of where we have been, we now turn to the potentially more interesting set of questions regarding where work in this area might proceed next.

EMPIRICAL WORK: WHERE MIGHT WE GO?

In reviewing the empirical work on behavioral risk in agriculture over the past twenty-odd years, several observations become clear. These are, in no particular order, roughly as follows:

- Dual results under uncertainty, at least in the case where price risk matters, have been developed although not widely adopted or applied, at least in the case where behavioral risk is potentially relevant.
- Only limited attempts have been made to incorporate wealth variables in risk-responsive production models and to otherwise impose and/or test the reciprocity relationships known to exist under price risk.
- The primal approach, while a potentially fruitful vehicle for obtaining considerable insights into the structure of (stochastic) production and the role of risk in decision making, has been utilized sparingly.

- The generalized method of moments (GMM) estimation framework, while gaining widespread acceptance as a procedure for modeling preferences under uncertainty in general economics, appears to have been an underutilized tool in agricultural risk analysis.
- Nonparametric estimation methods have generally not been widely explored or adopted as a possible means for generating expectations of relevant moments to include in risk-responsive output supply and input demand equations.
- New and potentially useful local and, for that matter, globally flexible functional forms such as the normalized quadratic, the Fourier form, or the asymptotically ideal model (AIM) have not been embraced in any serious way in agricultural risk and production analysis.
- Limited progress has been made to combine the programming approach to risk analysis, whether static or dynamic, with econometric estimation and simulation methods.
- Many studies still rely on aggregate data to infer estimates of key risk parameters. There is a need to develop more disaggregate analyses of risk effects.

In many ways the lack of serious progress on a number of these and related issues is discouraging. In reading Pope's 1982 article, one might have reasonably concluded nearly twenty years ago that the agricultural economics profession was poised to attack the issue of risk in agriculture on a broad front, with a broad array of tools. Appropriate theory seemed to be largely in place; econometric methods and computational abilities have only continued to improve; and there seemed to be broad agreement that risk was an important consideration in agricultural decision making. It is not our role to divine the underlying reasons for the evolution, trajectory, and maturation of the empirical analysis of risk in agriculture in recent years. In our view, there are several potentially fertile areas of future investigation, areas that might help rejuvenate efforts to examine the role of risk in agricultural production decisions.

GMM Estimation of Stochastic First-Order Conditions

One area that has received only limited attention in the area of agricultural risk analysis is the estimation of stochastic first-order conditions similar to (29) by use of GMM techniques. That is, GMM techniques provide a direct alternative to the FIML-numerical method of estimating first-order conditions from a primal specification such as (1) that, moreover, does not require reliance on any stylized duality results, or, for that matter, any need to assume away production risk. An early and notable effort by Antle (1987) to apply

GMM techniques to the estimation of producers' risk preferences (although not, it should be noted, explicitly to models based on first-order conditions similar to (29)) is the only known empirical example of this methodology in the agricultural economics literature.[11] This is surprising given that GMM estimation procedures have been in place for nearly twenty years (Hansen 1982, Hansen and Singleton 1982), and are now widely used in economics and in empirical finance to estimate parameters embedded in stochastic Euler equations and in dynamic asset pricing models (see, e.g., Hansen and Singleton 1983). Several potential advantages of the GMM approach include the following:

- Rational expectations assumptions may, as an integral part of the estimation procedure, be directly incorporated into the model and otherwise tested for without the need to estimate and solve explicitly for the underlying market supply-demand structure.
- Detailed stochastic specifications for *both* price and production risk may be included without an explicit *a priori* determination of their parametric form.
- Since direct numerical solution of the stochastic first-order conditions in (29) is not required to estimate "deep" technology and production function parameters, fairly general specifications for utility and production functions may be incorporated directly into estimation.
- Given an adequately flexible specification of the underlying utility function (i.e., the Expo-Power utility function of Saha, Shumway, and Talpaz 1994, or the utility function posited by Holt and Chavas 1996), direct tests of maintained hypotheses regarding the nature of risk preferences (i.e., CARA, CRRA, DARA, IRRA, etc.) may be performed.
- The modeling approach can utilize time series data, cross-sectional data, or, if available, panel data.
- The management of heteroscedasticity and autocorrelation within the model's error structure may be done in a reasonably efficient manner within the estimation framework.

Of course, to implement the GMM procedure, a set of relevant and reasonable instruments $\underline{z}$ that are deemed to be orthogonal to the error terms

[11] Chavas (2000) has recently utilized GMM estimation procedures to estimate stochastic Euler equations in the context of heterogeneous producer expectations in cattle supply response. No attempt was made, however, to incorporate behavioral risk response. Indeed, one potentially interesting extension of the study by Chavas (2000) is to consider a case where agents potentially have heterogeneous expectations and, as well, are assumed to react to price and production risk.

$\underline{\varepsilon}$ and e in (29) must be chosen.[12] In many instances, lagged market prices, both for inputs and outputs, as well as lagged observations on input and output levels might constitute a reasonable set of instruments. Moreover, if more instruments than first-order conditions are utilized in estimation, it is possible to perform an asymptotic test of the overidentifying restrictions, and therefore to perform a direct test of the validity of the model specification and the choice of instruments (Hansen 1982). Under fairly general conditions, the resulting GMM parameter estimates are consistent and asymptotically normal (Hansen 1982).

One potential drawback to the GMM approach, at least in the context of agricultural risk analysis, is that, to date, methods for evaluating expectations when the underlying distribution functions are truncated need to be refined. This is, at times, potentially important in agriculture, where various price support mechanisms have been used to limit downward price movements. Indeed, the incorporation of truncation effects for market prices was a central feature of the empirical models developed by Chavas and Holt (1990, 1996).

Combining Econometric and Programming Models

A traditionally important approach to investigating the implications of risk and risk response for agricultural decision making that is, nonetheless, not based on an econometric framework has been the Markowitz mean-variance efficiency model. The basic model is

$$\min v = \{\boldsymbol{x}^T \Sigma \boldsymbol{x} \mid \overline{\boldsymbol{R}}^T \boldsymbol{x} = \varepsilon, A\boldsymbol{x} \leq \boldsymbol{b}\}, \tag{36}$$

where $\boldsymbol{x}$ is a vector of activities, Σ is an estimate of the covariance matrix of (unit) revenues or net returns, $\overline{\boldsymbol{R}}$ is a vector of mean or expected revenues or net returns, A is a matrix of technical coefficients associated with resource constraint vector $\boldsymbol{b}$, v is total enterprise or farm variance, and e is expected (total) enterprise or farm returns. Solution of (36) yields the minimum variance portfolio conditional on e. Moreover, an efficient set, $e = \theta(v)$, is generated as e is varied (Hardaker, Huirne, and Anderson 1997). The optimal enterprise (farm) plan is determined by solving

$$\max_{e,v} EU = \{\varpi(e,v) \mid e = \theta(v)\}. \tag{37}$$

In empirical work when programming analysis is used, it is frequently the case that $\varpi(e,v)$ is approximated by a mean-variance utility function (e.g.,

---

[12] The basic idea is that the sample counterpart to the population moment condition $E[(\boldsymbol{\varepsilon}^T, e)^T z] = 0$ is assumed to hold at all observations.

Meyer 1987). In the simplest case, the utility function is assumed to be additive (corresponding to CARA),

$$\varpi(e, v) = e - (\varphi/2)v\,, \tag{38}$$

where φ is the absolute risk-aversion coefficient representing the trade-off between expected return and risk. Alternatively, one can let the decision maker choose his or her preferred (e,v) point on the efficiency frontier, $e = \theta(v)$, and deduce the implied risk aversion parameter φ. This latter approach appears attractive whenever it is difficult to assess more directly the nature of the decision maker's risk preferences.

A potential problem with the above approach is that the remaining elements in the model, specifically those embedded in Σ, $\overline{\boldsymbol{R}}$, A, and $\boldsymbol{b}$, must be somehow estimated or inferred. Small changes in any or all of these parameters can have a significant impact on the estimate of φ. One approach that makes sense is to specify econometric models to estimate the parameters in these vectors and matrices. These sub-models could, in turn, be single-equation OLS models, treated as a system and estimated by SUR, or some other form of linear or nonlinear estimator (i.e., frontier production function estimators). Once initial estimates of these technical coefficients are obtained, sub-sample bootstrap and/or sub-sample jackknife procedures, as developed for example by Politis and Romano (1994), could be used in combination with repeated solution of the mean-variance efficiency model. This in turn would allow the investigator to obtain standard error estimates for the risk-aversion coefficient, φ, as well as the solution vectors, $\boldsymbol{x}^*$. To date, the only known application along these lines reported in the agricultural economics literature is by Ziari, Leatham, and Ellinger (1997).

CONCLUSIONS

Nearly twenty years ago, Pope (1982) provided a summary of existing and emerging approaches to modeling the role of risk in agricultural production decisions. Since then, additional refinements to the theory have been made, with particular emphasis given to the potential for developing dual models of risk behavior – and the attendant reciprocity conditions. Additionally, much attention has been given to the role of production risk in cost minimization and, as well, in the estimation of behavioral models that include risk response. Headway has also been made in the adaptation of modern time-series techniques such as ARCH and GARCH models in the estimation of risk in a quasi-rational or rational expectations paradigm. Finally, some progress has been made in implementing primal formulations of the EU model, which in turn facilitates direct testing of various features of producer risk preferences and, as well, allows for fairly flexible specifications of underlying production technologies to be incorporated. Clearly, some progress has been made in our collective efforts to understand more thoroughly the role of risk in agriculture.

A central conclusion of this chapter, however, is that much remains to be done, at least with regards to the empirical investigation of behavioral response to uncertainty. While dual models under uncertainty have been developed, they have for the most part not been applied or explored. More simply, the Slutsky-like relationships that have been shown to exist in the case of price or revenue risk have received only limited attention in the empirical literature. No serious attempt has been made, for example, to estimate a consistent set of output supply and input demand equations – similar to those derived from, say, the ever-popular dual profit function relationships – that incorporate these reciprocity conditions. Without serious attempts at adequate empirical testing, we may never be able to deduce whether or not these relationships provide any additional explanatory power or structural insights.

While progress on the empirical investigation of risk over the past twenty years has been somewhat halting, there does appear to be significant scope for applying new and emerging econometric methods in our quest to learn more about behavioral risk response. GMM methods, widely adopted and utilized in mainstream economics, might serve as a logical vehicle for estimating stochastic first-order conditions resulting from an appropriately specified EU model. Indeed, it is somewhat surprising that this methodology has not been adopted more widely in agricultural economics research. As well, new and emerging bootstrap simulation procedures make it feasible to combine econometric estimation techniques with the specification and solution of Markowitz mean-variance programming models. Importantly, the ability to generate measures of precision for the solutions from models of this sort now exists, and should be exercised by those involved in the application of programming models to agricultural decision making.

Overall, it seems that much work has been done on risk, but much remains to be done, at least if we as professional agricultural economists are seriously interested in validating once and for all the role of risk in agricultural production decisions. The arsenal of tools that we now have available to explore various maintained hypotheses regarding risk continues to expand. Progress can occur only, however, if we continue to systematically adopt and test these new techniques, and if we approach the analysis of risk in a rigorous and scientific manner. Let us hope that such a path will be pursued.

REFERENCES

Antle, J.M. 1987. "Econometric Estimation of Producers' Risk Attitudes." *American Journal of Agricultural Economics* 69: 509-522.

Appelbaum, E., and U. Kohli. 1997. "Import Price Uncertainty and the Distribution of Income." *Review of Economics and Statistics* 79: 620-630.

Appelbaum, E., and A. Ullah 1997. "Estimation of Moments and Production Decisions under Uncertainty." *Review of Economics and Statistics* 79: 631-637.

Aradhyula, S.V., and M.T. Holt. 1989. "Risk Behavior and Rational Expectations in the U.S. Broiler Market." *American Journal of Agricultural Economics* 71: 892-902.

Arrow, K.J. 1971. *Essays in the Theory of Risk Bearing.* Amsterdam: North Holland Publishing Co.

Bar-Shira, Z., R.E. Just, and D. Zilberman. 1997. "Estimation of Farmers' Risk Attitude: An Econometric Approach." *Agricultural Economics* 17: 211-222.

Barten, A.P., and C. Vanloot. 1996. "Price Dynamics in Agriculture: An Exercise in Historical Econometrics." *Economic Modelling* 13: 315-331.

Batra, R., and A. Ullah. 1974. "Competitive Firm and the Theory of Input Demand under Price Uncertainty." *Journal of Political Economy* 82: 537-548.

Behrman, J.R. 1968. *Supply Response in Underdeveloped Agriculture: A Case Study of Four Major Crops in Thailand, 1937-63.* Amsterdam: North Holland Publishing Co.

Binswanger, H.P. 1981. "Attitudes Toward Risk: Theoretical Implications of an Experiment in Rural India." *Economic Journal* 91: 867-889.

Bollerslev, T. 1986. "Generalized Autoregressive Conditional Heteroscedasticity." *Journal of Econometrics* 31: 307-327.

Bollerslev, T., and H.O. Mikkelsen. 1996. "Modeling and Pricing Long Memory in Stock Market Volatility." *Journal of Econometrics* 73: 151-184.

Brorsen, B.W., J.-P. Chavas, W.R. Grant, and L.D. Schnake. 1985. "Marketing Margins and Price Uncertainty: The Case of the U.S. Wheat Market." *American Journal of Agricultural Economics* 67: 521-528.

Chambers, R.G., and J. Quiggin. 1998. "Cost Functions and Duality for Stochastic Technologies." *American Journal of Agricultural Economics* 80: 288-295.

___. 2000. *Uncertainty, Production, Choice and Agency: The State-Contingent Approach.* New York: Cambridge University Press.

Chavas, J.-P. 1985. "On the Theory of the Competitive Firm under Uncertainty When Initial Wealth Is Random." *Southern Economic Journal* 49: 818-827.

___. 2000. "On Information and Market Dynamics: The Case of the U.S. Beef Market." *Journal of Economic Dynamics and Control* 24: 833-853.

Chavas, J.-P., and M.T. Holt. 1990. "Acreage Decisions under Risk: The Case of Corn and Soybeans." *American Journal of Agricultural Economics* 72: 529-537.

___. 1996. "Economic Behavior under Uncertainty: A Joint Analysis of Risk Preferences and Technology." *Review of Economics and Statistics* 78: 329-335.

Chavas, J.-P., and R.D. Pope. 1985. "Price Uncertainty and Competitive Firm Behavior: Testable Hypotheses from Expected Utility Maximization." *Journal of Economics and Business* 37: 223-235.

Chavas, J.-P., R.D. Pope, and H. Leathers. 1988. "Competitive Industry Equilibrium under Uncertainty and Free Entry." *Economics Inquiry* 26: 331-344.

Coyle, B. 1999. "Risk Aversion and Yield Uncertainty in Duality Models of Production: A Mean-Variance Approach." *American Journal of Agricultural Economics* 81: 553-567.

Dalal, A. 1990. "Symmetry Restrictions in the Analysis of the Competitive Firm under Price Uncertainty." *International Economic Review* 31: 207-211.

Dickey, D.A., and W.A. Fuller. 1981. "Likelihood Ratio Statistics for Autoregressive Time Series with a Unit Root." *Econometrica* 49: 1057-1072.

Dillon, J.L., and P.L. Scandizzo. 1978. "Risk Attitudes of Subsistence Farmers in Northeast Brazil: A Sampling Approach." *American Journal of Agricultural Economics* 60: 425-435.

Duffy, P., K. Shalishali, and H.W. Kinnucan. 1994. "Acreage Response under Farm Programs for Major Southeastern Field Crops." *Journal of Agricultural and Applied Economics* 26: 367-378.

Engle, R. 1982. "Autoregressive Conditional Heteroscedasticity with Estimates of the Variance of United Kingdom Inflation." *Econometrica* 50: 987-1007.

Engle, R., and T. Bollerslev. 1986. "Modelling the Persestince of Conditional Variances." *Econometric Review* 5: 1-50.

Estes, E.A., L.L. Blakeslee, and R.C. Mittelhammer. 1981. "On Variances of Conditional Linear Least-Squares Search Parameter Estimates." *American Journal of Agricultural Economics* 63: 141-145.

Ezekiel, M. 1938. "The Cobweb Theorem." *Quarterly Journal of Economics* 52: 255-280.

Hansen, L.P. 1982. "Large Sample Properties of Generalized Method of Moments Estimators." *Econometrica* 50: 1029-1054.

Hansen, L.P., and K.J. Singleton. 1982. "Generalized Instrumental Variables Estimation of Nonlinear Rational Expectations Models." *Econometrica* 50: 1269-1286.

___. 1983. "Stochastic Consumption, Risk Aversion, and the Temporal Behavior of Asset Returns." *Journal of Political Economy* 91: 249-265.

Hardaker, J.B., R.B.M. Huirne, and J.R. Anderson. 1997. *Coping with Risk in Agriculture.* CAB International.

Hoffman, D.L. 1991. "Two-Step and Related Estimators in Contemporary Rational-Expectations Models: An Analysis of Small-Sample Properties." *Journal of Business and Economic Statistics* 9: 51-61.

Holt, M.T. 1993. "Risk Response in the Beef Marketing Channel: A Multivariate Generalized ARCH-M Approach." *American Journal of Agricultural Economics* 75: 559-571.

___. 1994. Price-Band Stabilization Programs and Risk: An Application to the U.S. Corn Market." *Journal of Agricultural and Resource Economics* 19: 239-254.

___. 1999. "A Linear Approximate Acreage Allocation Model." *Journal of Agricultural and Resource Economics* 24: 383-397.

Holt, M.T., and S.V. Aradhyula. 1990. "Price Risk in Supply Equations: An Application of GARCH Time-Series Models to the U.S. Broiler Market." *Southern Economic Journal* 57: 230-242.

___. 1998. "Endogenous Risk in Rational-Expectations Commodity Models: A Multivariate Generalized ARCH-M Approach." *Journal of Empirical Finance* 5: 99-129.

Holt, M.T., and G. Moschini. 1992. "Alternative Measures of Risk in Commodity Supply Models: An Analysis of Sow Farrowing Decisions in the United States." *Journal of Agricultural and Resource Economics* 17: 1-12.

Ishii, Y. 1977. "On the Theory of the Competitive Firm under Price Uncertainty: Note." *American Economic Review* 67: 768-769.

Just, R.E. 1974. "An Investigation of the Importance of Risk in Farmers' Decisions." *American Journal of Agricultural Economics* 56: 14-25.

Just, R.E., and R.D. Pope. 1978. "Stochastic Specification of Production Functions and Economic Implications." *Journal of Econometrics* 7: 67-86.

Just, R.E., and D. Zilberman. 1983. "Stochastic Structure, Farm Size and Technology Adoption in Developing Agriculture." *Oxford Economic Papers* 35: 307-328.

Krause, M.A., and W.W. Koo. 1996. "Acreage Response to Expected Revenues and Price Risk for Minor Oilseeds and Program Crops in the Northern Plains." *Journal of Agricultural and Resource Economics* 21: 309-324.

Krause, M.A., J.-H. Lee, and W.W. Koo. 1995. "Program and Nonprogram Wheat Acreage Responses to Price and Price Risk." *Journal of Agricultural and Resource Economics* 20: 1-12.

Lin, W. 1977. "Measuring Aggregate Supply Response under Instability." *American Journal of Agricultural Economics* 59: 903-907.

Lin, W., G.W. Dean, and C.V. Moore. 1974. "An Empirical Test of Utility vs. Profit Maximization in Agricultural Production. *American Journal of Agricultural Economics* 56: 497-508.

Love, A.H., and S.T. Buccola. 1991. "Joint Risk Preference-Technology Estimation with a Primal System." *American Journal of Agricultural Economics* 73: 765-773.

Machina, M.J. 1987. "Choice under Uncertainty: Problems Solved and Unsolved." *Journal of Economic Perspectives* 1: 121-154.

Markowitz, H. 1959. *Portfolio Selection: Efficient Diversification of Investments*. New York: John Wiley & Sons.

Meyer, J. 1987. "Two-Moment Decision Models and Expected Utility." *American Economic Review* 77: 421-430.

Moscardi, E., and A. de Janvry. 1977. "Attitudes Toward Risk among Peasants: An Econometric Approach." *American Journal of Agricultural Economics* 59: 710-716.

Moschini, G. 2001. "Production Risk and the Estimation of Ex Ante Cost Functions." *Journal of Econometrics* 100: 357-380.

Muth, J.F. 1960. "Optimal Properties of Exponentially Weighted Forecasts." *Journal of American Statistics Association* 55: 299-306.

___. 1961. "Rational Expectations and the Theory of Price Movements." *Econometrica* 29: 315-335.

Nerlove, M. 1958. "Adaptive Expectations and Cobweb Phenomena." *Quarterly Journal of Economics* 72: 227-240.

Nerlove, M., and I. Fornari. 1998. "Quasi-Rational Expectations, An Alternative to Fully Rational Expectations: An Application to U.S. Beef Cattle Supply." *Journal of Econometrics* 83: 129-161.

Nerlove, M., D.M. Grether, and J.L. Carvalho. 1979. *Analysis of Economic Time Series*. New York: Academic Press.

Pagan, A., and A. Ullah. 1988. "The Econometrics Analysis of Models with Risk Terms." *Journal of Applied Econometrics* 3: 87-105.

Politis, D.N., and J.P. Romano. 1994. Large Sample Confidence Regions Based on Subsamples under Minimal Assumptions." *Annuals of Statistics* 22: 2031-2050.

Pope, R.D. 1978. "Econometric Analysis of Risk Models – Some Explorations and Problems." Paper presented at the Western Agricultural Economics Association meetings, Bozeman MT.

___. 1980. "The Generalized Envelope Theorem and Price Uncertainty." *International Economic Review* 21: 75-86.

___. 1982. "Empirical Estimation and use of Risk Preferences: An Appraisal of Estimation Methods that use Actual Economic Decisions." *American Journal of Agricultural Economics* 64: 376-383.

___. 1988. "A New Parametric Test for the Structure of Risk Preferences." *Economic Letters* 27: 117-121.

Pope, R.D., and J.-P. Chavas. 1985. "Producer Surplus and Risk." *Quarterly Journal of Economics* 100: 853-869.

___. 1994. "Cost Functions under Production Uncertainty." *American Journal of Agricultural Economics* 76: 196-204.

Pope, R.D., and R.E. Just. 1991. "On Testing the Structure of Risk Preferences in Agricultural Supply Analysis." *American Journal of Agricultural Economics* 73: 743-748.

___. 1996. "Empirical Implementation of Ex Ante Cost Functions." *Journal of Econometrics* 72: 231-249.

Pratt, J.W. 1964. "Risk Aversion in the Small and in the Large." *Econometrica* 32: 122-136.

Saha, A., C.R. Shumway, and H. Talpaz. 1994. "Joint Estimation of Risk Preference Structure and Technology Using Expo-Power Utility." *American Journal of Agricultural Economics* 76: 173-184.

Sandmo, A. 1971. "On the Theory of the Competitive Firm under Price Uncertainty." *American Economic Review* 61: 65-73.

Traill, B. 1978. "Risk Variables in Econometric Supply Models." *Journal of Agricultural Economics* 29: 53-61.

Tronstad, R., and T.J. McNeill. 1989. "Asymmetric Price Risk: An Econometric Analysis of Aggregate Sow Farrowings." *American Journal of Agricultural Economics* 71: 630-637.

Ullah, A. 1988. "Nonparametric Estimation of Econometric Functionals." *Canadian Journal of Economics* 21: 625-658.

von-Massow, M., and A. Weersink. 1993. "Acreage Response to Government Stabilization Programs in Ontario." *Canadian Journal of Agricultural Economics* 41: 13-26.

von Neumann, J., and O. Morgenstern. 1944. *Theory of Games and Economic Behavior.* Princeton, NJ: Princeton University Press.

White, F.C., and R.F. Ziemer. 1982. "Farm Real Estate Pricing under Risk: An Empirical Investigation." *Southern Economic Journal* 49: 77-87.

Ziari, H.A., D.J. Leatham, and P.N. Ellinger. 1997. "Development of a Statistical Discriminant Mathematical Programming Model via Resampling Estimation Techniques." *American Journal of Agricultural Economics* 79: 1352-1362.

Chapter 12

AGRICULTURE AS A MANAGED ECOSYSTEM: IMPLICATIONS FOR ECONOMETRIC ANALYSIS OF PRODUCTION RISK

John M. Antle and Susan M. Capalbo
Montana State University–Bozeman

INTRODUCTION

Managed ecosystems are complex, dynamic systems with spatially varying inputs and outputs that are the result of interrelated physical, biological, and human decision making processes. There is a growing recognition by the scientific community that principles from the biological, physical, and social sciences must be integrated to understand and predict the behavior of complex biological and human systems such as managed ecosystems. This recognition is evidenced by recent federal government research initiatives on biocomplexity by the National Science Foundation, and on human dimensions of climate change by agencies such as the U.S. Environmental Protection Agency and Department of Energy. Agricultural ecosystems are arguably the most important and pervasive managed ecosystems. Understanding and predicting the behavior of agroecosystems is critically important for a number of leading public policy issues. These issues include the environmental and human health consequences of agroecosystems, and the impacts of climate change on the global food supply.

In this chapter we use the paradigm of agriculture as a managed ecosystem to investigate the implications for econometric analysis of production systems and production risk. The fundamental question we address is whether risk concepts can help us understand and predict the behavior of complex managed ecosystems such as agroecosystems. We premise our analysis on the commonsense view that farmers use a variety of strategies to manage the risks associated with the spatial and temporal variability in agroecosystems that is driven by both biophysical and economic processes. We argue that the prevailing analytical and empirical paradigm used by the agricultural economics profession, which largely abstracts from the spatial and temporal aspects of these complex systems, suffers from a number of significant limitations. These limitations explain why this paradigm has not been successful in improving the predictive power of economic models, and why models used in policy analysis do not typically incorporate risk features. How then can we develop a more useful quantitative approach to understand the role of risk in agricultural production systems?

We begin with a brief description of the paradigm of agriculture as a managed ecosystem and its implications for farm decision making processes and econometric model specification. Next we present a critical assessment of the literature on production risk – based largely on the paradigm of static neoclassical production models with risk aversion – and its usefulness in understanding and predicting the behavior of farmers' production decisions and the behavior of agroecosystems. The final section argues that the agro-ecosystem paradigm suggests a different approach to understanding how risk affects production decision making. This approach integrates biophysical and economic models into a simulation framework that accounts for the spatial and temporal variability of agricultural production systems. This type of approach can be used to investigate the spatial and temporal properties of agro-ecosystems and the role that production risk and risk management play in improving our ability to understand these systems and predict their behavior.

AGRICULTURE AS A MANAGED ECOSYSTEM: IMPLICATIONS FOR SPECIFICATION OF DECISION MAKING PROCESSES

We view agroecosystems as complex, dynamic systems with spatially varying inputs and outputs that are the result of interrelated physical, biological, and human decision making processes (Antle et al. 2001). This view is consistent with the rich literature on bioeconomic models in which economic decisions of farmers, foresters, and fishers are modeled as economic optimization subject to economic and biophysical constraints. However, most of the literature focuses on stylized theoretical models that abstract from empirical details needed to understand and predict behavior of these systems. The challenge we are addressing is how to use this paradigm to improve our ability to understand and predict the behavior of these systems across space and time. The agroecosystem paradigm can be represented using a diagram such as Figure 1. This paradigm implies two key characteristics of production systems, temporal variability and spatial variability.

Temporal Variability: Intra-Seasonal and Inter-Seasonal Dynamics

A key feature of agroecosystems is temporal variability. This temporal variability is driven by the dynamics of biological growth processes, and by

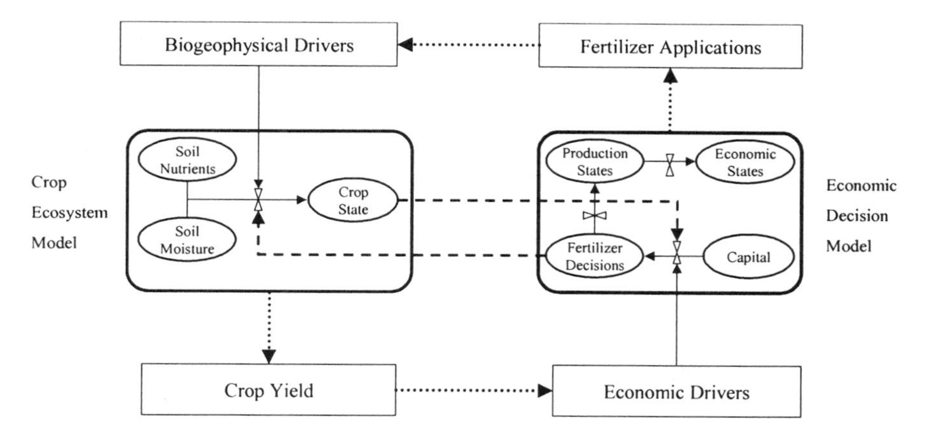

Figure 1. Agroecosystems Represented as Loosely or Closely Coupled Ecosystem and Economic Models

Source: Antle et al. 2001

Note: Dotted connectors represent feedbacks from system states to drivers in a loosely coupled model, dashed connectors represent feedbacks between processes in a closely coupled agro-ecosystem model.

the temporal variation of prices and other economic drivers of the system. As Figure 1 illustrates, the linkages between both biophysical and economic processes may involve various feedbacks in the system that give rise to various dynamic properties. This temporal variability occurs on short time steps within the growing season that can be represented as sequential intra-seasonal decision making (Antle 1983a, 1983b, Antle and Hatchett 1986, Antle, Capalbo, and Crissman 1994). In addition, feedbacks across growing seasons in both biophysical and economic dimensions give rise to inter-seasonal decision making processes (Antle and Stoorvogel 2001, Antle and Capalbo 2001).

A key factor in specification of dynamic production systems is the temporal unit of analysis. Biophysical processes operate in continuous time and are typically modeled in crop growth and environmental process models on a daily time step. Farm decision makers make intra-seasonal decisions on time steps that may be characterized according to stages of crop growth such as land preparation and planting, intermediate operations for fertilization, pest management, irrigation and cultivation, and harvest (Antle and Hatchett 1986). More generally, however, the time at which inputs are being applied may be endogenous to the decision making process (Antle, Capalbo, and Crissman 1994).

The intra-seasonal decision making process for crop production can be described using the scheme in Figure 2. Time t is defined as continuous on the nonnegative real line, and production activities occur at discrete points in

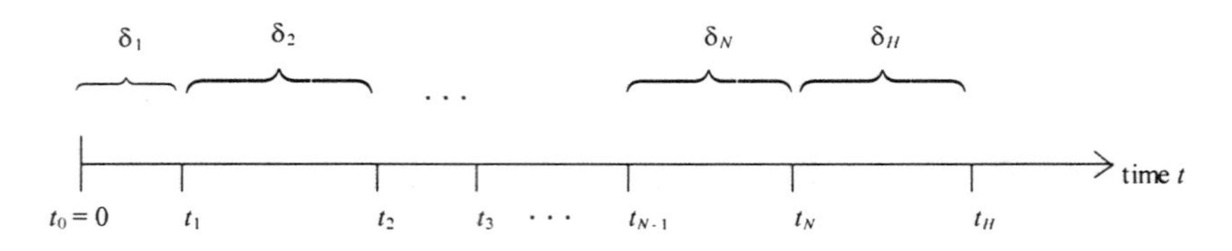

Figure 2. Decision Times (t_i) and Intervals (δ_i)

time. There are $N+2$ decisions occurring at times t_i, $i=0, 1,..., N, H$, with land preparation, planting, and related activities at time $t_0=0$, intermediate production activities at times $t_1,...,t_N$, and harvest at t_H. The intervals between decisions are defined as $\delta_i = t_i - t_{i-1}$ for $i=1, \ldots, N$ and $\delta_H = t_H - t_N$ so that $\sum_i \delta_i = t_H$.

Define a random vector ε_i on time interval δ_i to represent weather events on that interval (e.g., temperature, rainfall). For input x_t applied at time t (assumed here for simplicity to be a scalar), a general representation of a discrete, time-dependent production process can then be written as

$$q_0 = q_0(x_0, \varepsilon_0), \tag{1}$$

$$q_t = q_t(x_1, q_{t-1}, \varepsilon_1), \quad 0 < t < t_H,$$

$$q_H = q_H(x_H, q_t, \varepsilon_H),$$

where the subscripts on the functions indicate that the response of output to inputs depends on when the inputs are applied. If the time intervals δ_i between decisions are fixed exogenously, as in Antle (1983b) and Antle and Hatchett (1986), then this representation is useful. However, if the time intervals are endogenous, then this representation is not useful because in continuous time there are an infinite number of possible times at which input applications could occur on the (0, H) interval, and thus by implication there are an infinite number of possible production functions. To obtain an operational model, we represent the production process in each stage as a function of inputs employed and the time the activity occurs in relation to other activities in the production process. The ith production activity occurs at time $t_i = t_{i-1} + \delta_i$; and production q_i is a function of output from the previous stage, q_{i-1}, the time interval δ_i, the input vector x_i, and the random events ε_i that occurred during δ_i. Thus (1) becomes

$$q_0 = q_0(x_0, \varepsilon_0), \tag{2}$$

$$q_1 = q_1(x_1, q_{i-t}, \mathrm{t}_{i-1}, \delta_i, \varepsilon_i), \quad i = 1,..., N,$$

$$q_H = q_H(x_H, q_N, t_N, \delta_H, \varepsilon_H).$$

According to this model, parameters vary by stage of production rather than being explicit functions of time. The functions $q_i(\cdot)$ are assumed to be concave in x_i, q_{i-1}, and δ_i. The explanation for the concavity of the production function in δ_i is derived from the physiology of crop growth. As crop growth proceeds, there is a point in time where each operation such as cultivation, fertilization, pest control, etc., yields its greatest contribution to final output, given the state of crop growth and previous production activities.

Recursively substituting the stage functions q_i into q_H in (2) gives the composite production function

$$q_H = q_H(x_H, q_N(q_{N-1}(\ldots), t_{N-1}, \delta_N, \varepsilon_N), t_N, \delta_N, \varepsilon_H) \qquad (3)$$

$$\equiv q^c({}^Hx, {}^Nt, {}^N\delta, {}^H\varepsilon),$$

where ${}^Hx = (x_0,\ldots,x_H)$, and ${}^Nt$, ${}^N\delta$, and ${}^H\varepsilon$ are defined similarly. In conventional econometric analysis, intermediate products are not observed by the econometrician, hence the composite function q^c typically is estimated in econometric models. This is a critical point to which we shall return below.

Various sequential decision rules arise, depending on how the decision maker uses information, and the structure of these decision rules plays a key role in the econometric model. Here we assume the farm manager updates information when each production activity occurs. Thus, it is assumed that when the $(i\text{-}1)^{\text{th}}$ decision is implemented at time t_{i-1}, the manager updates information and plans the subsequent action x_i and its time of implementation $t_i = t_{i-1} + \delta_i$. Because t_{i-1} is known, the choice of t_i is equivalent to the choice of δ_i. An important implication of this model is that every decision corresponds to an observable input application, hence all observed values of the endogenous decision variables are positive.

In earlier work on dynamic production models (Antle 1983b, Antle and Hatchett 1986, Antle, Capalbo, and Crissman 1994), the farmer's objective function was assumed to be the maximization of expected returns. Antle (1983a) demonstrated that risk affects decisions in models with the structure presented here because expected returns is a nonlinear function of stochastic intermediate and final outputs. If the decision maker is assumed to be risk averse, the objective function can be specified as the maximization of expected utility of net returns over the relevant time horizon. While this changes (and complicates) the algebraic representation of the model and the derivation of the solution to the decision problem, it does not alter the general form of the solution in terms of the variables that are contained in the model. We define φ_i as the parameters of the output price distribution and θ^i as the vector of parameters of the distribution of $(\varepsilon_i,\ldots,\varepsilon_N,\varepsilon_N)$ at time t_i. Following the derivation in Antle, Capalbo, and Crissman (1994), we take input prices as known by the decision maker for simplicity, and it follows that the values of x_i and δ_i that maximize expected returns or expected utility of returns, conditional on information available at the time t_{i-1}, are generally of the form

$$x_0^* = x_0^*(\varphi_0, w^0, \theta^0) \tag{4}$$

$$x_i^* = x_i^*(\varphi_{i-1}, w^i, q_{i-1}, t_{i-1}, \theta^i), \qquad i = 1,\ldots,N,H$$

$$\delta_i^* = \delta_i^*(\varphi_{i-1}, w^i, q_{i-1}, t_{i-1}, \theta^i),$$

where we use the notation $w^i = (w_i,\ldots,w_N,w_H)$.

Antle, Capalbo, and Crissman (1994) note that the intermediate outputs, q_i, $i < H$, are not observed by the econometrician (but are observed by the farmer in the sense that crop growth is observed). They therefore recursively substitute the intermediate stage production functions (2) into (4) to obtain a reduced-form system of the form

$$x_0^* = x_0^r(\varphi_0, w^0, \theta^0) \tag{5}$$

$$x_i^* = x_i^r(\varphi_{i-1}, w^i, {}^{i-1}x, {}^{i-1}t, {}^{i-1}\delta, \theta^i, {}^{i-1}\varepsilon)$$

$$\delta_i^* = \delta_i^r(\varphi_{i-1}, w^i, {}^{i-1}x, {}^{i-1}t, {}^{i-1}\delta, \theta^i, {}^{i-1}\varepsilon), \qquad i = 1,\ldots,N,H\ ,$$

where we use the notation that ${}^{i-1}x = (x_0, x_1,\ldots,x_{i-1})$, and similarly for other variables.

For analysis of production risk we would like to be able to identify the risk structure of the model and investigate how risk affects decision making. Recall that the parameter vectors θ^i in system (5) represent the parameters of the production function errors that will be realized from t_i to harvest time t_H. In other words, at time t_i, the decision maker is making input decisions based on the anticipated distribution of final output, conditional on the state of the crop at that time. In equation (2), we can substitute out q_s for $s = i+1,\ldots,N, H$, to obtain the conditional composite production function for stage i,

$$q_H = q^{ci}(x^i, q_{i-1}, t^{i-1}, \delta^i, \varepsilon^i). \tag{6}$$

We can think of θ^i as the moments of the error ε^i in the conditional composite production function for stage i. Following the logic of the risk models in the literature, we can hypothesize that the moments of the distribution of output at time t_i in the growing season are functions of prospective input decisions x^i, input timing decisions δ^i, and the state of crop growth q_{i-1}, thus taking the form

$$\theta^0 = \theta^0(x^0) \tag{7}$$

$$\theta^i = \theta^i(x^i, q_{i\text{-}1}, t_{i\text{-}1}, \delta^i), \quad i = 1,\ldots,N,H\ .$$

Equation (7) represents the moments as functions of the intermediate outputs, inputs, and input timing decisions. Using the reduced-form input demand equations and the intermediate state production functions, we can substitute future outputs and future decisions out of the moment equations:

$$\theta^0 = \theta^{0r}(\varphi_0, w^0) \tag{8}$$

$$\theta^i = \theta^{ir}(\varphi_{i\text{-}1}, w^i, {}^{i\text{-}1}x, {}^{i\text{-}1}t, {}^{i\text{-}1}\delta, \theta^i, {}^{i\text{-}1}\varepsilon), \quad i = 1,\ldots,N,H\ .$$

These reduced-form moment functions have important implications for the analysis of production risk that we shall consider in detail below.

Inter-seasonal dynamics in the form of crop rotations play a critical role in maintenance of soil quality and productivity. The effects of crop rotations can be accurately modeled only on a site-specific basis, because their representation requires site-specific data on the history of land use. Aggregation across fields, even at the farm level, prevents the dynamics of soil quality from being accurately represented in both economic and biophysical process models. Modeling of some processes may require sub-field analysis, as when soil erosion occurs at different rates within a field (Antle and Stoorvogel 2001).

To illustrate, consider a simple crop-fallow rotation in which the effect of land use decisions on soil moisture carry over only for a single period, so farmers maximize the expected returns to each period's decision conditional on last period's decision (Antle and Capalbo 2001). The use of rotations in this case is an economic decision involving a tradeoff between the opportunity cost of the fallow and the productivity gains associated with the rotation. For each crop and location the production function takes the form $q_t = f(x_t, \lambda_{t\text{-}1})$, where $\lambda_{t\text{-}1} = 1$ if the previous use was a crop and equals zero if the field was fallowed, hence $f(x_t, 1) < f(x_t, 0)$. If a unit of land was previously cropped, the decision to fallow this season with the intent to crop again next season is based on net returns above variable cost calculated as $(p_{t+1}q_{t+1} - vc_{t+1})(1/1+r) - fc_t$, where vc_{t+1} is variable cost of crop production, r is the interest rate, and fc_t is the variable cost associated with fallow. The profit function takes the form $\pi_{\text{fal}}(p_{t+1}, w_{t+1}, r, \lambda_t, fc_t)$. The returns to growing a crop in period t after a crop was grown in period t-1 is equal to $p_t q_t - vc_t$, giving the profit function $\pi_{\text{crop}}(p_t, w_t, \lambda_{t\text{-}1})$. The farmer will use fallow if $\pi_{\text{fal}} > \pi_{\text{crop}}$. A similar analysis applies to the case where the field was fallowed in the previous period.

Spatial Variability: Site-Specific Production Decisions

Spatial variability is due to the fact that agricultural production processes are biological processes and therefore are dependent on site-specific soil and climate conditions. Implications for the structure of production decision making processes can be derived from a simple static representation of the production process. Following Antle and Capalbo (2001), the production process of activity j at site i in period t is defined by the production function $q_{ijt} = f_j\,(x_{ijt}, \mathrm{z}_{ijt}, \mathrm{e}_{it})$, where x_{ijt} is a vector of variable inputs, z_{ijt} is a vector of allocatable quasi-fixed factors of production and other fixed effects, and e_{it} is a vector of biophysical characteristics of the site. The vector e can contain variables such as soils, topography, and climate, or it could contain functions of these variables derived from biophysical process models. For example, a site-specific yield predicted by a crop growth model could be included as an explanatory variable and interpreted as an indicator of site-specific potential productivity, or a pest population variable derived from a pest population model could be included as a proxy for the farmer's expectations about pest pressure. For expected output price p_{ijt} and input price vector w_{ijt}, the profit function corresponding to the production function is $\pi_{ijt} = \pi_j(p_{ijt}, \mathrm{w}_{ijt}, z_{ijt}, e_{it})$. If a crop is not grown on a particular site, then the site is allocated to a conserving use with a return of π_{ict}. Define $d_{ijt} = 1$ if the jth crop is grown at location i at time t and zero otherwise, so that one of the d_{ijt} takes on a value of unity when a crop is grown. The land use decision on site i at time t is

$$\max_{(d_{i1t},\ldots,d_{\text{int}})} \sum_{j=1}^{N} d_{ijt}\pi_j(p_{ijt}, \boldsymbol{w}_{ijt}, \boldsymbol{z}_{ijt}, \boldsymbol{e}_{it}) + (1 - \textstyle\sum_{j=1}^{N} d_{ijt})\pi_{ict}\,. \tag{9}$$

The solution takes the form of a discrete step function,

$$d^*_{ijt} = d_j(\boldsymbol{p}_{it}, \boldsymbol{w}_{it}, \boldsymbol{z}_{it}, \boldsymbol{e}_{it}, \pi_{ict}), \tag{10}$$

where $\boldsymbol{p}_{it}$ is a vector of the p_{ijt}, $\boldsymbol{w}_{it}$ is a vector of the $\boldsymbol{w}_{ijt}$, and likewise for the other vectors. Using Hotelling's lemma, the quantity of the jth output on the ith land unit is given by

$$q_{ijt} = \frac{d^*_{ijt}\partial\pi_j(p_{ijt}, \boldsymbol{w}_{ijt}, \boldsymbol{z}_{ijt}, \boldsymbol{e}_{it})}{\partial p_{ijt}}\,. \tag{11}$$

Variable input demands are likewise given by

$$\boldsymbol{x}_{ijt} = \frac{d^*_{ijt}\partial\pi_j(p_{ijt}, \boldsymbol{w}_{ijt}, \boldsymbol{z}_{ijt}, \boldsymbol{e}_{it})}{\partial \boldsymbol{w}_{ijt}}\,. \tag{12}$$

Equations (10), (11), and (12) define the model of discrete, extensive margin (land use) decisions and continuous, intensive margin (supply and input use) decisions. This analysis assumes that expected net returns above variable cost are positive for at least one activity, otherwise the land is left idle as in the conventional analysis of the firm's shut-down decision.

The solution to (9) applies to a given land unit. Each land unit is managed separately under the assumptions that farmers are risk-neutral, sell their products into a well-functioning market (as opposed to subsistence farmers who produce for own consumption), and have access to well-functioning rental markets for land and capital. In this static framework, production decisions may be interrelated across land units because risk-averse farmers manage a portfolio of production activities to maximize expected utility. In that case, the decision problem (9) for each land unit would be nested into a decision problem for all land units, subject to an adding-up constraint on the land units.

IMPLICATIONS FOR ECONOMETRIC ANALYSIS OF PRODUCTION RISK

The models outlined above with spatial and temporal variability can be integrated into a single model, obviously at the cost of greater complexity (see, e.g., Antle and Stoorvogel 2001). The user of applied production modeling will be well advised to use the simplest model that captures the essential features of the system under study. Our goal here is not to derive the most general model, but rather to illustrate some key econometric implications of models with these features.

Input Endogeneity and Production Risk Measurement

Perhaps the most basic implication of the intra-seasonal decision model outlined above, and the implication emphasized by Antle (1983b), is that inputs are endogenous to final output. The earlier discussions of estimating moments of output as functions of inputs (Just and Pope 1978, Antle 1983c) were surprisingly silent on this issue. Love and Buccola (1991) observe that inputs are likely to be endogenous in their proposal to estimate risk preferences jointly with technology parameters. Shankar and Nelson (1999) observe that technology parameters can be estimated consistently without specifying risk preferences, under the well-known Zellner-Kmenta-Dreze argument that input decisions are made before output shocks are realized. Antle (1983b) showed, however, that this justification for treating inputs as exogenous is valid only in a single-period model; in a multi-period sequential model, as we derived above, inputs are endogenous.

Using the intra-seasonal decision model, we derived the primal moment functions (7) and the reduced-form moment functions (8). Due to the endogeneity of input quantities, the primal moment functions for periods $1,...,N,H$ cannot be estimated consistently using the Just-Pope (1978) or Antle (1983c) methods based on regressing functions of production function residuals on inputs. From (7) and (8), we can see that the *zero-period* primal moments are functions of the zero-period input quantities; these input quantities are determined before any realizations of the production errors are observed, so they can be estimated consistently using the standard single-equation methods. Similarly, the zero-period reduced-form moments are functions of expected output price and input prices and can be estimated consistently using the standard methods. However, the primal moment functions for periods $1,...,N,H$ (equation 7) depend on intermediate input quantities that are endogenous to output, and cannot be estimated consistently using the standard methods based on the production function residuals. The reduced-form moment functions also depend on lagged input quantities and lagged production errors, so they also cannot be estimated consistently using the standard methods.

One estimation strategy is to transform the reduced form to a final form by substituting out the lagged input quantities so that they are replaced by input prices. In most applications, input prices will not vary enough during the season to make estimation of this model possible. Because time-varying input prices are not likely to be available, the sequential model could be specified in a two-period model where inputs are grouped into predetermined (zero-period) inputs and endogenous (intermediate stage) inputs. Then the moment functions can be estimated in the form

$$\theta = \theta(\varphi_0, w^0, x_0). \tag{13}$$

Another estimation strategy would be to utilize conventional estimation procedures that account for the lagged endogenous variables in the sequential moment functions. This strategy would require input quantity and price data for each production stage.

Econometric Analysis of Discrete Land Use Decisions

The spatial production model outlined above also could be used in econometric analysis of production risk. As we noted, we can view farmers as managing a portfolio of land units, and spatial diversification can be one important means to manage risk. Econometric specification and estimation of disaggregate, site-specific production models must account for the discrete structure of land use decisions, the dynamics of crop rotations, the spatial variation in physical conditions, statistical properties of the spatial data, and features of the farmer's management behavior. A key problem with disaggregate data is that a cross-section of data will have an unbalanced property, i.e., not all production activities will be undertaken by all producers. Note that this problem does not typically arise with aggregated data as long as at least one farmer in each spatial unit produces each crop. This problem of unbalanced data means that a multi-product model that assumes positive values for all outputs on all land units, as is typically assumed in duality-based multi-output production models in the literature, is not appropriate.

Several econometric specifications are available for disaggregate models. A first choice would be to specify a multi-output production model along the lines in the literature (Huffman 1988). These models generally assume that all outputs are produced by all farms, thus they fail to account for discrete choice among production activities and the unbalanced data that result from these discrete choices.

Another approach would be to utilize a multinomial discrete choice model (e.g., logit or probit) to estimate the vector of reduced-form land use functions analogous to equation (10), and a system of factor demand equations such as (12) to represent management decisions conditional on land use decisions. While this approach provides an efficient means to estimate the probability of each land use alternative in a consistent manner, it has several practical limitations.

First, the discrete choice model is in reduced form and thus faces the limitations of models in which productivity is not represented explicitly. One solution to this problem could be to replace the vector $\boldsymbol{e}$ with a yield estimate from a crop model, under the assumption that the crop model yield represents the site's productivity potential. A second, related problem is that yields are observed only for the crop that is produced at the site, whereas the decision between alternative crops involves a comparison of productivity between all possible crops. This difficulty is magnified in an integrated assessment that involves simulation of unobserved conditions, e.g., a perturbed climate scenario, in which existing crops may be uncompetitive and new crops may become profitable. The productivity of alternative crops could be estimated using either a statistical model (e.g., an econometric supply function) or a biophysical simulation model.

Several other practical problems complicate the use of an explicit discrete choice model. The estimation and simulation of multi-dimensional probability distributions required for the discrete land use decision problem poses difficult computational problems, and available software limits this type of analysis to the logistic and normal distributions without any clear justification of these functional forms. Adaptation of these models to account for spatial dependence would be even more challenging than with conventional linear statistical models. A final limitation is that this approach would provide only reduced-form estimates of land use choices, but would not provide estimates of production, cost, and net returns that are of interest to policymakers.

Spatial and Temporal Aggregation

The dynamic and spatially explicit models discussed above suggest that farmers make decisions that are site-specific (i.e., specific to a field and in the case of precision farming specific even to points within a field) and temporally variable due to intra-seasonal and inter-seasonal management decisions. However, most data used for econometric analysis of production are aggregated across space and time. Even when data are collected by small spatial units such as the field or farm, input data are typically aggregated across time. For example, multiple fertilizer and pesticide applications are often made within a growing season, but data are collected as total application quantities (moreover, these inputs are rarely collected according to component active ingredients so there is also a problem with measurement units). Moreover, when data are collected at the farm level, inputs and outputs for a given crop are often aggregated across various spatial units (fields). And of course data that are reported for larger spatial units, such as counties, are aggregated across both space and time, and often in ways that are arbitrary and have unknown effects on the properties of the data.

An obvious implication, one that is largely ignored in most applied production analysis, is that spatial and temporal aggregation of data is likely to significantly reduce the information content of the data and bias inferences based on them. Antle and Stoorvogel (2001) provide an analysis of the effects of spatial aggregation on the measurement of the productivity effects of soil erosion that illustrates this point. A similar analysis could be made of the effects of temporal aggregation.

Our experience working with highly detailed data, collected on a site-specific basis according to individual production operations during the growing season, strongly confirms this implication. We offer our work from Ecuador (Crissman, Antle, and Capalbo 1998) as an example of how such data can be collected and the kinds of inferences that can be drawn from them. Perhaps the most egregious case in the agricultural economics literature is the research on pesticide productivity and pest management, where most researchers aggregate various types of active ingredients without regard to their greatly varying purposes and rates of effectiveness. It is little wonder that the econometric results are so widely varying and contradictory.

Does Risk Improve the Predictive Power of Economic Models?

The most fundamental question for risk analysis is whether it can improve our ability to predict behavior. Testing for significance of moments in econometric models (Just 1974) is a weak test in the sense that variables that are proxies for risk may be statistically significant but may not substantially improve the model's predictive capabilities. A stronger test of the importance of risk is to ask if it can significantly improve the ability of a model to predict either within-sample or out-of-sample behavior. Here we undertake to subject the spatial econometric-process model developed by Antle and Capalbo (2001) to this stronger test. As we discussed earlier (see equations 9-12), the motivation for the development of an econometric-process approach was the need to link economic analysis of production systems to site-specific biophysical simulation models. Site-specific data are used to estimate the econometric production models, and these data and models are then incorporated into a simulation model that represents the decision making process of the farmer as a sequence of discrete and continuous land use and input use decisions. This discrete/continuous structure of the econometric-process model is able to simulate decision making both within and outside the range of observed data in a way that is consistent with economic theory and with site-specific biophysical constraints and processes.

To implement this model, econometric production models (a system of supply functions and cost functions) were estimated using cross-sectional data from a sample of 425 farms and over 1,200 fields that are statistically representative of USDA's three Major Land Resource Areas (MLRA) in the grain-producing regions of Montana. The MLRAs were stratified into six zones (sub-MLRAs), based on high or low precipitation according to historical climate data. Log-linear production models consisting of a crop supply and a variable cost function were estimated using nonlinear three-stage least squares, for winter wheat, spring wheat, and barley. Data and parameter estimates for the supply and cost equations are reported and discussed in Antle and Capalbo (2001).

By operating at the field scale with site-specific data, the simulation can represent spatial and temporal differences in land use and management, such as crop rotations, that give rise to different economic outcomes across space and time in the region. Each field in the sample is described by area, location, and a set of location-specific prices paid and received by producers, and quantities of inputs. Using sample distributions estimated from the data, draws are made with respect to expected output prices, input prices, and any other site-specific management factors (e.g., previous land use). The econometric production models are simulated to estimate expected output, costs of production, and expected returns. The land use decision for each site is made by comparing the farmer's objective function for each production activity. The question we pose here is whether using an expected utility criterion based on mean and variance of expected returns will improve the model's ability to predict land allocation decisions relative to a model that uses expected returns to predict land allocation decisions.

In addition to the input endogeneity issue discussed above, it seems likely that methods for estimation of production moments based on residuals will be highly sensitive to model specification. To illustrate, we present here estimates of second-moment functions derived from an application of the Just-Pope (1978) model and the Antle (1983c) model to data for dryland grain production in Montana (see Antle et al. 1999 and Antle and Capalbo 2001 for details). We present two versions of these models: one is the conventional estimation procedure wherein moments are specified as functions of input quantities; the other is based on the sequential model discussed above wherein moments are specified as functions of input prices. In this dryland production system, the key inputs are machinery services, fertilizer, and insecticides. We assume that machinery operations are planned at the beginning of the season and are predetermined relative to output realizations, whereas fertilizer and insecticide applications are made during the growing season and are endogenous, as shown by the sequential model. The mean function was specified in the Cobb-Douglas form with an additive error, with explanatory variables as land, machinery, fertilizer, insecticides; dummy variables to indicate fallow, winter wheat, barley, and interactions between fallow and fertilizer, and between fallow and winter wheat and barley; and dummy variables for agroecological zones. The mean function was estimated using nonlinear least squares. The variance functions were specified in a similar way. The variance functions were estimated using the Just-Pope (1978) procedure (regressing the log of the absolute value of the mean function residual on the log of the Cobb-Douglas model), and using the Antle (1983c) procedure wherein the squared residual is regressed on the Cobb-Douglas function (estimated using nonlinear least squares).

Table 1 presents the variance function results. One notable feature is the poor fit of the Just-Pope model and low statistical significance of many of the parameters, and the much better fit of the Antle model. Another feature is the contrast between the primal models based on input quantities and the sequential models based on input prices. Particularly with the Antle models, there is a marked difference between the primal model parameters, which should be inconsistent, and the sequential model parameters, which should be consistently estimated.

Figure 3 presents the results of a within-sample test of the risk-neutral and risk-averse model predictions. The risk-averse model was specified with an Arrow-Pratt relative risk aversion coefficient value of 3 and a standard deviation of 0.5 (various values were tried and these values provided the best predictive power). Following Antle and Capalbo (2001), to test the predictive capability of the model the observed proportion of each land use (winter wheat, spring wheat, and barley in continuous or fallow rotations) was computed for each of six agroecological zones and compared to the simulated proportions. Note that the site-specific land use data follow a binomial distribution for each use (i.e., the data are coded 1 if use j occurs and zero otherwise). It follows that the sample proportions plotted in Figure 3 are sufficient statistics for the entire distribution (all of its moments are functions of this proportion). Thus, Figure 3 shows how the *distributions* of land use are represented by the econometric-process simulation models. Two facts are immediate from inspection of Figure 3. First, the plots of observed and simulated mean land use for both risk-neutral and risk-averse models fall along a 45-degree line, an indication that the simulation model does reproduce the observed

Table 1. Variance Function Elasticities for Montana Dryland Grain Production Based on Primal and Sequential Models

	Primal Model		Sequential Model	
	Just-Pope	Antle	Just-Pope	Antle
Land	0.906	3.701	0.966	4.974
	(4.14)	(12.68)	(4.84)	(11.64)
Machinery	-0.213	-0.988	-0.101	-2.115
	(-1.36)	(-6.29)	(-0.63)	(-8.96)
Fertilizer	0.298	-0.620	-0.450	-0.366
	(2.14)	(-2.79)	(-1.17)	(-0.39)
Fertilizer + Fallow	0.109	-0.716	-0.627	2.522
	(0.584)	(-5.21)	(-2.05)	(4.46)
Insecticide	-0.082	-0.511	-0.030	0.470
	(-0.95)	(-4.99)	(-0.43)	(6.39)
Spring Wheat Fallow	1.802	1.589	-0.098	0.350
	(1.32)	(0.65)	(-0.60)	(1.69)
Winter Wheat	-0.276	1.974	-0.388	0.474
	(-0.82)	(5.83)	(-1.14)	(0.67)
Winter Wheat + Fallow	0.162	0.837	-0.080	2.136
	(0.94)	(5.44)	(-0.46)	(8.51)
Barley	0.341	1.284	0.174	1.771
	(1.53)	(3.11)	(0.90)	(3.62)
Barley + Fallow	0.634	1.482	0.532	2.321
	(3.21)	(9.80)	(2.65)	(8.34)
R^2	0.254	0.531	0.240	0.525

Note: t-ratios in parentheses. Primal model uses input quantities, sequential model uses input prices. Fallow and crop effects are dummy variable coefficients.

data without a systematic bias. Second, it is clear that the risk-averse model does not perform better than the risk-neutral model in predicting land use distributions by agroecological zone. Indeed, in our attempts to choose values of risk-aversion coefficients for this exercise, we found that in most cases the risk-averse model performs worse than the risk-neutral model.

A NEW APPROACH TO THE ANALYSIS OF PRODUCTION RISK

We are not surprised by the finding that the conventional static expected utility formulation of a production model does not predict better than a risk-neutral model. We are not surprised because we do not believe that the static risk-aversion model captures the ways that spatial and temporal variability in the crop growth process interacts with farmers' land use and management

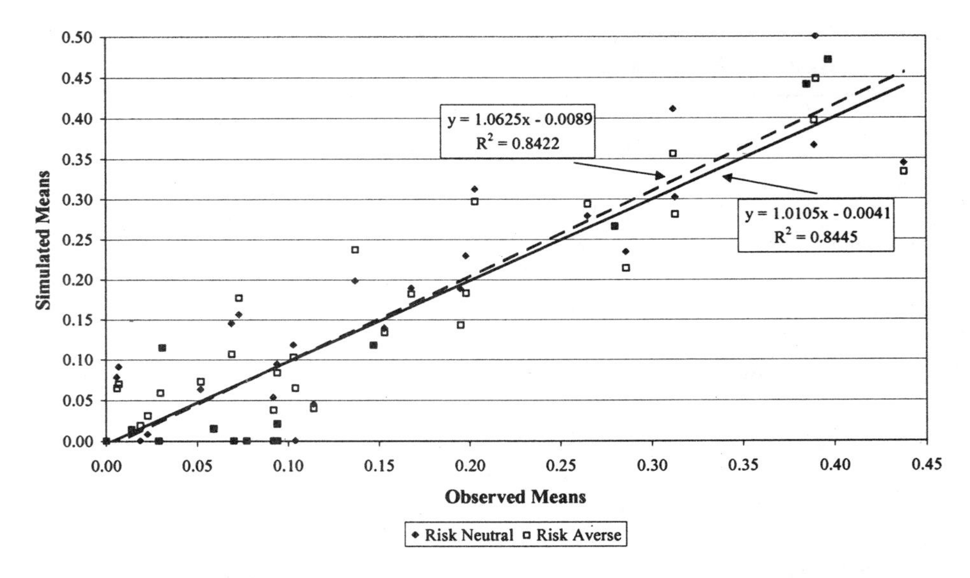

Figure 3. Observed vs. Simulated Mean Land Use in Montana Dryland Grain Production, Risk-Neutral and Risk-Averse Models

decisions. This behavior is in fact embedded into the parameters of the risk-neutral model, and cannot be effectively extracted using conventional econometric techniques due to the various problems associated with the spatial and temporal variation in behavior that we discussed above. In this section we return to the agroecosystem paradigm and discuss a new approach to the analysis of production risk. We hypothesize that this new approach could provide greater predictive power than conventional static risk-aversion models.

Agroecosystems are complex, spatially variable dynamic systems. To capture both the biophysical and economic dimensions of these systems, empirical models of agricultural production systems can be characterized as a set of linked sub-models, each with sets of drivers, state variables, flow variables, and processes (Antle et al. 2001). Figure 1 describes a simplified dryland crop agroecosystem that is composed of a crop ecosystem model and an economic decision model each with a set of drivers and outputs. We describe the modeling system as *loosely coupled* when it is constructed using state or flow variables from one sub-model as driving variables in the other sub-model. In Figure 1, the economic decision model determines fertilizer application rates as a function of economic drivers and crop yields. The crop ecosystem model determines yields as a function of exogenous biophysical drivers and fertilizer application rates. Under this structure the two models are loosely coupled by executing each model for a growing season sequentially, passing fertilizer application rates from the economic model to the ecosystem model, and crop yields from the ecosystem model to the economic model.

When states or processes from one sub-model are linked directly to processes in another sub-model, we describe the modeling system as *closely coupled*. Returning to Figure 1, the closely coupled structure is illustrated by the dashed lines linking the fertilizer decisions in the economic model to the crop growth processes in the ecosystem model, and by linking crop growth to the fertilizer decision making process in the economic model. In our recent applications linking agricultural ecosystem and economic simulation models, the only linkages between models involve land use and crop yields on an annual basis. To operate the crop ecosystem model, representative fertilizer, tillage, and other important management decisions are set as fixed boundary conditions and are not linked to the economic decision model. A more tightly coupled model would make these linkages between biophysical processes and management decision making. In many production systems, multiple fertilizer applications, pesticide applications, and tillage operations are made during the growing season in relation to weather events and crop growth, and these operations can have significant implications for biophysical processes which in turn feed back to decision making. Integrated pest management is a classic example of the farmer sequentially acquiring information during the growing season, updating information about pest populations, and making decisions based on that information. Another good example of the importance of these interactions is provided by recent research on the potential for agriculture to both emit greenhouse gases and sequester atmospheric CO_2 in the soil. Research shows that these processes could only be represented by a model that captured the interactions between management decisions such as fertilizer use and tillage operations (Robertson, Paul, and Harwood 2000, Watson et al. 2000). Management decisions across seasons, such as crop rotations, interact dynamically with weather events that determine important production constraints such as soil moisture and pest populations. Management decisions are also affected over time by the farmer's acquisition of information about crop and input prices.

Thus, with a closely coupled model it is possible to link processes in ways that more accurately reflect the interactions between biophysical and economic processes. In contrast to loosely and tightly coupled systems, an *integrated* system would have a single set of drivers and endogenous variables for all disciplinary components. Integration of the agroecosystem in Figure 1 would mean that the same set of biophysical and economic drivers would be inputs into a combined model of crop growth and economic decision making. For example, economic decisions would take account of all relevant information affecting the biophysical processes of crop growth as well as the exogenous economic drivers; likewise, management decisions such as fertilizer and pesticide applications and tillage operations would be incorporated into the crop growth processes. Thus, the key difference between the loosely or closely coupled systems is that an integrated system operates on temporal and spatial scales dictated by the processes within the model, not by the way that the disciplinary models were designed and coupled.

The development of tightly coupled or integrated agroecosystem models would provide the capability to investigate the role of risk in agricultural production systems in ways that represent the information used by farmers to make decisions in response to spatial and temporal variability. To illustrate, let us return to the intra-seasonal decision model discussed earlier. We noted in that discussion that a key implication of the sequential decision model is the endogeneity of inputs to output. This endogeneity arises because the farmer observes the crop growth process (and the effects of random shocks such as weather and pests) during the growing season and incorporates that information into decisions, but the econometrician does not observe crop growth. Consequently, the econometrician can only estimate the composite production function (or functions dual to it). We also showed that the stochastic properties of the production process (as represented by the condition moment functions in equation 7) also vary during the growing season as the farmer manages the crop in response to stochastic events. The lack of observability of crop growth makes it difficult to estimate econometrically a sequential decision model with risk.

These estimation problems associated with the conventional econometric approach to production modeling can be avoided by using the type of closely coupled or integrated models that we have described here. These problems are avoided because a biophysical model of crop growth provides estimates of the intra-seasonal information that the farmer uses for decision making. Thus, by coupling an economic decision model to a biophysical crop growth model, the production system can be both estimated and simulated taking intra-seasonal information into account. In addition, the use of this integrated simulation approach provides a way to bypass the estimation problems discussed above associated with the modeling of discrete land use decisions, as discussed in detail in Antle and Capalbo (2001).

A detailed scheme for this type of integrated modeling is presented in Figure 4. In this approach, linkages are made between biophysical models and economic models for both estimation and simulation. As represented in the upper half of Figure 4, these linkages can be made to facilitate the estimation of intra-seasonal and inter-seasonal dynamics on a given land unit, and they can be made to account for spatial variability across land units. Figure 4 represents these linkages in a loosely coupled format where information flows from crop growth models to economic models, but more closely coupled or integrated models would involve dynamic feedbacks between the model components as shown in Figure 1. The lower half of Figure 4 represents how biophysical crop growth models and other environmental process models can be linked to economic models for simulation. Again, the scheme in Figure 4 is based on the loosely coupled approach, and dynamic feedbacks could be incorporated. The critical point is that through coupled simulation of these models, the effects of spatial and temporal variability on farmer decision making could be investigated in ways that could not be investigated using conventional econometric models alone.

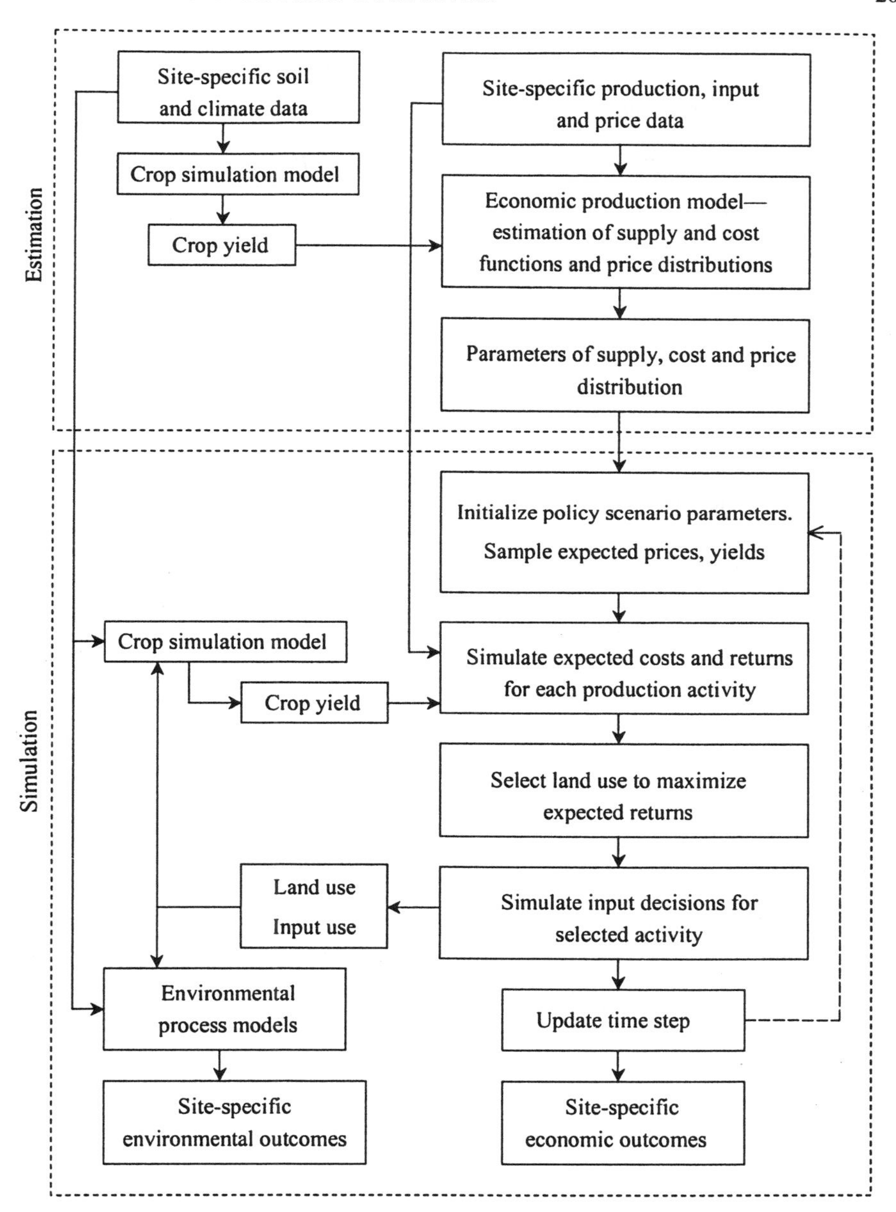

Figure 4. Structure of an Econometric-Process Simulation Model and Linkages to Biophysical Simulation Models in a Closely Coupled Model of an Agroecosystem
(Source: Antle and Capalbo 2001)

CONCLUSIONS

In this chapter we use the paradigm of agriculture as a managed ecosystem to investigate the implications for econometric analysis of production systems and production risk. Our goal is to critically assess how risk concepts can help us understand and predict the behavior of agroecosystems, i.e., complex spatially and temporally varying systems made up of interrelated biophysical and economic processes. We argue that the prevailing analytical and empirical paradigm used by the agricultural economics profession, which largely abstracts from the spatial and temporal aspects of these complex systems, suffers from a number of significant limitations. These limitations explain why this paradigm has not been successful in improving the predictive power of economic models, and why models used in policy analysis do not typically incorporate risk features.

We used the paradigm of agriculture as an agroecosystem to investigate the temporal and spatial structure of decision making processes. These characterizations give rise to a number of econometric problems, including the lack of observability of crop growth processes, which causes input endogeneity, and the site-specific character of decisions that introduces a discrete choice element.

We argued that a strong test of the value of the risk is the ability to improve predictive power of empirical models. We used a recent model of ours to show that, at least in that example of a spatially explicit dynamic model, adding a risk-aversion component did not increase the predictive power of the model.

In the final section of the chapter, we outlined a new approach to analysis of risk that utilizes the agroecosystem paradigm. In this approach, biophysical and economic models are coupled for both estimation and simulation. These models capture the spatial and temporal variability of agroecosystems that we hypothesize are essential to understanding how risk affects agricultural decision making.

REFERENCES

Antle, J.M. 1983a. "Incorporating Risk in Production Analysis." *American Journal of Agricultural Economics* 65: 1099-1106.

___. 1983b. "Sequential Decision Making in Production Models." *American Journal of Agricultural Economics* 65: 282-290.

___. 1983c. "Testing the Stochastic Structure of Production: A Flexible Moment-Based Approach." *Journal of Business and Economic Statistics* 1: 192-201.

Antle, J.M., and S.M. Capalbo. 2001. "Econometric-Process Models for Integrated Assessment of Agricultural Production Systems." *American Journal of Agricultural Economics* 83: 389-401.

Antle, J.M., S.M. Capalbo, and C.C. Crissman. 1994. "Econometric Production Models with Endogenous Input Timing: An Application to Ecuadorian Potato Production." *Journal of Agricultural and Resource Economics* 19: 1-18.

Antle, J.M., S.M. Capalbo, E.T. Elliott, H.W. Hunt, S. Mooney, and K.H. Paustian. 2001. "Research Needs for Understanding and Predicting the Behavior of Managed Ecosystems: Lessons from the Study of Agroecosystems." *Ecosystems* (in press).

Antle, J.M., S. Capalbo, J. Johnson, and D. Miljkovic. 1999. "The Kyoto Protocol: Economic Effects of Energy Prices on Northern Plains Dryland Grain Production." *Agricultural and Resource Economics Review* 28: 96-105.

Antle, J.M., and S.A. Hatchett. 1986. "Dynamic Input Decisions in Econometric Production Models." *American Journal of Agricultural Economics* 68: 939-949.

Antle, J.M., and J.J. Stoorvogel. 2001. "Integrating Site-Specific Biophysical and Economic Models to Assess Trade-offs in Sustainable Land Use and Soil Quality." In N. Heerink, H. van Keulen, and M. Kuiper, eds., *Economic Policy and Sustainable Land Use: Recent Advances in Quantitative Analysis for Developing Countries.* New York: Physica-Verlag.

Crissman, C.C., J.M. Antle, and S.M. Capalbo. 1998. *Economic, Environmental and Health Tradeoffs in Agriculture: Pesticides and the Sustainability of Andean Potato Production.* Boston: Kluwer Academic Publishers.

Huffman, W.E. 1988. "An Econometric Methodology for Multiple Output Agricultural Technology: An Application of Endogenous Switching Models." In S.M. Capalbo and J.M. Antle, eds., *Agricultural Productivity: Measurement and Explanation.* Washington, D.C.: Resources for the Future Press.

Just, R.E. 1974. "An Investigation of the Importance of Risk in Farmers' Decisions." *American Journal of Agricultural Economics* 56: 14-25.

Just, R.E., and R.D. Pope. 1978. "Stochastic Specification of Production Functions and Economic Implications." *Journal of Econometrics* 7: 67-86.

Love, H.A., and S.T. Buccola. 1991. "Joint Risk Preference-Technology Estimation with a Primal System." *American Journal of Agricultural Economics* 73: 765-774.

Robertson, G.P., E.A. Paul, and R.R. Harwood. 2000. "Greenhouse Gases in Intensive Agriculture: Contributions of Individual Gases to the Radiative Forcing of the Atmosphere." *Science* 289: 1922–1924.

Shankar B., and C.H. Nelson. 1999. "Joint Risk Preference-Technology Estimation with a Primal System: Comment." *American Journal of Agricultural Economics* 81: 241-244.

Watson, R.T., I.R. Noble, B. Bolin, N.H. Ravindranath, D.J. Verardo, and D.J. Dokken. 2000. "Land Use, Land-Use Change, and Forestry." A special report of the Intergovernmental Panel on Climate Change (IPCC), Cambridge, England: Cambridge University Press.

Chapter 13

SURVEY AND EXPERIMENTAL TECHNIQUES AS AN APPROACH FOR AGRICULTURAL RISK ANALYSIS

Brian Roe and Alan Randall
The Ohio State University

INTRODUCTION

The ubiquity and centrality of risk in agricultural processes helps define agricultural economics as a distinct field of study. Understanding agricultural agents' behavioral responses to risk, and the implications of these responses for market behavior and agricultural policy, has occupied a central position in agricultural economics research. Much of this research has relied upon the analysis of agents' behavior as observed and encoded by various public data collection agencies and private researchers. Simulation and programming techniques have also broadened our understanding of risk and extended the profession's ability to provide prescriptive analysis to both individual decision makers and policy advisors. Such techniques often rely upon the analysis of observed behavioral responses to calibrate key parameters of the underlying models. However, calibration of individual risk preferences and identification of agents' true behavioral objectives and constraints are often difficult, if not impossible, to obtain from observed behavior.

To augment and complement imperfect observation of behavior, economists traditionally have relied upon direct conversation with farmers and other agricultural agents to better understand the underlying objective function, the pertinent constraints upon behavior, and the details of the decision making process. Such conversations, however, are typically informal or, for many with direct agricultural experience, merely a matter of introspection.

However, a modest vein of the agricultural risk literature has turned to more formal and systematic methods that involve the direct questioning of agricultural agents.[1] A number of research efforts administer survey instru-

[1] See Young (1979) for an early summary of this literature. Bard and Barry (1999) review some more recent contributions. Our apologies to those who have contributed other research in this vein but are omitted in this brief overview.

ments that require agricultural agents to state intended behavior for impending real decisions (e.g., USDA's production intention surveys), to reveal subjective probabilities for relevant stochastic processes (e.g., Francisco and Anderson 1972, Pingali and Carlson 1985), to describe key attitudes and attributes surrounding the individual choice process via Likert-type scales (Howard, Brinkman, and Lambert 1997, Bard and Barry 1999), to state intended behavior in response to hypothetical scenarios (e.g., Officer and Halter 1968, Dillon and Scandizzo 1978, Gunjal and Legault 1995, Finkelshtain and Feinerman 1997), or to respond to real monetary incentives in an experimental setting (e.g., Binswanger 1980, Grisley and Kellogg 1983, Belaid and Miller 1987, Ward et al. 1996, Ward et al. 1999). Many of these studies borrow and augment survey techniques used by researchers involved in experimental economics and contingent valuation.

The goals of this chapter are to review the historical development of the experimental economics and contingent valuation literatures, to summarize a set of persistent anomalies to the neoclassical assumptions of agent rationality that these literatures have helped to identify, to forward several stylized facts concerning these anomalies, and to identify a set of intriguing questions that remain open to further debate. Finally, we will identify key discussions in the agricultural risk literature that might be informed by survey and experimental techniques.

EXPERIMENTAL AND SURVEY METHODS IN ECONOMICS

Applied economists are constrained by the natural experiments that occur in the economy and by their econometric ingenuity. These natural experiments must provide sufficient variation, and sufficient data collection must take place, before analysis can move forward and key hypotheses can be tested. However, recorded economic data cannot always confess the reality and nuance of the economic structures that interest us. Natural confounding of the data (i.e., a poor design of the natural experiment) often limits our ability to identify particular structural elements (e.g., the difficulty in identifying the separate effects of risk aversion, technical inefficiency and allocative inefficiency from observed production data sets). In other cases, the data may involve insufficient variation to allow for global identification of technical and behavioral parameters (e.g., projecting unregulated behavior from data recorded in regulated regimes). Sometimes the data simply may not exist (e.g., establishing values for public goods never before offered or for public goods with non-market amenities).

Each of these frustrations has motivated economists to take matters into their own hands and to formulate hypothetical or synthetic environments in which people react to the researcher's stimuli. These environments might vary from a simple survey with a contingent valuation question administered to respondents via telephone to experiments involving real monetary rewards for a group of subjects competing in a constructed market on a computer network.

The genesis of modern economic survey methods is often traced to L.L. Thurstone's (1931) research, in which he constructed an individual's indifference curve for shoes and hats by posing a series of hypothetical questions concerning the desirability of various bundles of these goods. Wallis and Friedman (1942) responded to this novel methodology in the following way:

> "It is questionable whether a subject in so artificial an experimental situation could know what choices he would make in an economic situation; not knowing, it is almost inevitable that he would, in entire good faith, systematize his answers in such a way as to produce plausible but spurious results. ..... The responses are valueless because the subject cannot know how he would react" (pp. 179-180).

This stern critique notwithstanding, such research continued over the following decades, albeit with more tightly designed surveys and experiments meant to address the Wallis and Friedman critique that such endeavors were unlike real economic choices both in the framing of the economic question and in the nature of payoffs and consequences. Economists still lob elements of this critique across conference rooms to this day in response to contingent valuation and experimental economics presentations. Elements of this critique also emerged in the critique of contingent valuation methods and helped shape the NOAA panel's guide (Arrow et al. 1993) for conducting relevant contingent valuation studies.

A Tale of Two Literatures

Economic research using surveys and experiments is divided into two principal literatures that feature less overlap than one might expect. The first concerns economic experiments (EE), which, prototypically, focus on repetitive exposure of college students to various economic stimuli in a laboratory setting and feature small but real monetary payouts commensurate with response. The second concerns surveys using the contingent valuation method (CVM), which, prototypically, provide detailed descriptions of a public good (or other goods not commonly transacted in markets) and pose questions to a representative sample of respondents in order to elicit the respondents' willingness to pay (accept) for the provision (removal) of the good.

As described above the genesis of EE is generally traced to Thurstone (1931). A number of influential experiments are typically cited from the 1940s (e.g., Chamberlin 1948) and 1950s (e.g., Allais 1953); the body of work greatly expanded during the 1960s and beyond. Roth (1995) provides a full treatment of the EE lineage.

The lineage of CVM skips the Thurstone experiment and, instead, cites Ciracy-Wantrup (1947) as the researcher who planted the seed for modern contingent valuation. Davis (1963) is generally cited as the first contingent valuation study while the modern thrust of the contingent valuation literature begins with the work of Randall, Ives, and Eastman (1974).

The contingent valuation method (CVM) emerged in response to the wide-spread adoption of benefit-cost analysis (BCA) by government administrators and to the recognition that certain non-market values, often associated with environmental amenities, were not being accurately counted in the BCA (Randall 1998). Other methods for valuing such non-market goods, such as travel cost and hedonics, could be applied only to a certain subset of goods and relied often on tenuous modeling assumptions. To measure these benefits economists formalized survey techniques used by other social scientists in a manner that could elicit individual willingness to pay for an alteration in the provision of a public good. After much trial and error to improve the survey process and after economists made the appropriate contributions in welfare theory and econometric methods, CVM emerged as another method in the toolbox of BCA practitioners.

As the usage and application of CVM grew within government evaluation circles, its critics became more vocal. The debate concerning the method's validity escalated after CVM was used to measure lost passive-use values in the calculation of corporate liability in the *Exxon Valdez* oil spill case in 1989. Sharp criticisms of the method and its validity were forwarded in a volume edited by Hausman (1993). In response to these damning criticisms the government assembled a blue-ribbon panel to evaluate the criticisms and to recommend actions concerning the future use of CVM (the NOAA panel, Arrow et al. 1993). The panel laid out practical guidelines for implementation of CVM studies that should be met if the results are to be deemed valid for deriving discrete values for public good offerings; the guidelines include benchmarks for survey method, survey response rates, question design, and question format.

Interestingly, the experimental economics literature largely fails to recognize the sizeable output and controversy of the CVM literature; e.g., *The Handbook of Experimental Economics* (Kagel and Roth 1995) mentions the term contingent valuation only once and does not even include the term in its index. Despite the historic lack of overlap and cross-citation between the experimental economics literature and the CVM literature, there emerges an interesting insight that will lead us to the next section of this chapter. The detractors of contingent valuation and stated preference methods largely focused their critical arrows at the adjectives "contingent" and "stated" as the reason for the methods' invalidity. Indeed they discovered and cited a number of anomalous findings from this literature that departed from the predictions of economic theory. Critics suggested that the removal of the "contingent," the "hypothetical," and the "stated" aspects of these methods would clear a path to methodological validity.

However, the growing body of EE research gave scant support to these suggestions. This literature shows that making incentives real rather than hypothetical reduces the magnitude but not the presence of reported anomalies, particularly when the null hypothesis is derived from the expected utility framework. Risk experiments tend on balance to support the conjecture that the fundamental axioms of rationality fail consistently to predict behavior.

Persistent Anomalies to Rational, Neoclassical Behavior

Do the Three P's of Rationality Hold? McFadden (1999) succinctly summarizes three separate neoclassical assumptions concerning human rationality that must carefully be examined when we, as researchers, turn directly to people in an effort to uncover elements of the economic decision making process and when we analyze the data that such questioning creates. The first assumption is that of *perception rationality*: agents behave as if information is perfectly processed to form perceptions and beliefs using strict Bayesian statistic principles. The second assumption is that of *preference rationality*: agents have preferences that are primitive, consistent, and immutable. The third assumption is that of *process rationality*: the agent's cognitive process is that of preference maximization subject to market constraints.

How does each tenant hold up under the scrutiny of experimentation? Each takes its share of abuse under the microscope of laboratory experiments. Again, following McFadden, we catalog the major categories of anomalies that appear as chinks in the armor of strict rationality.

Categories of Anomalies. The first category involves *context effects* in which the presentation of information influences how agents process information. Examples include issues of anchoring, framing, prominence, and saliency. This group of anomalies is deemed quite relevant for stated preference questions as well as probability judgments. However, certain anomalies, such as anchoring, have widespread concern for collecting any survey data, whether the data is gathered by hypothetical question, by experimental protocol, or by recall of real-world economic data. Anchoring anomalies occur when the choice of a cue presented by the interviewer affects the collected information. So suggested starting dollar bids in an experimental auction, suggested values on a bid card used in contingent valuation, or income bracket levels on a household consumption survey will likely impact the eventual answer encoded from the respondent.

Reference point effects refer to effects that are driven by manipulation or alteration of agents' baselines from which the relative benefits of potential changes are calculated. Well-known variations of reference point effects include Tversky and Kahneman's (1991) loss aversion theory, which postulates that the pain of marginal losses exceeds the pleasure of comparable gains, and Thaler's (1981) endowment effect, which postulates that the status quo holds a privileged position.

Availability effects refer to an agent's inherent tendency to draw upon certain types of information rather than the entirety of his information set when making a decision. Such effects include the representativeness effect, in which agents regularly fail to invoke Bayes' law; a regression effect, in which agents interpret recent observed changes as idiosyncratic changes in the underlying structure rather than random fluctuations; the primacy and recency effects, in which agents favor the first and last pieces of information they have received; and status quo effects, in which historical experience is more easily retrieved than hypothetical alternatives.

Superstition effects involve situations in which agents' subjective probabilities feature unwarranted interdependence (e.g., the gambler's fallacy of winning streaks across independent draws), in which agents attribute superior powers to other agents, and in which agents take apparently irrational actions to guard against the general possibility of being exploited by others.

Process effects arise from the way agents approach the decision making process. Examples include situations in which the agents derive benefits and losses from the decision making process itself, in which agents derive suboptimal rules to guard against unpredictable future lapses in willpower, and in which agents use inconsistent discounting of future events.

Projection effects arise when the researcher presents a choice task within a limited context but the respondent interprets the problem in a broader, strategic context. Here, the research typically fails to assess the context adopted by the subject; e.g., when agents misrepresent responses to contingent valuation questions in hopes of altering policy or when respondents try to project a positive image to the interviewer.

Stylized Facts and Open Questions

This set of robustly observed deviations from strictly rational behavior leads to a set of stylized facts and a larger set of open questions that are the subject of ongoing research. We share several of both below.

Losses and Gains Are Treated Asymmetrically. EE and CVM results consistently reveal that individuals treat losses and gains differently, even in carefully constructed situations that, in accordance to neoclassical theory, should be treated symmetrically. One well-documented manifestation of this tendency is that individuals' willingness to accept compensation for the removal of a wide variety of private and public goods greatly exceeds willingness to pay for the provision of the exact same good (see Horowitz and McConnell, forthcoming, for a review of such studies). While standard economic theory can explain this difference (Hanemann 1991, Randall and Stoll 1980), the empirical magnitudes recorded in many studies exceed the magnitude predicted by neoclassical theory; these findings hold across a wide variety of experimental and survey protocols and respondent classes.

Many researchers allude to non-neoclassical explanations for such results (e.g., loss aversion and other reference point explanations), though few formally test such a hypothesis against the competing neoclassical hypotheses. Mansfield (1999) provides one of the few published tests in this vein and finds evidence for the loss aversion explanation as well as evidence of inherent bias of the survey technique (in her case CVM). Using observed behavior, Collins, Musser, and Mason (1991) test for loss-gain asymmetry in the context of risk aversion (i.e., prospect theory) and find statistical evidence that the reference-point theory provides a better explanation of changes in farmers' risk aversion than does neoclassical theory. While more experimentation and testing remains to be done with loss-gain asymmetries, reference point explanations are becoming a stylized fact within economics literature.

People's Treatment of Small Probabilities Is Inconsistent. Perception rationality requires both that individuals act as perfect Bayesians and that they correctly comprehend and react to probabilities no matter what the absolute magnitude. Very small probabilities tend to be treated inconsistently across individuals and differently than expected utility theory would predict.

Evaluation of small probabilities is key to Ganderton et al.'s study (2000) which attempts to identify why people tend to underinsure against naturally occurring low-probability–large-loss events. Previous research (e.g., Camerer and Kunreuther 1989) reports a dichotomy of individual responses to such risks. Some people dismiss the possibility of such events by focusing on the fact that a very small number is essentially the same as zero (i.e., a threshold bias).[2] Such a response logically leads to insensitive responses to marginal changes in very small probabilities because the respondent deems all very small numbers to be essentially the same. Others recently exposed to such losses may tend to overestimate low probabilities (i.e., a recency bias) or focus on the magnitude of the loss (i.e., conjunction and availability biases). Why would anybody take a risk if such a terrible outcome could occur with any probability? Some experimental research (McClelland, Schulze, and Coursey 1993) found that such handling of small probabilities translated into a bimodal distribution of the willingness to pay for disaster insurance, with one group willing to pay prices above actuarially fair rates while the other group was not willing to buy insurance even with substantial subsidy. This would suggest that individuals inherently have a tough time dealing with very small probabilities and that such cognitive challenges lead to behavior inconsistent with perception rationality.

---

[2] Note these experiments are not dealing with problems associated with subjective probabilities that may legitimately vary across individuals (see Buschena, this volume, for additional discussion of subjective vs. objective probabilities) but rather exogenous, objective probabilities that are spelled out to respondents in the experimental session. The only way that the subjective vs. objective issue might enter is if respondents do not have faith in the experimenter's honesty or capability regarding the stated probabilities.

However, Ganderton et al. (2000) employ a more realistic experimental protocol that more closely mimics the features of real insurance markets (i.e., respondents buy insurance at posted prices rather than bidding for a fixed supply of insurance contracts). Such alterations yield a more uniform distribution of willingness to pay for disaster insurance and, in general, results are more closely in line with the predictions of expected utility theory than previous research. However, substantial deviation from expected utility remained. Respondents overvalued insurance for low-probability events, responded too strongly to changes in the posted probability of losses, and responded too weakly to changes in the size of potential losses. Many respondents also suffered the gambler's fallacy, i.e., they treated independent draws as if they were correlated. Continued experimentation could complement existing research of observed behavior in insurance markets and help forward our understanding of insurance purchases.

Embedding Is Not an Anomaly. One may be tempted to add the embedding effect to the list of cognitive anomalies. Often associated with CVM values, embedding occurs when the elicited value of a good changes depending upon whether the good is valued separately or is embedded as part of a more inclusive package of goods. However, as Hoehn and Randall (1996) show, the smaller values often associated with embedded goods are entirely consistent with standard utility-theoretic results. Given that households have limited discretionary budgets, the value placed on a single good should depend on the other goods available for valuation either prior to or simultaneously with it. Such results hold for public or private goods. For example, List and Shogren (1998) show that the winning bid for a rare baseball card was higher when it was auctioned alone than when it was auctioned as a package with other cards.[3]

Are Individual Risk Preferences Robust Across Institutions? While experimental techniques to uncover subjects' risk preferences have progressed greatly over the past several decades, the transferability of risk aversion elicited in experiments to risk aversion in real settings remains an open question. Questions of transferability arise on several fronts. As discussed by Binswanger (1980), most experiments deal with the resolution of single-period risk via the playing of lotteries. The generalizability of such risk preferences from a single-shot game, even one that is repeated numerous times in a lengthy laboratory session, to the relevant risk preferences that guide intra-season and inter-season choices in agriculture is not straightforward.

---

[3] List and Shogren (1998) uncover another interesting but unrelated feature of experimental auctions that warrants further consideration: while professional baseball card dealers made substantially higher average bids (willingness to pay, or WTP) for the rare card than did non-dealers, they tended not to offer the highest bid and, hence, would not outbid non-dealers for the card at auction. This deviation between average WTP and highest WTP at auction may be systematic and worthy of further experimentation.

Furthermore, there is some experimental evidence that the structure of a subject's risk preferences may not be stable across different types of experimental processes. Isaac and James (2000) report on a compelling series of experiments that identify a subject's risk preferences via two well-known experimental methods. Not only did most subjects' risk aversion coefficient change significantly from one method to the other, but also the ranking of individuals from most risk averse to most risk loving was significantly altered.

If such results are verified by other experiments, it suggests that extrapolation of risk preferences from one experimental situation to another may be ill advised; hence, one must also then question the extrapolation of risk aversion tendencies from risk experiments to any real-world situation. Indeed, if future research robustly rejects risk preference transfer across institutions, the usefulness of individual elicitation of risk preferences in any synthetic environment is highly questionable. Such a result would then motivate a research agenda that explores the robustness of observed risk preferences across time and decision situation.

Respondent Incentives: To Pay or Not to Pay? The start-up costs of performing experimental economics and administering surveys can be considerable. It typically requires careful thought with respect to experimental design, a subject to which most economists are not exposed. For experimental economics, it often requires a great physical coordination of respondents, laboratory space, laboratory staff, and the computer equipment now typically used to produce the stimuli and record respondent reactions. In addition to these costs, it can also be quite expensive to properly motivate respondents with the real monetary incentives deemed so crucial by many economists to remove the hypothetical bias thought to plague much of the contingent valuation and other stated preference literature. However, the efficacy of actual payments remains an open question.

Can real incentives eliminate observed anomalies? Not generally. Camerer and Hogarth (1999) review 74 published papers featuring experiments that varied the amount of real compensation afforded experimental subjects. They found that increasing monetary incentives rarely removes all deviations from strict rationality, either for naïve or experienced subjects, but usually decreases variation in responses. Increasing incentives will often decrease the percent of subjects who make choices in disagreement with standard rationality assumptions, however.

Furthermore, Horowitz and McConnell's (forthcoming) analysis of data from 45 studies that compare willingness to pay and willingness to accept suggests that real incentives have little effect on the degree of deviation between the two measures; i.e., mandatory incentives did not yield results more closely aligned with standard economic theory.

Will incentives alter risk preferences? Usually. In reviewing 13 studies concerning risk experiments, Camerer and Hogarth (1999) find mixed evidence on the effect of incentives. In three of the studies they reviewed, playing real gambles instead of hypothetical gambles had no effect on subjects' risk aversion, while in eight studies, risk aversion increased when real gambles were introduced. It is not clear if the typical response of greater risk aversion is a desirable feature of including incentives, or how even to test for the validity of any risk effect.

Do real incentives alter the absolute level of willingness to pay? Generally yes. A number of authors do find, when comparing respondents' stated willingness to pay for private goods to actual amounts later bid in auctions or exchanged in real markets, that the stated amount often exceeds the exchanged amount (e.g., Dickie, Fisher, and Gerking 1987, List and Shogren 1998, Fox et al. 1998). However, as Randall (1998) points out, while clean tests generally can be constructed for purely private goods, the power of such tests generally dissipates as the nature of the goods involves more non-market elements.

What Governs the Dynamic Choice Process? Economics research and, in particular, agricultural economics research has long embraced dynamic programming as an effective means for solving intertemporal policy issues (e.g., grain stockpiling) and for analyzing individual intertemporal investment and consumption choice. Slowly but surely, risk research in economics is moving away from models that impose static risk concepts to those that embrace intertemporal allocation as a means for self-insuring and guarding against risk.

In order for such analyses to be fully effective, however, the appropriate definitions of the intertemporal criteria used by farmers must be employed. The typical analysis assumes the objective to be maximization of the present value of a stream of expected utility or profit over time. In other words, additive separability of time is imposed upon preferences and future utilities or payouts are discounted exponentially. However, how people actually attack and solve dynamic decision problems is not well understood (Muller, forthcoming). A group of experimental economists have now begun to train their sights on understanding this rich and difficult area of economic behavior.

Researchers are critically assessing the reality of nearly all facets of the standard dynamic objective. Mounting experimental and revealed preference data suggest that individual preferences may deviate from such simple dynamic formulations. For example, Gigliotti and Sopher (1997) find that two-thirds of experimental subjects did not choose payment streams that maximized present value. When monetarily penalized for not choosing maximum present value, similar patterns persisted, though they were not as prevalent. Respondents often preferred increasing payment streams with lower present values rather than decreasing payment streams with higher present values, particularly when the payment streams were attached to the wages from a fixed-term employment rather than to payments derived from inheritance.

Exponential discounting in time-additive models is also coming under fire. Exponential discounting is often rejected in favor of hyperbolic discounting (Thaler 1981, Benzion, Rapaport, and Yagil 1989, Loewenstein and Elster 1992); i.e., respondents generally discount the distant future at lower rates than the near future. Azfar (1999) shows that such a pattern can be rationalized within a model featuring agents who are uncertain about future discount rates. Such rejection of exponential discounting may also be expected if agents hold separate preferences over ending values that are not adequately reflected in the original model. Furthermore, there is no reason to expect individuals to discount exponentially at positive rates, except that they face interest rates in the market, and should discount to the extent that utility is driven by such considerations.

A relatively untapped but exciting line of inquiry in experimental economics with significant relevance to agricultural investments is that of trying to understand how individuals actually tackle dynamic decision making problems. Carbone and Hey (1997) find that in the face of simple dynamic problems, most subjects attempt to use backward-induction methods postulated by most economic modeling; however, only a few subjects apply backward-induction in a consistent and thorough manner. When Anderhub et al. (2000) exposed subjects to more complex dynamic problems featuring consumption-savings decisions over finite but uncertain time horizons, they found that subjects chose strategies that, at least qualitatively, followed optimal strategies derived from stochastic dynamic programming, though subjects tended to consume less during early periods than would the optimal strategy whether preferences were additive or multiplicative through time. Muller (forthcoming) conducts similar experiments and obtains a similar pattern of under-consumption in early periods, as do Anderhub et al. (2000). By encoding the heuristics that each subject uses during the decision making process and noting how conditional responses differ from risk-neutral responses, Muller finds that risk aversion is a viable explanation for the under-consumption, though the degree of risk aversion for each subject can differ based on the exact position in the time horizon and the conditional probability of survival.

The problem of such experiments is that it is difficult to maintain experiments and delay actual payouts over a period long enough to correspond with actual possible time frames of relevant real-world decisions (e.g., it is tough to conduct an experiment with payouts over several years). Furthermore, people responding to complex dynamic decision making in business situations will likely have access to greater computing power and, hence, may be able to fully implement the principles of backward induction and dynamic optimality commonly invoked in dynamic programming solutions. Further experimentation in this field that allows for adequate computational time and costly consulting advice is warranted in the agricultural risk field.

Can Tournament Contracts Destabilize Markets? As agriculture continues its evolution toward integrated production systems, risk-absorbing integrators will consider tournament contracts like those used in the poultry industry as means to induce optimal levels of effort by risk-shedding agents. The firm-level implications of various tournament incentives have been widely assessed by economists (e.g., Nalebuff and Stiglitz 1983, Bull, Schotter, and Weigelt 1987, Knoeber 1989, Ehrenburg and Bognanno 1990). While these and other authors (Brown, Van Harlow, and Starks 1996) recognize that the introduction of tournaments can alter the riskiness of strategies adopted by individuals, less effort has been directed at understanding the possible aggregate effects that such tournaments might impose upon the market.

James and Isaac (2000) analyze possible aggregate effects of tournament contracts and show that convergence of an asset market toward the underlying asset's intrinsic value, which is among the more robust findings of experimental economics, can be derailed when an individual participant's compensation is tied to his or her relative performance against other traders. In such situations, traders who are trailing the market late in the tournament period undertake riskier positions in order to salvage any chance at winning the tournament. Such "beat the market" tournaments are fairly common among mutual fund managers and, in James and Isaac's experiments, it was shown that the underlying asset value was not uncovered during the progression of the market.

It is interesting to contrast James and Isaac's constructive approach – how does individual contract structure affect market stability? – to Tsoulouhas and Vukina's (1999) deconstructive approach – how does market volatility affect the type of contract structure that emerges? (Note Tsoulouhas and Vukina use observed behavior in livestock markets.) James and Isaac never assess whether the firms offering tournament contracts would realize their destabilizing effect and not offer such contracts. Tsoulouhas and Vukina point out that if bankruptcy is possible then the firms offering contracts in inherently volatile markets (e.g., hog integrators) would not find it optimal to offer tournament contracts. Such a difference in approach suggests that there remain many interesting questions about the relationship of individual contract structures and market stability that experimental techniques might be able to elucidate.

Future Directions of Experimental and Survey Methods

The net impact of the stylized facts presented in the previous sections and the progress made in addressing the many remaining open questions concerning human behavior under risk has been to enhance the credibility of survey and experimental methods. "Anomalous" results cannot be ascribed blithely to measurement error, and conflicts between theory and evidence can no longer be ascribed convincingly to weaknesses in the evidence. Furthermore, there are many interesting questions in agricultural risk management that can be illuminated by survey and experimental research, regardless of whether it is ever established satisfactorily that CVM is capable of producing reliable estimates of population willingness to pay for public goods (i.e., the question that faced the NOAA panel). We can be reasonably sure that experimental and stated preference methods provide robust marginal and relative measures of attitudes, preferences, and values, in a wide range of settings.

The emerging trend is toward proliferation of methods, blurring the boundaries among qualitative research, survey, and experimental methods, and eagerly adopting new information and communication technologies to expand the set of research possibilities. This proliferation of methods can be expected to dramatically increase the menu of methods for collecting stated preference data. Standing in the way of this proliferation of methods and procedures is the felt need of government to standardize CVM procedures for use in policy making and litigation, which provided the context for the NOAA panel. While one can sympathize with government's concerns, one must expect proliferation to win the day. Rather than a narrowly stylized and standardized CV method, we believe the future belongs to a broad-based research program of learning about preferences from what people tell us, whatever it takes.

The emerging trend toward developing and applying techniques that combine self-reported, experimental, and observed choice data will and should continue unabated. In a number of research contexts, the consistency of the various kinds of data has been established, typically after applying some kind of endogenous scaling method (e.g., Cameron 1992). Given our culture as economists, there is always a tendency to treat one of the data sets (typically observed choice data) as a benchmark against which to scale data generated by the other(s). Again, we would caution that we have less reason than we think for maintaining this premise of differential validity; it makes more sense to learn everything that can be learned by combining these kinds of data sets without imposing preconceived notions of their relative validity.

The agenda will turn toward mapping, across a broad front, the performance characteristics of various stated preference and experimental research methods. In this respect, the experimental and stated preference research programs will follow the lead of public opinion survey research. No one asks anymore whether public opinion surveys are valid and reliable. The question is essentially meaningless. Instead, an enormous amount of research has been accumulated, mapping the performance characteristics of alternative approaches and techniques of measuring attitudes and public opinion.

POTENTIAL APPLICATIONS TO AGRICULTURAL RISK

Given the vast array of stimulating research underway or already completed in the EE and CVM literatures, and the growing confidence in the ability of these methods to provide genuine insight into difficult questions, there exists a rich potential for transferring these methods to the agricultural risk arena or, in some cases, extending the agricultural risk studies that have already tapped these methodologies. We outline a few topics that seem particularly intriguing.

Eliciting Values for Farm Program Elements

Congress will, from time to time, change U.S. farm programs in a way that alters the risk faced by many farmers. The use of stated preference instruments, couched in the familiar terms of a referendum vote and describing various attributes of current and potential farm programs, could allow for the derivation of marginal trade-offs that farmers are willing to make. Inclusion of variables demarcated in simple dollars terms, such as policies affecting lump-sum payments, would allow for recovery of marginal and discrete welfare measures of key risk-related farm policy attributes. The results of such studies could provide a wealth of information concerning different types of farmers' and agribusinesses' marginal demands for various farm program attributes and perhaps help identify program priorities for various subsets of farmers.

Exploring Barriers to Crop and Revenue Insurance Participation

Low participation by farmers in U.S. crop insurance programs has long baffled researchers. Despite federal subsidization of the programs and indemnities outpacing premiums in many years, many farmers forgo crop insurance unless such participation is required in order to receive other farm program benefits. Just and Calvin (1994) outline several reasons that may drive low participation, including heterogeneity of risk, heterogeneity of average yields, other types of heterogeneity (e.g., farm size), adverse selection, moral hazard, and the expectation of disaster relief. Just, Calvin, and Quiggin (1999) analyze farm-level crop insurance participation using the farmers' subjective yield probabilities and find that farms with positive expected payouts tend to participate while those with negative expected payouts do not.

While informative, even analyses using observed farm-level and farmers' subjective probabilities leave much to be desired as the natural variation in program parameters and possible discrepancies between observed and subjective yield distributions can mask the marginal effects of various program parameters that govern participation or overlook various motivations for participation altogether. For example, there may exist a number of other sources of heterogeneity affecting participation that remain unobserved, such as administrative costs, risk aversion, and attitudes toward the government.

A better understanding of participation might be gained through a series of focus groups with farmers.[4] Focus group research provides a systematic though non-representative means for collecting anecdotes and refining stylized facts about behavior at the individual level. Such qualitative research is commonly used by political advisors, consumer marketing firms, and survey researchers to gain initial insights about the thought and decision making process of individuals and to calibrate quantitative data collection instruments. Output from such focus groups could be used to develop hypothetical choice experiments in which a representative sample of farmers choose among various insurance products that differ on a variety of attributes, to calibrate a laboratory experiment centered around insurance participation such that key institutional elements were replicated in the experiment, or to guide the collection of attitudinal and institutional covariates in future survey data so that they can be used as explanatory variables in econometric explanations of farm-level participation.

Assessing Non-Farm Support for Farm Programs

The continued legislative support and congressional funding of agricultural transfer and insurance programs depends to some degree on the underlying sentiment of the non-farm community. At least part of the continued funding of farm programs is dependent upon the stock of goodwill that voters hold for farmers and the agricultural sector. However, as most citizens become more distant from production agriculture and as popular press stories tend to paint agriculture as an industrializing sector, some support may be lost or more stipulations may be required in order for funding of agricultural programs to continue.

A regular set of survey instruments designed to elicit the general attitudes about and support for agriculture would serve as a means for gauging public support. The American Farm Bureau Federation currently conducts a biennial tracking poll of consumers' image of farmers and the farming sector and many other farm and non-farm organizations implement opinion polls on an ad hoc basis to assess consumer reaction to emerging topics (e.g., the use of biotechnology). However, a richer set of questions that probe the depth of support for agriculture and gauge potential reactions to various negative actions by farmers and the agricultural sector (e.g., pollution or industrialization) would provide better indicators of continued support. Particularly a set of surveys that elicit reaction to a suite of negative hypothetical scenarios

---

[4] Note that Schertz and Johnston's (1997) summary of discussions with panels of farmers concerning possible reactions to the 1996 Farm Bill does gather some interesting insights from farmers in a focus-group–like setting. These panel discussions covered barriers to crop insurance participation as just one of a myriad of topics, however. Furthermore, these panels consisted of exceptional farmers; while not meant to be representative, focus groups should attempt to include a broad range of participants.

(e.g., mad cow disease entering the U.S., a story of large farms polluting, stories of biotechnology mishaps, etc.) might serve as a systematic guide for which elements of agriculture could spoil the sentimentality or comprise the goodwill held by much of the general public.

Furthermore, a set of CVM or other stated preference surveys could be conducted to elicit willingness to pay or willingness to vote for policies that continue transfers into the agricultural community. Particularly, these instruments could probe how various conservation or environmental segments of farm programs affect consumers' willingness to expand or maintain transfers to the agricultural sector. An experimental design of both positive and negative information treatments prior to the implementation of such surveys could be used to gauge the public's willingness to support farm programs contingent upon swings in the image of the agricultural sector or swings in sectoral financial conditions.

Effects of Data Collection, Contracting, and Concentration on Markets

Agricultural markets are undergoing major changes as the effects of concentration and contracting have altered the number and size of transactions that take place in the market. These trends can affect the volatility of the marketplace and can increase the risk faced by its participants in many ways. Furthermore the government resources available to collect, verify, and disseminate information concerning transactions in any particular market are often decreasing, while the government's authority and motivation to compel market players to reveal pertinent information can change. The effects of concentration, contracting, and information collection, and the interaction of these elements, are examples of phenomena for which the natural variation in the market is not sufficient to allow for satisfactory structural identification. As such, the construction of experimental markets offers a complementary method for exploring the effects of such key elements for all market participants and society.

Experimental work has already explored issues arising in the fed cattle market (Anderson et al. 1998, Ward et al. 1996, Ward et al. 1999) and provides a taste of the types of issues that experimental markets can help illuminate: e.g., the impacts of reduced public information provision on price discovery, the effects of captive supply contracts on price volatility, and the effects of mandatory price reporting of privately transacted cattle on market dynamics. Many of these same questions could be explored for other key sectors with experiments that involve the institutional peculiarities of each sector.

Broader, cross-sector issues can also be explored, such as the implications of changes in farm and tax policy on land and other asset markets and general tests of the Lucas critique as applied to agricultural asset markets. Previous experimental research on such matters (Schmalensee 1976, Plott and Sunder 1982, Williams 1987, Forsythe and Lundholm 1990) finds mixed support for participants' ability to achieve rational expectations where support for rational expectations is greater when the market environment is simpler, traders are more experienced, and the productivity of assets is common knowledge. Adaptive expectations and least squares learning often provide more realistic explanations of the behavior observed in more complex laboratory market scenarios.

Swenson (1997) performs experiments to determine if subjects' forecasts of future policy changes conform to the rational expectations hypothesis and if subjects' purchases of assets are made in anticipation of such policy changes. He finds support for adaptive expectations but finds that subjects did learn to alter the timing of asset purchases in anticipation of policy changes, even as market environments were made quite noisy. Similar experiments might be formulated to investigate the potential effects of proposed farm policy, tax policy, and land zoning policy changes on the performance of land and other asset markets.

Identifying Farmers' Intertemporal Objectives

As emphasized in Just (2000), little is known about the exact nature of farmers' intertemporal preferences and, as discussed in the earlier section on what governs the dynamic choice process, deviations from exponentially discounted, additively separable objective functions commonly imposed in dynamic modeling can have major implications for derived optimal investment, savings, and risk management strategies. A set of experiments in which farmers are asked to rank the desirability of streams of payouts in realistic situations could be developed as a refinement of experiments conducted by Gigliotti and Sopher (1997). For example, farmers could be given the choice between two long-term net farm income insurance policies and provided with expected streams of payouts to each as well as summaries of the payout streams (e.g., the present value of each stream, expected ending addition to net worth, and maximum available income draws in each period). Repeated elicitation of choice among competing policies would allow for identification of subjects' intertemporal objective given an appropriate design of the stimuli. Adding uncertainty concerning the length of the time horizon, as in Anderhub et al. (2000), and adding the possibility of farm transfer motives would provide further information concerning optimal investment activities near retirement. A vast array of experiments could be envisioned that would help explore the effect of various tax policies on farm operators and the effect on late-life behavior of the farm sector.

SUMMARY AND CONCLUSIONS

Roth (1995) summarizes three main areas in which experimental and survey research can contribute. First, experimental and survey methods can inform economic theorists because these methods can use controlled environments to provide feedback about well-articulated theories that cannot always be cleanly analyzed with observed data. Second, if standard theories are refuted or consistent irregularities arise in both experimental and real data, experiments and surveys can be used to form a body of stylized facts from which new theory can be formed. Finally, surveys and experiments can provide rapid feedback to policymakers about issues that are not easily analyzed with observed data: e.g., the volatility implications of altering market structure, the dynamic effects of changing regulatory regimes, and the costs and benefits of altering the level of public goods offered. We suggest the field of agricultural risk analysis could benefit in each of these three areas from continued research using experimental and survey methods.

Experimental and survey methods are not a methodological panacea for answering all unresolved questions in the realm of agricultural risk. Rather these methods offer a set of research techniques that can complement the econometric and programming techniques that are more familiar to many agricultural risk practitioners and help us as a profession to improve our understanding of individual and aggregate responses to agricultural risk. The limited and prudent use of these techniques in previous agricultural risk work stands as a testament to the methods' usefulness, particularly given that many of those currently and previously using these techniques incurred substantial start-up costs to learn and apply them. As more economics and agricultural economics programs offer formal training in economic experiment and survey methods, we might imagine that the use of such methods will explode onto the agricultural risk scene and provide another means to forward the profession's knowledge base.

Finally, because of the extraordinary amount of control available in experimental and survey techniques, protocols can be explicitly formulated to test a wide array of behavioral hypotheses. Because these protocols are common knowledge, researchers can continue to build upon previous results in the literature and explore perceived irregularities from previous studies. Hence, research has great potential to be additive and, hence, more quickly and efficiently add to the stock of knowledge generated by the profession. This potential for additivity is particularly appealing because, as some have noted (Just 2000), past research in agricultural economics has often lacked such additivity.

REFERENCES

Allais, M. 1953. "Le Comportement de L'homme Rationnel Devant le Risque: Critique des Postulats et Axiomes de L'ecole Americaine." *Econometrica* 21: 503-546.

Anderhub, V., W. Guth, W. Muller, and M. Strobel. 2000. "An Experimental Analysis of Intertemporal Allocation Behavior." *Experimental Economics* 3: 137-152.

Anderson, J.D., C.E. Ward, S.R. Koontz, D.S. Peel, and J.N. Trapp. 1998. "Experimental Simulation of Public Information Impacts on Price Discovery and Marketing Efficiency in the Fed Cattle Market." *Journal of Agricultural and Resource Economics* 23: 262-278.

Arrow, K., R. Solow, P. Portney, E. Leamer, R. Radner, and H. Schuman. 1993. Report of the NOAA Panel on Contingent Valuation. *Federal Register* 58: 4601-4614.

Azfar, O. 1999. "Rationalizing Hyperbolic Discounting." *Journal of Economic Behavior and Organization* 38: 245-252.

Bard, S.K., and P.J. Barry. 1999. "Developing a Scale for Assessing Farmers' Risk Attitudes." Working Paper, Center for Farm and Rural Finance, University of Illinois, February 23.

Belaid, A., and S.F. Miller. 1987. "Measuring Farmers' Risk Attitudes: A Case Study of the Eastern High Plateau Region of Algeria." *Western Journal of Agricultural Economics* 12: 198-206.

Benzion, U., A. Rapaport, and J. Yagil. 1989. "Discount Rates Inferred from Decisions: An Experimental Study." *Management Science* 270-284.

Bergstrom, J.C., and J.R. Stoll. 1989. "Application of Experimental Economics Concepts and Precepts to CVM Field Survey Procedures." *Western Journal of Agricultural Economics* 14: 98-109.

Binswanger, H.P. 1980. "Attitudes Toward Risk: Experimental Measurement in Rural India." *American Journal of Agricultural Economics* 62: 395-407.

Bishop, R.C., and T.A. Heberlein. 1979. "Measuring Values of Extramarket Goods: Are Indirect Measures Biased?" *American Journal of Agricultural Economics* 61: 926-930.

Brown, K.C., W. Van Harlow, and L.T. Starks. 1996. "Of Tournaments and Temptations: An Analysis of Managerial Incentives in the Mutual Fund Industry." *Journal of Finance* 51: 85-110.

Bull, C., A. Schotter, and K. Weigelt. 1987. "Tournaments and Piece Rates: An Experimental Study." *Journal of Political Economy* 95: 1-33.

Buschena, D.E. 2001. "Non-Expected Utility: What Do the Anomalies Mean for Risk in Agriculture?" In R.E. Just and R.D. Pope, eds., *A Comprehensive Assessment of the Role of Risk in U.S. Agriculture.* Boston, MA: Kluwer Academic Publishers.

Camerer, C.F., and R. Hogarth. 1999. "The Effect of Financial Incentives in Experiments: A Review and Capital-Labor-Production Framework." *Journal of Risk and Uncertainty* 19: 7-42.

Camerer, C.F., and H. Kunreuther. 1989 . "Decision Processes for Low Probability Events: Policy Implications." *Journal of Policy Analysis and Management* 8: 565-592.

Cameron, T.A. 1992. "Combining Contingent Valuation and Travel Cost Data for the Valuation of Nonmarket Goods." *Land Economics* 68: 302-317.

Carbone, E., and J. Hey. 1997. "How People Tackle Dynamic Decision Problems." Mimeo, University of York.

Chamberlin, E.H. 1948. "An Experimental Imperfect Market." *Journal of Political Economy* 56: 95-108.

Ciracy-Wantrup, S.V. 1947. "Capital Returns from Soil-Conservation Practices." *Journal of Farm Economics* 29: 1181-1196.

Collins, A., W.N. Musser, and R. Mason. 1991. "Prospect Theory and Risk Preferences of Oregon Seed Producers." *American Journal of Agricultural Economics* 73: 429-435.

Davis, R.K. 1963. "The Value of Outdoor Recreation: An Economic Study of the Maine Woods." Unpublished Ph.D. dissertation, Harvard University, Cambridge, MA.

Dickie, M., A. Fisher, and S. Gerking. 1987. "Market Transactions and Hypothetical Demand Data: A Comparative Study." *Journal of American Statistical Association* 82: 69-75.

Dillon, J.L., and P.L. Scandizzo. 1978. "Risk Attitudes of Subsistence Farmers in Northeast Brazil: A Sampling Approach." *American Journal of Agricultural Economics* 60: 425-435.

Ehrenburg, R.G., and M.L. Bognanno. 1990. "Do Tournaments Have Incentive Effects?" *Journal of Political Economy* 98: 1307-1324.

Finkelshtain, I., and E. Feinerman. 1997. "Framing the Allais Paradox as a Daily Farm Decision Problem: Tests and Explanations." *Agricultural Economics* 15: 155-167.

Forsythe, R., and R. Lundholm. 1990. "Information Aggregation in an Experimental Market." *Econometrica* 58: 309-347.

Fox, J., J. Shogren, D. Hayes, and J. Kliebenstein. 1998. "CVM-X: Calibrating Contingent Values with Experimental Auction Markets." *American Journal of Agricultural Economics* 80: 455-465.

Francisco, E.M., and J.R. Anderson. 1972. "Chance and Choice West of the Darling." *Australian Journal of Agricultural Economics* 16: 82-93.

Ganderton, P.T., D.S. Brookshire, M. McKee, S. Stewart, and H. Thurston. 2000. "Buying Insurance for Disaster-Type Risks: Experimental Evidence." *Journal of Risk and Uncertainty* 20: 271-289.

Gigliotti, G., and B. Sopher. 1997. "Violations of Present-Value Maximization in Income Choice." *Theory and Decision* 43: 45-69.

Grisley, W., and E.D. Kellogg. 1983. "Framers' Subjective Probabilities in Northern Thailand: An Elicitation Analysis." *American Journal of Agricultural Economics* 65: 74-82.

Gunjal, K., and B. Legault. 1995. "Risk Preferences of Dairy and Hog Producers in Quebec." *Canadian Journal of Agricultural Economics* 43: 23-35.

Hanemann, W.M. 1991. "Willingness to Pay and Willingness to Accept: How Much Can They Differ?" *American Economic Review* 81: 635-647.

Hausman, J. (ed.) 1993. *Contingent Valuation: A Critical Assessment.* Amsterdam: Elsevier.

Hoehn, J.P., and A. Randall. 1996. "Embedding in Market Demand Systems." *Journal of Environmental Economics and Management* 30: 369-380.

Horowitz, J.K., and K.E. McConnell. Forthcoming. "A Review of WTP/WTA Studies." *Journal of Environmental Economics and Management.*

Howard, W.H., G.L. Brinkman, and R. Lambert. 1997. "Thinking Styles and Financial Characteristics of Selected Canadian Farm Managers." *Canadian Journal of Agricultural Economics* 45: 39-49.

Isaac, R.M., and D. James. 2000. "Just Who Are You Calling Risk Averse?" *Journal of Risk and Uncertainty* 20: 177-187.

James, D., and R.M. Isaac. 2000. "Asset Markets: How They Are Affected by Tournament Incentives for Individuals." *American Economic Review* 90: 995-1004.

Just, R.E. 2000. "Some Guiding Principles for Empirical Production Research in Agriculture." *Agricultural and Resource Economics Review* 29: 138-158.

Just, R.E., and L. Calvin. 1994. "An Empirical Analysis of U.S. Participation in Crop Insurance." In D.L. Hueth and W.H. Furtan, eds., *Economics of Agricultural Crop Insurance: Theory and Evidence.* Boston: Kluwer Academic Publishers.

Just, R.E., L. Calvin, and J. Quiggin. 1999. "Adverse Selection in Crop Insurance: Actuarial and Asymmetric Information Incentives." *American Journal of Agricultural Economics* 81: 834-849.

Kagel, J.H., and A.E. Roth. 1995. *The Handbook of Experimental Economics.* Princeton, NJ: Princeton University Press.

Knoeber, C.R. 1989. "A Real Game of Chicken: Contracts, Tournaments and the Production of Broilers." *Journal of Law, Economics and Organization* 5: 271-292.

List, J.A., and J.F. Shogren. 1998. "Calibration of the Difference between Actual and Hypothetical Valuations in a Field Experiment." *Journal of Economic Behavior and Organization* 37: 193-205.

Loewenstein, G., and J. Elster. 1992. *Choice Over Time.* New York: Russell Sage Publications.

Mansfield, C. 1999. "Despairing Over Disparities: Explaining the Difference between Willingness to Pay and Willingness to Accept." *Environmental and Resource Economics* 13: 219-234.

McClelland, G.H., W.D. Schulze, and D.L. Coursey. 1993. "Insurance for Low-Probability Hazards: A Bimodal Response to Unlikely Events." *Journal of Risk and Uncertainty* 7: 95-116.

McFadden, D. 1999. "Rationality for Economists?" *Journal of Risk and Uncertainty* 19: 73-105.

Muller, W. In press. "Strategies, Heuristics and the Relevance of Risk Aversion in a Dynamic Decision Problem." *Journal of Economic Psychology.*

Nalebuff, B.J., and J.E. Stiglitz. 1983. "Prizes and Incentives: Towards a General Theory of Compensation and Competition." *Bell Journal of Economics* 14: 21-43.

Officer, R.R., and A.N. Halter. 1968. "Utility Analysis in a Practical Setting." *American Journal of Agricultural Economics* 50: 257-277.

Pingali, P.L., and G.A. Carlson. 1985. "Human Capital, Adjustments in Subjective Probabilities, and the Demand for Pest Controls." *American Journal of Agricultural Economics* 67: 853-861.

Plott, C., and S. Sunder. 1982. "Efficiency of Experimental Security Markets with Insider Information: An Application of Rational-Expectations Markets." *Journal of Political Economy* 90: 663-698.

Randall, A. 1998. "Beyond the Crucial Experiment: Mapping the Performance Characteristics of Contingent Valuation." *Resource and Energy Economics* 20: 197-206.

Randall, A., B.C. Ives, and C. Eastman. 1974. "Bidding games for Valuation of Aesthetic Environmental Improvements." *Journal of Environmental Economics and Management* 1: 132-149.

Randall, A., and J.R. Stoll. 1980. "Consumer's Surplus in Commodity Space." *American Economic Review* 70: 449-455.

Roth, A.E. 1995. "Introduction to Experimental Economics." In J.H. Kagel and A.E. Roth, eds., *The Handbook of Experimental Economics.* Princeton, NJ: Princeton University Press.

Schertz, L.P., and W.E. Johnston. 1997. "Managing Farm Resources in the Era of the 1996 Farm Act." Market and Trade Economics Division, Economic Research Service, U.S. Department of Agriculture. Staff Paper No. AGES 9711, December.

Schmalensee, R. 1976. "An Experimental Study of Expectation Formation." *Econometrica* 44: 17-41.

Swenson, C.W. 1997. "Rational Expectations and Tax Policy: Experimental Market Evidence." *Journal of Economic Behavior and Organization* 32: 433-455.

Thaler, R. 1981. "Some Empirical Evidence of Dynamic Inconsistency." *Economics Letters* 201-207.

Thurstone, L.L. 1931. "The Indifference Function." *Journal of Social Psychology* 2: 129-167.

Tsoulouhas, T., and T. Vukina. 1999. "Integrator Contracts with Many Agents and Bankruptcy." *American Journal of Agricultural Economics* 81: 61-74.

Tversky, A., and D. Kahneman. 1991. "Loss Aversion in Riskless Choice: A Reference-Dependent Model." *Quarterly Journal of Economics* 106: 1039-1061.

Wallis, W.A., and M. Friedman. 1942. "The Empirical Derivation of Indifference Functions." In O. Lange, F. McIntyre, and T.O. Yntema, eds., *Studies in Mathematical Economics and Econometrics in Memory of Henry Schultz.* Chicago: University of Chicago Press.

Ward, C.E., S.R. Koontz, T.L. Dowty, J.N. Trapp, and D.S. Peel. 1999. Marketing Agreement Impacts in an Experimental Market for Fed Cattle." *American Journal of Agricultural Economics* 81: 347-358.

Ward, C.E., S.R. Koontz, D.S. Peel, and J.N. Trapp. 1996. "Price Discovery in an Experimental Market for Fed Cattle." *Review of Agricultural Economics* 18: 449-466.

Williams, A.W. 1987. "The Formation of Price Forecasts in Experimental Markets." *Journal of Money, Credit and Banking* 19: 1-18.

Young, D.L. 1979. "Risk Preferences of Agricultural Producers: Their Use in Extension and Research." *American Journal of Agricultural Economics* 61: 1063-1070.

Part 4

SOURCES AND CONSEQUENCES OF AGRICULTURAL RISK: HOW FARMERS MANAGE RISK

Chapter 14

MODELING PRICE AND YIELD RISK*

Barry K. Goodwin and Alan P. Ker
The Ohio State University and University of Arizona

INTRODUCTION

Agricultural producers face a wide array of risks that influence their production and marketing decisions. Though a precise and comprehensive taxonomy may be difficult, risk is typically assumed to originate from the random nature of prices (for both inputs and outputs) and yields (for both animal and plant production). In addition, producers may face other less tangible sources of risk, including those risks associated with liability issues and the potential for capital gains and losses in asset values that may arise from exogenous shocks such as policy changes.[1]

A precise definition of "risk" may be elusive. Economists generally consider risk to involve unanticipated movements in prices, yields, revenues, or other variables of economic importance. Knight (1921) made a now famous distinction between *risk* and *uncertainty*. Risk was taken to pertain to situations where the mathematical odds associated with an outcome are known, though the outcome is not. In contrast, uncertainty involves those situations where neither the outcome nor the odds associated with the outcome are known. An example of the former includes the odds associated with a poker hand or a casino game while an example of the latter is the case of parimutuel gambling. The latter case is generally more pertinent to agricultural risk where the probabilities associated with various outcomes are unknown and thus must be estimated. In Knight's own words, "...the probability in which the student of business risk is interested is an estimate." The economics

---

* This research was supported by the North Carolina Agricultural Research Service and Arizona Agricultural Research Service. We are grateful to Shiva Makki, Matt Roberts, and Keith Coble for their helpful comments. Goodwin was at North Carolina State University when this work was completed.

[1] Liability issues have become increasingly important factors affecting the risks associated with agricultural production. Animal waste spills, chemical contamination, and improper handling of genetically modified organisms are examples of liability issues that may be important to the risk of a farm operation.

literature does, in some cases, tend to reveal some confusion about how risk should be characterized and measured.

In many cases, the variance associated with a variable of interest is taken to be the ideal measure of risk in empirical work. This can be misleading, however, when comparisons of the variances of variables with different means is undertaken. For example, such logic might lead one to conclude that corn is necessarily a more risky crop than wheat because the measured variance of corn is almost always higher than that of wheat. An alternative measure of risk that addresses this point is the coefficient of variation, which scales the variance by the mean in order to provide a relative measure of risk (variability) that accounts for differences in means. However, comparisons based on such a measure of risk only address comparative differences in the first two moments of the distribution of interest. Many economic variables have interesting distributional characteristics involving higher moments that may be important to risk. For example, the conventional wisdom maintains that crop yields tend to be negatively skewed while commodity prices tend to be positively skewed. An adequate representation of risk may therefore require an estimate of the probability density function (pdf) or the cumulative distribution function (cdf), such that all relevant moments of the distribution may be characterized.

In most cases, a workable definition of risk implies some assessment of the probabilities associated with certain events. For example, yield risk generally pertains to the probability that yields below some threshold will be observed. Likewise, price risk generally considers some assessment of the likelihood that prices below (or above) some level will be observed. Thus, it is the properties of the distribution of the random variable of interest that must generally be assessed in an analysis of risk.

Estimation of distributional properties of a random variable may be pursued in several different ways. One may estimate the moments from empirical observations of the variable of interest. Of course, knowledge of the moments is of limited use without knowledge of the parametric distribution. Alternatively, one may attempt to evaluate distributional properties of a random variable by estimation of the parameters characterizing a specific parametric specification, which is assumed to be known a priori. For example, one may assume that the distribution of prices is lognormal; in which case, two parameters (the mean and variance) are sufficient to characterize the entire distribution. Other familiar distributions include the beta distribution, which is often and perhaps incorrectly assumed to be an appropriate representation of the distribution of crop yields. In a method of moments sense, estimation of a number of moments equal to the number of parameters characterizing the distribution is sufficient to characterize the entire distribution, provided one has a priori knowledge of the *parametric* family characterizing the random variable.

A shortcoming of parametric approaches involves the fact that one rarely has a priori knowledge of the appropriate parametric family. As Horowitz

(1993) has noted, there is seldom sufficient justification for assuming that the distribution of a random variable belongs to an assumed parametric family. Despite this obvious shortcoming of parametric techniques, the standard approach is to assume a particular parametric form, usually without explicit specification testing, and to push ahead with inferences without looking back. Recent research has attempted to address such concerns through the adoption of semiparametric and nonparametric methods. Such methods mitigate the need to specify a distributional family. Of course, such flexibility may not come at a cost. Nonparametric methods may be more demanding in terms of the data required to accurately assess a probability distribution. Essentially, assuming a parametric model during the estimation process is perfectly analogous to imposing a restriction. If that restriction is correct, estimation efficiency is increased. Conversely, if that information is incorrect, estimation efficiency is generally decreased and possibly significantly so.

The objective of this chapter is to outline and review issues underlying the measurement of risk for agricultural product yields and prices. We give careful attention to issues underlying parametric and nonparametric techniques for evaluating risk. Our analysis is broken into three sections. The following section discusses issues underlying the measurement of crop yield risk. The third section of the chapter addresses the measurement of price risk, where many of the same issues arise. We then address issues underlying the measurement of revenue risk. In this case, one must deal with modeling multiple sources of non-independent risk – prices and yields – which may have different marginal distributions. For example, a common situation in the measurement of revenue risk involves beta-distributed yields and lognormal prices, which tend to be negatively correlated. The final section of our chapter briefly reviews our main points and offers some concluding remarks.

MODELING YIELDS

Difficulty in empirically modeling any phenomenon stems from one aspect – a lack of data. Ideally, available data are sufficiently rich to reduce via specification testing the set of potential models such that subsequent economic analysis based on the estimated models is invariant. We have never been so fortunate in our empirical analyses involving yield data and the estimation of yield densities. Perhaps not surprisingly, we were unable to find any empirical analysis in the literature where the authors undertook a sensitivity analysis of the economic findings to alternative modeling assumptions.

This section intertwines the customary literature review with a general discussion of modeling issues germane to estimating yield densities. We start by discussing the nature of yield data to provide an underlying framework for the remainder of the section. Subsequently, we discuss the common parametric models advanced in the literature, our view of those models, and our view of the parametric approach. This will be followed by a discussion of the nonparametric methods forwarded in the literature, our view of those methods, and our view of the nonparametric approach. Finally, we will discuss in greater depth more advanced methods – combined parametric and nonparametric estimators and nonparametric estimators that make use of extraneous data – and our view of these methods.

The Nature of Yield Data

Understanding the nature of the data-generating process of yields is crucial as it can provide guidance in modeling yields. This is a necessary first step because completely data-driven evaluation methods such as specification testing tend to be uninformative; the power of such tests against relevant alternatives is negligible because of the lack of yield data (see the section on parametric methods for a complete discussion). Yields follow some spatio-temporal process. By averaging over some spatial region and conditioning on some $X \in R^d$, one recovers the conditional mean yield density for that given space. As the spatial region or conditioning vector X changes, the underlying conditional yield density will change, and perhaps significantly. For example, the mean yield density for all-practice corn in a given Iowa county conditioned on $X = \{\text{weather variables, time}\}$ may be approximately normal. Conversely, if $X = \{\text{time}\}$ the literature suggests the conditional mean yield density would not be approximately normal.

In most empirical analyses time is the only conditioning variate and thus we restrict attention to mean yield densities conditioned on the temporal process of yields. Prior to forging ahead with the spatial process of yields, we make three points regarding the temporal process of yields. First, all the modeling issues hereafter discussed with respect to estimating the conditional mean yield density are relevant issues for modeling the temporal process. Second, the estimated residuals derived from the temporal model are conditional on that model and as such are vulnerable to misspecification problems, which in turn may lead to poor inferences regarding the conditional mean yield density (see Just and Weninger 1999 for an example). Third, various methods have been used in the literature including IMA models (Goodwin and Ker 1998, Ker and Goodwin 2000, and Bessler 1994), polynomial models (Just and Weninger 1999), robust splines (Skees, Black, and Barnett 1997), non-linear models (Atwood and Watts 2000), and locally robust nonparametric smoothing methods (Ker and McGowan 2000, and Ker and Coble 1998).

The reader may have noticed that we are discussing the density of a *mean* and question why central limit theorems (CLTs) are not at play. Strong spatial dependence, an empirical stylized fact, negates appealing to CLTs for dependent processes when considering mean yields. These theorems require that spatial dependence dies off at a sufficiently quick rate or that spatial dependence disappears after some finite distance. While this is certainly true for yield data, it is almost never true for the spatial region – such as unit-, farm-, county-, or crop-reporting district – over which the mean is being taken. Therefore, we argue that while mean yields may be normal for some crop-region specifications, they are not normal because of spatial averaging (more on this will be discussed in the next section, on parametric methods).

A major difficulty in estimating conditional mean yield densities is lack of yield data for the region of interest, typically the farm. Farm-level yield data are often at most 10-15 years in length (if one is fortunate), which is rarely adequate to estimate a farm-level yield density with a sufficient degree of confidence. As such, most have attempted to combine farm-level yield data with the associated, but significantly longer, county-level mean yield data (see for example AgRisk[2] and the Income Protection Rating Methodology). Others have used the longer county-level mean yield data as a proxy for individual farm-level data. Unfortunately, without sufficient farm-level yield data, it is difficult to determine which of the three methods (using small farm-level yield series, using a pseudo farm-level yield series generated from both the available farm-level and county-level yield series, or using county-level yield series) most minimizes the appropriate loss function for a given problem.

Ker and Goodwin (2000) first inferred about possible distributional structures based on conjectures about the underlying data-generating process of yields. They suggested that yields come from one of two distinct sub-populations: a catastrophic sub-population and a non-catastrophic sub-population. Thus, they conjectured that conditional yields may best be modeled as a mixture of two unknown distributions where the secondary distribution (from catastrophic years) lives on the lower tail of the primary distribution (from non-catastrophic years) and has significantly less mass. The secondary distribution would be expected to have less mass because catastrophic events are realized with far less frequency than their complement. The secondary distribution would also be expected to live on the lower tail of the primary distribution because realized yields tend to be far less in catastrophic years. They further suggested that yields are highly dependent across space with respect to which sub-population they are drawn from. On the other hand, depending on which sub-population is realized, they hypothesized that yields are only mildly dependent across space.

While it could be easily argued that yields may be a mixture of more than two (possibly infinite) distributions, this simplified structure is sufficiently rich to garner some insights into the possible distributional structures of mean yields. Yields may have a unimodal symmetric density (mass of catastrophic distribution is negligible), a negatively skewed density (the mass of catastrophic distribution is non-negligible and distribution is relatively flat), or a negatively skewed bi-modal density (the mass of catastrophic distribution is

[2] AgRisk is a program designed to assist crop producers in managing harvest-time revenue risk using forward, futures, options, and crop insurance contracts.

non-negligible and distribution is relatively peaked). We note that most of the common parametric families discussed in the next section cannot accommodate all three of the distributional structures arising from this possibly over-simplified view of the data-generating process of yields.

Parametric Methods

The space of parametric models is dense and thus the probability of assuming the correct parametric model is zero. Therefore, asymptotically, parametric estimators converge at $O(1)$; that is, they do not converge to the true model. This is neither surprising nor damaging to proponents of parametric methods. Consider the behavior of parametric estimators when the model is false: parameters converge at the parametric rate, $O(n^{-1})$, such that the Kullback-Leibler distance between the true density and the best parametric approximate in the assumed parametric family is minimized. Hence, it is possible that an incorrect parametric model may have greater efficiency, in terms of an appropriate metric such as mean integrated squared error (MISE), than a correctly specified parametric model for a given finite sample. Also, it is possible that an incorrect parametric model may have greater efficiency than a nonparametric estimator. Therefore, parametric models should not be so easily dismissed because they may be incorrect. The main problem with parametric models was expressed in Ker and Coble (2001) – yield data is not sufficiently rich to reduce the set of potential parametric models to the point of economic invariance. Therefore, one is left grasping for some degree of confidence.

Most in the literature have postulated the beta distribution; see for example Hennessy, Babcock, and Hayes (1997), Babcock and Hennessy (1996), Coble et al. (1996), Borges and Thurman (1994), Kenkel, Busby, and Skees (1991), C.H. Nelson (1990), and Nelson and Preckel (1989). These authors found sufficient evidence of skewness and/or kurtosis in their yield data and opted to use the beta distribution in lieu of the normal distribution. Interestingly, none of these authors reported any statistical test justifying the use of the beta distribution. Gallagher (1987) used a gamma distribution. While in general it is considered a stylized fact that yields are not normally distributed, recently Just and Weninger (1999) called this into question. More complex methods have been postulated by Moss and Shonkwiler (1993), who made use of an inverse hyperbolic sine transformation to accommodate deviations from normality in third and fourth moments. Finally, Ker (2000) used a mixture of two normals to model yield densities. We will discuss below the two most commonly used parametric distributions.

The Normal Distribution. Just and Weninger (1999) attempt to resurrect the normal distribution as a reasonable candidate for modeling yields. They base their arguments on (i) the use of aggregate time-series data; (ii) misinterpretation of statistical significance; and (iii) misspecification of the temporal process of yields. While we feel that they make some important and valid arguments, we disagree with the flavor of their article.

First and most important, Just and Weninger (1999) put far too much reliance on the inability to reject normality. Table 1 presents the results of a simulation analysis based on 25,000 random samples of size n=25 from the following mixture of two normal distributions: $\lambda N(\mu_1,\sigma_1^2)+(1-\lambda)N(\mu_2,\sigma_2^2)$ with parameters $\mu_1=120$, $\sigma_1=20$, $\mu_2=60$, and $\sigma_2=20$, with λ ranging from 0.95 to 0.80. The probability of Type II error with a common Wald-type test based on the third and fourth moments ranges from 0.81 to 0.91, while the resulting insurance rates at the 65 percent coverage level based on the normal are biased downwards from 52 percent to 39 percent. Therefore, while the statistical test will tend to fail to reject the normal, the economic consequences of using the normal to estimate yield densities and derive crop insurance rates can be disastrous. That is, *the inability to reject normality (or any parametric form) based on a relatively small sample should never be used as grounds to assume normality.* An alternative way to proceed would be to design more powerful tests given the economic analysis at hand. For example, if yield densities are estimated in attempts to recover crop insurance rates, a more powerful test may be designed by focusing on departures from normality in the lower tail only. Ker and Coble (2001) restrict the region of interest to the lower tail to increase the power of their sup-type test in testing the beta distribution against a nonparametric alternative.

Table 1. Simulations from a Mixture of Two Normals (Type I Error = 0.05)

λ	*P* (Type II Error)	True Rate	Normal Rate
0.95	82.38%	1.016	0.491
0.90	81.22%	1.761	0.894
0.85	84.93%	2.370	1.311
0.80	91.60%	2.852	1.752

Second, given that farm-level yields may be decomposed into a region-wide random component and a farm-specific random component, the appropriate test is not on the farm-specific random component but rather on the region-wide random component. The region-wide random component determines which of the two sub-populations the farm-level yield is drawn from and thus is more likely to represent departures from normality. Therefore, testing the aggregate time series yield data for departures for normality is reasonable. Also, given that the convolution of a normal with any non-normal distribution results in a non-normal distribution, it is sufficient to test only the region-wide random component at the appropriate significance level. It is unlikely that the farm-specific random component deviates significantly from normality because it is a sum of numerous independent random events.

Third, while we agree with Just and Weninger about the effect of strong spatial correlation and the impact it has on testing multiple time-series, we disagree that one can then also appeal to CLTs for dependent processes to support the assumption of normality. Spatial correlation does not both die off sufficiently quick within the region of interest to appeal to these CLTs and die off insufficiently quick to cause correlation among the statistical tests across different regions.

Beta Distribution. The beta distribution is a common alternative to the normal when estimating yield densities. However, the beta distribution can not accommodate one of the main possible distributional structures: bi-modality or near bi-modality. Additionally, with the exclusion of Borges and Thurman (1994), all applications of the beta distribution have assumed a support for the distribution, thereby using only the first two moments of the yield data. This is surprising because the motivation for using the beta is to account for the skewness and/or kurtosis in the yield data, yet the third and fourth moments of the sample data are not used. Also, empirical analysis suggests (Ker and Coble 2001) that estimating the three or four parameter beta distributions may be more inefficient in the size of samples generally used to estimate yield distributions than nonparametric or semiparametric methods.

Nonparametric Methods

Whether an underlying yield density is normal, beta, or a complex mixture of various parametric distributions can never be determined. This is commonly referred to as the model selection problem. Hendry (1995) discusses this by stating

> "...that no realistic sufficient conditions can be established which ensure the discovery of a 'good' empirical model, nor are any required for empirical econometrics to progress. However, there are a number of necessary conditions which can rule out many poor models, allowing us to focus on the best remaining candidates" (p. 5).

These rules – which cannot hope to uncover the true model – reject models that are deemed statistically inconsistent with the realized data. Unfortunately yield data are not sufficiently rich to reduce the set of parametric models to the point of economic invariance.

Turvey and Zhao (1993), Goodwin and Ker (1998), and Ker and Goodwin (2000) mitigated the model selection problem by using nonparametric kernel methods. Despite this very attractive feature, these methods are not widely employed. Although Yatchew (1998) suggests, among other things, complexity (theoretically, not the application thereof), perhaps inefficiency may be a reason for lack of use. Nonparametric methods tend to be inefficient relative to maximum likelihood methods when the assumed parametric model is correct. In fact, it is possible, and even likely for very small samples such as those corresponding to farm-level yield samples, that an incorrect parametric form – say, normal – is more efficient than the standard nonparametric kernel estimator. Therein lies the exigency in modeling yields. First, yield data are rarely sufficient to reduce the set of potential parametric models such that subsequent economic analysis is invariant among that set. Second, it is unknown which if any of the non-rejected parametric models are efficient for the given sample size relative to a nonparametric estimator. Third, nonparametric estimators, although capable of mitigating the model selection problem, require more yield data because no additional information – in the form of a parametric structure – is used. That is, standard nonparametric kernel estimators are asymptotically inefficient relative to maximum likelihood when the correct parametric model is assumed ($O\,[n^{-4/5}]$ versus $O\,[n^{-1}]$). [3]

Goodwin and Ker (1998) and Turvey and Zhao (1993) used standard nonparametric kernel methods. The kernel density estimator places a bump or individual kernel at each sample realization from the density of interest. The estimate of the density at any given point in the support is simply the sum of the individual kernels at that point. Although this explanation is oversimplified, it serves to illustrate the intuitive nature of the nonparametric kernel density estimator.

The kernel estimate of a density function can be represented as a convolution of the sample distribution function with the chosen kernel, and thus

$$\hat{f}(x) = \int K_h(x-u)\,dF_n(u)\,, \tag{1}$$

where h is the bandwidth or smoothing parameter, $K_h(u) = 1/hK(u/h)$, K is the kernel function, and $F_n(u)$ is the sample distribution function. Throughout, K is assumed to be a square integrable symmetric probability density function with a finite second moment and compact support. Denoting $\mu_2(K) = \int u^2 K(u)du$ and $R(K) = \int K(u)^2 du$ while letting f be the unknown density of interest, standard properties for second order kernels are

$$\begin{aligned} E\hat{f}(x) - f(x) &= \int K(u)[f(x-hu) - f(x)]du \\ &= 1/2h^2\mu_2(K)f''(x) + O(h^4) \\ Var[\hat{f}(x)] &= (nh)^{-1} f(x)R(K) + o[(nh)^{-1}], \end{aligned} \tag{2}$$

[3] We note however that nonparametric kernel estimators can be constructed that achieve a convergence rate as close as desired to the parametric rate by using increasingly higher order kernels.

and thus

$$MSE\{\hat{f}(x)\} = (nh)^{-1} f(x)R(K) + 1/4h^4[\mu_2(K)]^2[f''(x)]^2 + o[(nh)^{-1} + h^4],$$
$$MISE\{\hat{f}\} = (nh)^{-1} R(K) + 1/4h^4[\mu_2(K)]^2 R[f''(x)] + o[(nh)^{-1} + h^4]. \qquad (3)$$

Ker and Goodwin (2000) considered more advanced nonparametric kernel methods. First, they employed a variable smoothing approach to estimate the conditional yield densities, noting that a variable smoothing approach significantly decreases the dependency of estimated tail probabilities on the specific location of the tail realizations. Second, they restricted their estimated densities to have variance equal to the sample variance. They show that the variance of the kernel density estimate is greater than the sample variance almost surely for an order two kernel. For yield samples the additional variance can be significant, which in turn can have a profound effect on the derived insurance rates.

Advanced Methods

Here we discuss two approaches that may offer significant improvements in modeling yields. The first approach combines the advantages of both the parametric and nonparametric estimators. The second approach makes use of extraneous yield data to improve the efficiency of the nonparametric estimator.

Semiparametric Density Estimation. An estimator that mitigates the dependency on the assumed parametric model if the parametric assumption is incorrect yet behaves as efficient as a parametric estimator if the parametric assumption is correct would be most promising. Semiparametric density estimators have been introduced that, with diminutive cost, encapsulate the benefits of both parametric and nonparametric methods, while mitigating their disadvantages. Most important, if the assumed parametric model is incorrect but sufficiently close to the true density, the semiparametric estimator utilizes this information to increase efficiency relative to the standard nonparametric kernel estimator.

Semiparametric density estimators can be characterized into three categories. The first estimator, introduced by Olkin and Spiegelman (1987), is a convex combination of a parametric and a nonparametric estimator. The second estimator, introduced by Hjort and Jones (1996), constructs a local likelihood for densities and subsequently performs locally parametric nonparametric density estimation. The final estimator, introduced by Hjort and Glad (1995), starts with a parametric estimate and nonparametrically corrects it. Ker and Coble (2001) sampled from both pilot yield densities as well as the Marron and Wand (1992) test densities for various small to moderate sample sizes. According to mean integrated squared error (MISE), the latter two estimators performed equally admirably while the Olkin and Spiegelman estimator performed poorly. Ker and Coble chose – as we do – to concentrate on Hjort and Glad's semiparametric estimator because of its relative ease of application and intuitive nature.

Hjort and Glad's Semiparametric Density Estimator. The idea of Hjort and Glad's (1995) semiparametric density estimator is to start with an initial parametric estimate and then multiply a nonparametric kernel-type correction factor. In this sense, the estimator has an empirical Bayes flavor, with the parametric start acting as a prior. If we start with the parametric estimate $f(x,\hat{\theta})$, the correction factor, given the true density $f(x)$, is necessarily $r(x) = f(x)/f(x,\hat{\theta})$. This is estimated by $\hat{r}(x) = \int K_h(x-u)/f(u,\hat{\theta})dF_n(u)$, a nonparametric kernel-type estimator. Combining the start with the correction factor yields the semiparametric estimator

$$\tilde{f}(x) = f(x,\hat{\theta})\hat{r}(x) = \int K_h(x-u)\frac{f(x,\theta)}{f(u,\theta)}dF_n(u). \tag{4}$$

Prior to outlining the properties we must first consider the behavior of the parametric start when the parametric model is incorrect. In this case, $f(x,\theta) \to f(x,\theta_0) = f_0(x) \neq f(x)$. If the maximum likelihood estimator is employed, then $\hat{\theta} \to \theta_0$, where θ_0 minimizes the Kullback-Leibler metric $\int f(x)\log[f(x)/f(x,\theta)]dx$ from the true f to the approximate $f(.,\theta)$. The properties of the semiparametric estimator are

$$\begin{gathered} E\tilde{f}(x) - f(x) = 1/2h^2\mu_2(K)f_0(x)r''(x) + o(h^2) \\ Var(\tilde{f}(x)) = (nh)^{-1}f(x)R(K) + o((nh)^{-1}), \end{gathered} \tag{5}$$

and thus

$$\begin{aligned} MSE[\hat{f}(x)] &= (nh)^{-1}f(x)R(K) + 1/4h^4[\mu_2(K)]^2[f_0(x)r''(x)]^2 + o[(nh)^{-1} + h^4], \\ MISE[\tilde{f}] &= (nh)^{-1}R(K) + 1/4h^4[\mu_2(K)]^2R(f_0r'') + o[(nh)^{-1} + h^4], \text{ and} \\ AMISE[\hat{f}] &= (nh)^{-1}R(K) + 1/4h^4[\mu_2(K)]^2R(f_0r''). \end{aligned} \tag{6}$$

At first glance it is not very intuitive how any gains in efficiency arise. First consider when the assumed parametric model is correct that $r(x) = 1$ for all x. Since a constant can be estimated nonparametrically at the parametric rate of $O(n^{-1})$ and the parametric start is $O(n^{-1})$, $\tilde{f}$ attains the parametric rate $O(n^{-1})$. When the assumed parametric model is incorrect, efficiency gains may still arise but are less transparent. Consider that

$$AMISE(\hat{f}) - AMISE(\tilde{f}) = 1/4h^4[\mu_2(K)]^2[R(f'') - R(f_0 r'')], \qquad (7)$$

where recall that $R(g(x)) = \int g(x)^2 dx$ and therefore $R(f'')$ is a measure of the global curvature. That is, the more curvature in the underlying density, the more difficult it is to estimate nonparametrically. Therein lies the potential efficiency gains with the semiparametric estimator. If the parametric model is close, $r(x)$ will gently oscillate around one, thereby yielding less curvature than $f(x)$. In turn, $R(f_0 r'')$ will tend to be smaller than $R(f'')$, and thus $MISE(\tilde{f}) < MISE(\hat{f})$. In fact, the standard kernel estimator is equivalent to the semiparametric estimator using a uniform start. However, the uniform distribution is very conservative and can be vastly improved upon in most empirical settings. For estimating yield densities, a normal or beta distribution would, in almost all circumstances, be closer to the true density than the uniform distribution, and thus the semiparametric estimator would tend to outperform the standard kernel estimate.

Ker and Coble (2001) investigated the Hjort and Glad (1995) semiparametric estimator by undertaking two simulations. In the first simulation they sampled from pilot estimates of all-practice corn county mean yield densities in Illinois, and compared the relative performance of the competing estimators for small to moderate sample sizes. The results suggested that for very small samples (< 15), the normal maximum likelihood estimator would tend to be preferred given its relatively low estimation error and ease of use. Conversely for samples above 15, the semiparametric estimator with the normal start was preferred to the competing parametric and nonparametric estimators.

While their first simulation considered the general estimation efficiency of the semiparametric estimator relative to the competing estimators, their second simulation focused on economic implications of estimation error. The competing estimators are used to estimate a set of yield densities and derive the associated premium rates. They evaluated the competing estimators by calculating out-of-sample loss ratios based on decision rules for retaining or ceding crop insurance contracts. These results confirmed their results from the first simulation: the semiparametric estimator with a normal start is more efficient than the parametric models (normal, beta), the standard nonparametric kernel estimator, and the current Federal Crop Insurance Corporation / Risk Management Agency rating methodology.

Extraneous Yield Data. Lack of yield data is the single greatest impediment to reducing estimation error. Two methods have been introduced that make use of extraneous yield data and hold promise at reducing estimation error. The first was introduced by Ker (1998) and applied to rating crop insurance contracts by Ker and Goodwin (2000). This estimator uses empirical Bayes methods and the pointwise limiting distribution of kernel estimators to recover a nonparametric empirical Bayes posterior density estimate. While the first method shrinks each kernel density towards the average, the second method starts with a kernel density estimate for the pooled data and estimates a correction factor for each individual density. This was introduced by Ker (2001a) and applied to rating crop insurance contracts in Ker (2001b).

Empirical Bayes Nonparametric Density Estimation. Ker and Goodwin (2000) considered a set of conditional yield densities, one for each county. Denote the number of counties as Q and the adaptive kernel estimate at support point y_j for county i as $\hat{f}_{ij}$. Ker (1998) proposed the following hierarchical model:

$$\begin{aligned} \hat{f}_{ij} \mid \mu_{ij} &\sim N(\mu_{ij}, \sigma_{ij}^2) \\ \mid \mu_{ij} &\sim N(\mu_j, \tau_j^2) \end{aligned}, \tag{8}$$

where $\mu_{ij} = f_{ij} + \beta_{ij}, f_{ij}$ is the unknown density value for county i at support point y_j, β_{ij} is the bias for county i at support point y_j, σ_{ij}^2 is the variance of the adaptive kernel density estimate for county i at support y_j, μ_j is the mean value of the densities across counties at support y_j, and τ_j^2 is the variance across counties at support y_j. The intuition behind the hierarchical model is that even though the μ_{ij}'s are mutually independent for a given j, they are tied together in that there is one loss function for estimating the Q densities at support point y_j. Thus, in a flavor similar to Stein's paradox, an estimator (the posterior) which is a function of the $\{\hat{f}_{1j},...\hat{f}_{Qj}\}$ is constructed which *may* be preferable to the standard kernel estimate. If the bias term β_{ij} was zero or asymptotically zero, we could say that the resulting posterior dominates the standard kernel estimate, thus making it inadmissible. Unfortunately such a statement cannot be made, and dominance – either finite or asymptotically – cannot be asserted regarding either estimator. The posterior estimate is

$$\tilde{f}_{ij} = \hat{f}_{ij}\left(\frac{\tau_j^2}{\tau_j^2 + \sigma_{ij}^2}\right) + \mu_j\left(\frac{\sigma_{ij}^2}{\tau_j^2 + \sigma_{ij}^2}\right), \tag{9}$$

where the unknowns $(\mu_j, \tau_j^2, \sigma_{ij}^2)$ must be estimated. Ker and Goodwin used bootstrap methods to estimate the variance σ_{ij}^2, while estimates of the mean and variance across counties was obtained using the following method of moments estimators: $\hat{u}_j = 1/Q\Sigma_{i=1}^Q \hat{f}_{ij}$ and $\hat{\tau}_j^2 = \hat{s}_j^2 - 1/Q\Sigma_{i=1}^Q \hat{\sigma}_{ij}^2$, where $\hat{s}_j^2$ = $1/Q - 1\Sigma_{i=1}^Q (\hat{f}_{ij} - \hat{\mu}_j)^2$. Therefore, the empirical Bayes nonparametric kernel density estimator at support y_j for county i is

$$\tilde{f}_{ij} = \hat{f}_{ij}\left(\frac{\hat{\tau}_j^2}{\hat{\tau}_j^2 + \hat{\sigma}_{ij}^2}\right) + \hat{\mu}_j\left(\frac{\hat{\sigma}_{ij}^2}{\hat{\tau}_j^2 + \hat{\sigma}_{ij}^2}\right). \quad (10)$$

The resulting posterior or empirical Bayes nonparametric kernel estimate $\tilde{f}_{ij}$ is very intuitive. As the estimated variance of the kernel estimates across counties increases $\hat{\tau}_j^2 \uparrow$, the less the set of adaptive kernel estimates $(\hat{f}_{1j}, \hat{f}_{2j}, ..., \hat{f}_{Qj})$ will shrink towards the overall mean $(\hat{\mu}_j)$. Conversely, the larger the estimated variance of the kernel estimate for a given county $(\hat{\sigma}_{ij}^2)$, the more the given adaptive kernel estimate $(\hat{f}_{ij})$ shrinks toward the overall mean $(\hat{\mu}_j)$. As expected with many shrinkage or Stein-type estimators, the greater the variance within the experimental units relative to the variance across the experimental units, the greater the shrinkage and the greater the potential improvements in efficiency. Ker (1998) indicates that the empirical Bayes nonparametric kernel estimator may offer the largest efficiency gains in small samples, where the variance within counties tends to be relatively high as compared to the variance across counties.

The simulations undertaken in Ker and Goodwin (2000) suggest that this estimator may provide very significant efficiency gains in estimating conditional yield densities. They found that 61 years of data were required for an adaptive kernel to estimate the shape of the conditional yield densities (up to a location-scale transformation) as accurately, on average and with respect to L_2 norm, as the empirical Bayes nonparametric kernel density estimator given only 35 years of data.

One aspect of the empirical Bayes nonparametric kernel density estimator that is of concern is estimation of τ_j^2. Spatial correlation will bias the estimate seriously downward. Ker and Goodwin (2000) attempt to circumvent this by pooling into groups of non-contiguous counties. While this certainly reduces the effect, it does not eliminate the bias, and as such there is greater shrinkage to the mean than would otherwise be expected. The estimator in the following section circumvents this problem.

Nonparametric Density Estimation from a Nonparametric Start. This estimator, introduced in Ker (2001a), was designed to have superior performance, relative to the standard kernel estimator applied separately to the individual samples from the set of unknown densities $f_1, ..., f_Q$, when those densities are identical or similar, while not losing much if they are dissimilar. If the densities were in fact identical, the efficient estimator would pool the Q samples and estimate a single density. However, if the densities are not identical, this estimator is inconsistent. Ker's idea is to combine a kernel estimate based on the pooled data with a kernel estimate based on the individual data in much the same fashion as a parametric estimate is combined with a nonparametric estimate.

This estimator starts with a nonparametric estimate based on the pooled data, which we denote $\hat{g}(x)$, and then multiplies a nonparametric smooth of the individual correction function $r_i(x) = f_i(x)/\hat{g}(x)$. The idea being that if the densities are identical or similar, the pooled estimate represents a reasonable start – in the flavor of Hjort and Glad (1995) – from which to estimate a correction factor function for each individual density. The correction factor function is estimated by $\hat{r}_i(x) = \int K_h(x-u)/\hat{g}(u)dF^i_{n_i}(u)$, thus leading to

$$\tilde{f}_i(x) = \hat{g}(x)\hat{r}_i(x) = \int K_h(x-u)\frac{\hat{g}(x)}{\hat{g}(u)}dF^i_{n_i}(u), \tag{11}$$

where $F^i_{n_i}(u)$ is the sample distribution function corresponding to density f_i. The properties of the estimator are such that

$$\begin{aligned} E\tilde{f}(x) &= f(x) + 1/2g(x)h^2\mu_2(f/g)''(x) + o(h^2) \\ Var\,\tilde{f}(x) &= (nh)^{-1}R(K)f(x) + 2(Nh_p)^{-1}\left[\int\left(K^2(u) - K(u)(K\cdot K)(u)\right)du\right]f(x) \\ &\quad + (Nh_p)^{-1}\left[\int(K(u) - K\cdot K(u))^2\,du\right]f(x) \\ &\quad + o\left((nh)^{-1} + (Nh_p)^{-1}\right) \end{aligned} \tag{12}$$

and

$$\begin{aligned} AMISE(\tilde{f}) &= 1/4h^4R[g(f/g)''] + (nh)^{-1}R(K)f(x) \\ &\quad + 2(Nh_p)^{-1}\left[\int\left(K^2(u) - K(u)(K\cdot K)(u)\right)du\right]f(x) \\ &\quad + (Nh_p)^{-1}\left[\int(K(u) - K\cdot K(u))^2\,du\right]f(x), \end{aligned} \tag{13}$$

where n and h correspond to the individual sample, while N and h_p correspond to the pooled sample.

The idea behind this estimator is to reduce the global curvature of the underlying function being estimated in a nonparametric setting. Generally, the correction factor function will have less global curvature if the start is anywhere close to the unknown density. Unlike the combined parametric and nonparametric estimator, this start is nonparametric, which begs the question: Where do any possible efficiency gains come from? This approach makes use of extraneous data in the estimation of the initial start density. As a result, the total curvature that is being estimated with the individual sample data tends to be significantly reduced.

Ker (2001a) found that the estimator performs well even if the densities are quite dissimilar. Therefore, the pooled estimate, $\hat{g}(x)$, need not have to provide a close approximation to each or any of the individual densities. This is a testament to the idea of Hjort and Glad (1995) that the standard kernel estimate, which corresponds to a start with the uniform distribution over the support, is a conservative start for most densities and can be improved upon. Also, if the densities are equal, the estimator is more efficient, asymptotically, than the standard nonparametric kernel estimator based on the pooled data because of the inherited bias reduction properties of the estimator. That is, if the true densities are identical, then $\widetilde{f}$ behaves like the Jones, Linton, and Nielsen (1995) estimator in that the bias is $O(h^4)$ if $h_p = ch$.

Ker (2001b) illustrates that the estimator out-performs the government's current rating methodology, standard kernel methods, standard parametric methods (beta, normal), and finally the semiparametric estimator of Hjort and Glad. Most important, the estimator circumvents the spatial correlation by choosing both smoothing parameters (pooled estimate and individual correction) using likelihood or least squares cross-validation methods based on the individual data only.

Conclusions for Modeling Yields

With respect to modeling yields, lack of data is a double-edged sword. On the one hand, adding information such as a parametric model has greater benefits when data is lacking. On the other hand, the data is insufficient to provide guidance in choosing a parametric model. As illustrated, the inability to reject a candidate model should never be used to justify the use of that parametric model in small samples. A minimax solution would be to employ nonparametric methods. While these mitigate the model selection problem, no additional information is used in the estimation process that "may" reduce efficiency.

The methods just discussed offer the greatest hope. The semiparametric density estimators mitigate the errors associated with incorrect assumptions while allowing potentially useful information in terms of an appropriate parametric model to be expressed. Simulations by Ker and Coble (2001) suggest significant efficiency gains are possible in estimating yield densities. Simulation results using the nonparametric estimators that make use of extraneous yield data are also very encouraging. The bias reduction properties as well as the ease with which spatial correlation is circumvented suggests that the nonparametric density estimation from a nonparametric start may offer more promise in estimating yield densities than the empirical Bayes nonparametric kernel density estimator.

MODELING PRICES

An enormous literature exists addressing issues relating to the measurement of price risk. In general, most farmers consider output prices to be their main source of risk, followed by yields and then input prices.[4] The conventional wisdom holds that the combination of random output shocks (due to weather) and generally inelastic demands tends to lead to greater volatility and risk for prices of agricultural products relative to nonagricultural commodities. Thus, agricultural commodity prices tend to be highly variable and this variability may be state-dependent (i.e., conditional upon market conditions or information at a point in time).

Much of this research has involved deriving techniques for pricing options contracts and other derivative securities.[5] Options are most simply viewed as insurance contracts written on price. Thus, in an efficient market with risk-neutral investors, their price should be equivalent to the actuarially fair insurance premium, which should equal the expected payout of the contract. This general technique for pricing assets, often referred to as the "no-arbitrage" approach, was developed by Merton (1973) and Cox and Ross (1976). For example, a European call option with a strike price of S that expires at time T should have a premium of

$$c(P_T) = e^{rt} \int_0^{\infty} f(P_T) \max(0, P_T - S) dP_T \, , \tag{14}$$

where $f(P_T)$ is the probability density function (pdf) of the price of the asset at expiration. Thus, the particular form of the pdf for prices is key to appropriately assessing risk from a consideration of options prices and this relationship can be used to extract a measure of $f(P_T)$ from observed options prices. A closely related approach to deriving the form of $f(P_T)$ from observed options premia data was developed by Breeden and Litzenberger (1978), who noted that $f(P_T)$ could be derived by twice differentiating a function representing the pricing of call options: [6]

$$\frac{\partial^2 c(P)}{\partial P^2} = e^{-rt} f(P_T) \, . \tag{15}$$

This relationship is frequently used to derive the risk-neutral pdf from call prices observed over different strikes. The vast empirical literature that has

---

[4] In a survey of Kansas farmers, Goodwin and Kastens (1993) found that 67 percent indicated that their main source of risk is commodity prices.

[5] To be more precise, most research has worked in the opposite direction – determining the probability implications of observed options prices.

[6] Alternatively, a measure of the cdf of the price is obtained by differentiating the call pricing function a single time.

derived risk-neutral pdf's for asset prices has generally followed one or the other approach.[7]

Lognormality, Smiles, and Smirks

Observation of commodity prices as well as considerable anecdotal evidence suggest that price distributions tend to be positively skewed. This evidence has led financial researchers to the widespread adoption of the assumption that prices tend to be lognormally distributed. In particular, the Black-Scholes option valuation formula, which is based upon the assumption of lognormally distributed prices, has gained widespread acceptance.[8] Black-Scholes essentially derives parameters of the (lognormal) price distribution by equating the expected rate of return on the asset to the return on a risk-free bond and solving for the unknown parameters in equation (14).

Lognormality and the Black-Scholes model have become widely accepted standards for representing and modeling price risk. For example, current revenue insurance rating procedures assume lognormality for agricultural prices and utilize the Black-Scholes model to model price variability (and thus risk) from options price data. Implied volatilities calculated from the Black-Scholes model are widely reported and utilized to represent price risk.

In spite of the widespread acceptance of the Black-Scholes model and its concomitant assumption of lognormality, anecdotal evidence and considerable recent research has yielded results that lead one to question the assumption of lognormality. In particular, volatility patterns that are inconsistent with a lognormal distribution for prices are often revealed. Under the assumptions inherent in the Black-Scholes option pricing model (i.e., lognormality), the implied volatility of an asset should be the same across all strike prices of the asset having the same maturity date. That is, a plot of implied volatilities against different strike prices should be flat if prices are lognormally distributed. However, the implied volatility across different strikes for an asset often rises as one moves far in or out of the money (i.e., as one moves into the tails of the price distribution). As Bahra (1997) notes, such patterns suggest that market participants assign higher probabilities to tail events than does a lognormal distribution – meaning that the underlying price distribution has fatter tails than does the lognormal.[9]

Figure 1 illustrates the implied volatilities for May 2001 corn and July 2001 wheat. A rise in the implied volatility as one moves toward the tails of the distribution is apparent, suggesting that the implied distribution of corn and wheat prices is inconsistent with lognormality. These patterns are often

---

[7] Comprehensive reviews of this literature are contained in Bahra (1997), Campbell, Lo, and MacKinlay (1997), and Chang and Melick (1999). Our discussions follow these reviews closely.

[8] To be more precise, the Black-Scholes model assumes that the price of an underlying asset evolves according to geometric Brownian motion.

[9] Such evidence of skewness or "fat-tails" (leptokurtosis) is frequently noted in the literature.

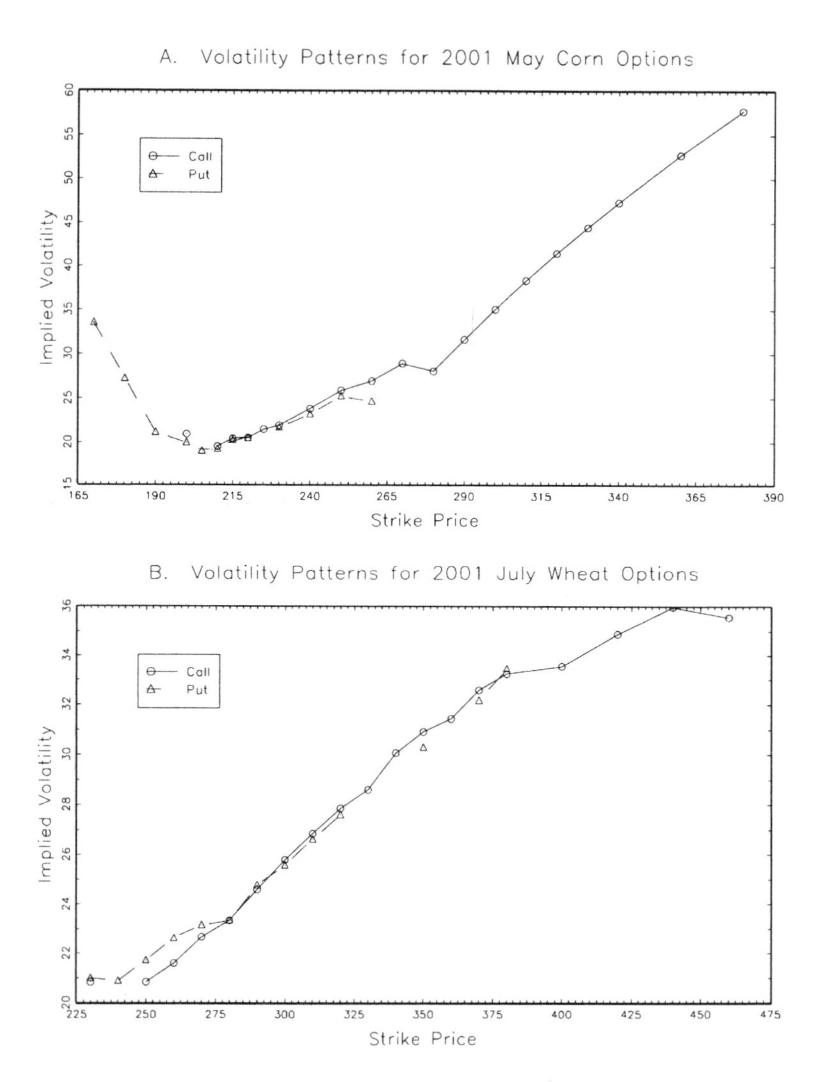

Figure 1. Volatility Smiles: July Corn and May Wheat

referred to as "smiles" and, when the patterns are non-symmetric, as in these cases, as "smirks." These patterns are more the rule than the exception – suggesting that the widespread assumption of lognormality may not be consistent with the evidence.

The assumption of lognormality has been questioned by a wide range of studies. Fama (1965) first noted the existence of leptokurtotic behavior in prices that was inconsistent with lognormality. Early studies noting price behavior that was inconsistent with lognormality include Black (1975),

MacBeth and Merville (1980), and Rubinstein (1985). More recent studies have pursued various specification tests to determine the appropriateness of lognormality and alternative specifications and have considered alternative parametric approaches to evaluating the distribution of prices. Again, the bulk of this research has involved the utilization of observed options prices to extract a measure of the implied probability density function for prices.

Parametric Alternatives to Lognormality

The general approach to deriving an estimate of the parameters of the risk-neutral pdf implied in an option price involves either directly solving equation (14) for the parameters of the underlying pdf or using numerical optimization techniques to derive parameter estimates that minimize the sum of squared differences between actual options prices and those predicted by the pdf. Such an approach is entirely general and can be applied to any parametric specification of the pdf. The goal of much recent research has been to specify parametric alternatives that are more general in terms of their flexibility than the lognormal pdf or that possess features that more closely mimic what is implied by observed options prices. Candidate distributions considered in the literature include the stable Paretian, Burr, gamma, Weibull, and exponential distributions as well as various mixture distributions.

Sherrick, Garcia, and Tirupattur (1996) evaluate the lognormality assumption using a number of specification tests applied to soybean options price data. Lognormality is rejected in approximately one-third of the cases they evaluate. Their results also reveal considerable variability in the skewness and kurtosis implicit in the distributions implied by their sample of options prices. They recommend selecting a flexible parametric form that is capable of representing a wide range of skewness and kurtosis. Their choice is the three parameter Burr III distribution, which they note is capable of covering all of the regions of support implicit in the Pearson types IV, VI, and the bell-shaped curves of the Pearson type I, the gamma, the Weibull, normal, lognormal, exponential, and logistic distributions. The Burr III distribution pdf is given by

$$f(P)=\frac{\alpha\gamma P^{\alpha\gamma-1}(\tau^{a}+P^{\alpha})-\gamma\alpha P^{\alpha(\gamma-1)}}{(\tau^{\alpha}+P^{\alpha})^{\gamma+1}}. \tag{16}$$

Their results indicate that the Burr III distribution provides a more accurate representation of observed option prices for contracts closer to maturity, while the lognormal distribution provides a better fit at long maturities.

Buschena and Ziegler (1999) consider various parametric specifications of the pdf implied by option prices for their suitability in modeling price risk for corn and soybean revenue insurance contracts. In addition to the Burr III, they also consider the Burr XII and the lognormal distributions. Their results indicate that lognormality receives relatively strong support and that deviations from lognormality are common only close to the expiration of the option contract. Their results also indicate that the use of market-based options generally provides a more accurate assessment of risk than what is implied by historical data.

Recent research has adopted densities that are comprised of mixtures of standard distributions, such as the normal and the lognormal.[10] In particular, the density is represented as a weighted sum of k components,

$$f(P) = \sum_{i=1}^{k} \lambda_i \phi_i(P), \tag{17}$$

where the λ_i terms are mixing parameters that are constrained to lie on the open interval (0,1) and to satisfy $\sum_{i=1}^{k} = 1$ and $\lambda_i > 0$ for all i. Such an approach adds considerable flexibility since mixtures are capable of capturing considerable variation in terms of skewness and kurtosis and other distributional features inherent in the data. For example, a mixture of two normals nests the conventional normal case (obtained when the means and variances of each component density are identical), skewness (when the variances differ), and kurtosis (when the means differ – even allowing two modes). Mixture models have much in common with the nonparametric models discussed below in that standard nonparametric density estimation techniques generally use an n-component mixture of component densities to estimate an unknown density.[11]

Ritchey (1990) proposed using a mixture of lognormals to generalize the standard Black-Scholes option pricing model. Ritchey showed that this approach has analytical simplicity in that the implied options premia are simply weighted averages of the premia implied by each lognormal component where the weights are given by the mixing parameters. Ritchey finds that lognormality results in serious mispricing of out of the money options close to expiration. Melick and Thomas (1997) also adopted a mixture of three lognormal distributions to model oil prices. Similar mixture models have been applied by Mizrach (1996), who also used a mixture of lognormals, and Söderlind and Svensson (1997), who used a mixture of normal densities.

---

[10] An excellent discussion of finite mixture distributions is contained in Everitt and Hand (1981).

[11] For example, use of a normal kernel in nonparametric density estimation is analogous to using a mixture of n normals, centered on each point with variances determined by the bandwidth parameter, and mixing parameters equal to 1/n.

Goodwin, Roberts, and Coble (2000) used a mixture of normals and lognormals to evaluate techniques for measuring price risk in constructing revenue insurance contracts. As they note, the utilization of mixture distributions provides a ready means for specification testing. Consider a simple mixture of a normal (ϕ) and lognormal (ℓ) densities:

$$f(P) = \lambda\phi(P) + (1-\lambda)\ell(P). \tag{18}$$

The value of λ may be helpful in evaluating whether the underlying density is consistent with normality (for $\lambda = 1$) or lognormality (for $\lambda = 0$). Because parameters of one of the components may be unidentified under either null, the testing procedures involve nonstandard distributions. McLachlan (1987) and Feng and McCulloch (1994, 1996) have developed appropriate inferential procedures for specification testing of mixture densities.

Semiparametric and Nonparametric Approaches

Many recent studies have pursued an alternative route to evaluating the risk-neutral pdf involving equation (15) above. These studies often utilize a direct application of the Breeden and Litzenberger (1978) result discussed above. Essentially, this approach involves estimation of an interpolated call option pricing function which can then be twice differentiated (usually using numerical differentiation) to yield the risk-neutral pdf. The specific approach sometimes uses the Black-Scholes formula to derive a smile relationship (between strike prices and implied volatilities) and then to fit a parametric or nonparametric function relating the implied volatility to the strike price. This allows the strike price to be related back to the call option price.

Shimko (1993) used a quadratic function to relate volatility to the strike price. It is standard to extrapolate outside the observed range of strike prices using lognormal distributions to model the tails, with the restriction that the entire pdf must integrate to one. Alternatively, one may interpolate the call price function directly. The estimated call price function must satisfy certain properties – namely, it must be convex, monotonic as one moves in or out of the money, and twice-differentiable. Estimation of the call price function may use parametric functional relationships, such as quadratic function, or nonparametric techniques. Bates (1991) used a cubic spline to interpolate call prices. Neuhaus (1995) used numerical derivatives, evaluated only at observed call prices, to derive probability implications of observed call option prices.

Aït-Sahalia and Lo (1998) followed a completely nonparametric procedure by using nonparametric kernel regression to estimate the entire call price function. In light of the significant data requirements typically associated with nonparametric estimation, they used a cross-sectional, time-series panel of options prices and strikes. Their results demonstrate that standard Black-

Scholes techniques fail to account for the skewness and kurtosis that is captured in the nonparametric estimates. Explicit specification tests comparing the parametric Black-Scholes model and the nonparametric alternatives reject the restrictions inherent in the Black-Scholes model. Aït-Sahalia and Lo (2000) expanded these methods to make inferences regarding subjective risk aversion. A similar line of research was proposed by Coutant (1999), who used Hermite polynomial expansions to model the relationship between volatility and strike price.

Modeling Time-Varying Risk

Standard parametric approaches, such as the geometric Brownian motion assumption of the Black-Scholes model, are generally based on the assumption that the variance of the price is constant or varies over time in a deterministic fashion. A variance that varies in a deterministic fashion is a minor complication, as standard parametric models can simply replace the variance with an integral that evaluates it over the life of the option. Considerable empirical evidence has recently been presented to question the assumption that volatility is deterministic.[12] An early approach to measuring variability was to consider the variance observed over a certain period of time. Campbell, Lo, and MacKinlay (1997) note that "...however, [it is] both logically inconsistent and statistically inefficient to use volatility measures that are based on the assumption of constant volatility over some period when the resulting series move through time" (p. 481). A very large literature has developed to address this limitation in terms of stochastic volatility models.[13] A stochastic variance term has important implications for modeling risk since uncertainty about the variance adds another source of uncertainty to the asset's price.

The theory underlying asset pricing dynamics is generally expressed in terms of continuous dynamic processes. Empirical applications of time-varying volatility models are almost always forced to work with discretely sampled data. Researchers have observed that variance may vary according to market conditions. For example, periods of poor weather or low stocks may result in much more volatile crop prices. It has also been widely observed that periods of high price variability tend to be followed by more high price variability and vice versa. This conditional and autoregressive nature of variability has led to a very large literature that has developed and estimated conditional heteroscedasticity models.

The overwhelming majority of these studies have pursued autoregressive conditional heteroscedasticity models of price variance. The basic autoregres-

[12] See, for example, Amin and Ng (1993) and Hull and White (1987).

[13] See Campbell, Lo, and MacKinlay (1997) for a detailed discussion and review of this literature.

sive conditional heteroscedasticity model (ARCH) was introduced by Engle (1982). If e_t represents a normally distributed innovation in a price series, where the innovation has a zero mean but a time-varying variance (i.e., $e_t \sim N[0,\sigma_t^2]$), an ARCH model is generally written as

$$\sigma_t = \alpha_0 + \sum_{i=1}^{k} \alpha_k e_{t-i}^2 . \tag{19}$$

Of course, the α_i terms must be restricted to be positive to ensure positive predicted variances. A parsimonious generalization of the ARCH model was presented by Bollerslev (1986):

$$\sigma_t = \gamma_0 + \sum_{i=1}^{p} \gamma_i \sigma_{t-i}^2 + \sum_{i=0}^{q} \alpha_i e_{t-i}^2 . \tag{20}$$

Again, the parameters of equation (20) must be constrained to ensure positive variances. The standard approach to ARCH and GARCH modeling is to assume a parametric distribution for innovations (again, lognormality dominates) and to choose parameters by maximizing the appropriate likelihood function. Extensions of the GARCH modeling techniques have been pursued in several directions. Cases where the parameters on the lagged variance and square innovation terms sum to one imply a unit autoregressive root in the variance process such that variance shocks permanently alter the future time path of variances. These models have been termed integrated GARCH or IGARCH models (see Nelson 1990). It is straightforward to generalize GARCH models to a multivariate framework where variance shocks to one price may have an effect on other prices. The number of parameters to be estimated grows considerably in multivariate models since covariance effects must be accounted for. In addition, cross-equation restrictions are also necessary in order to restrict predicted variance terms to be positive. Kroner and Ng (1998) survey multivariate GARCH models. Models permitting the variance to exert influences on the first moment of prices have also been developed. These are commonly referred to as GARCH in mean models, or GARCH-M models.

The applied ARCH and GARCH modeling literature is immense and its sheer size precludes a comprehensive review. These models are useful in terms of identifying the dynamic process that the evolution of a price's variance may follow. However, the models may provide limited information about the factors actually causing the autoregressive effects in variances. Fackler (1986) pointed out that autoregressive patterns in variances typically reflect the autoregressive nature of the arrival of new information. For example, information about weather and crop growing conditions tends to arrive in an autocorrelated fashion during the growing season. These exogenous shocks will be revealed in terms of autoregressive variance effects.

Kenyon et al. (1987) evaluated factors affecting grain price variability. Their results suggested that a higher stocks-to-use ratio was correlated with lower price variability and that a higher ratio of prices to loan rates was correlated with a higher level of price variability. Similar research was carried out by Streeter and Tomek (1992). Their results found that the level of supply did not appear to affect price variability for soybeans and that higher stocks actually had a positive influence on price variability for soybeans. Hennessy and Wahl (1996) found that growing conditions, represented by temperature and rainfall measures, had a significant influence on price variability for agricultural commodities. This line of research has also concluded that variables related to the structure of futures markets may affect the variability of prices in these futures markets. Streeter and Tomek (1992) found that more speculative activity tended to result in less soybean price variability. Day-trading (or "scalping") was found to be correlated with more price variability by Streeter and Tomek (1992) and Goodwin and Schnepf (2000).

Goodwin and Schnepf (2000) used standard conditional heteroscedasticity, ARCH, and GARCH models to evaluate the variability of corn and wheat prices. Their results indicated that information about growing conditions, stocks, total use, and other market factors tended to explain changes in the variance of prices. Significant seasonal effects in variability of agricultural prices were revealed. Their results also suggested that it was difficult to identify the effects of these factors when the models also included autoregressive variance effects (i.e., ARCH and GARCH effects). This is consistent with Fackler's (1986) assertion that ARCH and GARCH models tend to reflect the autocorrelated nature of the arrival of new information.

As we have noted, the assumption of lognormality in asset prices has remained widespread. However, the body of research that has developed methods for evaluating the time-varying price variability has been cognizant of the departures from lognormality discussed above. In particular, the presence of a higher degree of kurtosis than is consistent with lognormality has been frequently identified. A simplistic approach to this problem is to continue to assume normality in log prices but to interpret the estimates as quasi-maximum likelihood and thus make adjustments for the standard errors. Alternatively, one may wish to model the distribution directly. Bollerslev (1987) introduced a t-distribution into the GARCH modeling framework. The t-distribution allows one to model kurtosis by making the degrees of freedom of the t-distribution a parameter that is estimated.[14] Alternative distributions have also been considered. D. Nelson (1991) used a generalized error distribution (GED). Engle and Gonzalez-Rivera (1991) used a nonparametric estimate of the error density. Distribution-free estimation procedures, such as the generalized method of moments (GMM), may also permit nonparametric

---

[14] Of course, as the degree of freedom increases, the density converges to a normal. This parameter is usually estimated in its inverse form, such that a convenient test of normality is available in a consideration of the null that the inverted parameter is zero.

estimation of GARCH models. Linton and Perron (1999) assume that standardized residuals from a GARCH model have a distribution within the exponential power family. Li and Turtle (2000) adopt an "estimating function" (EF) approach as an alternative to least squares or maximum likelihood to address potential non-normality in GARCH models. They find that such an approach offers efficiency gains over maximum likelihood and quasi maximum likelihood techniques when the underlying distribution is non-normal.

Finally, much recent research has been directed at capturing the discrete jumps in underlying parameters describing the stochastic process that may occur as new information becomes available. These jumps, when ignored, may manifest themselves in terms of implied non-normality of log prices. Mixture distributions represent one approach to capturing such changes. GARCH models based upon mixture distributions have been developed by Roberts (2000). An alternative approach involves representing the underlying stochastic process as a mixed diffusion-jump process. Merton (1976) suggested a jump-diffusion for stock prices where the jump occurs at a Poisson rate and the jump size is normally distributed. If prices are lognormal and as the probability of jumps goes to zero, these models reduce to standard lognormal models. Jorion (1988) incorporated a jump process into a GARCH model of exchange rates. Similar application of mixed diffusion-jump processes was done in a study of exchange rates by Akgiray and Booth (1988). Their results indicated that the diffusion-jump estimator provided a better fit to observed exchange rate data than did a mixture of normals.

Conclusions for Modeling Prices

In all, our rather selective survey of the literature on modeling price risk leads us to several conclusions. First, a model of price risk will generally need to evaluate the entire density of prices. To the extent that the parametric family describing the distribution of prices is known, parameters of the distribution may be estimated from observed prices. In this context, a rather standard assumption regarding prices has been that they are lognormally distributed, in which case the mean and variance of prices describes the entire distribution. Our survey suggests that the overwhelming majority of empirical results concludes against this finding. In particular, the probabilities associated with extreme prices are generally greater than what is implied by a lognormal distribution. Evidence of this leptokurtosis is abundant across a wide range of applications. Thus, methods capable of accommodating departures from normality are needed. A wide range of alternative probability density estimators are available, including flexible parametric distributions such as the Burr, nonparametric estimators, and finite mixture distributions. Although lognormality will likely continue to be a standard assumption, evaluators of price risk should be cognizant of the fact that lognormality is generally not supported by the available data.

MODELING REVENUE RISK

Analysis of agricultural risk, though usually comprised of considerations of price and/or yield risk, are generally concerned with the effects of yield and price uncertainty on revenues. One may have precise measures of yield risk (the yield density) and price risk (the price density), and yet making the next step to a consideration of the risk associated with revenue (the product of yields and prices) may not be obvious. A fundamental complicating factor is the fact that yields and prices are rarely independent. To the extent that yields are spatially correlated, yield shortfalls on an individual farm or in a particular area such as a county or state are likely to be correlated with higher prices. That is, yields and prices tend to be negatively correlated. Of course, weather patterns tend to be spatially correlated, and thus yield shortfalls do tend to occur over a wide area. Any consideration of revenue risk that utilizes measures of price and yield risk must also consider the degree of correlation between yields and prices. To the extent that the suspected negative correlation is indeed present, a "natural hedge" exists such that revenue shortfalls associated with low yields are naturally offset by higher prices.

Measurement or simulation of revenue risk typically proceeds by first measuring the marginal distributions of yields and prices and the degree of correlation between the two. To the extent that yields and prices are drawn from a common parametric family, a joint pdf can be used to generate correlated draws for simulation or Monte Carlo integration purposes. However, in that the marginals are likely to be from different parametric families, some method for drawing correlated random variables from different marginal distributions is needed.[15]

Three procedures have been utilized in the literature to accomplish random sampling of correlated random variables from specified marginal distributions. The first two of these procedures are essentially equivalent, though the latter is more flexible in terms of being applicable to large sets of variables. The first two of these procedures is often called the "weighted linear combination" approach, developed by Johnson and Tenenbein (1981). This approach essentially involves a translation scheme whereby functions of uncorrelated random draws from two different marginals (say, for price and yield) are combined to form a linear combination that yields bivariate distributions with the appropriate degree of correlation. The weight chosen to form the linear combination is determined by the extent of correlation and the distribution used in the translation process. The idea is to generate a pair of random variables (X,Y) from a given pair of marginals $F_1(X)$ and $F_2(Y)$ that have a given degree of correlation. How this correlation is to be measured is discussed by Johnson and Tenenbein (1981), and they describe the method for

[15] For example, consider the common case of prices (often assumed to be lognormal) and yields (often assumed to be beta distributed). Specification of the appropriate joint pdf would be challenging to say the least.

two common measures of correlation – Spearman's rank correlation coefficient (ρ_S) and Kendall's tau (τ) coefficient. Let U' and V' be independent random draws from a common probability distribution. We then consider

$$U = U' \quad \text{and} \quad V = cU' + (1-c)V', \tag{21}$$

where c is a constant on the interval (0,1) that is determined by the extent of the correlation between X and Y. Let $X' = H_1(U)$ and $Y' = H_2(V)$, where $H_1(U)$ and $H_2(V)$ are the cumulative distribution functions of U and V, respectively. Then, draws from the desired marginals possessing the desired degree of correlation are given by

$$X = F_1^{-1}(X'), \tag{22}$$

and, if ρ_S is positive,

$$Y = F_2^{-1}(Y'), \tag{23}$$

and, if ρ_S is negative,

$$Y = F_2^{-1}(1-Y'). \tag{24}$$

Essentially, this procedure is analogous to taking random draws from a known distribution (such as a normal), constructing a weighted linear combination that will result in variables with a known correlation, translating these to a uniform using the cdf, and then using these uniform random deviates in the inverse cdf for each variable to generate random draws from the desired marginals. Johnson and Tenenbein (1981) have tabulated values of the weighting factor c that should be used for different levels of correlation and different distributions.

The Johnson and Tenenbein (1981) procedure is limited in that it is applicable only to the bivariate case. A simple multivariate extension developed by Fackler (1991) can be applied to any number of correlated random variables. This procedure essentially involves the same intuition. Random numbers are drawn from a known multivariate distribution, such as the normal, that has the desired correlation structure. These values are then translated to uniform random variates using the normal cdf and then are translated to the appropriate marginals using the inverse cdf applicable to each marginal.[16] A slight complication associated with this approach involves measuring the

[16] This is completely analogous to the weighted linear combinations (WLC) approach of Johnson and Tenenbein (1981), where the weights are derived as those values implied by the Choleski decomposition of the covariance matrix underlying the desired correlation structure.

correlation structure. In particular, the transformations involved in this approach may not preserve the correlation structure as represented by a Pearson correlation coefficient. However, Fackler (1991) has shown that a rank measure of correlation, such as Spearman's correlation coefficient, will be preserved through the transformation associated with this technique.

A limitation associated with these approaches is that they are fundamentally parametric – a priori knowledge of the appropriate marginal distributions and the degree of correlation is necessary in order to implement the procedures. Thus, the aforementioned limitations associated with parametric modeling of price and yield distributions are again relevant. Measurement of correlation is especially important as the joint distribution will be significantly influenced by the extent of correlation among the individual variables. For example, the Johnson and Tenenbein (1981) procedure is currently used to draw correlated yields (from a beta marginal distribution) and prices (from a lognormal marginal distribution) in constructing rates for crop revenue insurance. These rates are quite sensitive to assumptions regarding the degree of correlation between yields and prices. In particular, rates fall substantially as this correlation increases, and thus accurate rates are critically dependent upon measurement of this correlation. Unfortunately, the amount of data available for measuring this correlation is often quite limited, and thus measures of this correlation may be inaccurate.

A more fundamental limitation of these methods involves the fact that, although the techniques are able to match first and second moments, including covariances, there are no assurances that higher ordered "cross-moments" are matched. These moments are difficult to interpret on an intuitive basis, though they certainly could be important characteristics of a joint distribution that could influence insurance premium rates and other measures of risk. An example of the techniques using simulated data is presented in Figure 2. Forty thousand beta-distributed yields (Panel A) and normally-distributed prices (Panel B), having an *average* Spearman correlation coefficient of -0.30, were generated and used to construct a simulated revenue series (Panel C).[17] The Johnson and Tenenbein (1981) procedures were then used to generate an alternative simulated distribution from the same marginal distributions (Panel D). The simulated revenue distributions were used to estimate the probabilities associated with revenues below 150 and above 500. For the first set of simulated data, the associated probabilities were 0.017 and 0.034, respectively. In the second case, the respective probabilities are 0.022 and 0.014, representing differences of 25-93 percent. The implied revenue distributions are clearly different, though the prices and yields underlying both have the same means, variance, and correlations.

[17] This simulation reflects a suspicion that correlation may be state-dependent. The correlation coefficient varies according to the sign of the yield residual for each observation, though on average it is equal to -0.30.

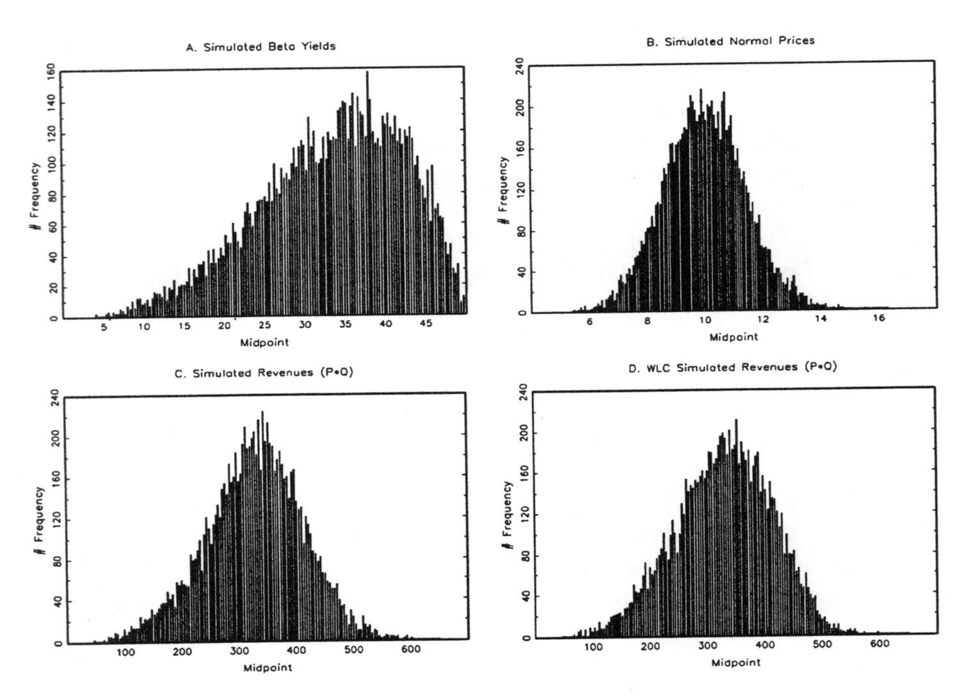

Figure 2. Simulated Yields, Prices, and Revenues

Taylor (1990) proposed approximating joint distribution functions by using the identity

$$f(p_1, p_2) = f(p_1) f(p_2 | p_1) . \tag{25}$$

Such an approach holds promise only if sufficient data exist to measure the conditional distribution. As Taylor notes, this may be feasible if p_1 takes only a limited number of values. Thus, this approach is likely to be operational only in limited cases where a large amount of data is available, thus permitting accurate measurement of the conditional distributions.

Finally, the method of copulas may hold some promise for measuring risk in multivariate relationships. A copula $C[\cdot]$ is a function that relates marginal distributions $F(\cdot)$ to a joint distribution function:

$$C[F(X_1), F(X_2),..., F(X_N)] = F(X_1, X_2,..., X_N) . \tag{26}$$

Frees and Valdez (1998) discuss techniques for identifying copulas from a sample of data. These techniques generate approximations to an unknown copula, which can then be used to represent the joint distribution function. Frees and Valdez (1998) offer several examples from the insurance literature. The techniques associated with approximating copulas share many of the limitations outlined above – they are of a parametric nature and thus are sensitive to misspecification, and they are approximations, which may suffer from approximation errors.

CONCLUDING REMARKS

The objective of this chapter was to provide an overview of issues underlying the measurement of yield, price, and revenue risk in agricultural markets. A voluminous literature addressing each aspect of risk modeling exists. We have attempted to provide a selective and critical review of this literature. Of course, our review reflects our own biases. We argue that modeling of "risk" usually must involve more than simply considering variance terms or the like. A full comprehension of risk requires knowledge of higher moments of a distribution.

Difficulty in modeling yields results because yield data is not sufficiently rich to reduce the set of candidate parametric models to the point of economic invariance. The semiparametric and advanced nonparametric methods appear to offer the best solution to this difficult empirical problem.

The assumption of lognormality has prevailed in models of price risk. This assumption is popular because of its analytical simplicity and because prices often (but not always) exhibit behavior that is consistent with lognormality. However, as we discuss, an abundance of empirical research and anecdotal information suggests that lognormality may be a flawed assumption, especially when modeling low probability events. A wide range of alternatives, including flexible parametric specifications and nonparametric methods, are discussed.

Finally, we review methods used to evaluate revenue risk. Revenue is the product of correlated yields and prices. These methods generally attempt to provide approximate measures of the joint distribution of prices and yields. However, as we note, these methods generally only match first and second moments and thus may yield flawed measures of a joint distribution.

REFERENCES

Aït-Sahalia, Y., and A.W. Lo. 1998. "Nonparametric Estimation of State-Price Densities Implicit in Financial Asset Prices." *Journal of Finance* 53: 499-547.

___. 2000. "Nonparametric Risk Management and Implied Risk Aversion." *Journal of Econometrics* 94: 9-51.

Akgiray, V., and G.G. Booth. 1988. "Mixed Diffusion-Jump Process Modeling of Exchange Rate Movements." *Review of Economics and Statistics* 70: 631-637.

Amin, K., and V. Ng. 1993. "Option Valuation with Systematic Stochastic Volatility." *Journal of Finance* 48: 881-910.

Atwood, J., and M. Watts. 2000. "Normality of Crop Yields." Unpublished manuscript, Montana State University.

Babcock, B.A., and D.A. Hennessy. 1996. "Input Demand under Yield and Revenue Insurance." *American Journal of Agricultural Economics* 78: 416-427.

Bahra, B. 1997. "Implied Risk-Neutral Probability Density Functions From Option Prices: Theory and Application." Bank of England, ISSN 1368-5562.

Bates, D.S. 1991. "The Crash of '87: Was It Expected? The Evidence from the Options Markets." *Journal of Finance* 46: 1009-1044.

Bessler, D. 1994. "Aggregate Personalistic Beliefs on Yields of Selected Crops Estimated Using ARIMA Processes." *American Journal of Agricultural Economics* 76: 801-809.

Black, F. 1975. "Fact and Fantasy in the Use of Options." *Financial Analyst Journal* 31: 684-701.

Bollerslev, T. 1986. "Generalized Autoregressive Conditional Heteroscedasticity." *Journal of Econometrics* 31: 307-327.

___. 1987. "A Conditional Heteroscedastic Time Series Model for Speculative Prices and Rates of Return." *Review of Economics and Statistics* 69: 542-547.

Borges, R., and W. Thurman. 1994. "Marketing Quotas and Subsidies in Peanuts." *American Journal of Agricultural Economics* 76: 801-809.

Breeden, D., and R. Litzenberger. 1978. "Prices of State-Contingent Claims Implicit in Option Prices." *Journal of Business* 51: 621-651.

Buschena, D., and L. Ziegler. 1999. "Reliability of Options Markets for Crop Revenue Insurance Rating." *Journal of Agricultural and Resource Economics* 24: 398-423.

Campbell, J.Y., A.W. Lo, and A.C. MacKinlay. 1997. *The Econometrics of Financial Markets*. Princeton, NJ: Princeton University Press.

Chang, P.H.K., and W.R. Melick. 1999. "Background Note for Workshop on Estimating and Interpreting Probability Density Functions." Unpublished manuscript, Bank of International Settlements, Basel, Switzerland (June 14).

Coble, K.H., T.O. Knight, R.D. Pope, and J.R. Williams. 1996. "Modeling Farm-Level Crop Insurance Demand with Panel Data." *American Journal of Agricultural Economics* 78: 439-447.

Coutant, S. "Implied Risk Aversion in Option Prices Using Hermite Polynomials." Manuscript presented at Workshop on Estimating and Interpreting Probability Density Functions, Bank of International Settlements, Basel, Switzerland (June 14).

Cox, J., and S. Ross. 1976. "The Valuation of Options for Alternative Stochastic Processes." Journal of Financial Economics 3: 145-166.

Engle, R. 1982. "Autoregressive Conditional Heteroskedasticity with Estimates of the Variance of UK Inflation." *Econometrica* 50: 987-1008.

Engle, R., and G. Gonzalez-Rivera. 1991. "Semiparametric ARCH Models." *Journal of Business and Economic Statistics* 9: 345-359.

Everitt, B., and D.J. Hand. 1981. *Finite Mixture Distributions*. New York: Chapman and Hall.

Fackler, P.L. 1986. "Futures Price Volatility: Modeling Nonconstant Variance." Paper presented at the 1986 AAEA Meetings, Reno, Nevada.

___. 1991. "Modeling Interdependence: An Approach to Simulation and Elicitation." *American Journal of Agricultural Economics* 73: 1091-1097.

Fama, E. 1965. "The Behavior of Stock Market Prices." *Journal of Business* 38: 34-105.

Feng, Z.D., C.E. McCulloch. 1994. "On the Likelihood Ratio Test Statistic for the Number of Components in a Normal Mixture." *Biometrics* 50: 1158-1162.

___. 1996. "Using Bootstrap Likelihood Ratios in Finite Mixture Models." *Journal of the Royal Statistical Society* (Series B) 58: 609-617.

Frees, E.W., and E.A. Valdez. 1998. "Understanding Relationships Using Copulas." *North American Actuarial Journal* 2: 1-25.

Gallagher, P. 1987. "U.S. Soybean Yields: Estimation and Forecasting with Nonsymmetric Disturbances." *American Journal of Agricultural Economics* 69: 798-803.

Goodwin, B.K., and T. Kastens. 1993. "Adverse Selection, Disaster Relief, and the Demand for Insurance." Unpublished manuscript, Kansas State University.

Goodwin, B.K., and A.P. Ker. 1998. "Nonparametric Estimation of Crop Yield Distributions: Implications for Rating Group-Risk (GRP) Crop Insurance Contracts." *American Journal of Agricultural Economics* 80: 139-153.

Goodwin, B.K., M.C. Roberts, and K.H. Coble. 2000. "Measurement of Price Risk in Revenue Insurance: Implications of Distributional Assumptions." *Journal of Agricultural and Resource Economics* 25: 195-214.

Goodwin, B.K., and R. Schnepf. 2000. "Determinants of Endogenous Price Risk in Corn and Wheat Futures Markets." *Journal of Futures Markets* 20: 753-774.

Hendry, D.F. 1995. *Dynamic Econometrics*. New York: Oxford University Press.

Hennessy, D.A., B.A. Babcock, and D.J. Hayes. 1997. "Budgetary and Producer Welfare Effects of Revenue Insurance." *American Journal of Agricultural Economics* 79: 1024-1034.

Hennessy, D.A., and T.I. Wahl. 1996. "The Effects of Decision Making on Futures Price Variability." *American Journal of Agricultural Economics* 78: 591-603.

Hjort, N.L., and I.K. Glad. 1995. "Nonparametric Density Estimation with a Parametric Start." *Annals of Statistics* 23: 882-904.

Hjort, N.L., and M.C. Jones. 1996. "Locally Parametric Nonparametric Density Estimation." *Annals of Statistics* 24: 1619-1647.

Horowitz, J. 1993. "Semiparametric Estimation of a Work Trip Choice Model." *Journal of Econometrics* 58: 49-70.

Hull, J., and A. White. 1987. "The Pricing of Options on Assets with Stochastic Volatilities." *Journal of Finance* 42: 281-300.

Johnson, M., and A. Tenenbein. 1981. "A Bivariate Distribution Family with Specified Marginals." *Journal of the American Statistical Association* 76: 198-201.

Jones, M.C., O. Linton, and J.P. Nielsen. "A Simple Bias Reduction Method for Density Estimation." *Biometrika* 82: 327-338.

Jorion, P. 1988. "On Jump Processes in the Foreign Exchange and Stock Markets." *Review of Financial Studies* 1: 427-445.

Just, R.E., and Q. Weninger. 1999. "Are Crop Yields Normally Distributed?" *American Journal of Agricultural Economics* 81: 287-304.

Kenkel, P.I., J.C. Busby, and J.R. Skees. 1991. "A Comparison of Candidate Probability Distributions for Historical Yield Distributions." Presented paper at SAEA meetings, Fort Worth, February.

Kenyon, D., K. Kling, J. Jordan, W. Seale, and N. McCabe. 1987. "Factors Affecting Agricultural Futures Price Variance." *Journal of Futures Markets* 7: 73-91.

Ker, A.P. 1998. "Empirical Bayes Nonparametric Kernel Density Estimation." Unpublished manuscript, Department of Agricultural and Resource Economics, University of Arizona.

___. 2000. "Estimating Crop Insurance Rates with Mixture Distribution and Seminonparametric Maximum Likelihood Methods." Unpublished manuscript, University of Arizona.

___. 2001a. "Nonparametric Estimation of Possibly Similar Densities." Unpublished manuscript, University of Arizona.

___. 2001b. "Nonparametric Estimation of Possibly Similar Densities with Application to Rating Crop Insurance Contracts." Unpublished manuscript, University of Arizona.

Ker, A.P., and K. Coble. 1998. "On Choosing a Base Coverage Level for Multiple Peril Crop Insurance Contracts." *Journal of Agricultural and Resource Economics* 23: 427-444.

___. 2001. "Modeling Yields." Unpublished manuscript, University of Arizona.

Ker, A.P., and B.K. Goodwin. 2000. "Nonparametric Estimation of Crop Insurance Rates Revisited." *American Journal of Agricultural Economics* 83: 463-478.

Ker, A.P., and P. McGowan. 2000. "Weather Based Adverse Selection: The Private Insurance Company Perspective." *Journal of Agricultural and Resource Economics* 25: 386-410.

Knight, F.H. 1921. *Risk, Uncertainty, and Profit.* Boston: Houghton Mifflin Co. Available online from http://www.econlib.org/library/Knight/knRUPtoc.html. (Accessed June 26, 2001.)

Kroner, K., and V. Ng. 1998. "Multivariate Garch Modelling of Asset Returns." In R. Jarrow, ed., *Volatility: New Estimation Techniques for Pricing Derivatives.*

Li, D.X., and H.J. Turtle. 2000. "Semiparametric ARCH Models: An Estimating Function Approach." *Journal of Business and Economic Statistics* 18: 174-186.

Linton, O., and B. Perron. 1999. "The Shape of the Risk Premium: Evidence from a Semiparametric Garch Model." Cahier 9911, Universite de Montreal, Faculte des arts et des sciences.

MacBeth, J., and L. Merville. 1980. "Tests of the Black-Scholes and Cox Call Option Valuation Models." *Journal of Finance* 35: 285-303.

Marron, J.S., and M.P. Wand. 1992. "Exact Mean Integrated Squared Error." *The Annals of Statistics* 20: 712-736.

McLachlan. 1987. "On Bootstrapping the Likelihood Ratio Test Statistic for the Number of Components in a Normal Mixture." *Journal of the Royal Statistical Society* (Series C: Applied Statistics) 36: 318-324.

Melick, W.R., and C.P. Thomas. 1997. "Recovering an Asset's Implied PDF from Option Prices: An Application to Crude Oil During the Gulf Crisis." *Journal of Financial and Quantitative Analysis* 32: 91-115.

Merton, R.C. 1973. "Theory of Rational Option Pricing." *Bell Journal of Economics and Management Science* 4: 141-183

___. 1976. "Option Pricing When Underlying Stock Returns Are Discontinuous." *Journal of Financial Economics* 3: 125-144.

Mizrach, B. 1996. "Did Option Prices Predict the ERM Crises?" Rutgers University Economics Working Paper No. 1996-10.

Moss, C.B., and J.S. Shonkwiler. 1993. "Estimating Yield Distributions with a Stochastic Trend and Nonnormal Errors." *American Journal of Agricultural Economics* 75: 1056-1062.

Nelson, C.H. 1990. "The Influence of Distribution Assumptions on the Calculation of Crop Insurance Premia." *North Central Journal of Agricultural Economics* 12: 71-78.

Nelson, C.H., and P.V. Preckel. 1989. "The Conditional Beta Distribution As a Stochastic Production Function." *American Journal of Agricultural Economics* 71: 370-378.

Nelson, D. 1990. "Stationarity and Persistence in the GARCH(1,1) Model." *Econometric Theory* 6: 318-334.

___. 1991. "Conditional Heteroscedasticity in Asset Returns: A New Approach." *Econometrica* 59: 347-370.

Neuhaus, H. 1995. "The Information Content of Derivatives for Monetary Policy." Discussion Paper No. 3/95, Deutsche Bundesbank.

Olkin, I., and C.H. Spiegelman. 1987. "A Semiparametric Approach to Density Estimation." *Journal of the American Statistical Association* 82: 858-865.

Ritchey, R.J. 1990. "Call Option Values for Discrete Normal Mixtures." *Journal of Financial Research* 13: 285-296.

Roberts, M. 2000. "A Mixture of Normals GARCH Model." Unpublished manuscript, North Carolina State University.

Rubinstein, M. 1985. "Nonparametric Tests of Alternative Option Pricing Models Using All Reported Trades and Quotes on the 30 Most Active CBOE Option Classes from August 23, 1976, through August 31, 1978." *Journal of Finance*, pp. 455-480.

Sherrick, B.J., P. Garcia, and V. Tirupattur. 1996. "Recovering Probabilistic Information from Option Markets: Tests of Distributional Assumptions." *Journal of Futures Markets* 16: 545-560.

Shimko, D. 1993. "Bounds of Probability." *Risk* 6: 33-37.

Skees, J.R., J.R. Black, and B.J. Barnett. 1997. "Designing and Rating an Area Yield Insurance Contract." *American Journal of Agricultural Economics* 79: 430-438.

Söderlind, P., and L.E. Svensson. 1997. "New Techniques to Extract Market Expectations from Financial Instruments." NBER Working Paper No. 5877 (January).

Streeter, D., and W. Tomek. 1992. "Variability in Soybean Futures Prices: An Integrated Framework." *Journal of Futures Markets* 12: 705-728.

Taylor, C.R. 1990. "Two Practical Procedures for Estimating Multivariate Nonnormal Probability Functions." *American Journal of Agricultural Economics* 72: 210-217.

Turvey, C.G., and C. Zhao. 1993. "Parametric and Nonparametric Crop Yield Distributions and their Effects on All-Risk Crop Insurance Premiums." Working Paper, Department of Agricultural Economics and Business, University of Guelph (May).

Yatchew, A. 1998. "Nonparametric Regression Techniques in Economics." *Journal of Economic Literature* 36: 669-721.

Chapter 15

AGRICULTURAL TECHNOLOGY AND RISK*

Michele C. Marra and Gerald A. Carlson
North Carolina State University

INTRODUCTION

On the face of it, it would seem that agricultural technology might be thought of and modeled in the same way we would other agricultural inputs. As with any other productive input, such as labor, we can think of a derived demand for a particular technology that changes with output price, technology price, and the technology's marginal product. So, why devote a separate chapter to technology? Webster's New American Dictionary (1995) defines technology as "a manner of accomplishing a task using special knowledge of a mechanical or scientific subject." There are two parts of this definition that support the tendency to differentiate a new technology from other agricultural inputs or older technologies: "special knowledge" and "scientific." Special knowledge implies an initial period where the knowledge has yet to be acquired. This period may be characterized by uncertainty about how the technology works and how to apply it in a particular situation. As the special knowledge is acquired, experimentation and learning take place. The special knowledge may not be applicable to any other manner of accomplishing the task, so that if a newer technology is introduced, the learning must take place again. This implies an initial opportunity cost of adopting a new technology that is higher than using an older, conventional input. The "scientific" part of the definition, especially in the case of agricultural technologies, implies that the development of the technology probably takes place off the farm. It may take some time between the technology development and dissemination of sufficient information about it to be seriously considered by the farmer. Ultimate usefulness of the technology on the farm depends on its applicability to the unique growing conditions and other characteristics of the farm and the farmer. Again, this implies some additional level of initial uncertainty. A useful distinction, then, between conventional agricultural inputs and tech-

* We thank Mitch Renkow and the participants in the SERA-IEG 31 annual meeting at Gulf Shores, Alabama (especially Paul Mitchell), for valuable comments and suggestions. We accept responsibility for any remaining errors.

nology might be the degree of "newness" and special knowledge requirements of the input or production technique. Therefore, we leave aside conventional input choices and land allocation to different crops (except *newly* developed cultivars or varieties).

On a more practical level, we want to study technology to be able to provide better predictions of agricultural supply for policymakers. We want to understand how decision makers' adoption decisions are made not only for scholarly purposes, but also to improve the efficiency of the information dissemination process. Technology-producing industries and public organizations (e.g., universities and international research groups) also benefit from a clearer understanding of the technology adoption and diffusion process. Understanding constraints to rapid and more extensive adoption of new technologies is important in formulating subsidy and research priorities, particularly in developing countries.

How Have We Studied Technology?

There have been several approaches to the study of technology in agriculture, some by agricultural economists, some by sociologists or geographers. On a macro-level, Ruttan (1996) and others have studied how repeated adoption of new technologies translates into changes in supply over time, or technological change. They developed the notion of new technology "embodied" in a particular category of inputs, such as seeds or equipment, or "disembodied," essentially affecting all productive inputs in the same manner (e.g., human capital). Griliches (1957) also studied the aggregate economic behavior associated with hybrid corn in his seminal article characterizing the diffusion path of a new technology across areas and over time. Geographers have studied the aggregate spatial aspects of the diffusion of new technologies, but have not generally considered economic characteristics, the technology, or the adopters as explanatory variables (Rogers 1995). Sociologists have studied the diffusion of technologies as they have affected the welfare of communities as well as public and private attitudes toward new technologies such as biotechnology (Rogers 1995).

Rosenberg (1976), in a review of research on adoption of innovations, suggested that an important part of the process is interpreting and waiting for technical improvements:

> "Practical businessmen tend to remember what social scientists [presumably economists are the exception] often forget: that the very rapidity of the overall pace of technological improvement may make a postponed adoption decision privately (and perhaps even socially) optimal. ...decisions to postpone the adoption of an innovation are often based upon well-founded and insufficiently appreciated expectations concerning the future time-flow of further improvements. Even the most widely accepted justification for

> postponement, the elimination of conspicuous but not overwhelmingly serious technical difficulties, or 'bugs', can reasonably be interpreted as merely a special case of expectations of future technological improvement" (pp. 533-534).

This points to one of three observations that have given rise to explicit consideration of risk and information in the study of the microeconomics of technology adoption and diffusion, namely that adoption of, or investment in, potentially profitable new technologies has, in many cases, lagged behind the pace predicted by neoclassical theory. Partial adoption of new technologies by individual decision makers (either as selecting only some components or applying the technology package to only part of the eligible farming area) is another empirical observation that has given rise to renewed attempts to develop alternative hypotheses to explain adoption behavior. The third is an observed reluctance to disinvest in an old technology in favor of a superior new one.

Chapter Road Map

This chapter is not meant to be a complete review of the technology adoption/investment under uncertainty literature. Rather, it focuses on the main developments in the theory of technology choices in the presence of uncertainty or risk, tests of these theories in both developed country and developing country agricultural settings, and some illustrative empirical examples. This section contains the introductory material. In the next section we discuss how the literature dealing with divisible technologies and associated inputs and risk has progressed using research related to three types of divisible technologies as representative of the literature. We consider also the farm scale effects on adoption and farmer learning about modern technologies in this section. Then we review recent developments in the long-term investment literature and discuss how they relate to agricultural technology and risk. We take up the topics of asset fixity, sunk costs, irreversiblity, and hysteresis in this section. The fourth section of the chapter summarizes what we have learned and outlines what we think are the promising future avenues for technology and risk research.

We acknowledge that the manner in which we decided to divide this chapter into its sections is, in many ways, arbitrary. There is really a continuum of technology characteristics, such as "newness," useful life, or degree of sunk costs or fixity associated with technology adoption/investment, not an arbitrary "divisible" and "long-term" distinction. So far, the economic modeling has differentiated between the two to a large degree, so for expositional purposes, it seems reasonable to make the distinction. We hope that, on balance, the chapter divisions help the reader's progress through the literature.

DIVISIBLE TECHNOLOGIES AND RISK

Technologies embodied in the variable inputs or new production techniques such as integrated pest management (IPM) are usually considered to be divisible. There is usually no concern about resale of assets, so the multi-period interactions of the kind prevalent in the real option or investment literature considered below are not present. Other uncertainties associated with multi-period decisions can apply, however. (See Antle and Hatchett 1986 and Zacharias and Grube 1986 for treatment of these issues for new fertilizer and pesticide technologies, respectively.)

Economic aspects of these technologies without special attention to risk have been evaluated at various points in time. Perrin and Winkleman (1976), Feder, Just, and Zilberman (1985), and Byerlee (1996) review modern variety technologies, while reviews of pest control technologies and pesticides can be found in Carlson and Wetzstein (1993) and Fernandez-Cornejo, Jans, and Smith (1998). Adoption of transgenic crops (modern varieties created through use of biotechnology) is reviewed by Marra (2001), and many theoretical and empirical studies are reported in the proceedings of the annual conferences of the International Consortium on Agricultural Biotechnology Research (Evenson et al. 1999, 2000, Santaniello et al. 2000).

Not all issues, models, and results involving agricultural technology and risk can be considered here. We will concentrate on empirical and theoretical developments that occur when new, commercial use of technologies is imminent or just beginning. As we stated in the introduction to this chapter, this is the stage of development and implementation of innovations when there are frequently high levels of uncertainty about the relative profitability of the new technology. We also narrow our focus to three technology types – modern conventional crop varieties (MCVs), transgenic crops (TCs), and new pest management technologies (PMs). There are, of course, other divisible technologies that have been the focus of adoption and diffusion studies that include aspects of risk (there is a separate chapter in this volume on precision farming, for example), but these three can serve to illustrate the important developments in this line of research, while avoiding too much repetition.

Sources of Technology-Related Risk

There are many sources of uncertainty for farm operators when allocating land and resources to any technology, and it may be helpful to enumerate these as they relate to the adoption of modern conventional crop varieties, transgenic crops, and innovations in pesticides and pest management.

***Modern Conventional Crop Varieties* (MCVs).** Usually, adoption of these technologies requires more purchased inputs such as fertilizer and water, which means that there can be more resources at stake. In some cases, MCVs when first available are more agroclimatic-specific than local varieties, and they will have lower returns in marginal soil, temperature, and pest areas. The modern rice varieties did not have insect and disease resistance built in initially, as did the modern wheat varieties developed for some regions that contained genes for rust resistance.

Transgenic Crops **(TCs).** For transgenic crops with herbicide tolerance such as Roundup Ready® (RR) soybeans, there seem to be uncertainties about relative profitability compared to conventional weed control systems. Initially, they have had lower yields in some varieties and locations. For insect-resistant varieties the uncertainty comes primarily from variable pest infestations (for example Bt corn faces variable European corn-borer densities that are linked to location, insect egg-laying weather, and time of planting factors). Bt cotton uncertainty depends upon insect infestation levels, the mixture of pest species, and whether the pests in a given location are resistant to conventional insecticides. Marketing costs and prices of transgenic crops are also random variables, especially with identity preservation requirements and uncertain premiums for non-TC products.

New Pesticides and Pest Management Technologies **(PMs).** The major unknowns at pest management decision time for farmers are pest density, pesticide efficacy, damage per pest, pest type, pest monitoring efficacy, crop yield, and crop value (price, quality) without pest damage. Weather related to application timing and duration of a pest management activity is sometimes critical. There are policy risks such as cancellation or new restrictions on methods of use for particular pesticides. There are also long-run uncertainties such as entry of new pests, buildup of resistance to particular pesticides, and the timing of the discovery and approval of new pesticide products including pesticides contained in biotech crops. In addition there can be uncertainty about future revelation of new health and environmental risks, both on and off the farm.

Technology Policies and Risk

Specifying the policy issues surrounding technology adoption and risk also may help narrow our focus to the research purposes and findings that are most critical. The technology evaluations that can assist policymakers include those that help in identifying, quantifying, and understanding the following:

- *availability* of the technology by geographical area, which involves technology supply uncertainties including infrastructure and government policies on permitting technologies to be tested and sold;

- constraints to initial *adoption rates* such as lack of credit, farm scale, information gaps, risk aversion, or mean income effects; and
- factors that explain *ceiling* or equilibrium adoption levels by area, which include the same factors that affect adoption rates.

Griliches (1957) first identified the availability, rate of adoption, and technology use ceiling descriptors for aggregate technology studies in agriculture. These correspond to his origin, slope, and ceiling parameters of the S-shaped curves for hybrid corn adoption over time. Griliches and many other researchers of aggregate technology use (Dixon 1980, Knudson 1991, Mundlak 2000) have not included risk as a focal point. However, others (Hazell 1984, Anderson and Hazell 1989, and Traxler et al. 1995) have attempted to consider risk as a factor in aggregate studies since there is disagreement on the change in income variability that has or has not occurred as a result of adoption of the MCVs.

The major policy issues in developing countries in enhancing the use of modern seed varieties have been on questions of deployment of scarce public research funds, income distribution effects (Renkow 1993), and use of credit, information, and other input (e.g., fertilizer and seed) subsidies. Recently, sustainability and risks associated with environmental damage related to agricultural technology have become more prominent. Concern over the proliferation of confined livestock feeding technologies in the U.S. is one example of this. In the U.S. and developed countries there is much more attention paid to health, safety, and environmental damage risks. There is little attention paid to input subsidies or credit limits, but information dissemination (extension) and research funds allocation under uncertainty are major issues. We use pest management and transgenic crops to illustrate risk research in developed countries and adoption of modern conventional varieties to illustrate progress in developing countries.

Observable Features of Technologies, Research Questions, and Findings

A major difficulty in technical change research by economists has been defining what it is. Technology is often embodied in conventional agricultural inputs, it sometimes involves multiple components, and frequently it will modify the crop production environment.

The observable, decision-theoretic features of MCVs have been:

- the use or non use of MCVs on a particular land parcel, usually measured at a particular time (*adoption*);
- *partial adoption* of an MCV or some proportion of available land area; and
- use of the MCV in combination with other inputs on the same decision unit (*component adoption*).

Two often-posed research questions (theoretical or empirical) related to observations on adoption are why there is not full adoption on target farms and why the technology does not fit particular farm areas. A related question is why there is adoption delay by some farmers. The adoption of components of technology "packages" such as an MCV or complementary inputs such as fertilizer and irrigation rather than adoption of the package has also been investigated. Risk aversion or increased income variability with use of the modern varieties is a hypothesis to be compared to others (Feder 1982, Leathers and Smale 1991, Smale, Just, and Leathers 1994, and Foster and Rosenzweig 1995). For micro-studies the analysis is usually based on a set of cross-section or time-series, cross-section choices by individual farmers. For aggregate studies, yield variation over time and across modern and traditional varieties is examined.

Modern Conventional Varieties. Modern variety studies at the aggregate level (diffusion of a technology for an area) are not too common today. However, serious policy issues remain. One critical policy issue is whether the release of the MCVs since the 1960s has increased or decreased income variability in the long run. Using state- or country-level data, Hazell (1984, 1989) examined yield variability about time trends for major cereals in most countries of the world, and concludes that there was higher variability during the 1970s than during the 1960s. He hypothesizes that this is due to fewer cereal varieties being grown following the introduction of the modern varieties so that there is higher positive correlation across fields and regions. The researchers reporting in Anderson and Hazell (1989) present arguments and some evidence that the higher regional correlation might be due to factors other than release of modern varieties, such as more homogeneous production practices, more covariant patterns of rainfall (especially when examined over short time intervals), more irrigated area leading to synchronized production across locations, more covariant shortages of complementary inputs, movement of cereals to more marginal environments, and higher correlation in cereal prices in the 1970s than in the 1960s. The policy question of the long-term variability effects of the initial release of MCVs has not been resolved and may not be as important as the nature of mean and variability effects of MCVs after they have been available for some time.

Another study that sheds light on the long-term yield variability of modern varieties is the Mexican wheat study of Traxler et al. (1995). They use the Just and Pope (1979) production model to estimate mean yield and variance of yield equations utilizing experimental plot data. They include vintages of wheat varieties that were released over the 1955-1985 period, some of which were widely planted worldwide. They find that mean yield increased from 1955 to 1980, and then reached a plateau. Variance of yield rises until about the 1970 vintages and then declines steadily (Figure 1 shows these effects over time). This tradeoff is common in breeding where each variance-reducing factor added to a breeding line will call for some sacrifice in mean yield (Carlson and Main 1976). Heisey et al. (1997) have estimated mean-genetic diversity trade-off for wheat growers of the Punjab of Pakistan. They find considerable costs of growing the wheat cultivar mixes that are recommended for disease control. This is partly due to the partial public good nature of plant diseases for farmers who share the disease. (See Figure 2 for the average, actual portfolio, potential, and recommended tradeoff curves.)

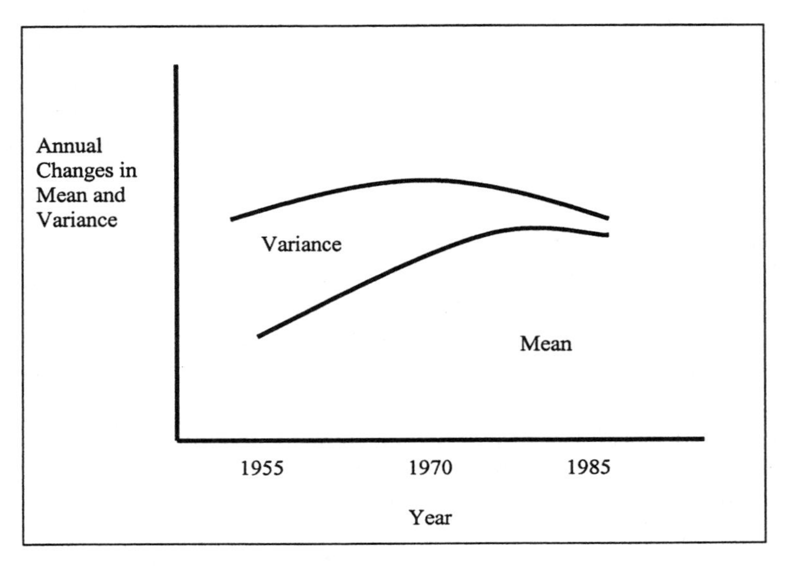

Figure 1. Mean Yield and Yield Variability Changes from Wheat Cultivars of Various Vintages
Source: Traxler et al. 1995

There is a growing number of farm-level adoption studies for modern varieties that include a risk framework. Two successful studies are reviewed here, both based on extensive data sets collected in a developing country. The Foster and Rosenzweig (1995) study was based on a 3-year panel of data on about 1,500 farmers from 101 Indian villages. The profit functions and MCV adoption equations show that there is both learning from doing and learning from neighbors in the same village who have planted the modern rice.

Increased area planting and experience increase adoption. There is evidence of free-riding from neighbors, especially if the neighbor is predicted to be accurate in assessing the technology. The prediction of accuracy is based on neighbors' asset holdings (farm animals, irrigation assets, and farm equipment). The simulations based on estimated coefficients can give adoption delays of about 2 years for small (as measured by asset holdings) farmers with larger farmer neighbors. (Figure 3 shows the nature of their estimated adoption curves with rich and poor neighbors.)

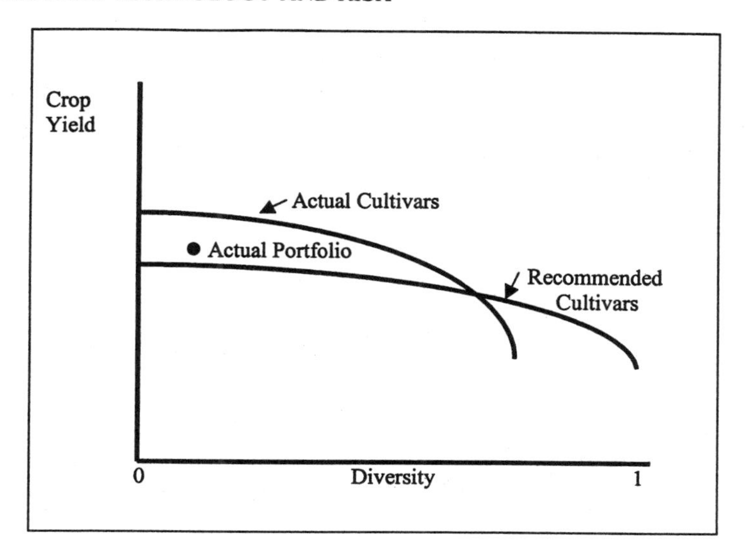

Figure 2. Diversity – Yield Frontiers, Actual and Recommended Cultivars
Source: Compiled from Heisey et al. 1997

Another of the major issues is explaining why ceiling levels of adoption of modern varieties are "relatively low" in some regions. The four competing hypotheses according to Smale, Just, and Leathers (1994) are (1) input fixity for fertilizer measured by asset levels and predicted entry in a credit club, (2) risk aversion (portfolio) measured by ratios of modern to local variety yield variances, (3) learning or experimentation measured by years of experience with modern varieties, and (4) safety-first preferences expressed by family consumption requirements relative to land area available for maize production. The authors estimate the full model and then perform nested, likelihood ratio tests on 2- and 3-way groups of the four factors using maize planting data from 420 Malawi farmers. Their main finding is that the full model is more likely to explain land allocation patterns than any single approach or combination of two or three approaches. This study implies that it is difficult to have meaningful research in the MCV adoption area unless studies allow for competing hypotheses, use extensive data, and apply appropriate econometrics methods.

Transgenic Crops. There are fewer studies for TCs, but adoption, partial adoption, and component adoption in the form of adoption with changes in pesticide use have been the observable variables of interest. There are empirical studies involving information and risk hypotheses for Bt corn (Hyde et al. 1999, Darr and Chern 2000), Bt cotton (Hubbell, Marra, and Carlson 2000, Marra, Hubbell, and Carlson 2001), and herbicide-tolerant corn and soybeans (Darr and Chern 2000).

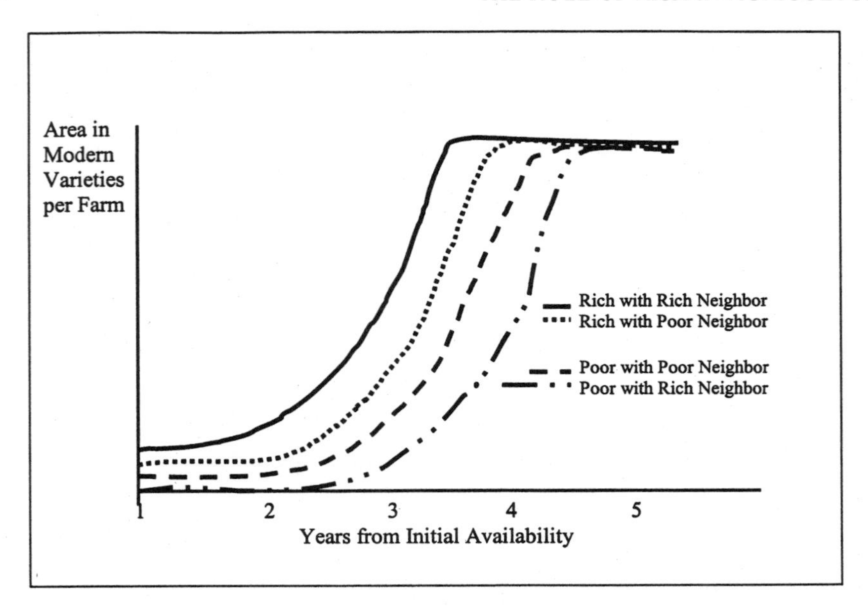

Figure 3. Predicted Adoption of Modern Varieties as a Function of Own and Neighbors' Asset Levels
Source: Compiled from Foster and Rosenzweig 1995

Most of the studies we review here are based on farmer surveys of actual adoption of the seeds. However, the simulation study by Hyde et al. (1999) is based on data from expert opinions concerning cornborer infestations in Indiana, and the bST milk study by Saha, Love, and Schwart (1994) is based on an ex ante farmer survey. The Saha study found no influence of either risk attitudes or risk perceptions on the prospective adoption decisions of 150 Texas farmers. However, the expressed intentions on *degree* of adoption increase with herd size and these same risk factors.

The Hyde et al. (1999) study found that mean profitability estimates varied systematically with European cornborer (ECB) infestation levels. They found that high levels of absolute risk aversion could make Bt corn attractive if probabilities of ECB infestation of zero are 0.6 or lower. It is unclear why the researchers did not examine long-run data that is available on relative frequency of ECB infestations in Indiana.

Marra, Hubbell, and Carlson (2001) examined the effect of information quality and depreciation in the old technology in explaining patterns of Bt cotton adoption over two years for a sample of southeastern U.S. cotton growers. They find that information from nearby sources on yield and cost changes can be outweighed by more distant estimates with more precision because the latter are estimated from larger samples. Popularity of adoption as measured at the state level does have a weak influence on adoption by individual farmers in the second year. Also, first-year adoption is increased by findings of profitability from experimental plot studies available from the industry at the time of seed purchase.

Darr and Chern (2000) surveyed Ohio growers about their adoption of Bt corn and RR soybeans in 1998-2000. There are consistent findings of higher adoption rates in 2000 for farmers with more acres and for those in higher ECB infestation areas. They found higher probabilities and intensities of adoption with farmers who routinely "sell grain forward." Farmers who said they were "concerned about price premiums" for conventional corn had lower adoption intensities. None of these "risk aversion measures" were found to be associated with changes in the adoption or adoption intensity of RR soybeans. These indirect measures of risk aversion may change over time, or be correlated with unobservable factors. This can yield unstable and biased parameter estimates in these adoption models.

Alexander, Fernandez-Cornejo, and Goodhue (2000) surveyed Iowa growers about their corn and soybean biotech crops. Only cross tabulations are available to date from the study, but they do show that those farmers who planted neither Bt corn nor RR soybeans in either 1999 or 2000 were older and had farms smaller in size (gross income and acres) than either the continued adopters or previous adopters who have since disadopted. The RR partial adopters seem to have more formal education than do full adopters. Disadopters tend to be like adopters except they have smaller farms and less human capital. More work is needed to relate this work to the random factors that affect profitability and variability in income from ECB and weed damage.

Pest Management Technologies. Studies related to PM adoption have focused on rates of use, efficient indicators for use (economic thresholds), information, human capital, and various pesticide substitutes (IPM programs, insurance, labor, equipment). There has been much less emphasis on partial adoption. There has been considerable attention paid to dynamic models of pesticide use as a way of dealing with uncertainties that unfold over time. Finally, some attention has been given to the general question of the relationship of pesticide use to income variability (Carlson 1984, Pannell 1991, and Horowitz and Lichtenberg 1993, 1994).

Pesticide productivity studies with respect to risk have been common for some time (Hildebrant 1960 and Carlson 1970). New pesticide products become available, new pests arrive in particular areas, and evolving pest biotypes (strains resistant to particular pesticides, crop rotations, or cultivars) contribute to an environment in which farmer decisions are made under both dynamic and stochastic uncertainty.

In developing countries, pesticide adoption is relatively new and its contribution to both mean return and variability of returns fits the mold of a new technology, whose net benefits are directly influenced by sporadic and changing pest pressure, weather, and variable pesticide prices and availability. Some of the modern variety adoption work has contributed to our understanding of the effect of pesticides, and pest monitoring (IPM) on the expected value and variance of income (Feder and Slade 1984, Herdt, Castillo, and Jayasuriya 1984, and Flinn and Garrity 1989). We see partial adoption and sequential adoption of pesticides as components of packages with new varieties (Byerlee and Hesse de Polanco 1986).

Some of the most significant biological technologies for pest management have come from the increased disease and insect resistance that has been incorporated into modern seed varieties. The main risk feature is catastrophic loss if some major varieties lose their ability to control key pests. This is most prominent for wheat rusts (Carlson and Main 1976, Heisey et al. 1997), but also is prevalent in nematode control (Zacharias and Grube 1986). Understanding which aspects of genetic diversity will help prevent resistance development is critical for long-term disease management. Smale et al. (1998) found that in rain-fed wheat districts in Pakistan the genealogical characteristics of varieties are associated positively with mean yield and negatively with yield variance, but the same is not true for irrigated wheat where regional concentration among varieties was most critical.

In developed country agriculture, economists have made contributions examining the value of information (Carlson 1970 and Swinton and King 1994) and incorporating risk aversion in normative models for farmers facing both single and multiple pest species (Musser, Tew, and Epperson 1981, Moffitt et al. 1986, and Cochran, Robinson, and Lodwick 1985). In most cases the recommended control strategies are risk reducing (Carlson 1984 and Table 1). Feder (1979) developed a model that incorporated risk aversion and multiple sources of variability (pest numbers, N, pesticide effectiveness, k, and damage per pest, d). However, there are frequently more random components than these three. Carlson (1984), in a model with land allocation across crops and with multiple pest species, considers pest-free yield (Y), crop prices correlated with pest damage, direct damage of pesticides, and unmeasured pest control as other sources of uncertainty. It is quite possible for random, pest-free yield, positive pest damage-crop price correlation, and direct damage to crops from pesticides to result in variability increases from pesticide use (Horowitz and Lichtenberg 1994 and Pannell 1991). Lazarus and Swanson (1983) show this for insecticide use in combination with rotation of soybeans and corn for corn rootworm control. Pannell (1991, 2000) and Swinton and King (1994) find that optimum weed control strategies can give small income variability increases or decreases with more herbicide use. Many of the rain-fed crops of the world (including many semi-arid cereals in both developed and developing countries) have highly variable, pest-free crop values relative to pesticide treatment costs. In this setting adding pesticide treatments can increase income variability.

There have been several estimates of pesticide use behavior under uncertainty at the farm level (Burrows 1983, Pingali and Carlson 1985, Antle 1988, Horowitz and Lichtenberg 1993, and Smith and Goodwin 1996). Burrows finds that IPM adoption is simultaneously determined with pesticide use and

Table 1. Sources of Risk, Utility Formulation, and Evidence on Marginal Risk Effects of Pest Control Inputs

	Measured Sources of Risk[a]	Utility Formulation(s)[b]	Crop(s)	Marginal Risk Effects Found	
				Pesticides	Other
Carlson (1970)	N, K	EV, DA	peaches	-	monitoring (-)
Carlson (1979)	N	EV, SD	cotton		monitoring (-)
Cochran, Robinson, and Lodwick (1985)	$\underline{N}$ Y	EV, SD	apples	-	monitoring (?)
Feder (1979)	N d k	M-P-S	none	-	
Hall (1977)	N	EV	citrus, cotton		monitoring (-)
Lazarus and Swanson (1983)	N d P_y Y $P_s Y_s$	EV	corn, soybeans	+	rotation (-)
Moffit et al. (1986)	N_1 N_2	EV, SD	soybeans		monitoring (-)
Musser, Tew, and Epperson (1981)	$\underline{N}$	EV, SD	vegetables	-	monitoring (?)
Burrows (1983)	N	A	cotton, citrus	-	IPM (-)
Carlson (1984)	$\underline{N}$ $\underline{d}$ $\underline{k}$ $\underline{Y}$ $\underline{P}$	EV	none	- +	
Pingali and Carlson (1985)	$\underline{N}$ $\underline{k}$	EV, PY	apples	-	monitoring (-), pruning (-)
Antle (1988)	$\underline{N}$ k	EV, CARA (3rd)	tomatoes	-	IPM (-)
Flinn and Garrity (1989)	$\underline{N}$	EV	rice	-	IPM (-)
Leathers and Quiggin (1991)	N	DARA, CARA, MS, LS	none	+ -	
Pannell (1991)	Y N	EV	wheat	+ -	monitoring (-)
Horowitz and Lichtenberg (1993, 1994)	Ndk Y P	EV	corn	+ -	monitoring (+, -), insurance (-)
Smith and Goodwin (1996)	$\underline{N}$ Y	R	wheat	-	insurance (-)

[a] N = pest density ($\underline{N}$ refers to multiple pest species, N_1 is scout observation, N_2 is area forecast), d = damage per pest (N), k = percent pest reduction, P_y = crop prices, Y_s = yield of substitute crop, Y = pest-free yield, P_s = substitute crop price, Ndk = damage risk. [b] EV = expected profit-variance of profit, SD = stochastic dominance analysis, M-P-S = mean-preserving-spread, DA = disaster avoidance, PY = potential revenue, LS = location and spread, MS = monotone spread 3rd = third moment, A = crop acres/total farm size, R=farmer-reported risk preferences.

both are associated with lower income variability for California citrus and cotton growers. Pingali and Carlson include insect and disease damage variability along with crop scale as risk measures in demand equations for a panel of apple growers. They find that predicted variance of insect damage increases insecticide use. The scale of the crop value at risk of pest damage (potential yield times expected price) is significant in increasing both insecticide and fungicide use. Antle includes the first three moments of insect damage to tomatoes to estimate the insecticide, IPM, and other input use behavior of tomato growers. He finds that increased insecticide use is associated with lower second and third moments of crop damage. Use of insecticides in an IPM program is more risk reducing. "These results suggest that the IPM technology does enhance the effectiveness of pesticide use, and the increased effectiveness comes about primarily through the risk attributes of the technology and not through an increase in the mean marginal product of insecticides" (Antle 1988, p. 109).

Several other demand studies for pesticides have tried to find the effect of pesticide use on the demand for crop insurance. Generally, these expenditures are thought to be substitute methods for risk reduction. However, as indicated above pesticides can be risk increasing, and Horowitz and Lichtenberg (1993), using a recursive model of input purchases, find higher levels of use of pesticides among those corn and soybean farmers who buy crop insurance. Smith and Goodwin (1996) cannot reject simultaneity in insurance and pesticide purchases for a set of Kansas wheat farmers, and find that crop insurance and pesticide use are substitutes.

Information and Learning

One of the major avenues of research in technology adoption has been trying to understand which model or models of learning best fit the data. Both the gathering of information and the processing of information take resources and time. Bayesian learning postulates, which combine prior information with experimental or new information by weighting each according to their perceived precision (Stoneman 1981, Feder and O'Mara 1982, and Lindner, Fischer, and Pardey 1979), applied the Bayesian model to the time of initiation of adoption. The passive learning by observing the results of own-farm planting of the new technology is thought of as a version of "learning by doing." Active testing of new technologies by farmers is a task that takes time and resources that is consistent with partial adoption (Feder and Slade 1984) and sequential adoption of components (Leathers and Smale 1991). Several researchers have pointed out that on-farm testing can also tell managers how efficient they are about learning new technologies (Lindner, Pardey, and Jarrett 1982, Tsur, Sternberg, and Hockman 1990, Foster and Rosenzweig 1995).

Clearly, formal education and technology testing experience can help in both the collection of information and in the processing phase. Numerous authors have included human capital variables to account for heterogeneity across farmers in their use of information. Rahm and Huffman (1984) call this efficiency in allocation, while Pingali and Carlson (1985) and others have evaluated how human capital affects differences between objective and subjective probabilities.

There often seem to be delays in use of information relative to that implied by Bayesian learning (Fisher, Arnold, and Gibbs 1996). Individual observations on technologies may be correlated or biased because of unmeasured site-specific factors. Additionally, decision makers may use simplified methods of processing or collecting information. Ellison and Fudenberg (1993) refer to these as "rules of thumb." One of the important shortcuts in evaluating new technologies might be their "popularity" as measured by aggregate adoption of technologies in nearby areas. Marra, Hubbell, and Carlson (2001) compare the popularity hypothesis with other information quality measures of the Fisher, Arnold, and Gibbs type for adoption of Bt cotton and found that farmers will use these indirect sources of information for Bt cotton adoption decisions.

As indicated above, attention has also been given to the importance of farmers learning from their neighboring farmers (Foster and Rosenzweig 1995, Besley and Case 1993, and Cameron 1999). Foster and Rosenzweig propose a model of learning from village farmers that makes the scale of operation endogenous. Foster and Rosenzweig focus on learning about uncertain input use (fertilizer) for the MCVs, while Besley and Case treat the profitability of the MCV technology as uncertain and exogenous. Both models are dynamic and require panel data to estimate.

Farm Size and Adoption

Scale of the farm or a farm enterprise usually is included in all evaluations of adoption. The most common view is that the cost of acquiring information for a large farm is similar to that of a smaller farm, hence there will be a lower cost per unit of area on the large farm (Perrin and Winkleman 1976 and Feder and O'Mara 1981). Larger farms may also be located in areas with better information sources or growing conditions more favorable for the new varieties. Feder (1980) showed that larger farm size might indicate less credit or capital limitations on adopting new technologies.

Just and Zilberman (1983) developed a portfolio model in which a minimum farm size without adoption reflected credit limits or fixed costs of adoption. Additionally, a concave adoption-size relationship was consistent with risk aversion and positively correlated returns to the old and new technologies. Marra and Carlson (1987), in an empirical test of the Just and Zilberman hypotheses, found a concave relationship between degree of adoption and farm size for the new, double-cropped soybean-wheat technology.

They also found evidence of minimum size with no adoption when corrections were made for the effects of management quality and special equipment. (See Figure 4 for the graphical depiction of the Just and Zilberman model and the estimated relationships from Marra and Carlson.) These and other studies (Smale, Just, and Leathers 1994) show that there may be other reasons besides risk aversion for farmers to diversify across technologies or to exhibit partial adoption.

LONG-TERM INVESTMENT IN AGRICULTURAL TECHNOLOGIES

The neoclassical theory of investment is centered on the result that a profit-maximizing firm will commit scarce resources to a new project if the sum of the present value of the future net cash flows from the project and the present value of the project's salvage value is greater than the project cost. It has long been observed that firms appear to require a marginal benefit larger than this theory would suggest before they undertake some investments. In other words, the investment *threshold* seems to be higher than the sum of the discounted benefits stream for some types of investments. Investment lags are also observed in the sense that neoclassical theory tends to over-predict the rate of diffusion of new technology, even after accounting for the time it takes for information of the availability of the new technology to reach all the decision makers.

Firms are also observed to disinvest more slowly than neoclassical theory would suggest. This also can be thought of in terms of the difference between a neoclassical exit threshold value of an asset and the observed threshold value. In the case of disinvestment, the expected marginal benefit must be significantly below the investment cost before the firm will disinvest.

Asset Fixity

The observed disinvestment lag in agriculture was termed "asset fixity" by Johnson and Quance (1972). They were trying to explain why, when agricultural prices fall, the consequent reductions in quantities supplied are not as great as are the increases in quantities supplied when prices rise. They called this phenomenon the "overproduction trap" and held it responsible for persistent low returns in agricultural production. The phenomenon was blamed on the fact that many agricultural investments have resale values far below their purchase prices (less any physical depreciation); i.e., there are some *sunk costs* associated with the investments. Farmers were reluctant to pay the sunk costs if they thought there was a decent probability that higher future prices would increase the resale value of their investment. Johnson and Pasour

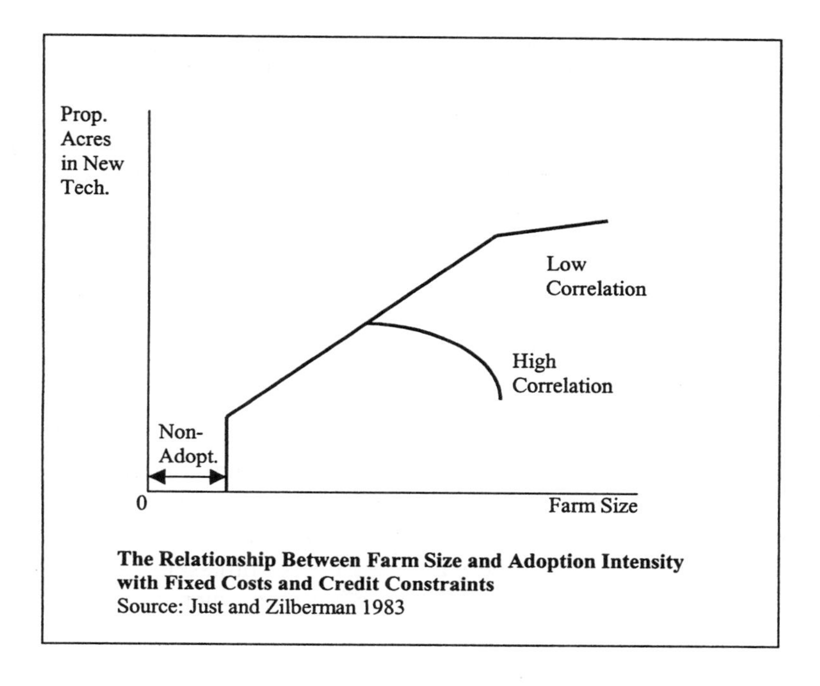

The Relationship Between Farm Size and Adoption Intensity with Fixed Costs and Credit Constraints
Source: Just and Zilberman 1983

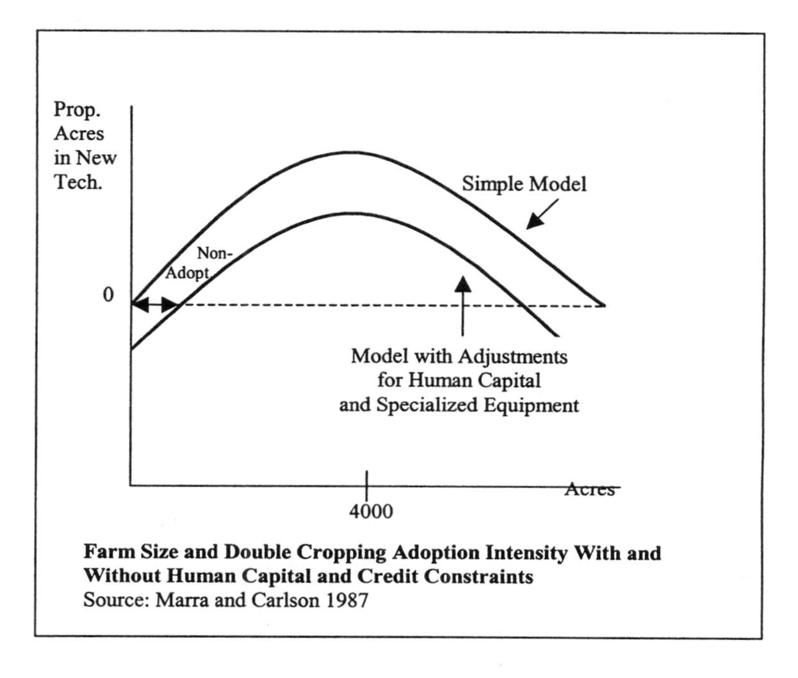

Farm Size and Double Cropping Adoption Intensity With and Without Human Capital and Credit Constraints
Source: Marra and Carlson 1987

Figure 4. The Relationship Between Farm Size and Technology Adoption: Theoretical and Empirical Evidence

(1981) argued that producers were reacting to the fact that these assets' *value in use* was greater than their salvage values and, thus, the low disinvestment was the result of rational behavior on the part of agricultural producers, not the hope of better times to come. Either way, both camps acknowledged that the measured supply response resulting from falling prices was less than the one resulting from rising prices. While the notion of sunk costs explained the observed disinvestment lags, it did not address the delays in making the initial investment until much later.

Irreversibility, Uncertainty, and Investment Lags

Around the same time as the Rosenberg (1976) review, resource economists began to borrow several pertinent findings from the field of finance, namely the Nobel Prize winning idea of a market for contingent claims (Black and Scholes 1973, Merton 1973). The thrust of this research was to show that there is a *real option value* of waiting to invest when there is uncertainty about future returns, and that there are sunk costs in the sense that one cannot recoup all investment costs when disinvesting, and when future information might lead to a better decision. This sunk cost was called the *irreversibility* of the investment. The term irreversibility was initially applied to investment projects that would alter a natural or unique area in some way so that it would be very costly to return it to its natural state. Henry (1974) used the term to describe the following problem:

> "...a new circumferential highway is now being planned around Paris as a direct connection between the various suburbs located ten kilometers beyond the city limits. It may cut through public forests (Versailles, Malmaison, etc.), ancient royal estates..." (p. 1006).

– an irreversible investment indeed! Arrow and Fisher (1974) also considered problems of natural resource depletion such as transforming old growth forests into farmland. Theirs was a two-period model in which the true state of demand is revealed in the second period.

Using dynamic models developed by Bellman (1961) and Samuelson (1964) and stochastic calculus (Dixit and Pindyck 1994), the effects of output and input price uncertainty on the investment decision then began to be explored. A series of papers in the resource economics literature developed these notions as they apply to irreversible investment decisions, particularly of extractive industries. (See Dixit and Pindyck [1994] for a comprehensive review and synthesis of this literature.) Even back as far as Arrow and Fisher (1974), the definition of irreversibility began to take on a less restrictive meaning. Today, the term is used in the context of any investment decision involving sunk costs.

Only recently has this line of inquiry been applied to agricultural investment problems. While dynamic optimization under risk has been in agricultural economists' toolbox for more than twenty years (see, for example, Taylor and Talpaz 1979, or Burt, Koo, and Dudley 1980), Hertzler (1991) was the first to describe the methods of stochastic calculus and Ito control in the agricultural economics literature in 1991 and to suggest several examples of where it could be used to analyze agricultural decisions under risk.

Chavas (1994) combines the techniques of irreversible investment theory and sunk costs to derive optimal investment, and entry and exit rules. He is able to encompass both the option value approach to explain investment lags and sunk costs to explain disinvestment lags under the assumption of risk neutrality. Taking the Johnson and Quance (1972) view of asset fixity, Chavas (1994) concludes that there may be under-investment in some agricultural technology and that government-provided risk-reducing programs, such as price supports, may be optimal under some circumstances. A similar unified model of investment was derived concurrently by Abel and Eberly (1994). In their model investment is a function of the shadow value, q, of the investment. There are two action threshold values for q and between these two values there is no incentive to invest or to disinvest. This interval is termed the *range of inaction*. They take the Johnson and Pasour (1981) approach to asset fixity by focusing on the shadow value, or value in use, of the investment as the relevant decision variable.

Figure 5 illustrates some of the main points derived from Chavas (1994), Abel and Eberly (1994), and a later paper by Barham and Chavas (1997). Consider a firm's investment/disinvestment decision for a particular investment (a) that is characterized by significant sunk costs so that, once the investment is made, its purchase price cannot be fully recouped when it is sold; (b) that is used in a production process where there is revenue uncertainty so that the investment's future marginal value product is uncertain; and (c) where the decision maker is risk neutral. If the decision maker is deciding whether or not to make the initial investment, the expected marginal value product (MVP) of the investment (Abel and Eberly's "q") is the additional benefit, and the additional cost of investing (MC^+) is the investment's purchase price *plus* the value of the option to wait. The greater the uncertainty about future returns, the larger is the option value of waiting to gather more information. So as long as MVP > MC^+, the decision maker will decide to make the investment. For values of MVP < MC^+, the decision is whether or not to disinvest. Now the MVP (the shadow value, q) becomes the marginal cost of disinvesting (the value in use foregone if the investment is sold) and the investment's salvage value is the marginal benefit, MB^-, of the decision to disinvest. If the MVP of the investment falls between the option value and the salvage value in Figure 5, then the optimal decision is to keep the investment in production since its value in use is greater than its salvage value. If the MVP falls below the salvage value, then the marginal benefit of disinvesting (the salvage value) is greater than the marginal cost (the MVP

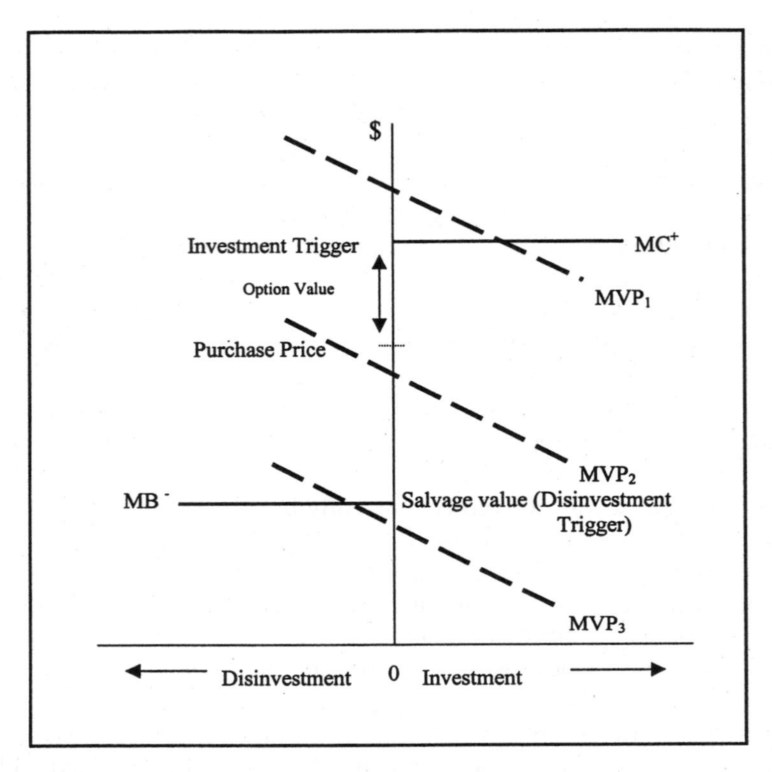

Figure 5. The Investment/Disinvestment Decision with Uncertainty and Sunk Costs

Source: Adapted from Barham and Chavas (1997) and from Abel and Eberly (1994)

foregone), so the optimal decision is to disinvest. MVP_1 in Figure 5 illustrates the case where the optimal decision is to invest. Notice that there is zero investment unless MVP_1 is well above the purchase price. This illustrates the investment lags that have been observed. MVP_2 is the case of asset fixity, where the value in use is greater than the salvage value. MVP_3 illustrates the case of disinvestment, or for some types of investments, industry exit. One of the important points here is that the purchase price is not a trigger for action in either direction. It is irrelevant for the disinvestment decision (a true sunk cost) and only a sub-component of the investment trigger.

Purvis et al. (1995) applied this idea to an ex ante analysis of Texas dairy farmer investment in new waste management technology and found that, compared to the neoclassical net present value approach to the investment decision, the option value approach resulted in a significantly higher expected income stream before it would make economic sense to invest in the new technology. This study was one of the first empirical works in agricultural economics that considered the uncertainty/irreversibility features of the decision explicitly.

Richards (1996) brought the term *hysteresis* to the agricultural economics literature. In economic terms, hysteresis is the tendency for a phenomenon (action, investment, etc.) to persist after the initial cause for it has disappeared. Asset fixity is an example of economic hysteresis. He examines the problem of investing in production quota licenses by Canadian dairy farmers. He considers the fact that the licenses are related to other quasi-fixed inputs in production, providing an additional, indirect avenue for hysteresis to occur. He finds that, indeed, hysteresis is present and results in slower changes in herd sizes than would be optimal under the neoclassical rule.

Price and Wetzstein (1999) use the same principles to calculate empirical entry and exit thresholds for Georgia peach producers. They assume price and yield both follow a stochastic process and calculate revenue, price, and yield thresholds for entry and exit. They find that there is as much as a $4,000 per acre expected revenue difference between the entry threshold and the exit threshold, depending on the degree of uncertainty involved.

Pietola and Myers (2000) extend the application of investment theory under uncertainty and irreversibility to the dual problem in their study of the Finnish pork industry. They find that because of increasing uncertainty, Finnish hog farmers have been slow to adjust to larger herd size to take advantage of significant economies of scale. Interestingly, they find that the fixity of labor appears to be one of the major reasons for the slow adjustment, rather than capital asset fixity.

In an interesting recent article, Zhao and Zilberman (1999) extend the theory to allow endogeneity of the degree of irreversibility of the investment. They include the fact that decision makers may not only delay investments because of the option value of waiting for more information but explicitly account for the fact that decision makers will be willing to pay more (or be willing to accept less output) for more flexible technologies in the face of uncertainty.

Zhao (2000), using a game-theoretic approach and option value, considered the case where the option value of waiting to adopt is related to the opportunity to observe earlier adopters' experience with the technology. This is a finding similar to those found in the Foster and Rosenzweig (1995) and Besley and Case (1993) studies discussed in the previous section on divisible technologies.

Technological Change and Investment under Uncertainty

Zilberman, Wetzstein, and Marra (1993) consider the case of an investment decision where there is a current technology choice and uncertain availability of new, more modern technology in the future. They consider a two-period model with new technology available in the second period with probability less than one. They find that ignoring the potential availability of new technology will result in over-investment in the current technology in the first period. The optimal amount of investment in the current technology in the first period depends on the discount rate, the probability of new technology availability in the second period, and the welfare- (profit or resource-saving) increasing potential of the new technology relative to the current technology. The option value of waiting for the new technology in their model does not imply that there is no investment in the first period, just that there is less than total investment to preserve the option of acquiring the new technology in the second period.

Another interesting effect of the rate of technological progress on the uncertain investment decision is considered in Woo and Wright (2000). They analyze the problem of the optimal timing of genetic evaluation and varietal development to combat a particular disease, whose onset is uncertain, from germplasm held in gene banks. They find that technological progress that more or less steadily reduces the search and development costs favors later (ex post as opposed to ex ante) evaluation and development in many circumstances because the costs of doing so will be lower with newer technology. This finding is consistent with conventional real option theory. However, they show that the possibility of a future *technological breakthrough* that substantially reduces development costs implies *earlier* investment in the evaluation phase so that the gains from the breakthrough in development technology can begin immediately upon its arrival. This finding is, on the face of it, somewhat contrary to the Zilberman, Wetzstein, and Marra (1993) result and real option theory. However, the problem Woo and Wright (2000) address is one of technological progress in a later phase of production and its effects on timing of investments in an earlier phase, not in the later phase, itself.

CONCLUSIONS AND PROMISING AREAS FOR FUTURE RESEARCH

Progress has been made in explaining adoption, partial adoption, and adoption delays by more complete learning models that include learning from neighbors and adjusting for information quality. These learning models should prove to be useful for other agricultural technology adoption problems. The role of neighbors' assets and own assets as explanations for the size of risk exposure and information quality also has helped in explaining adoption. More scale effect research is warranted. In developed countries farm sizes are continuing to increase over time, and transgenic crops may have some scale effects. For developing countries, finding technologies for small farm sizes is a critical policy objective.

Subjective perceptions of risk and of risk aversion can be of some help in adoption models. However, there are many "risk aversion" variables obtained from surveys that have been of questionable usefulness because they are so ambiguous (e.g., purchase of crop insurance, use of forward sales contracts).

With new pesticides and transgenic crops there is additional uncertainty.

There is also sharp downside risk when the modern conventional varieties break down, pest resistance occurs, or new pests move into an area. Models that allow tests for the effects of higher moments of the revenue distribution need more attention in the context of adoption models.

The study of new technologies and ignoring the changing nature of currently used technologies is a fairly standard approach. Describing and developing models with depreciation rates of current technologies can be important in explaining new technology adoption. Transgenic crops with pest resistance build-up and rust susceptibility of the pre-modern variety era are examples where biological depreciation is clearly applicable and have not been explored adequately.

Spatial diffusion or supply and adoption demand studies are needed for the new transgenic crops both in the U.S. and in developing countries. There is also a major opportunity to study why there is regional and individual farmer partial adoption of transgenic crops. Also, the reasons for adoption of RR soybeans have not been thoroughly studied. Is it management input saving, convenience, risk reducing, or something else?

It is possible to get variability-increasing and -decreasing effects of pesticides, especially when pest-free yield is relatively variable to begin with. The question of whether risk management tools, such as crop insurance and IPM, result in more or less input use is still an open question and should be explored further.

In many cases of repeated decisions with pesticide products that are not new, there is evidence that risk aversion is not very important. Moreover, many recent studies related to new technologies have shown that behavior that was once thought to be the effects of risk aversion could be explained as well or better with other factors such as credit constraints, learning, or asset fixity. Even so, risk, or uncertainty is still very important in explaining behavior, and there is more work to do in agricultural economics, especially in the long-term investment models and applications. Even with divisible technologies, where there is learning and partial adoption, there can be multi-period effects.

Replacement (or backstop) technology research should be an active area for economists. This line of research is relevant today as more and better transgenic crops come to market. The general theoretical implications of technology adoption or investment in the face of ongoing technical change have not been developed sufficiently as yet.

There are many, still untapped, opportunities to apply the option value/sunk cost models to agricultural technology problems. Examples are

- precision farming adoption,
- biotech innovation problems,
- agricultural technologies that are environmentally friendly (or environmentally hazardous), and
- technologies requiring large investments in human capital.

In sum, there is still a lot of work to do. One of the major constraints we face in meeting the challenges in the study of agricultural technology and risk is one of sufficient data availability. A geographically broad panel data set at the farm level is needed to test some of the newer hypotheses. We should either secure sufficient funding to collect such data and/or work with USDA's National Agricultural Statistics Service to ensure that the surveys they conduct have the sampling properties and farm-level information we need to continue this work.

REFERENCES

Abel, A., and J. Eberly. 1994. "A Unified Model of Investment Under Uncertainty." *The American Economic Review* 84: 1369-1384.

Alexander, C., J. Fernandez-Cornejo, and R. Goodhue. 2000. "Determination of GMO Use: A Survey of Iowa Corn-Soybean Farmer's Acreage Allocation." In R. Evenson, L. Paganetto, V. Santaniello, P.L. Scandizzo, and D. Zilberman, eds., *Fourth International Conference on the Economics of Agricultural Biotechnology*. Rome, Italy: International Consortium on Agricultural Biotechnology Research, Tor Vergata University.

Anderson, J., and P. Hazell (eds). 1989. *Variability in Grain Yields*. Baltimore: Johns Hopkins University Press.

Antle, J. 1988. *Pesticide Policy, Production Risk and Producer Welfare*. Washington, D.C.: Resources for the Future.

Antle, J., and S. Hatchett. 1986. "Dynamic Input Decisions in Econometric Production Models." *American Journal of Agricultural Economics* 68: 937-949.

Arrow, K., and A. Fisher. 1974. "Environmental Preservation, Uncertainty and Irreversibility." *Quarterly Journal of Economics* 88: 312-319.

Barham, B., and J.-P. Chavas. 1997. "Sunk Costs and Resource Mobility: Implications for Economic and Policy Analysis." Department of Agricultural and Applied Economics Staff Paper Series No. 410, University of Wisconsin-Madison.

Bellman, R. 1961. *Adaptive Control Processes: A Guided Tour*. Princeton, NJ: Princeton University Press.

Besley, T., and A. Case. 1993. "Modeling Technology Adoption in Developing Countries." *A.E.R. Papers and Proceedings* 83: 396-402.

Black, F., and M. Scholes. 1973. "The Pricing of Options and Corporate Liabilities." *Journal of Political Economy* 81: 637-659.

Burrows, T. 1983. "Pesticide Demand and Integrated Pest Management: A Limited Dependent Variable Analysis." *American Journal of Agricultural Economics* 65: 806-810.

Burt, O., W. Koo, and N. Dudley. 1980. "Optimal Stochastic Control of U.S. Wheat Stocks and Exports." *American Journal of Agricultural Economics* 62: 172-187.

Byerlee, D. 1996. "Modern Varieties, Productivity, and Sustainability: Recent Experience and Emerging Challenges." *World Development* 24: 697-718.

Byerlee, D., and E. Hesse de Polanco. 1986. "Farmers' Stepwise Adoption of Technology Packages: Evidence from the Mexico Altiplano." *American Journal of Agricultural Economics* 68: 519-527.

Cameron, L. 1999. "The Importance of Learning in the Adoption of High-Yielding Variety Seeds." *American Journal of Agricultural Economics* 81: 83-94.

Carlson, G. 1970. "A Decision-Theoretic Approach to Crop Disease Prediction and Control." *American Journal of Agricultural Economics* 52: 216-223.

___. 1979. "The Role of Pesticides in Stabilizing Agricultural Production." In J. Sheets and D. Pimentel, eds., *Pesticides: Contemporary Roles in Agriculture, Health and the Environment*. Clifton, NJ: Humana Press.

___. 1984. "Risk Reducing Inputs Related to Agricultural Pests." In *Risk Analysis of Agricultural Firms: Concepts, Information Requirements and Policy Issues*, Proc. Regional Research Project S-180, Department of Agricultural Economics, University of Illinois, Urbana, IL.

Carlson, G., and C. Main. 1976. "Economics of Disease Loss Management." *Annual Review of Phytopathology* 14: 381-403.

Carlson, G., and M. Wetzstein. 1993. "Pesticides and Pest Management." In G. Carlson, D. Zilberman, and J. Miranowski, eds., *Agricultural and Environmental Resource Economics*. New York: Oxford University Press.

Chavas, J.-P. 1994. "Production and Investment Decisions." *American Journal of Agricultural Economics* 76: 114-127.

Cochran, M., L. Robinson, and W. Lodwick. 1985. "Imposing the Efficiency of Stochastic Dominance Techniques Using Convex Set Stochastic Dominance." *American Journal of Agricultural Economics* 67: 289-295.

Darr, D., and W. Chern. 2000. "Estimating Adoption of GMO Soybeans and Corn: A Case Study of Ohio, U.S.A." In R. Evenson, L. Paganetto, V. Santaniello, P.L. Scandizzo, and D. Zilberman, eds., *Fourth International Conference on the Economics of Agricultural Biotechnology*. Rome, Italy: International Consortium on Agricultural Biotechnology Research, Tor Vergata University.

Demers, M. 1991. "Investment Under Uncertainty, Irreversibility and the Arrival of Information Over Time." *The Review of Economic Studies* 58: 333-350.

Dixit, A., and R. Pindyck. 1994. *Investment Under Uncertainty*. Princeton, NJ: Princeton University Press.

Dixon, R. 1980. "Hybrid Corn Revisited." *Econometrica* 48: 1451-1461.

Ellison, G., and D. Fudenberg. 1993. "Rules of Thumb for Social Learning." *Journal of Political Economy* 101: 612-643.

Evenson, R., W. Lesser, V. Santaniello, and D. Zilberman (eds.). 1999. In R. Evenson, L. Paganetto, V. Santaniello, P.L. Scandizzo, and D. Zilberman, eds., *The Shape of the Coming Agricultural Biology Transformation: Strategic Investment and Policy Approaches from an Economic Perspective*. Rome, Italy: International Consortium on Agricultural Biotechnology Research, Tor Vergata University.

Evenson, R., L. Paganetto, V. Santaniello, P.L. Scandizzo, and D. Zilberman (eds). 2000. *Fourth International Conference on the Economics of Agricultural Biotechnology*. Rome, Italy: International Consortium on Agricultural Biotechnology Research, Tor Vergata University.

Feder, G. 1979. "Pesticides, Information and Pest Management Under Uncertainty." *American Journal of Agricultural Economics* 61: 97-103.

___. 1980. "Farm Size, Risk Aversion and the Adoption of New Technology Under Uncertainty." *Oxford Economic Papers* 32: 263-283.

___. 1982. "Adoption of Interrelated Agricultural Innovations: Complementarity and the Impacts of Risk, Scale and Credit." *American Journal of Agricultural Economics* 64: 94-101.

Feder, G., R.E. Just, and D. Zilberman. 1985. "Adoption of Agricultural Innovations in Developing Countries: A Survey." *Economic Development and Cultural Change* 33: 225-298.

Feder, G., and G. O'Mara. 1981. "Farm Size and the Adoption of Green Revolution Technologies." *Economic Development and Cultural Change* 30: 59-76.

___. 1982. "On Information and Innovation Diffusion: A Bayesian Approach." *American Journal of Agricultural Economics* 64: 145-147.

Feder, G., and R. Slade. 1984. "The Acquisition of Information and the Adoption of New Technology." *American Journal of Agricultural Economics* 66: 312-320.

Fernandez-Cornejo, J., S. Jans, and M. Smith. 1998. "Issues in the Economics of Pesticide Use in Agriculture: A Review of the Empirical Evidence." *Review of Agricultural Economics* 20: 462-488.

Fisher, A., A. Arnold, and M. Gibbs. 1996. "Information and Speed of Innovation Adoption." *American Journal of Agricultural Economics* 78: 1073-1081.

Flinn, J., and D. Garrity. 1989. "Yield Stability and Modern Rice Technology." In J. Anderson and P. Hazell, eds., *Variability in Grain Yields*. Baltimore: Johns Hopkins University Press.

Foster, A., and M. Rosenzweig. 1995. "Learning by Doing and Learning from Others: Human Capital and Technical Change in Agriculture." *Journal of Political Economy* 103: 1176-1209.0

Griliches, Z. 1957. "Hybrid Corn: An Exploration Into the Economics of Technical Change." *Econometrica* 25: 501-525.

Hall, D. 1977. "The Profitability of Integrated Pest Management: Case Studies of Cotton and Citrus in the San Joaquin Valley." *Bulletin of the Entomological Society of America* 23: 267-274.

Hazell, P. 1984. "Sources of Increased Instability in Indian and U.S. Cereal Production." *American Journal of Agricultural Economics* 66: 302-311.

Heisey, P., M. Smale, D. Byerlee, and E. Souza. 1997. "Wheat Rusts and the Costs of Genetic Diversity in the Punjab of Pakistan." *American Journal of Agricultural Economics* 79: 726-737.

Henry, C. 1974. "Investment Decisions Under Uncertainty: The 'Irreversibility Effect'." *The American Economic Review* 64: 1006-1012.

Herdt, R., L. Castillo, and S. Jayasuriya. 1984. "The Economics of Insect Control on Rice in the Philippines." In *Judicious and Efficient Use of Insecticides in Rice*. Los Banos, Philippines: International Rice Research Institute.

Hertzler, G. 1991. "Dynamic Decisions Under Risk: Applications of the Ito Stochastic Control in Agriculture." *American Journal of Agricultural Economics* 73: 1126-1137.

Hyde, J., M.A. Martin, P.V. Preckel, and C.R. Edwards. 1999. "The Economics of Bt Corn: Valving Protection From the European Corn Borer." *Review of Agricultural Economics* 21: 442-454.

Hildebrant, P. 1960. "The Economic Theory of the Use of Pesticides, Part II: Uncertainty." *Journal of Agricultural Economics* 14: 52-61.

Horowitz, J. and E. Lichtenberg. 1993. "Insurance, Moral Hazard and Chemical Use in Agriculture." *American Journal of Agricultural Economics* 75: 926-935.

___. 1994. "Risk-Reducing and Risk-Increasing Effects of Pesticides." *Journal of Agricultural Economics* 45: 82-89.

Hubbell, B., M. Marra, and G. Carlson. 2000. "Estimating the Demand for a New Technology: Bt Cotton and Insecticide Policies." *American Journal of Agricultural Economics* 82: 118-132.

Johnson, M., and E. Pasour, Jr. 1981. "An Opportunity Cost View of Fixed Asset Theory and the Overproduction Trap." *American Journal of Agricultural Economics* 63: 1-7.

Johnson, G., and C. Quance. 1972. *The Overproduction Trap in U.S. Agriculture*. Baltimore, MD: Johns Hopkins University Press.

Just, R., and R. Pope. 1979. "Production Function Estimation and Related Risk Considerations." *American Journal of Agricultural Economics* 61: 276-284.

Just, R., and D. Zilberman. 1983. "Stochastic Structure, Farm Size, and Technology Adoption in Developing Countries." *Oxford Economics Papers* 35: 307-328.

Knudson, M. 1991. "Incorporating Technological Change in Diffusion Models." *American Journal of Agricultural Economics* 73: 724-733.

Lazarus, W., and E. Swanson. 1983. "Insecticide Use and Crop Rotation Under Risk: Rootworm Control in Corn." *American Journal of Agricultural Economics* 65: 738-747.

Leathers, H., and J. Quiggin. 1991. "Interaction Between Agricultural and Resource Policy: The Importance of Attitudes Toward Risk." *American Journal of Agricultural Economics* 73: 757-764.

Leathers, H., and M. Smale. 1991. "A Bayesian Approach to Explaining Sequential Adoption of Components of a Technical Package." *American Journal of Agricultural Economics* 73: 734-742.

Lindner, R., A. Fischer, and P. Pardey. 1979. "The Time to Adoption." *Economics Letters* 2: 187-190.

Lindner, R., P. Pardey, and F. Jarrett. 1982. "Distance to Information Source and the Time Lag to Early Adoption of Trace Elements Fertilizers." *Australian Journal of Agricultural Economics* 26: 98-113.

Marra, M. 2001. *Farm Level Benefits of Transgenic Crops: A Critical Review of the Evidence to Date.* IFPRI Report. Washington, D.C.: International Food Policy and Research Institute. In press.

Marra, M., and G. Carlson. 1987. "The Role of Farm Size and Resource Constraints in the Choice Between Risky Technologies." *Western Journal of Agricultural Economics* 12: 109-118.

Marra, M., B. Hubbell, and G. Carlson. 2001. "Information Quality, Technology Depreciation, and Bt Cotton Adoption in the Southeast." *Journal of Agricultural and Resource Economics.* In press.

Merton, R. 1973. "The Theory of Rational Option Pricing." *Bell Journal of Economics and Management Science* 4: 141-183.

Moffitt, J., L. Fansworth, R. Zavaleta, and M. Kogan. 1986. "Economic Impact of Public Pest Information: Soybean Insect Forecasts in Illinois." *American Journal of Agricultural Economics* 68: 274-279.

Mundlak, Y. 2000. *Agriculture and Economics Growth: Theory and Measurement.* Cambridge, MA: Harvard University Press.

Musser, W., B. Tew, and J. Epperson. 1981. "An Economic Examination of an Integrated Pest Management Production System with a Contrast Between E-V and Stochastic Dominance Analysis." *Southern Journal of Agricultural Economics* 13: 199-124.

Pannell, D. 1991. "Pests and Pesticides, Risk and Risk Aversion." *Agricultural Economics* 5: 361-383.

Pannell, D., B. Malcolm, and R. Kingwell. 2000. "Are We Risking Too Much? Perspectives on Risk in Farm Modelling." *Agricultural Economics* 23: 69-78.

Perrin, R., and D. Winkleman. 1976. "Impediments to Technical Progress on Small Versus Large Farms." *American Journal of Agricultural Economics* 58: 888-894.

Pietola, K., and R. Myers. 2000. "Investment Under Uncertainty and Dynamic Adjustment in the Finnish Pork Industry." *American Journal of Agricultural Economics* 82: 956-967.

Pingali, P., and G. Carlson. 1985. "Human Capital, Adjustments in Subjective Probabilities, and the Demand for Pest Controls." *American Journal of Agricultural Economics* 67: 853-861.

Price, J., and M. Wetzstein. 1999. "Irreversible Investment Decisions in Perennial Crops with Yield and Price Uncertainty." *Journal of Agricultural and Resource Economics* 24: 173-185.

Purvis, A., W. Boggess, C. Moss, and J. Holt. 1995. "Technology Adoption Decisions Under Irreversibility and Uncertainty: An Ex Ante Approach." *American Journal of Agricultural Economics* 77: 541-551.

Rahm, M., and W. Huffman. 1984. "The Adoption of Reduced Tillage: The Role of Human Capital and Other Variables." *American Journal of Agricultural Economics* 66: 405-413.

Renkow, M. 1993. "Differential Technology Adoption and Income Distribution in Pakistan: Implications for Research Resource Allocation." *American Journal of Agricultural Economics* 75: 33-43.

Richards, T. 1996. "Economic Hysteresis and the Effects of Output Regulation." *Journal of Agricultural and Resource Economics* 21: 1-17.

Rogers, E. 1995. *Diffusion of Innovations.* New York: The Free Press (Macmillan & Co.).

Rosenberg, N. 1976. "On Technological Expectations." *Economics Journal* 86: 523-535.

Ruttan, V. 1996. "What Happened to Technology Adoption–Diffusion Research?" *Sociologia Ruralis* 36: 51-73.

Saha, A., H. Love, and R. Schwart. 1994. "Adoption of Emerging Technologies Under Output Uncertainty." *American Journal of Agricultural Economics* 76: 836-846.

Samuelson, P. 1964. "Tax Deductibility of Economic Depreciation to Insure Invariant Valuation." *Journal of Political Economy* 72: 571-573.

Santaniello, V., R. Evenson, D. Zilberman, and G. Carlson (eds.). 2000. *Agriculture and Intellectual Property Rights*. New York: CABI Publishing.

Smale, M., J. Hartell, P. Heisey, and B. Senaver. 1998. "The Contribution of Genetic Resources and Diversity to Wheat Production in the Punjab of Pakistan." *American Journal of Agricultural Economics* 80: 482-493.

Smale, M., R. Just, and H. Leathers. 1994. "Land Allocation in HYV Adoption Models: An Investigation of Alternative Explanations." *American Journal of Agricultural Economics* 76: 535-546.

Smith, V., and B. Goodwin. 1996. "Crop Insurance, Moral Hazard, and Agricultural Chemical Use." *American Journal of Agricultural Economics* 78: 428-438.

Stoneman, P. 1981. "Intra-Firm Diffusion, Bayesian Learning and Profitability." *Economics Journal* 91: 375-388.

Swinton, S., and R. King. 1994. "The Value of Information in a Dynamic Setting: Case of Weed Control." *American Journal of Agricultural Economics* 76: 36-46.

Taylor, C., and H. Talpaz. 1979. "Approximately Optimal Carryover Levels for Wheat in the United States." *American Journal of Agricultural Economics* 61: 32-40.

Traxler, G., and J. Falck-Zepeda, J. Ortz-Monasterio, and K. Sayre. 1995. "Production Risk and the Evolution of Varietal Technology." *American Journal of Agricultural Economics* 77: 1-7.

Tsur, Y., M. Sternberg, and E. Hockman. 1990. "Dynamic Modeling of Innovation Process Adoption with Risk Aversion and Learning." *Oxford Economics Papers* 42: 336-355.

Webster's New American Dictionary. 1995. New York: Merriam-Webster, Inc.

Woo, B., and B. Wright. 2000. "The Optimal Timing of Evaluation of Genebank Accessions and the Effects of Biotechnology." *American Journal of Agricultural Economics* 82: 797-811.

Zacharias, T., and A. Grube. 1986. "Integrated Pest Management Strategies for Approximately Optimal Control of Corn Rootworm and Soybean Cyst Nemotode." *American Journal of Agricultural Economics* 68: 704-715.

Zhao, J. 2000. *Information Externalities and Strategic Delay in Technology Adoption and Diffusion*. Unpublished manuscript, Department of Economics, Iowa State University.

Zhao, J., and D. Zilberman. 1999. "Irreversibility and Restoration in Natural Resource Development." *Oxford Economics Papers* 51: 559-573.

Zilberman, D., M. Wetzstein, and M. Marra. 1993. *Economics of Nonrenewable and Renewable Resources*. In G. Carlson, D. Zilberman, and J. Miranowski, eds., *Agricultural Resource and Environmental Economics*. New York: Oxford University Press.

Chapter 16

QUALITY AND GRADING RISK*

Ethan Ligon
University of California, Berkeley

INTRODUCTION

Many (perhaps most) sorts of agricultural commodities are not homogeneous, but instead vary according to a set of quality characteristics. Some of these characteristics may be easy to inexpensively measure – many sorts of fresh fruit, for example, are sorted into different sizes by using a sizing belt. However, others may be difficult to measure non-destructively – think of trying to measure the color of flesh of a whole melon. In between these extremes (characteristics that are easily measured and those that are difficult or impossible to non-destructively measure) lie the sorts of characteristics that consumers may place a high value on, yet which can be easily measured only by the prospective consumer, such as the smell of a tomato.

Choosing to purchase a commodity that bundles a variety of observable characteristics can be thought of as the hedonic choice of Lancaster (1966). Here we want to consider the case in which an indivisible commodity can be regarded as a bundle of characteristics, but where not all of these characteristics are observed prior to consumption.

Not only might it be the case that some quality characteristics are not observed by consumers, but it may also be the case that those same characteristics may be unobserved by the producer and any intermediaries. Indeed, actors upstream from the consumer may be able to observe even less than the consumer. The way a fruit ripens over time adds some additional uncertainty to the problem faced by a wholesaler, for example. Other characteristics (think again of fragrance) may be easily observed by any actor, but may be difficult to communicate.

To some extent the producer or intermediary may have some control over the distribution of characteristics they cannot directly observe – the timing of

* My thanks to Brent Hueth, my collaborator on much of the research that has informed my thinking on this subject. This research has been sponsored in part by the NRI Competitive Grants Program/USDA Award No. 00-35400-9123 and by the Giannini Foundation; however, the views expressed in this article are not endorsed by either of these sponsors.

harvest will influence the ripeness perceived by the final consumer, for example. This control may mitigate the problems associated with the fact that some quality characteristics cannot be observed, but may also add an interesting twist – if the actions taken by the producer influence eventual outcomes, but these actions are hidden from intermediaries and consumers, then this asymmetry can have a profound influence on the organization of an industry.

To this point we have talked about "quality characteristics" without referring to "grades." We will think of a grade as a particular categorization of some kind of produce according to its measurable quality characteristics – by assumption, the grade cannot convey more information than a complete description of characteristics, and will typically convey less. Often grading information is compiled by the government, or by some third party established under the auspices of a state or federal marketing order.[1]

The importance of grades varies considerably across different kinds of commodities. For example, Kansas wheat is graded by the state according to characteristics including weight per bushel, dockage and defects, protein content, and water content. These characteristics explain about 24 percent of the variation in prices paid for wheat; additional characteristics that do not influence the grade but which are often privately measured (mostly having to do with the behavior of the grain as it is milled) account for an additional 17 percent of the variation in price (Espinosa and Goodwin 1991, Table 4). In this case grading information appears to be relevant to purchasing decisions, but incomplete. To the extent that buyers must rely on private measurement for more subtle characteristics, it is not clear that the state-sponsored grading system has much value in this case. A more extreme case in which state-sponsored grading appears to have little usefulness is that of many fresh-market horticultural commodities. Grading information for most fresh fruits and vegetables (usually based on size and defects) is of such little use that information on grade is seldom communicated to the final consumer, who can see for himself the tomato size and whether it is bruised.[2] Yet the consumer cares more about ripeness, information which is not included in the grade (Bierlen and Grunewald 1995). In this case the relatively crude information contained in the grade may be of use farther upstream, to intermediaries who may need to exchange lots of fruits or vegetables without directly observing them. This case contrasts with the grading system for many sorts of meat. For example, USDA mandates a grade for cuts of beef that depends on the weight of the entire carcass and fat cover, information which bears on the fat

[1] Dimitri, Horowitz, and Lichtenberg (1996) observe that USDA uses grades to set quality standards for 92 distinct horticultural commodities, while grading is ubiquitous for meat in the U.S.

[2] Of course, this grading information may be of use to the firm insofar as it makes it simpler to communicate with (or provide incentives to) other firms in the supply chain. We abstract from this with our stylized focus on a three-actor chain (grower, firm, consumer); however, see Hennessy (1995) or Hollander, Monier, and Ossard (1999) for a different stylized focus.

content of the steak, but may not be easily inferred from the casual observation of a single steak. As a consequence, information on beef grades usually *is* communicated to the consumer ("Select," "Choice," etc.), though some retail chains in the U.S. see fit to assign more informative grades to the beef they sell (Considine et al. 1986).

To sum up, grades are meant to summarize a variety of quality characteristics; however, these summaries differ in their value according to both the commodity and place in the production chain. This variation across commodities may explain differences in the conclusions drawn by different authors evaluating the welfare consequences of grading services. For example, Dupré (1990) praises the effects of grading on the Canadian dairy industry, while Freebairn (1973) argues that uniform grading of Australian beef ought to reduce the uncertainty faced by consumers, and Marette, Crespi, and Schiavina (1999) argue that olive oil ought to be graded since magnetic resonance imaging is necessary to detect the fraudulent use of non-olive oils. These are all cases in which grade arguably conveys information to the final consumer above and beyond what that consumer might be able to observe at the supermarket. In regards to horticultural commodities, Bockstael (1984) and Hollander, Monier, and Ossard (1999) argue that a competitive industry will, by itself, provide the efficient amount of grading.[3]

In this chapter we will consider the case of a grading system as a special case of a system in which producers observe some vector of characteristics, intermediaries observe a second vector, and consumers a third. These different vectors need not be exclusive; for example, consumers' information may be a superset of an intermediary's information. In a similar spirit, Bockstael (1984) assigns existing research on grading and quality standards into four distinct categories, according to whether or not consumers or producers make decisions that influence the quality of produce they consume or produce. The models presented in this chapter transcend Bockstael's categorization by introducing uncertainty (consumers and firms may be able to draw imprecise inferences about quality) and an additional set of actors (employees or contractors to the firm), but the spirit of her typology remains intact: what firms and consumers observe about quality and when they observe it is of crucial importance.

The remainder of the chapter proceeds as follows. In the following section, we sketch some simple models which differ only in their information structure, and make some simple observations regarding the implications of these different models for welfare and industry structure, with each model accompanied by an illustrative example. The final section concludes.

[3] This raises the issue of how to design an optimal grading system, a topic considered by Berck and Rausser (1982) and Bockstael (1987) but which is beyond the scope of this chapter, which will henceforth assume that the information provided by grade is determined exogenously.

MODELS

In this section we develop a short sequence of models, designed to illustrate the different issues having to do with both the consumers and producers of a good variable quality. As indicated above, the chief distinction between these models has to do with who observes what. We imagine three different kinds of actors: producers, who actually cultivate crops, grow livestock, and so on; intermediaries, who obtain agricultural commodities from the producers; and consumers, to whom the intermediaries sell. We denote by a some set of actions taken, investments made, or information observed by the producer; this is, in some sense, the information *generated* by the producers, which may or may not be shared with intermediaries and consumers. Similarly, we denote by b the information generated by the intermediary; think of this as the measurement of quality characteristics, when this occurs. Finally, the information generated by the consumer, c, may include informal quality measurement (squeezing apples, for example), and also includes the utility actually derived from consumption.

Our first model, entitled *Observable Quality*, serves as a sort of benchmark. In the benchmark model, we imagine that while production is uncertain, in the sense that a given set of inputs yields an output of variable quality, subsequent to harvest all relevant quality attributes are observable to all parties: when shopping for apples at a supermarket, for example, a consumer specifies a particular set of quality characteristics that completely capture all "payoff-relevant" information – given this information, the consumer knows exactly what utility he will derive from later consuming the apple. We think of this as the actual quality of the commodity, c. All parties costlessly observe production-related information a, as well as final quality c. Because all parties costlessly observe all relevant attributes of the foodstuff, no grading system can provide any additional useful information in this model. We characterize the solution to the problems facing producers, intermediaries, and consumers, and construct a competitive equilibrium for a particular specification of consumer preferences.

For contrast, in our second model, *Unobservable Quality*, we suppose that no quality attributes are observable: every consumer necessarily buys a "pig in a poke." Although consumers know how much utility they derive from their purchase ex post (that is, consumers know c), these "preference shocks" are assumed to be unobservable. Further, though consumers may be able to observe production practices a, they cannot associate different values of a with the resulting product, a situation that many authors (e.g., Bockstael 1984, Hennessy 1995) liken to Akerlof's (1970) "lemons" problem. As a consequence, there is only one price for any given kind of foodstuff, regardless of quality, with obvious consequences for firms' incentives to produce higher quality goods.

Our third model, entitled *Grading*, maintains the assumption that quality (and subsequent utilities) are unobservable, but that some *grading character-*

istics can be observed, and that these characteristics can be used to infer quality, albeit with imperfect precision. These observable characteristics can be interpreted as grades or (perhaps better yet) as the kinds of measurements that might be used in grading.

The fourth model we construct, *Consumer Grading*, extends our notion of grading to permit firms and consumers to observe different sets of grading characteristics. Returning to our example of a consumer purchasing apples, we have in mind that consumers may be privy to some information that may be difficult for the supermarket to observe. For example, consumer A, who has previously eaten an apple from a particular bin, might provide some word-of-mouth information on the sweetness of the apple to consumer B, or consumer B might have previously eaten an apple from the bin himself. Alternatively, the consumer may engage in a very detailed inspection of a small number of apples, the sort that may be too costly for the firm to engage in. This reduces the uncertainty faced by consumers, of course, and in equilibrium provides incentives to produce goods of higher quality.

Finally, in the model entitled *Contractor Investment* we suppose that the production characteristics a cannot be directly observed by either intermediaries or the consumer, but are observed by producers. As a consequence, though intermediaries and producers cannot directly observe c, it is possible to infer something about the characteristics consumers observe by looking either at the quantities demanded or realized prices. This final model delivers the setting considered by Hueth and Ligon (1999), in which (if arrangements are efficient) these employees or contractors bear some "price risk," since prices are a signal associated with consumers' private perception of some quality characteristics.

Observable Quality

In this section we develop a benchmark model of quality, in which firms make decisions that influence the quality of goods supplied to consumers, and in which consumers observe the quality of goods prior to making a purchase. In spirit, this benchmark model is very close to that of Bockstael (1984), though in her model quality is assumed to be discrete, and there is no uncertainty.

Consumers. Begin by considering the problem facing a person who consumes a single, indivisible unit of some foodstuff of variable quality, $c \in C \subseteq \Re$, and who also values consumption of some other composite commodity $x \in X$. The consumer's preferences over these two goods are given by a utility function $U : X \times C \to \Re$, assumed to be strictly increasing in either argument, weakly concave, and continuously differentiable.

Now, imagine our consumer at the grocery store who is trying to decide what produce to purchase. We take the price of the composite good to be numeraire, and imagine that every quality of produce can be found at the grocery store, with produce of quality c sold at a price of $p(c)$. Given this environment, the consumer can reduce expenditures on other goods in order to purchase higher quality produce, solving the problem

$$\max_{c \in C} U[x - p(c), c]. \tag{1}$$

Here we can interpret x as the expenditures on the composite commodity when the consumer makes no purchases of the foodstuff. Now, if the price function $p(c)$ is continuously differentiable, then the first-order condition associated with (1) is simply

$$p'(c) = \frac{U_c[x - p(c), c]}{U_x[x - p(c), c]},$$

where p' denotes the marginal change in price with a change in quality, U_x denotes the marginal utility of x, and U_c of c, so that the consumer equates the marginal increase in price with his marginal rate of substitution between x and c. Note, however, that care must be taken in interpreting this condition: if p is not convex, then the first-order condition may not characterize the consumer's optimum. Note that an increase in the price of "quality" is best thought of as an increase in the derivative $p'(c)$. Also note that so long as the cross-partial $U_{xc}(\cdot,\cdot)$ is not too large and negative, then quality will be a normal good for this consumer: increases in income translate into increases in the quality of the goods demanded.

Producers. Producers are taken to be the actors who make the management decisions and investments, and take actions necessary to actually produce some agricultural commodity. Producers have the same preferences as consumers, but we will find it convenient to work with the indirect utility function, denoted $V(w, \vec{p})$, where w is the producer's income, and $\vec{p}$ is a vector of prices, which are taken as a given by the producer. In addition to consuming, the producer takes some costly action $a \in A \subseteq \Re$, which determines the joint distribution of quality characteristics (c), and the information generated by the firm (b), given by $F(b, c \mid a)$. The wage received by producers is determined via negotiations with a firm, but in general may be contingent on any of the variables a, b, or c. Since the producer controls only a, his problem is to solve

$$\max_{a \in A} \int V[w(a,b,c) - a, \vec{p}] dF(b, c \mid a). \tag{2}$$

Firms. We turn next to the problem facing a representative firm in this environment of observable quality. Firms in this environment serve as intermediaries, hiring or contracting farmers to produce agricultural commodities, and then marketing the produce to consumers. The firm orders the producer to take action a, observes quality measurements b, and also observes quality characteristics c, just as consumers do. We denote by $f(b,c\,|\,a)$ the conditional probability density function (pdf) of (b,c), and assume for simplicity that $f(b,c\,|\,a) > 0$ for all $(b,c) \in B \times C$ and $a \in A$. Firms are assumed to take prices $p(c)$ as given, so that a profit-maximizing firm solves the problem

$$\max_{a \in A, \{w(a,b,c)\}_{(a,b,c) \in AxBxC}} \int [p(c) - w(a,b,c)] dF(b,c\,|\,a), \tag{3}$$

subject to offering a set of wages guaranteeing the producer a utility of at least $\underline{U}$,

$$\int V[w(a,b,c) - a, \vec{p}] dF(b,c\,|\,a) \geq \underline{U}. \tag{4}$$

If b and c are conditionally independent, this problem yields the first-order conditions

$$\int p(c) \frac{f_a(c\,|\,a)}{f(c\,|\,a)} dF(c\,|\,a) = \lambda V'[w(a,b,c) - a\,|\,\vec{p}] \tag{5}$$

and

$$1 = \lambda V'[w(a,b,c) - a\,|\,\vec{p}], \tag{6}$$

where $f(c\,|\,a)$ is the marginal conditional density of c, and where λ is the Lagrange multiplier associated with the participation constraint (4). Note from the first of these first-order conditions that the farmer's compensation $w(a,b,c)$ turns out to depend optimally only on a and $\underline{U}$, and from that combination of the two constraints the firm will instruct the farmer to choose a such that $Ep(c) f_a(c\,|\,a) / f(c\,|\,a) = 1$.

We briefly examine the relationships that emerge between farmers and intermediaries in this environment. First, since quality c is directly observable by all parties, there turns out to be no role for quality measurements b (mathematically, this is revealed by the fact that b integrates out of the firm's first-order conditions). Second, producers in this environment bear no risk and make no real decisions; they simply follow directions from the firm, and receive a non-contingent salary in exchange. Thus, in an environment with observable actions a and observable quality c, we expect something like perfect vertical integration to emerge as the most efficient organizational form.

Equilibrium. We define equilibrium in qualities as follows.

Definition. *A competitive equilibrium in qualities is a distribution of incomes across households* $G(x)$, *a price function* $p(c)$, *an investment level for firms of* a, *and an allocation of qualities* c *to a consumer with income* x *of* $d(c,x)$ *such that*

1. *Given the price function* $p(c)$ *and any income* x, $d(c,x)$ *solves* (1), *the problem facing a consumer with income* x*;*
2. *Given the price function* $p(c)$, a *solves* (3), *the firms' problem of choosing investments to maximize profits;*
3. *Firms' expected profits are zero, or*

$$\int p(c)dF(c \mid a) = w(a)\text{; and} \tag{7}$$

4. *Markets clear for commodities of almost every quality, or*

$$\int\int_{\underline{c}}^{c} d(\hat{c},x)dF(\hat{c} \mid a)dG(x) = F(c \mid a) \tag{8}$$

for all $c \in C$.

Example. Here we assume a particular parametric form for consumer preferences, and construct an equilibrium for an economy with these preferences. First, note that if preferences are Gorman-aggregable, then one can construct preferences for a representative consumer, of the form $\overline{U}[\overline{x}-p(c),c]$, where $\overline{x}$ is per capita income. Then the demand function for the representative consumer will be equal to per capita demand for individual consumers in the economy, with qualities demanded satisfying

$$p'(c) = \frac{\overline{U}_c[\overline{x}-p(c),c]}{\overline{U}_x[\overline{x}-p(c),c]}.$$

Market clearing implies that prices must be such that this first-order condition is satisfied for almost all c. Now, we select a particular parametric form of Gorman-aggregable preferences, assuming that $U(x,c) = \log x + \alpha \log c$. The preferences of the representative consumer take an identical form, so that any continuous differentiable price function must satisfy the differential equation

$$p'(c) = \alpha \frac{\overline{x}-p(c)}{c}.$$

Solutions to this equation take the form

$$p(c) = \overline{x} - \frac{k}{c^a} \tag{9}$$

for some constant k, determined by the requirement that markets clear. Accordingly, while the exact value of k depends on the pdf $f(c \mid a)$ = $\overline{x} - k / c^a$, for any pdf market clearing implies $k > 0$. A complete description of an equilibrium would involve specifying a particular pdf, and then finding a pair (k, a) to solve (7) and (3).

Unobservable Quality

To this point, we have made the extreme assumption that the quality of agricultural produce can be costlessly assessed, so that a different price $p(c)$ can be charged for every possible quality. This may be a reasonable assumption for some kinds of commodities – bananas in a U.S. supermarket, for example, differ in quality mainly according to their ripeness, which is easily evaluated by even casual inspection. However, it is not difficult to think of other kinds of quality characteristics that are not easily observed. Now let us suppose that the opposite is true: that when the consumer purchases a unit of this indivisible foodstuff, he does so in complete ignorance of its quality; the quality of the purchased good is revealed only when the foodstuff is actually consumed. Some version of this model lies at the heart of much earlier research on the welfare effects of minimum grading standards (Price 1967). The consumer knows the distribution of different quality outcomes, given actions and investments a taken by producers, the probability distribution $F(c \mid a)$ introduced in the section on observable quality. The consumer now has no real choice to make, since there is no way to choose among different qualities. So long as he consumes any of the foodstuff, his expected utility is now given by

$$\int U(x - p, c) dF(c \mid a).$$

Since U is concave by assumption, the problem facing a social planner will also be concave. As a consequence, the addition of uncertainty unambiguously reduces welfare, despite the fact that the firm can now charge only a single price p, which may be smaller than the price paid when quality is observed. If minimum quality standards could somehow be enforced despite the unobservability of c, then the imposition of such standards could be welfare improving, as in Price (1967). However, it is hard to imagine a useful characteristic that could be observed by a third party grader but not by firms or consumers.

Grading

Next we consider a type of hybrid of the two previous models. Quality is not directly observable, but some other random variable (the *grade*, denoted by $b \in B$) can be observed by both producers and consumers. Furthermore, while grade does not generally allow one to perfectly infer quality, it may provide some information. In particular, let $f(c,b \mid a)$ denote the joint pdf of quality and grade conditioned on action a taken by producers, while $f(c \mid b,a)$ denotes the distribution of unobserved quality conditioned on both grade and action, and $f(b \mid a)$ denotes the pdf of grade conditioned only on action. With this new notation in place, analysis proceeds much as in the *Observable Quality* model, with the difference that prices and ex ante consumer payoffs depend only on b. In particular, the consumer's problem becomes

$$\max_{b \in B} \int U[x - p(b), c] dF(c \mid b, a), \tag{10}$$

while firms solve

$$\max_{b \in B} \int p(b) dF(b \mid a) - w(a), \tag{11}$$

where wages $w(a)$ paid to producers are determined as in the *Observable Quality* model. A competitive equilibrium in this setting is analogous to the equilibrium defined by the *Observable Quality* model, but with the consumer's problem given by (10), the firm's problem by (11), and with prices indexed only by b, rather than by c. Note that if b and c are conditionally independent, then the model in fact reduces to the *Unobservable Quality* model, as grading yields no information regarding quality beyond that already revealed by knowledge of a.

Relative to the *Observable Quality* model, this problem involves greater uncertainty for consumers (what is the utility that will be derived from consuming foodstuff of grade b?). The concavity of the firms' and consumers' problems implies that this additional uncertainty results in a welfare loss. Two consequences of interest stem from this potential welfare loss. First, it may be worthwhile to make investments to improve grading. If consumers are better able to infer the quality of foodstuffs, then this extra information has value. Second, the strong conclusions drawn by Bockstael (1984) in an environment with quality being perfectly observable (minimum quality standards are always welfare-reducing) also apply to this more general setting, since consumers can always choose not to purchase goods which are *likely* to be of low quality, based on their observable grade.

Example. A recent example of this sort of grading problem is given by Chalfant et al. (1999), who consider the problem of sizing prunes. Whole prunes are often packaged in such a way that consumers cannot choose individual prunes based on size; instead they must select a package of prunes, drawn from the distribution of graded prunes. The chief element involved in prune grading is size. Prunes (and many other kinds of fresh produce) are sized by conveying them over a surface with holes of increasing size; once a prune reaches a hole larger than the diameter of the prune, it (usually) falls through into a bin of similarly sized prunes. Occasionally, however, a prune of fixed size fails to fall into the smallest hole through which it could fit, and instead falls through a hole meant to capture larger prunes. Note that the converse never happens.

In its design of a grading scheme, the prune industry appears to take the position that the quality characteristic valued by consumers is the weight of the prune, not its diameter. Accordingly, in this environment, we can think of the relevant quality characteristic, c, as just reflecting prune weight, and can think of the grade b as the radius of the prunes that are supposed to fall into a particular bin. If all prunes were perfect spheres of identical density, and if all prunes fell into the correct bin, then c and b would be related by $c = kb^3$, where the constant k is just $4/3\pi$ times the density of the prune. Since neither of these premises is exactly correct, c and b are actually related by $c = kb^3 + \varepsilon$, where ε is a grading error.

A nice feature of the Chalfant et al. (1999) paper is that the authors actually conduct experiments that allow them to construct precise engineering estimates of the distribution of ε; producers are assumed to be able to affect the size distribution of prunes by engaging in the costly practice of shaking trees, which tends to dislodge particularly small fruit. Using their estimates of grading error, the authors argue that this error leads to both lower consumer welfare and under investment in tree-shaking.[4]

Consumer Grading

Here we suppose that the intermediary can engage in quality measurement and grading, as in the previous section, but also that the consumer has some independent information that helps him to predict the utility derived with the consumption of some commodity. To model this, we modify the notation above, writing $c = h(c_1, c_2)$. The idea is that realized quality c is a function of

[4] Chalfant et al. (1999) claim that this is due to an "adverse-selection" problem. Because there is no asymmetry of information in their model, this seems an unusual use of the term. A sensible extension of this example might be to suppose that tree-shaking is an unobservable action, leading to an information asymmetry of the sort considered below in the section on contractor investment, but even this would be an example of moral hazard, not adverse selection.

characteristics c_1 observed by the consumer *prior* to purchase (think of a shopper checking the firmness of apples) and other characteristics (think of sweetness) observed only upon consumption.[5]

In this environment, prices can depend on measured characteristics b and *also* characteristics observed by consumers in advance of purchase c_1, as the firm sets retail prices to clear markets. As a consequence, the problem facing consumers is

$$\max_{(b,c_1)\in B\times C_1} \int U[x-p(b,c_1),h(c_1,c_2)]dF(c_2 \mid b,c_1,a), \tag{12}$$

while firms solve

$$\max_{a\in A} \int p(b,c_1)dF(b,c_1 \mid a)-w(a). \tag{13}$$

The additional information observed by the consumer reduces uncertainty, *ceteris paribus*. However, since only characteristics c_1 affect prices received by producers, quite perverse outcomes are possible when c_1 and c_2 are not independent.

Example. Here we report on a small experiment conducted by the author in the spring of 1999. We randomly selected four different kinds of Red Delicious apples from a local produce market ("small," "large," "extra large," and "organic"). Though we purchased a fairly large amount of each kind of apple, we carefully selected them so that there was much more apparent variation across types than within types. We gave 48 experimental subjects the opportunity to visually examine one of each of the four different types of apples, and asked them to rank the apples according to their *prediction* of utility derived from consuming the apple, based on their visual inspection (their "visual ranking"). We then gave the subjects a blind taste-test of the apples they had ranked, and asked them to rank the apples by utility derived from actual consumption (their ex post ranking).

There was considerable agreement among subjects regarding both visual and ex post rankings, with the average Spearman correlation coefficient between subjects' visual rankings (Friedman's statistic, which is distributed approximately chi-squared with three degrees of freedom for our experiment) equal to 25.7, while the same statistic for the ex post ranking was 10.69, both

[5] This twofold distinction is similar to one drawn by Nelson (1970), though he goes farther. Our c_1 corresponds to what he terms "search" characteristics, and our c_2 to what he calls "experience" characteristics. He also describes "credence" characteristics, which are valued by the consumer, but never observed. Examples of these last might include beliefs about production characteristics that directly enter consumer utility, such as whether or not a fruit was produced using organic methods.

highly significant. *However*, despite this agreement across subjects, visual rankings were a poor predictor of ex post rankings, the average Spearman correlation coefficient between the two rankings being equal to 0.23 (not significantly different from zero). This average hides the considerable heterogeneity in the apparent ability of subjects to use visual characteristics to predict ex post rankings. The modal Spearman correlation coefficient was 1, but if one sets aside these "perfect" predictors, then the average Spearman correlation coefficient drops to –0.12. Fourteen of the subjects gave predictions nearly the *reverse* of the ex post outcome (i.e., the apple they thought would be their favorite was in fact their least favorite; the apple they thought would be their second favorite was in fact their second least favorite, and so on). A firm marketing apples to these latter subjects would have done well to encourage growers to produce apples with appealing visual characteristics (large and particularly red in color) which actually turn out to be negatively related to appealing to taste in this sample.[6]

Contractor Investment

In each of the models considered so far, production information a is observed not only by the producer, but also by the firm. As a consequence, the contractual arrangements made between the firm and producers have looked very much as though the grower is simply a salaried employee of the firm – all risk in quality outcomes is borne by the intermediary, who simply pays the grower for doing his job (taking production decisions a, chosen by the firm).

In this section we turn our attention to the case in which the firm can suggest to the farmer that he take actions a, but may not be able to verify that the farmer actually follows those suggestions. This treatment is similar to the environment described by Hueth and Ligon (1999), but extends that treatment by being more explicit about the source of variation in prices received by the firm for a particular agricultural commodity. The producer retains considerable autonomy, but as a consequence must also bear some of the risk associated with the decisions he makes. The risk here stems from the possibility that a influences the probability of quality outcomes c in ways that cannot be easily measured by the firm. Thus, in designing the contract, the firm solves

$$\max_{a\in A,\{w(a,b,c_1)\}_{(a,b,c_1)\in A\times B\times C_1}} \int [p(b,c_1)-w(a,b,c_1)]dF(b,c_1\mid a), \quad (14)$$

to offering a set of wages guaranteeing the producer a utility of at least $\underline{U}$,

---

[6] This experiment seems to bear out the anecdotal conclusions drawn from interviews with several Washington State growers of Red Delicious apples reported in *The New York Times* (see Egan 2000).

$$\int V[w(a,b,c_1) - a, \vec{p}]dF(b,c_1 \mid a) \geq \underline{U}, \tag{15}$$

and subject also to the requirement that a, recommended by the firm, be consistent with the incentives facing the farmer, so that

$$a \in \arg\max_{\hat{a} \in A} \int V[w(a,b,c_1) - \hat{a} \mid \vec{p}]dF(b,c_1 \mid \hat{a}). \tag{16}$$

Now, if the firm offers the grower a constant compensation $w(a)$, the grower will respond by choosing the least expensive action $\hat{a}$; this is clearly inefficient. However, if the firm is able to observe only b and the price $p(b,c_1)$, the efficient contract will typically expose the grower to some price risk. Assume for simplicity's sake that $p(b,\cdot)$ is invertible, so that knowledge of b and $p(b,c_1)$ permits the firm to infer c_1. Then, so long as the production problem is suitably concave and the grower is sufficiently risk averse (Jewitt 1988), any interior solution to the contracting problem will satisfy

$$\frac{1}{V'[w(a,b,c_1) - a \mid \vec{p}]} = \lambda + \mu\left[\frac{f_a(b,c_1 \mid a)}{f(b,c_1 \mid a)} + \eta(a,b,c_1)\right], \tag{17}$$

where λ is the Lagrange multiplier associated with the participation constraint (15), μ is the multiplier associated with the incentive compatibility constraint (16), and $\eta(a,b,c_1)$ is the relative risk aversion of the producer. Note that when the incentive compatibility constraint is not binding, then we recover the constant compensation for growers seen in previous sections; when (16) is binding, then compensation depends on the market price via the likelihood ratio $f_a(b,c_1 \mid a)/f(b,c_1 \mid a)$.

Thus, in this environment it emerges that farmers will bear some of the risk associated with uncertain production of quality; this risk is necessary in order to provide growers with the appropriate incentives to make investments leading to higher quality output. This risk manifests itself in two ways: first, the grower's compensation will depend on the prices ultimately paid by consumers, since these prices allow one to infer what the quality characteristics c_1 are, which in turn provide some information on the unobserved action a. Second, quality measurements b will be of value only when these quality measures improve inference regarding a. In a survey of the contracts offered by intermediaries to producers of fruits and vegetables in California, the author has found that 59 percent of fresh market handlers write contracts that expose producers to this sort of price risk, contrasted with only 14 percent of processors, suggesting that a model with unobservable contractor investment is appropriate for many fresh market commodities.

CONCLUSIONS

In this chapter we interpret "grading risk" to be the kind of uncertainty associated with quality measurement b. Risk associated with these measures can affect both consumers and producers, but evaluation of this risk is complicated by the fact that these quality measures also influence prices and the form of the compensation firms offer producers.

We have presented a sequence of models with different assumptions regarding what is observed, and by whom; some stylized results from this sequence are summarized in Table 1. Our first model presumes that producers' actions and final quality characteristics can be observed by all parties. In this model consumers bear no risk – they only purchase commodities of known quality. Less obvious is the result that producers should also bear no risk – firms simply employ agents involved in production, telling them precisely what to do, and awarding them a non-contingent salary. An industry structure of complete vertical integration would deliver efficient outcomes in this environment.

Table 1. Information and Grading Risk

Model	Firm Observes	Consumer Observes/ Prices Depend On	Consumer Utility
Observable Quality	a,b,c	c	$U[x-p(c),c]$
Unobservable Quality	a	---	$E[U(x-p,c)\mid \underline{a}]$
Grading	a,b	b	$E(U[x-p(b),c]\mid a,b)$
Consumer Grading	$a, b, p(b, c_1)$	b,c_1	$E(U[x-p(b,c_1),c]\mid a,b,c_1)$
Contractor Investment	$b, p(b,c_1)$	b,c_1	$E(U[x-p(b,c_1),c]\mid \hat{a},b,c_1)$

continued....

Model		Firm Revenue	Example Commodities
Observable Quality	...	$E[p(c)\mid a]$	banana
Unobservable Quality	...	p	pig-in-a-poke
Grading	...	$E[p(b)\mid a]$	prunes
Consumer Grading	...	$E[p(b,c_1)\mid a]$	Delicious apples
Contractor Investment	...	$E[p(b,c_1)\mid \hat{a}]$	mature green tomatoes

Our second model modifies the first by assuming that quality characteristics cannot be observed by any party. Hennessy (1995) likens this environment to the "lemons" model of Akerlof (1970) (though Akerlof's model is one of hidden types, rather than of hidden actions); an immediate result is that producers will take the least costly action, which will typically mean that realized quality will be very low. Consumers now face uncertainty having to do with the quality of the goods they purchase; notably, producers still bear no risk, since it is not necessary to provide any incentives to improve quality.

Our third model maintains the assumption that quality is unobservable, but now supposes that the firm can do some kind of grading, which (imperfectly) reveals some information regarding quality. In the U.S., it seems likely that beef is a commodity produced largely in accord with this model. In this model, consumers continue to face risk, but the typical quality of the commodity is apt to be much higher. Producers continue *not* to face risk – an integrated packer, for example, may continue to simply compensate the producer according to the actions and investments made by labor.

Our fourth model adds onto the third the possibility that consumers may be able to do some informal grading of their own prior to purchase, observing some characteristics that allow them to predict the utility they will derive if they actually consume the good. An experiment with Red Delicious apples provides some evidence that this sort of informal consumer grading is useful to consumers, but that some uncertainty regarding quality remains.

Our fifth and final model finally relaxes the assumption that the characteristics of the producer are observable. The result is that efficient arrangements between the producer and firm no longer look like an employer-employee relationship, but rather like a contingent contract between two independent parties. In particular, the producer now bears some share of the risk associated with grading and in the price ultimately paid by consumers. Something like this set of arrangements is consistent with the majority of agricultural contracts in California.

REFERENCES

Akerlof, G.A. 1970. "The Market for 'Lemons': Quality Uncertainty and the Market Mechanism." *The Quarterly Journal of Economics* 84: 488-500.

Berck, P., and G.C. Rausser. 1982. "Consumer Demand, Grades, and Margin Relationships." In G.C. Rausser, ed., *New Directions in Econometrics Modeling and Forecasting in U.S. Agriculture*. New York: Elsevier.

Bierlen, R., and O. Grunewald. 1995. "Price Incentives for Commercial Fresh Tomatoes." *Journal of Agricultural and Applied Economics* 27: 138-148.

Bockstael, N.E. 1984. "The Welfare Implications of Minimum Quality Standards." *American Journal of Agricultural Economics* 66: 466-471.

___. 1987. "Economic Efficiency Issues of Grading and Minimum Quality Standards." In R.E. Kilmer and W.I. Armbruster, eds., *Economic Efficiency in Agricultural and Food Marketing*. Ames, IA: Farm Foundation.

Chalfant, I.A., I.S. James, N. Lavoie, and R.I. Sexton. 1999. "Asymmetric Grading Error and Adverse Selection: Lemons in the California Prune Industry." *Journal of Agricultural and Resource Economics* 24: 57-79.

Considine, I.I., W.A. Kerr, G.R. Smith, and S.M. Ulmer. 1986. "The Impact of a New Grading System on the Beef Cattle Industry: The Case of Canada." *Western Journal of Agricultural Economics* 11: 184-194.

Dimitri, C., J.K. Horowitz, and E. Lichtenberg. 1996. "Grading Services as a Mechanism for Dispute Resolution in Fruit and Vegetable Markets." Unpublished manuscript, Department of Agricultural and Resource Economics, University of Maryland, College Park.

Dupré, R. 1990. "Regulating the Quebec Dairy Industry, 1905-1921: Peeling Off the Joseph Label." *The Journal of Economic History* 50: 339-348.

Egan, T. 2000. "'Perfect' Apple Pushed Growers into Debt." *The New York Times* (November 4, 2000, p. A1).

Espinosa, J.A., and B.K. Goodwin. 1991. "Hedonic Price Estimation for Kansas Wheat Characteristics." *Western Journal of Agricultural Economics* 16: 72-85.

Freebairn, J.W. 1973. "The Value of Information Provided by a Uniform Grading System." *Australian Journal of Agricultural Economics* 17: 127-139.

Hennessy, D.A. 1995. "Microeconomics of Agricultural Commodity Grading: Impacts on the Marketing Channel." *American Journal of Agricultural Economics* 77: 80-89.

Hollander, A., S. Monier, and H. Ossard. 1999. "Pleasures of Cockaigne: Quality Gaps, Market Structure, and the Amount of Grading." *American Journal of Agricultural Economics* 81: 501-511.

Hueth, B., and E. Ligon. 1999. "Producer Price Risk and Quality Measurement." *American Journal of Agricultural Economics* 81: 512-524.

Jewitt, I. 1988. "Justifying the First-Order Approach to Principal-Agent Problems." *Econometrica* 56: 1177-1190.

Lancaster, K.J. 1966. "A New Approach to Consumer Theory." *Journal of Political Economy* 74: 132-157.

Marette, S., J.M. Crespi, and A. Schiavina. 1999. "The Role of Common Labelling in Context of Asymmetric Information." *European Review of Agricultural Economics* 26: 167-178.

Nelson, P. 1970. "Information and Consumer Behavior." *Journal of Political Economy* 78: 311-329.

Price, D.W. "Discarding Low Quality Produce with an Elastic Demand." *Journal of Farm Economics* 49: 622-632.

Chapter 17

FINANCE AND RISK BEARING IN AGRICULTURE

Peter J. Barry
University of Illinois

INTRODUCTION

Ongoing structural change in agriculture is significantly affecting the nature of risks faced by agricultural producers, the menu of available risk management practices, and the distributions of risk and returns within the food system (Boehlje and Lins 1998). A tri-model structure has emerged in production agriculture characterized by the co-existence of large industrialized units, commercial-scale family operations, and small, part-time or limited resource farms (see Economic Research Service 2000) for a more extensive typology). The industrialized component is characterized in part by large size, high levels of vertical coordination between agricultural production and other stages of the food system, and internalization of numerous risk management functions and other services (e.g., legal, accounting, communications, government relationships). Small farms, while numerous, make limited contributions to overall economic activity, are heavily dependent on non-farm income, and engage in minimal risk management.

The commercial-scale family farms represent the traditional component of a viable production sector, but family farms seem to be dividing and transitioning toward the other two categories. In responding to risks, commercial-scale farmers are especially proficient in dealing with production risks, they are improving significantly in addressing market risks, and they have subtle, yet proven abilities to handle policy risks through farm organizations and other influential means. In finance, those debt-using farmers (about half of all farmers) still are heavily dependent on risk management contributions from lenders, which is logical to expect from financial partners. They do not yet give the same attention to the "value of the firm" under risk as is found in other economic sectors. Finally, perhaps the most significant area of risk for many farmers nowadays is in the stability of their relationships with other contracting parties – ensuring access to markets through contractual opportunities with input suppliers and processors, competing for farmland leasing contracts, working with insurance agents and other providers of risk management contracts, and further developing the financial contracts that characterize lender-borrower relationships. Dealing with incentive and information issues is central to stabilizing and enhancing these types of contractual relationships.

My purpose in this chapter is to conduct a contemporary appraisal of the role of finance in agricultural risk management – from the vantage points of agricultural firms, financial markets and institutions, and public policy.[1] Finance inherently involves a long-term perspective on risk management. Valuing firms, evaluating new investments under risk, adjusting capital structures, managing liquidity, developing financial relationships, and creating new financial instruments and institutions are long-term, dynamic elements of strategic risk planning. Their effective use is clearly evident in the time paths of a farm's economic value and business performance.

AGRICULTURAL FIRMS: VALUATION UNDER RISK

In corporate finance, one of the standard approaches for characterizing a firm's risk position is to employ capital asset pricing model (CAPM) (or arbitrage pricing theory) concepts in order to estimate risk premiums on equity capital shares traded in well-developed financial markets. Systematic risk is measured and priced in terms of the investment's returns relative to a well-diversified market portfolio where much of the investment's stand-alone risk has been diversified away. The resulting risk premiums are used in estimating the cost of equity capital (with adjustments in financial leverage) for evaluating new investments or valuing the firm itself based on the corporate firm's weighted average cost of debt and equity capital (Robison and Barry 1996).

The basic question about the measurement of an investment's (or other activity's) risk is whether all of the project's risk should be considered or whether some of the risk can be diversified away (Robison and Barry 1996, Brigham and Gapenski 1991). The question hinges on the level of risk that the project *adds* to an economic unit. At one extreme, a firm that invests in only one asset and whose owner only invests in the firm would have the same risk at the asset, firm, and ownership levels. All of the asset's stand-alone risk would be considered in economic valuation analysis. At the other extreme, a firm (generally large, incorporated, and with publicly traded common stock) that invests in many assets, none of which is very large relative to total assets, and whose owners have portfolios that are well diversified, need only consider an investment's net contribution of risk to the investor's portfolio. Diversification at the firm level, then, has little or no value in terms of the firm's own risk-return position. In this case, the net contribution to risk is

[1] For previous appraisals of the relationships between finance and risk, see Barry and Robison (2001) and Lee and Baker (1984).

represented by the investment's systematic (or non-diversifiable) risk measured by the CAPM.

An intermediate point in this range is a firm that holds a relatively large number of assets (or enterprises), but whose owners concentrate their investments in the firm. This situation characterizes many farms and other small businesses in which ownership is concentrated in the hands of one or more families whose assets are heavily committed to the business. Under these conditions the investment analysis occurs at the firm level and the risk of a new investment or risk management practice is represented by its net contribution to the firm's risk.

With limited diversity and non-marketability of equity claims, the investment's contribution to the firm's risk probably is less than the investment's total risk, but likely involves some degree of non-systematic risk. This risk might be measured by employing the single-index model to compute a beta for the new investment in terms of a farm's total returns or a group of farms' total returns, or some other relevant index rather than the market portfolio (Collins and Barry 1988b, Tauer 2000, Hussain and Pederson 1994). The beta value then could be used to estimate the farm's cost of equity capital for use in determining the net present value of new investment earnings or risk management effects over time. Alternatively, a utility-based approach to measuring risk premiums could occur, although information about the investor's risk attitude then would be needed (Collins and Barry 1988a).

The basic point is to consider the amount of risk that is added to or subtracted from the farm's overall risk, recognizing that some of the stand-alone risk of the risky activity would be diversified away. These additions or subtractions can occur as the result of new investments or expansion plans, new contractual arrangements (i.e., contract production), or as a result of the utilization of various risk management techniques (enterprise diversity, marketing instruments, insurance, etc.) employed on a sustained basis as part of a long-term risk management strategy. Trade-offs between business risk and financial risk may also be involved (Gabriel and Baker 1980). The common basis for measurement is the effects of these risk changes on the farm's economic value resulting from the capitalization of its earnings, using either risk-adjusted earnings or a risk-adjusted discount rate.

The absence of trading markets for farmers' equity capital hampers applications of these approaches in agriculture. No clear pricing mechanisms are available to value risks and the effectiveness of farmers' risk management practices, or to gauge the rewards for risk bearing (Collins 1993). In Illinois, for example, it is well known that corn production has lower and more variable yields in southern Illinois than in northern Illinois. The market for number two yellow corn, however, is oblivious to farmers' risk positions. It delivers no "risk premium" to southern Illinois producers. Rather, the southern Illinois producers must seek returns to their risk bearing in other ways. Other work has attempted to identify risk premiums in futures and forward contracting markets (e.g., Townsend and Brorsen 2000). Even then, however, the risk premiums are not established within the context of a farm's full range of production, marketing, and finance activities.

Perhaps the best candidate for considering risk pricing in agriculture is the market for farmland. Because land values primarily reflect the characteristics of the anticipated future returns to land, one would expect riskier regions to have lower land values and higher risk-adjusted rates of return to land. Prior studies using CAPM concepts have demonstrated risk pricing of farmland relative to broader indices of market portfolios with results consistently showing relatively low systematic risks of farmland investments (e.g., Barry 1980, Bjornson and Innes 1992). Moreover, Chavas and Jones (1993) found a link between land values and the variability of farm income. Little evidence, however, is available about whether other costs and returns to risk bearing are reflected through differences in land values. More generally, we know little about the risk-adjusted costs of equity capital employed in agriculture, and how these costs would differ across risk regimes in agriculture.

FARM FINANCIAL STRUCTURE UNDER RISK

Distinguishing risk effects is further complicated by a dichotomy of financial structure behavior exhibited by farmers. Recently, we tested whether farm financial structure is explained by a partial adjustment theory (i.e., moving toward a target or equilibrium financial structure partially determined by risk considerations) or a pecking order theory of financial structure (Barry, Bierlen, and Sotomayor 2000).[2] The pecking order reflects an ordered preference to financing sources based on lower costs for internal sources of funds and higher costs for external funds, namely debt and leasing of farmland in this case. The premise is that the cost of debt capital may exceed the cost of a firm's internal sources of funds due to the lender's efforts to resolve asymmetric information problems and misaligned incentives in lender-borrower relationships. Such risk-reducing efforts create screening, monitoring, enforcement, and other agency costs associated with adverse selection and moral hazard problems, that generally are passed on to the borrower through higher interest rates or are incurred by the borrower through his or her own efforts to comply with the terms of the debt contract. Leasing arrangements between landlords and farmers may experience similar agency costs, although the longer-term nature of these relationships may reduce these costs relative to those in debt contracts.

In contrast, under the partial adjustment hypothesis, farmers are believed to adjust their debt, equity, and leasing levels towards an optimal, cost-minimizing financial structure. Debt costs generally are lower than equity costs because of the equity holder's higher financial risk (and reflecting the

[2] Also see Ahrendsen, Collender, and Dixon (1994) and Collins and Karp (1993) for related work on farm financial structure.

tax deductibility of interest). Both costs tend to increase when leverage increases as the result of greater financial risks and related administrative and agency costs. Risk premiums on equity capital and the resulting financial structure may also reflect the decision maker's level of risk aversion (Collins and Barry 1988a, Collins and Karp 1993, Gwinn, Barry, and Ellinger 1992, Jensen and Langemeier 1996).

An empirical application tested these theories by fitting a set of simultaneous financial equations with farm panel data from Illinois. Model results showed support for both theories, in that farms follow a pecking order in adjusting to changes in, or deviations from, long-run targets on financial structures, with a stronger pecking order effect for farms with greater asymmetric information problems. Especially important are the relationships between cash flow and investment, debt financing, and leasing. Strong cash flows lead crop farmers to expand through leasing land and other investment expenditures while paying down debt or refraining from borrowing. Weak cash flows or deficits may lead to lower investments, potential disinvestments, and increased borrowings. The implications for lenders and farmers are important, because farmers' financing demands (and even go-for-broke behavior) could increase substantially at the same time that their creditworthiness is declining. Moreover, the dichotomous approach to financial structure hampers its use as a direct indicator of farmers' risk attitudes or risk response behavior.

FARM REAL ESTATE LEASING AND RISK

Leasing of farmland by farmers is extensive, and recent evidence indicates a shift in rental arrangement away from crop share leases toward cash leases (although share leases still dominate in Illinois and other highly productive land areas) (Barry et al. 2000). Share leases have been highly risk-efficient financial arrangements because they reflect a perfect, positive correlation between the farmer's production and rental obligations, thus stabilizing the farmer's income position. Cash rents and debt obligations do not reflect such formal correlation relationships. Cash leases have other liquidity and risk implications because part of the cash rent is due in advance and the farmer must pay all of the crop's variable costs of production rather than sharing them with landlords. Relative to fixed cash leases, farmers can transfer production risk to landowners in the same proportion that crops are shared, with evidence of little if any reduction in expected returns due to the rigidities of conventional sharing levels (e.g., ½ - ½; ⅓ - ⅔) in different production regions (Barry et al. 2000). While the relationship between farmland leasing and risk is an old topic, further work is likely needed to better delineate and value the contemporary risk effects of the respective leasing choices.

FINANCIAL MARKETS AND RISK BEARING

The conventional view of financial leveraging is that greater reliance on external capital increases a firm's financial risk, as reflected by a greater probability of loss, diminished liquidity, and increased transaction costs of dealing with external parties. Most financial reporting systems in agriculture, indeed, indicate that, at some point, higher financial leverage is associated with diminished farm financial performance (e.g., Ellinger et al. 2000). At the same time, however, financial markets and institutions can contribute extensively to the effectiveness of risk bearing in agriculture. These contributions reflect the financial partnering role of lenders with borrowers and the incentive arrangements and information positions lenders experience as financial leveraging intensifies. Debt capital not only reflects financial claims on a firm's assets and earnings, but control rights as well.

In light of the inherent incompleteness of financial contracts and unanticipated contingencies, control allocation rules are needed to determine who is best suited to exercise control in various situations and the types of signals needed for control transfer (Berglof 1990). Under normal conditions, equity holders own and control the firm's assets. Under increasing adversity, however, loan performance can quickly deteriorate and indebtedness may reach levels at which the consequences of the borrower's actions that lead to default are increasingly borne by the lender. At some point, it is logical for the residual rights of ownership and control to shift to lenders so that they have the appropriate incentives to exercise the degree of effort needed to protect their debt claim. In this light, debt capital is a form of "contingent ownership" of the borrowing firm's assets and enterprises, and the control process is strongly influenced by the risk-bearing mechanisms contributed by the lender and arising from the lender-borrower relationships.

Financial institutions and markets can contribute in numerous ways to risk management and risk bearing in agriculture. Included are the following.

Pooling and Spreading Risk

By virtue of the large size and diversity of loan portfolios, financial institutions can significantly reduce the stand-alone credit risks of individual loans, similar to the diversifying of non-systematic risks discussed earlier for farm businesses. As a result, the average risk premium contained in borrowers' loan rates can be reduced to reflect the risk added by loans to the overall loan portfolio, and the premiums then can be allocated among borrowers based on risk-based adjusted interest rates that reflect the respective levels of systematic risk. Many lenders do utilize risk-adjusted pricing schemes, although they tend to group borrowers into several credit risk classes for monitoring and pricing, rather than customizing credit arrangements to each borrower. Generally, the respective risk premiums are developed in an ad hoc fashion, rather than precisely reflecting specific differences in probabilities of loan losses across borrowers.

Larger sizes and geographic scope of institutions further add to potential gains in risk efficiency from diversifying, although the marginal gains likely diminish significantly as these expansions continue. These patterns were confirmed by Sherrick, Barry, and Ellinger (2000) in the simulation of portfolio credit risks for securitized agricultural mortgage loans with relatively strong underwriting characteristics. Our results indicate that actuarial insurance costs against credit risk are initially highly sensitive and then become relatively insensitive as pool size increases. Much of the gains in risk efficiency are realized by pools in the $50 million to $100 million range and above. The savings in interest rates appear low for individual borrowers, but are of considerable magnitude when viewed in terms of aggregate farm debt and in terms of their accumulation through time.

Risk Rating and Credit Scoring of Borrowers

A major function of financial intermediation is the collection and processing of information about the creditworthiness of borrowers. The goals are to reduce the likelihood of adverse selection and moral hazard problems, minimize probabilities of loss, tailor financing terms and requirements to the credit risk characteristics of individual borrowers, and to determine sufficient capital reserves to hold against expected and unexpected loss. The result is a rich set of information about the risk positions of borrowers and how a farmer's financial performance, business practices, and financing arrangements may influence these risk positions.

Lenders accomplish these ratings through a combination of judgmental and quantitative analysis. The quantitative analysis may be based on worksheets reflecting the lender's own experience or on scoring models estimated using statistical regression procedures applied to farm- or loan-level financial data (Splett et al. 1994, Turvey 1991, Miller and LaDue 1989). Either approach involves the identification of key variables reflecting different attributes of the borrower's loan performance, selection of measures for the variables, placing weights on the variables, collecting appropriate data, and combining this information into a weighted average credit score.

For feasibility of implementation, lenders generally group their borrowers into a designated number of risk classes (e.g., three to five for smaller institutions, ten to twenty for larger institutions) and give extended attention to borderline cases. In some instances (e.g., smaller loans), the credit score or risk rating determines the loan decisions. Generally, however, the rating approach yields information that is combined with other judgmental factors to make loan decisions, price loans, monitor borrowers, evaluate portfolio risks, and communicate among lending personnel.

For internal institutional management and regulatory capital purposes, larger financial institutions and regulators are giving increasing attention to the use of comprehensive credit risk models, which determine probabilities of loss, and which also consider interest rate and operational risks (Saunders 1999, Caouette, Altman, and Narayanan 1998). Improved conceptualization of loss probabilities, new computational technologies, enhanced marketability of loans through securitization, and new derivative products for managing credit risk are some of the motivating factors. Probabilistic concepts are now rigorously employed to measure expected and unexpected loan losses, correlations among loans, and risk tolerances. Econometric models are receiving wider use in estimating loan-level relationships to credit risk for application to projecting potential losses and risk ratings on all loans comprising loan portfolios. A typical result of these new models is "value-at-risk" information that helps to make capital adequacy decisions at all levels of the institution – loan level, customer level, loan department level, loan portfolio level, and institutional level (Matten 2000).[3] These developments should lead to improvements in the capacities of financial markets to identify and carry credit risks.

Financial institutions (and specialized financial rating companies) are unique in generating these comprehensive risk ratings, credit scores, and other measures of credit risk for borrowers. Agricultural lenders are no exception to this process. No other entities provide comparable, comprehensive risk assessments (even country risk ratings are compiled by international lenders). The risk ratings compiled by insurance companies generally focus on specific contingencies rather than overall performance. Even the new "revenue insurance" tools for farmers are based on ratings of revenue rather than net income or wealth positions of asset holders.

The generality of lenders' risk ratings is limited, however, by the different perspectives of lenders and borrowers. In entering into the loan contract, the lender is primarily concerned about adverse selection of borrowers (whether the borrower is riskier than believed at loan origination) and moral hazard (whether the borrower generates greater credit risk during the term of the loan contract). These problems are caused by misaligned incentives as well as asymmetric information (Stiglitz and Weiss 1981, Jensen and Meckling 1976, Baker 1968). The borrower is motivated by profitability and wealth accumulation, because he or she shares directly in the returns (favorable or unfavorable) earned by the loan proceeds. In contrast, the lender is restricted to the fixed return of the loan funds plus interest. Thus, in evaluating a borrower's creditworthiness, the lender emphasizes loan repay ability and safety while the borrower focuses more on profitability and wealth. The result of the

[3] The recent and rapid adoption of "value-at-risk" concepts represents a breakthrough in applying the older "safety first" paradigm to bridge the gap between theory and application in risk analysis.

lender's evaluation is extensive information about the borrower's risk position, although it is oriented toward the lender's goal attainment.

Performance Monitoring

Closely related to the risk rating and scoring procedures is performance monitoring by lenders. Monitoring is intended to reduce asymmetric information between the lender and borrower throughout the loan contract, and to guard against potential opportunistic behavior. Monitoring, therefore, is part of the cost of managing credit risks. Lenders attempt to minimize these costs by tailoring the extent and style of monitoring to the borrower's characteristics. An extreme example is the case of the very large, multi-national, highly leveraged firm that has lender personnel located on site in the borrower's business, and where the lender has on-line access to the borrower's accounting system. An opposite extreme is the case of micro finance in developing countries in which monitoring helps ensure the lender that the borrower is "still there."

Agricultural lenders typically conduct monitoring through the traditional "farm visit," and other types of contact with borrowers. In some cases, the lender's monitoring function is out-sourced through the use of, for example, warehousing companies to monitor collateral and field servicing companies to handle the borrower's payment transactions. Out-sourcing then reflects the cost effectiveness of utilizing specialized monitoring services. In general, monitoring is a natural part of lender-borrower relationships and contributes importantly to the risk-bearing capacity of financial markets.

Providing Liquidity

The concept of a credit reserve, maintained as unused borrowing capacity, has had a central focus as a means of providing liquidity to farmers (Baker 1968, Barry, Baker, and Sanint 1981). Farmers reflect this focus in terms of having a "friendly banker." The idea is to fall back on the credit reserve in times of adversity when financial obligations are difficult to meet with cash flows or holdings of financial assets.

A credit reserve can be a highly efficient form of liquidity, as long as adversity is not excessive and adverse incentives do not lead to go-for-broke behavior by borrowers. Utilization of such reserves is reflected by the establishment of formal lines of credit and by lenders' flexibility in providing forbearance, carry-over loans, and refinancing when farmers experience adversity. Lenders, therefore, can contribute significantly to risk bearing in agriculture through these liquidity channels and the related smoothing of farmers' cash flows.

Considerable empirical work over the years has addressed how an agricultural borrower's credit capacity and financial strategies respond to actions, performance levels, and structural characteristics of farm businesses. The typical approach has been to elicit lenders' responses to case loan situations and to evaluate their implications for financial performance using farm-level simulations or optimization models. Applications include forward contracting and credit (Barry and Willmann 1976), crop insurance and credit (Pflueger and Barry 1986), income variability and credit (Barry, Baker, and Sanint 1981), split lines of credit (Baker 1968), asset structure (Sonka, Dixon, and Jones 1980), leased vs. owned land and credit (Baker 1968), mortgage vs. contract financing of farmland (Baker 1968), and financing independent vs. contract hog production (Barry et al. 1997). The empirical results of these studies have consistently documented the linkages between borrowing capacities (and other financing terms) from lenders and the borrower's decision making. The pecking order theory of farm financial structure also highlights the importance of credit reserves, because access to external debt capital under cash shortfall conditions depends on the lender's willingness to respond.

Financial Constraints on Debt and Equity Capital

External credit rationing by lenders is suggested as a plausible outcome to informational asymmetries that hamper the lender's distinction between high-risk and low-risk borrowers, either through credit denials to some potentially creditworthy borrowers or through limitations on loan sizes to all potentially creditworthy borrowers (Stiglitz and Weiss 1981, Gale and Hellwig 1985).[4] Credit rationing may also arise under other conditions in which the demand for credit exceeds the supply of credit. Moreover, any type of external credit rationing may arise whether or not borrowing actually occurs, through constraints placed on the capacity to borrow.

Such external constraints largely reflect uncertainties about loan performance, and thus constrain the borrower's financial risks, at least from the lender's point of view. Borrowers, themselves, may internally ration the use of their borrowing capacity due to the effects of their risk and/or time attitudes (Barry and Baker 1971, Barry, Robison, and Nartea 1996). Holding some credit in reserve creates a valuable source of liquidity, especially when other methods of risk management are limited. Distinguishing between external and internal credit rationing can be difficult, in light of the wide range of lenders' non-price provisions in financial contracts (see the following section).

Ultimately, the primary constraint for borrowers is equity capital rather than debt capital. Under asset-based lending or when limits are placed by

[4] Effective collateral pledges have been shown to partially alleviate these types of credit rationing (Bester 1985).

lenders on the borrower's financial leverage (i.e., the debt-to-equity ratio), increases in equity capital are needed to expand access to debt capital. For most farmers, retained earnings and long-term capital gains on farmland are the only feasible ways of generating equity capital. Family or other partnerships can pool equity to expand credit capacity. Other forms of external equity capital seldom are employed, except for the case of large, integrated, industrialized agricultural firms that can directly access external capital markets.

Financial Contracting and Constraints on Decision Making

Besides the financial constraints just cited, lenders may also exercise various types of risk controls over the borrower's decision making. The "contingent ownership" nature of debt capital, discussed above, implies that the residual rights of ownership and control should increasingly shift to lenders as their stake in the borrower's unit increases and as more of the costs of the borrower's actions are incurred by lenders through increased credit risks. Such shifts provide lenders with the appropriate incentives to effectively protect their debt claim. Most debt contracts, indeed, are written in this fashion. That is, under extreme adversity, ownership and control revert to debt holders according to the seniority and size of their claims.

Debt contracts can accomplish this through an increasingly extensive set of contract provisions, as adversity increases and as loan situations become more complex. Included among the terms and covenants in debt contracts may be the following:

- loan maturity
- repayment schedule
- collateral requirements
- third party guarantees
- loan commitments for future fund availability
- loan repayment on demand
- late fees and penalties
- foreclosure procedures
- inspections
- reporting requirements
- required accounting formats
- maintained performance standards
- sales restrictions on assets and products
- withdrawal limitations on dividends and family consumption
- limitations on additional debt and leasing
- insurance requirements
- compliance certification

The length and scope of this list illustrates the strong influence that lenders may have on a borrower's range of decision making and risk position.

These contract provisions are intended to forestall, correct, or constrain the actions of borrowers that cause adverse selection and moral hazard problems, thus improving the prospects of loan performance. The greater are the complexity and potential adversity of loan situations, the more extensive are the provisions in loan contracts. As debt instruments, these contracts do not directly involve lenders in the borrower's decision making. At the same time, however, the provisions may increasingly constrain the range of choices available to borrowers, thus resembling the effects of managerial participation. The end results are to control the borrower's risk position and, thus, safeguard the lender's debt claim.

Encouraging Other Risk Management Practices

In addition to these control mechanisms in financial contracts, lenders may also influence the borrower's risk management in less direct ways. By encouraging farmers to employ forward contracts for crop sales, hedging in futures markets, crop insurance, and other risk management practices, lenders will improve the prospects of the borrower's successful loan performance and protect more fully against credit risk. Such encouragement may not be formally expressed in a loan contract, but it may occur more subtly through expanded access to credit, lower costs of borrowing, and greater flexibility by lenders in exercising forbearance. Borrowers learn about these potential influences and may employ risk management more rigorously to more strongly solidify the lender-borrower relationship.

Anecdotal evidence of these lender influences came to light in recent focus groups with Illinois farmers about risk management. The intensity of use of crop insurance, marketing instruments, and other risk management practices clearly increased as farmers' obligations to meet fixed financial payments on debt and leased acreage became greater. The safety motivations for both farmers and lenders clearly converged under these conditions (none of these farmers appeared to [yet] be in a go-for-broke mode).

Credit Guarantees

Financial markets may enhance the access to credit of high-risk borrowers through the loan "guarantee" concept of public programs, through co-signees on notes, or through the use of commercial loan insurance. Public loan guarantees for commercial lender financing of young, limited resource, or emergency borrowers are made explicit through the credit programs of the Farm Services Agency of the U.S. government and many state credit programs. Implied government guarantees also stand behind the Government-Sponsored Enterprise programs of the Farm Credit System and Farmer Mac. Many of the new securitization instruments also include credit enhancements in the form of guarantees or insurance that add another safety element for investors and reallocate risk bearing in a cost-effective fashion. Thus, guarantees are another means by which financial markets can defuse risks and add to overall risk-carrying capacity in agriculture.

Financial Discipline and Borrowers' Efforts

Michael Jensen's (1986) free cash flow concepts may also apply to the relationship between credit risk and external financial obligations. The general idea is that businesses with surplus cash and small, if any, external obligations may become lax in management, squander resources, and be inattentive to effective performance. The creation of external obligations then may have beneficial effects through the motivation of managers to expend the effort needed to successfully meet the obligations and, thus, signal their reputations for high performance.

Jensen used this argument to rationalize a high incidence of leveraged buy-outs and takeovers in corporate finance. Business performance could quickly increase as a result of the need to meet these new obligations. Nasr, Barry, and Ellinger (1998) found similar evidence of a positive relationship between the production efficiency of farms and their financial leverage positions, with other farm and operator characteristics accounted for. Greater debt obligations were associated with higher levels of production efficiency. Thus, the involvement of financial institutions and markets with farm businesses could heighten some performance attributes and have risk-reducing effects, at least until conditions for borrowers' adverse incentives and go-for-broke behavior arise.

FINANCIAL POLICIES AND RISK BEARING

Credit programs and other financial policies may also significantly contribute to risk bearing in agriculture. Lenders of last resort, such as the Farm Services Agency and various state credit programs in the U.S., have the targeted mission of providing financial products and services to agricultural borrowers whose credit risks are high enough to disqualify them for commercial finance, but who have reasonable prospects for eventual graduation to commercial lenders. Many other countries have similar programs targeted to agriculture, including the role of agricultural development banks in developing countries. The recent emphasis on partial governmental "guarantees" of loans by commercial lenders utilizes the skills of these lenders in credit evaluation, provides incentives for the lender's risk controls through loss sharing, reduces the costs and subsidies of direct public credit, and is believed to facilitate the borrower's transition to non-government support.

In this process, the government guarantee (and direct loans as well) relieves the commercial lender from fully bearing the credit risk, although various forms of adverse selection and moral hazard still may characterize the public credit relationship (LaDue 1990). The magnitude of this public credit support is evident from the experiences of farm financial stress in the 1980s. During this time, commercial banks and the Farm Credit System in the U.S. experienced combined loan losses of $8.37 billion. In contrast, the Farm Services Agency (and its predecessor, the Farmers Home Administration) has experienced approximately $21 billion in losses since the early 1980s.

Most debt-using commercial-scale family farms rely on lenders who have specialized knowledge about, dedication to, and perhaps vulnerability from agricultural lending. Specialization helps to overcome the information problems and relatively high transaction costs of agricultural lending. Competitive yet still relationship-based lending remains the favorite approach to resolving adverse selection and moral hazard problems, and allows the lender to exert stronger controls during downturns in economic cycles.

In providing risk-bearing services, the role of farm credit programs is closely related to the use of other agricultural policy instruments that influence the level and variability of farm income. If these other elements of agricultural policy play a strong role in risk bearing and income enhancement, the types of financial institutions serving agriculture matter much less. If, however, agricultural policy as well as farmers' own risk management play a reduced role in countering the effects of agricultural risks, then the nature of agricultural finance can matter a great deal.

The higher-risk case emphasizes the importance of lenders who are specialized in and dedicated to agricultural finance through all phases of the economic cycle. The rub in this case, however, is that specialized, dedicated agricultural lenders themselves may need some level of government or institutional backup and support to offset the risk effects of their concentrated and sustained lending programs.

The U.S. example for this situation is the case of financial institutions that are government-sponsored enterprises (GSEs). Through their federal charters, the agency status of the debt securities they sell in the financial markets, and the implied (but not mandated by statute) government backing, the Farm Credit System and Farmer Mac are able to provide specialized and dedicated financial services to agriculture. Moreover, both of these GSEs have experienced statutory and regulatory changes throughout their histories, in part intended to enhance their risk-carrying capacities to the benefit of agricultural borrowers through expanded credit availability and lower interest rates.

The risk-bearing function of financial markets clearly makes credit programs an inherently important part of agricultural policy. Targeted assistance and filling market gaps in risk bearing are major motivations for public credit programs (Bosworth, Carron, and Rhyne 1987). At the same time, however, the integrity of credit markets is vulnerable to the political popularity of credit programs and to over-reliance on credit as a means of channeling subsidies to targeted groups. Thus, the larger the subsidy needed to achieve the public purpose, the less the assistance should be channeled through public credit programs.

RESEARCH IMPLICATIONS

This chapter has emphasized the identification of key issues and problems associated with finance and risk bearing in agriculture. The empirical focus reflects the diverse, changing structural characteristics of agriculture and finance. The time focus is long-term, so that the economic value of agricultural businesses can reflect the risk-return effects of strategic risk management behavior in which short-term production and market responses to risk are part of a long-term risk management strategy. The farm-level focus is also comprehensive in that a firm's overall risk position (i.e., risk adjusted wealth or expected utility) is evaluated in terms of the marginal effects of selected risk management practices, net of their diversifiable risks. The risk-bearing dimensions of financial markets and policies also are considered.

This empirically motivated, long-term, and comprehensive perspective has several key implications for how we conceptualize our work, the data that are employed, and the methods of analyzing these data. The importance of information, incentives, contractual relationships, and transaction costs in risk analysis gives value to the insights provided by organizational economics (or the new institutional economics) as a complement to the traditional production function characterization of the firm. Elements of agency theory, incomplete contracting, property rights, and transaction cost economics are included.

These vantage points help to identify the kinds of costs and control relationships that not only affect the identification of a firm (i.e., independent vs. integrated stages), but help in determining data needed to estimate the organization's economic value under risk conditions. Agency and transaction costs, including those incurred to safeguard against adverse incentives, incomplete contracts, and opportunistic behavior, may be especially difficult to quantify. Experimental approaches, case studies, or other elicitation procedures often are needed to generate primary, micro-analytic data in these cases (Lajili et al. 1997, Moss 2000). These approaches are similar to past efforts in eliciting risk attitudes, conducting contingent evaluations, and measuring lenders' credit responses to borrowers' managerial practices and financial strategies.

The economic value of the firm perspective utilizes costs of financial capital and measures of systematic risk. Estimating farmers' costs of equity capital under risk remains problematic, as is information about the pace and magnitude of changes in intertemporal risk (and time) attitudes that reflect decreasing, constant, or increasing absolute or relative risk aversion. Considerable short-term risk analyses have employed these longer-term theoretical concepts of risk attitudes, without much validation of their empirical characteristics.

The costs of equity capital contain risk premiums representing the level of risk (both business risk and financial risk) and risk attitudes. Some demonstrations of single index and related approaches to estimating marginal changes in risk and returns have occurred. However, we have not made much headway in mapping demand and supply schedules of risk-bearing services for different types of agricultural firms and linking this schedule to the decision maker's cost-of-equity for use in estimating changes in the firm's economic value as new and different combinations of risk and control actions are implemented.[5]

Quantification of systematic risks associated with the demand and supply schedules of risk-bearing services remains a high priority topic, in contrast to the tendency toward micro-level case studies where the stand-alone risk of various management alternatives is the primary measure. Also promising is further exploration of the risk-pricing of agricultural assets, at a level below that of a global market portfolio.

Employment of multi-period, stochastic farm-level models to analyze risk management alternatives has a long history, and the relative advantages of traditional simulation versus optimization approaches have been well established (e.g., Mapp and Helmers 1984). Simulation approaches likely have made more headway in recent times, thanks to enhanced computational technologies, greater flexibility in accommodating empirical characteristics, inter-relationships among model components, varying shapes of probability distributions, and a broader range of decision maker objectives. Still, however, optimization models have advantages as well, especially in identifying the role of various types of constraints on goal attainment.[6]

It seems advisable to choose analytical methods that best fit the goals of the analysis, the empirical characteristics of the situation under study, and the availability of usable data. As stated in the chapter's introduction, the ongoing structural changes in the farming sector and increasing concerns by farmers about contractual relationships may lead to new or different methods

[5] See Gardner (1978) for a characterization of the demand and supply schedules of risk-bearing services.

[6] Recently, we applied conventional risk (quadratic) programming to analyze, among other issues, cash vs. share leasing of farmland where share leasing was "more costly, but less risky" for the farmer because of the need to compensate the landowner for risk bearing. The appearance (i.e., dominance) of "more costly" share leasing in the risk-neutral solution (where risk reduction had no value, but liquidity was constraining) was then explained by the favorable liquidity consequences of share leasing. Under cash leasing, the farmer must make rental payments in advance and bear all of the operating expenses, thus depleting financial resources more quickly. These cost-risk-liquidity relationships would be more challenging to depict in a simulation approach.

of generating research data and to more emphasis on conceptual linkages among organizational alternatives, risk and other performance attributes, control over decisions and returns, and related policy issues. In finance, we have no shortage of relevant risk questions, but a significant step remains one of effectively linking finance to the other areas of risk analysis.

REFERENCES

Ahrendsen, B.L., R.N. Collender, and B.L. Dixon. 1994. "An Empirical Analysis of Optimal Farm Capital Structure Decisions." *Agricultural Finance Review* 54: 108-119.

Baker, C.B. 1968. "Credit in the Production Organization of the Firm." *American Journal of Agricultural Economics* 49: 507-521.

Barry, P.J. 1980. "Capital Asset Pricing and Farm Real Estate." *American Journal of Agricultural Economics* 62: 41-46.

___. 1995. *The Effects of Credit Policies on U.S. Agriculture.* Washington, D.C.: The AEI Press.

Barry, P.J., and C.B. Baker. 1971. "Reservation Prices on Credit Use: A Measure of Response to Uncertainty." *American Journal of Agricultural Economics* 53: 222-227.

Barry, P.J., C.B. Baker, and L.R. Sanint. 1981. "Farmers' Credit Risks and Liquidity Management." *American Journal of Agricultural Economics* 63: 216-227.

Barry, P.J., R. Bierlen, and N. Sotomayor. 2000. "Financial Structure of Farm Businesses Under Imperfect Capital Markets." *American Journal of Agricultural Economics* 82: 920-933.

Barry, P.J., L.M. Moss, N.L. Sotomayor, and C.L. Escalante. 2000. "Lease Pricing for Farm Real Estate." *Review of Agricultural Economics* 22: 2-16.

Barry, P.J., B. Roberts, M. Boehlje, and T. Baker. 1997. "Financing Capacities of Independent Versus Contract Hog Production." *Journal of Agricultural Lending* 10: 8-14.

Barry, P.J., and L.J. Robison. 2001. "Agricultural Finance: Credit, Credit Constraints and Consequences." In B. Gardner and R. Rausser, eds., *Handbook of Agricultural Economics.* Amsterdam: North-Holland. Forthcoming.

Barry, P.J., L.J. Robison, and G. Nartea. 1996. "Changing Time Attitudes in Intertemporal Analysis." *American Journal of Agricultural Economics* 78: 972-981.

Barry, P.J., and D.R. Willmann. 1976. "A Risk Programming Analysis of Forward Contracting With Credit Constraints." *American Journal of Agricultural Economics* 58: 62-70.

Berglof, E. 1990. "Capital Structure as a Mechanism of Control: A Comparison of Financial Systems." In M. Aoki, B. Gustafson, and O. Williamson, eds., *The Firm as a Nexus of Treaties.* London: Sage Publications

Bester, H. 1985. "Screening Vs. Rationing in Credit Markets With Imperfect Information." *American Economic Review* 75: 850-855.

Bjornson, B., and R. Innes. 1992. "Another Look at Returns to Agricultural and Non-Agricultural Assets." *American Journal of Agricultural Economics* 74: 109-119.

Boehlje, M.D., and D.A. Lins. 1998. "Risks and Risk Management in an Industrialized Agriculture." *Agricultural Finance Review* 58: 1-16.

Bosworth, B., A. Carron, and E. Rhyne. 1987. *The Economics of Federal Credit Programs.* Washington, D.C.: The Brookings Institution.

Brigham, E.F., and L.C. Gapinski. 1991. *Financial Management: Theory and Practice* (6th ed.). Chicago: Dryden Press.

Caouette, J.B., E.I. Altman, and P. Narayanan. 1998. *Managing Credit Risk.* New York: John Wiley & Sons.

Chavas, J.P., and B. Jones. 1993. "An Analysis of Land Pricing Under Risk." *Review of Agricultural Economics* 15: 351-366.

Collins, R.A. 1993. "The Robustness of Single Index Models in Crop Markets: Comment." *Journal of Agricultural and Resource Economics* 18: 131-134.

Collins, R.A., and P.J. Barry. 1988a. "Beta-Adjusted Hurdle Rates for Proprietary Firms." *Journal of Economics and Business* 40: 139-145.

___. 1988b. "Risk Analysis With Single Index Models: An Application to Farm Planning." *American Journal of Agricultural Economics* 68: 152-161.

Collins, R.A., and L.S. Karp. 1993. "Lifetime Leverage Choice for Proprietary Farmers in a Dynamic, Stochastic Environment." *Journal of Agricultural and Resource Economics* 18: 225-238.

Economic Research Service, USDA. 2000. "Farm Resource Regions." Agricultural Information Bulletin No. 760 (September).

Ellinger, P.N., C.L. Escalante, P.J. Barry, and D. Raab. 2000. *Financial Characteristics of Illinois Farms*. University of Illinois: The Center for Farm and Rural Business Finance.

Gabriel, S.C., and C.B. Baker. 1980. "Concepts of Business and Financial Risk." *American Journal of Agricultural Economics* 62: 560-564.

Gale, D., and M. Hellwig. 1985. "Incentive-Compatible Debt Contracts: The One-Period Problem." *Review of Economic Studies* 52: 647-663.

Gardner, B.L. 1978. "Farm Policy and Research on Risk Management." In *Market Risks in Agriculture: Concepts, Methods, and Policy Issues*, Technical Report No. 78-1, Texas Agricultural Experiment Station, College Station, TX.

Gwinn, A.S., P.J. Barry, and P.N. Ellinger. 1992. "Farm Financial Structure Under Uncertainty: An Application to Grain Farms." *Agricultural Finance Review* 52: 43-56.

Hussain, A., and G. Pederson. 1994. "Application of the Single Index Model to Minnesota's Agriculture." *Journal of the ASFMRA* 58: 102-108.

Jensen, F.E., and L.N. Langemeier. 1996. "Optimal Leverage With Risk Aversion: Empirical Evidence." *Agricultural Financial Review* 56: 85-97.

Jensen, M.C. 1986. "Agency Costs of Free Cash Flow, Corporate Finance, and Takeovers." *American Economic Review* 76: 323-329.

Jensen, M.C., and W. Meckling. 1976. "Theory of the Firm: Managerial Behavior, Agency Costs, and Ownership Structure." *Journal of Finance Economics* 3: 305-360.

LaDue, E. 1990. "Moral Hazard in Federal Farm Lending." *American Journal of Agricultural Economics* 72: 774-779.

Lajili, K., P.J. Barry, S.T. Sonka, and J.T. Mahoney. 1997. "Farmers' Preferences for Crop Contracts." *Journal of Agricultural and Resource Economics* 22: 264-280.

Lee, W., and C.B. Baker. 1984. "Agricultural Risks and Lender Behavior." In P.J. Barry, ed., *Risk Management in Agriculture*. Ames: Iowa State University Press.

Mapp, H., and G. Helmers. 1984. "Methods of Risk Analysis for Farm Firms." In P.J. Barry, ed., *Risk Management in Agriculture*. Ames: Iowa State University Press.

Matten, C. 2000. *Managing Bank Capital: Capital Allocation and Performance Measurement* (2nd ed.). New York: Wiley.

Miller, L.H., and E.L. LaDue. 1989. "Credit Assessment Models for Farm Borrowers: A Logit Analysis." *Agricultural Finance Review* 49: 22-36.

Moss, C.B., R.N. Weldon, and R.P. Muraro. 1991. "The Impact of Risk on the Discount Rate for Different Citrus Varieties." *Agribusiness* 7: 327-338.

Moss, L.E. 2000. "A Transaction Cost Economics and Property Rights Theory Approach to Farmland Lease Preferences." Ph.D. thesis, University of Illinois.

Nasr, R., P.J. Barry, and P.N. Ellinger. 1998. "Financial Structure and Efficiency of Grain Farms." *Agricultural Finance Review* 58: 33-46.

Pflueger, B.W., and P.J. Barry. 1986. "Crop Insurance and Credit: A Farm Level Simulation Analysis." *Agricultural Finance Review* 46: 1-14.

Robison, L.J., and P.J. Barry. 1996. *Present Value Models and Investment Analysis*. East Lansing: Michigan State University Press.

Saunders, A. 1999. *Credit Risk Management*. New York: John Wiley & Sons.

Sherrick, B., P.J. Barry, and P.N. Ellinger. 2000. "Valuation of Credit Risk in Agricultural Mortgages." *American Journal of Agricultural Economics* 82: 71-81.

Sonka, S., B.L. Dixon, and B.L. Jones. 1980. "Impact of Farm Financial Structure on the Credit Reserve of the Farm Business." *American Journal of Agricultural Economics* 62: 565-570.

Splett, N., P.J. Barry, B.L. Dixon, and P.N. Ellinger. 1994. "A Joint Experience and Statistical Approach to Credit Scoring." *Agricultural Finance Review* 54: 39-135.

Stiglitz, J.E. 1985. "Credit Markets and the Control of Capital." *Journal of Money, Credit, and Banking* 17: 133-152.

Stiglitz, J.E., and A. Weiss. 1981. "Credit Rationing in Markets With Imperfect Information." *American Economic Review* 71: 393-411.

Tauer, L.W. 2000. "Estimating Risk-Adjusted Interest Rates for Dairy Farms." Paper presented at the NC-221 Annual Meeting, Federal Reserve Bank of Minneapolis, October 2-3, 2000.

Townsend, J.P., and D.W. Brorsen. 2000. "Cost of Forward Contracting Hard Red Winter Wheat." *Journal of Agricultural and Applied Economics* 32: 89-94.

Turvey, C.G. 1991. "Credit Scoring for Agricultural Loans: A Review With Applications." *Agricultural Finance Review* 51: 43-54.

Chapter 18

DOES LIQUIDITY MATTER TO AGRICULTURAL PRODUCTION?*

Michael J. Roberts and Nigel Key
Economic Research Service, U.S. Department of Agriculture

INTRODUCTION

Farmers have a variety of tools with which to cope with risk. These tools include shifting their purchases of producer or consumer durable goods between periods, accumulating or depleting savings or buffer stocks, borrowing or restructuring debt, purchasing insurance, and participating in futures markets. Given the variety of methods farmers have for coping with risk, we first set out to answer the fundamental question: Does risk coping affect agricultural production decisions? To answer this we must address the issue of why risk matters. That is, what are the market failures that make risk matter and how do they affect producer behavior?

The dominant method of explaining agricultural behavior under price or yield uncertainty assumes that farmers choose inputs in order to maximize a static, concave (risk-averse) utility function. The seminal paper by Sandmo (1971) predicts that in the presence of price uncertainty, a risk-averse producer will produce less than if prices were known. Pope and Kramer (1979) offer one of the first models examining the effect of production risk on input use. They consider a stochastic production function, a constant relative risk aversion utility function, and allow inputs to increase or decrease risk. In the single-input case, they show that a risk-averse agent uses more (less) of an input depending on whether the input marginally decreases (increases) risk. Loehman and Nelson (1992) extend Pope and Kramer's model to include multiple inputs in which all inputs are either risk-increasing or -decreasing and all pairs are classified as either risk substitutes or complements. This characterization allows them to draw several general conclusions that hold for

* The views expressed are those of the authors and do not necessarily represent those of the U.S. Department of Agriculture. We thank Bruce Andersen, John Antle, Daniel Hellerstein, Jeffrey LaFrance, Rulon Pope, Meredith Soule, Brian Wright, and David Zilberman for helpful comments and suggestions. All remaining errors are our own.

relative risk aversion and exponential utility functions. Leathers and Quiggin (1991) develop further implications of the effect of uncertainty on input use. They derive implications for input response to prices and yield risk under increasing, decreasing, and constant absolute risk aversion. As early as Freund (1956), this method was used to explain how risk influences crop allocations and land use. Modern permutations include Chavas and Holt (1990, 1996) and Rambaldi and Simmons (2000).

Within the framework of the static model with risk-averse producers, risk affects production choices because of farmers' preferences. At its essence, this approach does not allow *any* kind of dynamic risk-coping. That is, the approach assumes autarkic behavior with respect to risk trading, forcing farmers to consume all individual shocks, idiosyncratic or otherwise, within the period in which a shock is experienced. However, risk-averse farmers have various means at their disposal for smoothing consumption including the accumulation and depletion of liquid assets, formal and informal insurance markets, or borrowing (Deaton 1992). Researchers have also recognized that households can alter their production decisions in order to reduce the variability of income (Alderman and Paxson 1994, Morduch 1995, Kochar 1999). In general, the consumption-smoothing literature provides evidence that households are, to a large extent, able to maintain consumption in the face of unexpected income shocks (Paxson 1992, Townsend 1995). Hence, in a *dynamic* context, risk (price or yield shocks) may be important to farmers only if they are unable to take advantage of available intertemporal risk-coping methods. Consequently, it may not be risk preferences, but rather intertemporal constraints that are crucial to understanding how risk affects production choices. We turn to the capital market literature to provide a theoretical basis for why risk might matter to agricultural decisions in a dynamic context.

With perfect markets, it is inconsequential whether capital investment is financed with retained earnings, equity, or debt (Modiglianni and Miller 1958). However, credit market imperfections (resulting perhaps from adverse selection or moral hazard in financial markets) could make liquidity and internal net worth important for capital investment (Jaffee and Russell 1976, Stiglitz and Weiss 1981). The importance of market imperfections can be estimated with a model wherein firms maximize the expected present value of all future profit flows and finance capital investment with retained earnings and debt, but face a borrowing constraint. The typical empirical strategy is to test for the importance of this constraint by measuring the correlation of cash flows with investment decisions while attempting to control for investment opportunities – changes in the marginal value of capital investment (see Hubbard 1998 for a review of this literature).

Because this literature has found liquidity important for investment decisions of large manufacturing firms, it suggests that liquidity constraints could be especially important in agriculture, which is both a risky industry and one comprised largely of partnerships and sole-proprietorships. Farm-households normally hold a disproportionate share of wealth in farm capital; and on average they tend to consume a smaller portion of their income than do non-farm households (Hubbard and Kashyap 1992, Carroll 1997). These patterns suggest that precautionary saving, self-finance, and consequently liquidity constraints are important features in farm business decision making. These constraints may impose risk-coping costs to farmers regardless of their preferences and ability to smooth consumption. The empirical challenge, therefore, is to discern how much of the change in farmers' production behavior is due to an inability to manage risk versus an efficient response to new information in their economic environment.

Although the empirical evidence of the importance of liquidity constraints is compelling, a skeptic might argue that these results are spurious due to an inadequate account of investment opportunities. After all, a positive relationship between investment opportunities and cash flow follows naturally if firms' revenue shocks persist over time – if good times now are an indication of good times to come in the future. For example, if a competitive firm's product price follows a random walk (i.e., all price shocks are *permanent*), then price shocks would simultaneously influence both cash flows and investment opportunities, whether or not firms are liquidity constrained. To control for this convoluting factor, the econometrician must estimate the liquidity effect simultaneously with the marginal value of investment. The skeptic would argue, however, that such controls are imperfect, and what the econometrician does not know about the value of investment is partially reflected by cash flows.

Uncertainty about persistent shocks can have important and sometimes complicated effects on economic decisions that hinge on the sequential nature of decisions and uncertainty resolution.[1] These information effects on choices may convolute estimates of risk effects, whether the information effects are attributed to liquidity constraints, risk aversion, or other proxies of market incompleteness. The difference between these two kinds of effects is fundamental: where risk effects imply a measure of inefficiency, information effects do not.

One way to separate risk effects from information effects is to isolate a *transitory* source of cash flow, which should bear no relation to investment opportunities – for example, by using a natural experiment in which all except internal net worth is held the same. To find such an experiment in most industries is difficult, because most shocks show a large degree of autocorrela-

[1] Note that *uncertainty* about persistent shocks also affects the profitability of investment. See the book by Dixit and Pindyck (1994) for a thorough analysis of this more subtle kind of information effect. In Chapter 15, Marra and Carlson (this volume) explain how information effects are important for technology choice. These information effects make it especially difficult to interpret empirical results based on static production models with risk-averse objectives. Also, in Chapter 12, Antle and Capalbo (this volume) show how explicit modeling of the sequential decision process appears to matter more than inclusion of risk preferences.

tion.[2,3] However, in agriculture yield shocks possess just the right elements for such an experiment. After removing time trends in yields, the residuals have a large variance and almost no autocorrelation. Yield shocks, like the year-to-year variations in weather that underlie them, are about as close to an ideal naturally occurring independent and identically distributed random variable as an econometrician could hope to use as a source of identification.

The remainder of this chapter provides an example of how yield shocks can be used to identify risk effects that result from a constraint on borrowing. This research assembles National Agricultural Statistics Service (NASS) county-level annual data on yields and planted acreage and NASS state-level annual data on prices.[4] The acreage and yield data include up to 81 years for each of over two thousand U.S. counties.[5] For each county having 20 or more observations, a locally linear, non-parametric yield trend curve is estimated. The residuals from these trend curves are used as a proxy for unexpected yield shocks. The estimated yield shocks are used in two ways, the first quasi-experimental and the second structural. First, the shocks are used to divide the data into six mutually exclusive and collectively exhaustive groups according to how many of the five immediately preceding yield shocks are negative (from zero to five). Average acreage plantings are compared between these six sub-samples, both with and without controls for differences between counties and county-level planting trends. The results provide compelling evidence of the importance of past yield shocks in the planting decision.

Second, a stochastic Euler equation is derived that corresponds to the planting decision of each county-representative farm. The equation includes a multiplier, which is non-zero when a borrowing constraint binds. The Euler equation is estimated, allowing for separate marginal revenue and marginal cost functions for each county and for different estimated multipliers for sub-samples that have experienced either good or bad prior shocks. The procedure effectively tests whether marginal profits differ significantly between counties that have just had two positive yield shocks, two negative yield shocks, and one positive and one negative shock. Findings show that acreage plantings respond positively to past yield shocks and that marginal profits respond negatively. The results, which estimate that an average yield shock causes a planted-acreage response of about 2 percent and a change in the

---

[2] Hubbard, Kashyap, and Whited (1995) use variation in tax payments as an independent source of variation in firms' cash flow. This natural experiment differs from yield shocks in that tax payments are mostly anticipated as opposed to unanticipated and are likely to retain a correlation with investment profitability (albeit an "imperfect" one).

[3] Note that simply trying to control for information effects by modeling them also entails certain pitfalls. In particular, estimated information effects are very sensitive to the specified autoregressive process (Schwartz 1997).

[4] These data are publicly available from the NASS internet site at http://www.nass.usda.gov.

[5] Missing values for prices in some states reduces the number of counties to 1,952 for the structural estimates.

marginal cost of being liquidity-constrained of between $1.50 and $2.50 per acre not planted, are robust across estimates that incorporate different controls and functional forms.[6]

In addition to using a compelling natural experiment and a wealth of data to assess the importance of imperfect markets, to our knowledge this is the first study to measure the effect of liquidity on production choices other than capital investment. The chapter also shows how, why, and (by some measures) how much imperfect markets make idiosyncratic profit variability costly to farmers in a way that does not hinge upon risk aversion. By relating risk to agricultural land use, the methodology used in the chapter may be useful in the analyses of government programs, such as subsidized crop insurance and the conservation reserve program, that aim to both stabilize farm income and protect environmentally sensitive land. The results also point toward future work to examine how responses and responsiveness to income shocks have changed with changes in government farm programs.

OPTIMAL PLANTING DECISIONS UNDER UNCERTAINTY

Suppose that the representative farmer of county i in each period t chooses a vector of inputs $\boldsymbol{x}_{it}$ so as to maximize a value function $V_{it}(\boldsymbol{x}_{it}|\Theta_{it})$, which equals the expected present value of all future profit flows given the information Θ_{it} available at time t. The objective is described mathematically as

$$\max_{\boldsymbol{x}_{it}} V_{it}(\boldsymbol{x}_{it}|\Theta_{it}) = E_{it}\left[\sum_{k=0}^{T=t}\left(\frac{1}{1+r}\right)^{k} \pi_{i,t+k}(\boldsymbol{x}_{i,t+k}, s_{i,t+k}) \,|\, \Theta_{it}\right], \tag{1}$$

where $\pi_{i,t}(\boldsymbol{x}_{it}, s_{it})$ is a function relating the flow of profits as a function of the input vector $\boldsymbol{x}_{it}$ and a state vector s_{it}, which depends on past inputs and random variables such as prices, weather, and so on, and r is the interest rate. If the random fluctuations included in s_{it} are independent of aggregate economic fluctuations, then the unconstrained solution to this maximization problem solves the first-best perfect-markets equilibrium, regardless of farmers' risk preferences. With perfect markets, all individual idiosyncratic shocks are pooled, effectively averaging to zero, so in a first-best world these risks "should" be costless.

Consider a special case of the profit function for which production of a single crop costs $c(l_{it})m_i(\boldsymbol{x}_{it}', s_{it})$ for planting l_{it} acres. The vector $\boldsymbol{x}_{it}'$ includes production inputs besides land. Revenues equal $l_{it}p_t y_{it}$, where p_t equals the

[6] In particular, the results change very little if the time interaction terms and/or adjustment costs are removed from the model.

realized crop price in period t and y_{it} equals farm i's per-acre yield in period t. Price depends on $\boldsymbol{s}_{it}$ and yield depends on $\boldsymbol{s}_{it}$, $\boldsymbol{x}_{it}'$, and l_{it}. Lastly, there is an acreage-adjustment cost equal to $(a_i / 2)(l_{it} - l_{i(t-1)})^2$. These assumptions imply that

$$\pi_{i,t}(\boldsymbol{x}_{it}, \boldsymbol{s}_{it}) = l_{it} p_{it}(\boldsymbol{s}_{it}) y_{it}(l_{it}, \boldsymbol{x}_{it}', \boldsymbol{s}_{it}) - c_i(l_{it}) m_i(\boldsymbol{x}_{it}', \boldsymbol{s}_{it}) - \frac{a_i}{2}(l_{it} - l_{i(t-1)})^2 . \quad (2)$$

The first-order condition (or stochastic Euler equation) associated with the land allocation decision l_{it} is

$$E_{it}\left[p_{it}(\boldsymbol{s}_{it}) y_{it}(l_{it}, \boldsymbol{x}_{it}', \boldsymbol{s}_{it}) + l_{it} p_{it}(\boldsymbol{s}_{it}) \frac{\partial y_{it}}{\partial l_{it}} - c'(l_{it}) m(\boldsymbol{x}_{it}', \boldsymbol{s}_{it}) \right.$$

$$\left. + \frac{a_i}{1+r}(l_i(t+1) - l_t) - a_i(l_{it} - l_{i(t-1)}) \mid \Theta_{it} \right] = 0 , \quad (3)$$

which should hold if the farmer makes value-maximizing decisions. Equation (3) effectively says that the expected marginal revenue equals expected marginal costs.

OPTIMAL CREDIT-CONSTRAINED DECISIONS UNDER UNCERTAINTY

If the farmer faces a borrowing constraint then the Euler equation is slightly different from equation (3). During periods when this constraint binds, the Euler equation will include a positive multiplier associated with the additional value of having more credit or cash on hand to finance the cost of planting an additional acre. Mathematically, the objective is the same as in equation (1), with the additional constraint

$$B_i + A_{it} - c_i(l_{it}) m_i(\boldsymbol{x}_{it}', \boldsymbol{s}_{it}) \geq 0 \quad \text{for all } i \text{ and } t, \quad (4)$$

where A_{it} denotes collaterizable assets and B_i denotes the limit on additional funds available to borrow. The assets in period t do not include revenues earned in period t because these revenues are realized after expenditures on production are made.

The Lagrangian associated with this optimization problem is

$$L_{it}(\boldsymbol{x}_{it} \mid \Theta_{it}) = E_{it}\left[\sum_{k=0}^{T-t}\left(\frac{1}{1+r}\right)^k \pi_{i,t+k}(\boldsymbol{x}_{i,t+k}, \boldsymbol{s}_{i,t+k}) \mid \Theta_{it}\right] + \lambda_{it}[B_i + A_{it} - c_i(l_{it})\, m_i(\boldsymbol{x}_{it}', \boldsymbol{s}_{it})], \tag{5}$$

which has the following first-order condition associated with the acreage choice l_{it},

$$E_{it}\left[p_{it}(\boldsymbol{s}_{it})y_{it}(l_{it}, \boldsymbol{x}_{it}', \boldsymbol{s}_{it}) + l_{it}p_{it}(\boldsymbol{s}_{it})\frac{\partial y_{it}}{\partial l_{it}} - c'(l_{it})m(\boldsymbol{x}_{it}', \boldsymbol{s}_{it}) + \frac{a_i}{1+r}(l_{i(t+1)} - l_t) - a_i(l_{it} - l_{i(t-1)}) \mid \Theta_{it}\right] = \lambda_{it}[c'(l_{it})m(\boldsymbol{x}_{it}', \boldsymbol{s}_{it})] \geq 0 . \tag{6}$$

The parameter λ_{it} is the multiplier associated with the liquidity constraint for the ith farmer in period t.[7] The first-order condition (6) differs from (3) in that the expectation may be greater than zero if the constraint contemporaneously binds. It is important to recognize, however, that this constraint will not necessarily bind contemporaneously even if it does bind globally. The constraint may cause the entire anticipated path of choices to shift; the Euler equation, however, only balances expected marginal tradeoffs between periods. A positive λ_{it} therefore implies the existence of a borrowing constraint, but not the other way around; a borrowing constraint may be binding through its effect on the transversality conditions (but never contemporaneously binding) even if equation (3) is always true. Put more plainly, firms recognize the constraint and therefore may build up a buffer stock of assets so as to avoid ever being contemporaneously constrained. This point is emphasized by Hubbard and Kashyap (1992), who use a similar approach to examine liquidity-constrained capital investment choices, and by Zeldes (1989), who examines liquidity-constrained consumption and savings decisions.

It is also worth noting that, under certain limited conditions, a relaxation of the credit constraint may not reduce the size of λ_{it}. If the production function is concave, continuous, and smooth with respect to all input choices, then the constraint will bind for all choices simultaneously and λ_{it} will decrease as the constraint is relaxed. If, however, another input choice such as capital investment is "lumpy," then constraint may not bind marginally with respect to acreage even if it does bind with respect to the "lumpy" choice. For example, partial relaxation of the constraint could therefore induce greater use of the lumpy input and cause the acreage constraint to bind marginally. A similar result occurs if there are increasing returns to scale. Unless there exists a

[7] For simplicity, we have assumed that the borrowing constraint does not limit adjustment costs.

"lumpy" input choice that is also a technical substitute for acreage, then holding all else equal, contemporaneously constrained acreage plantings should be less if unconstrained.

A change in liquidity can affect the value λ_{it} in several different ways. Three plausible scenarios are depicted in Figures 1a–1c. Each plot shows the marginal profitability of increasing acreage l_{it}. Figure 1a shows the most plausible and intuitive case with diminishing marginal profitability of land; as the liquidity constraint is relaxed, land use increases from l_{it}^C to an unconstrained level l_{it}^* and the marginal value of liquidity λ_{it} decreases to zero. In Figure 1b, there is increasing returns to scale, all farms are liquidity-constrained, and λ_{it} increases as the constraint eases. In Figure 1c, the production function includes at least one "lumpy" input, so a relaxation of the liquidity constraint may cause another input or inputs to increase by a discrete amount, shift the marginal productivity of land, and cause the change in λ_{it} to be ambiguous. These different possibilities, together with the fact that liquidity constraints could still affect the global value even if they do not contemporaneously bind, mean that empirical results need to be interpreted with care.

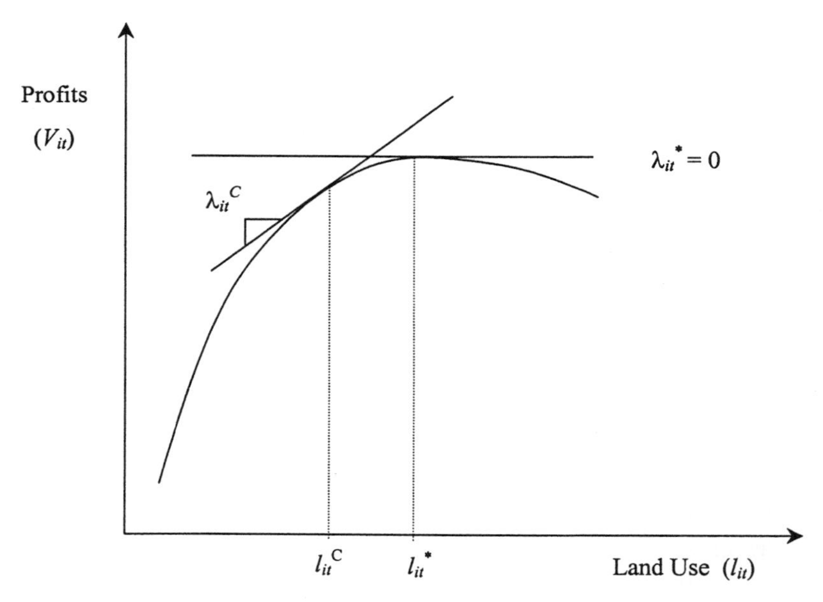

Figure 1a. Land Use and Profits: Concave Production

EMPIRICAL ANALYSIS

In an experiment where randomly assigned treatment groups are given different quantities of cash, and therefore have different levels of

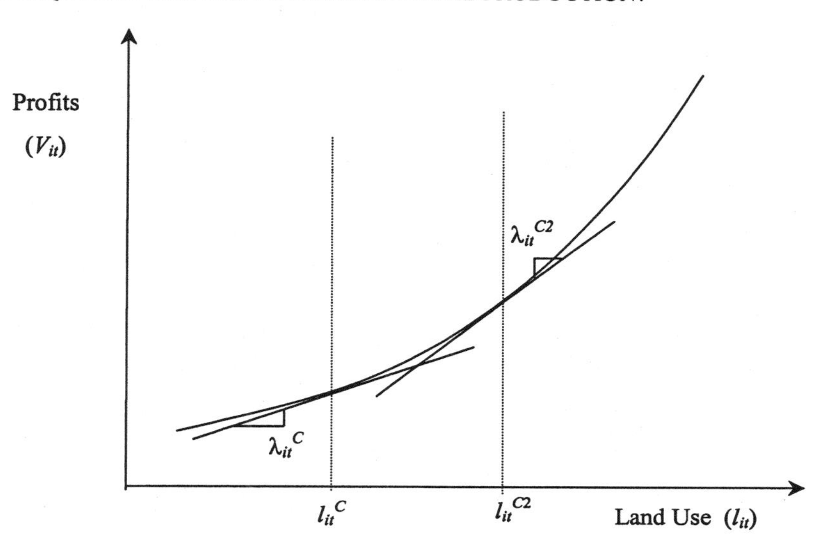

Figure 1b. Land Use and Profits: Convex Production

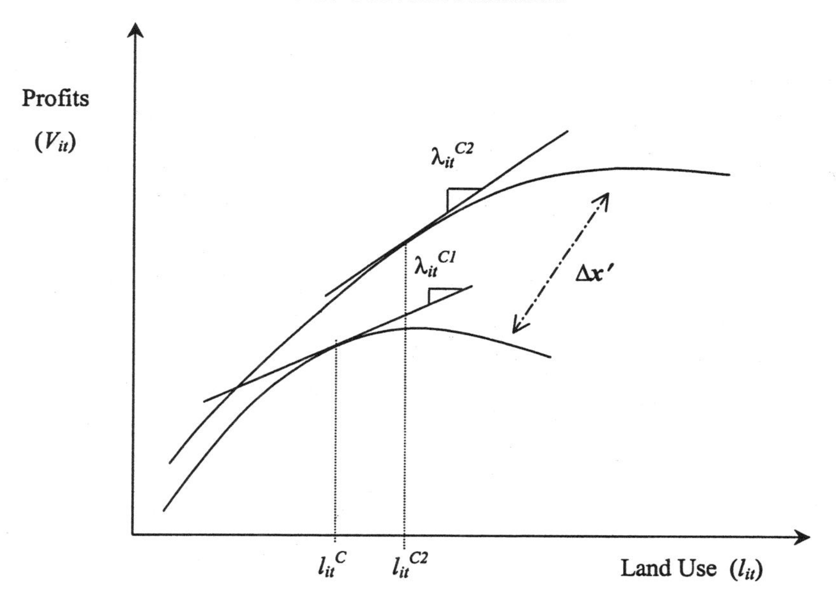

Figure 1c. Land Use and Profits: Lumpy Capital Investment

collaterizable assets (A_t), the liquidity constraint will have a different likelihood of contemporaneously binding; or, if the constraint always binds, it will be relaxed (or tightened) more for some groups than for others. Given such an experiment, there are two testable implications of the liquidity-constrained model verses the non-liquidity-constrained model. First, treatment groups with more cash should, on average, plant greater acreages. Second, marginal revenue less marginal cost should equal zero more frequently for treatment groups with greater levels of cash. It is therefore logical to conclude that marginal revenue less marginal cost will, on average, be greater for groups with smaller levels of cash on hand.

Although such an experiment does not exist, yield shocks constitute a close substitute for such an experiment. Most year-to-year yield variation constitutes random year-to-year variation in weather or pest damage. Because this random variation is transitory, it should affect cash on hand but not the expected profitability of choices made in subsequent years. Moreover, yield fluctuations vary widely and randomly over both space and time, and so are unlikely to be correlated with missing or unobserved fundamental variables that are necessarily missing from almost every econometric model.

The empirical analysis is thus comprised of three steps presented in the following subsections: (a) estimation of the yield shocks, (b) comparison of planted acreages between sub-samples delineated using past estimated yield shocks, and (c) structural estimation of equation (3) (or 6) to identify changes in the average value of λ_{it} between similarly delineated sub-samples.

Estimation of Yield Shocks

The first step of the analysis is to estimate the yield shocks, which are used in this study as the exogenous source of variation in liquidity. We also demonstrate that yield shocks have the appropriate properties to serve this purpose, which requires that they have little or no persistence and have a reasonably large variance. Although price shocks, due to their large degree of persistence, may be more important to farmers, the goal here is to separate risk effects (or liquidity effects) from information effects. Because persistent shocks include information about future profitability, they make the task of separating liquidity effects from information effects more challenging.

The empirical methods are illustrated using data from a representative county. Figure 2 presents a plot of wheat yields and a fitted yield trend curve for Cowley County, Kansas. The fitted trend curve was estimated using a non-parametric procedure called "loess," shorthand for "local-polynomial regression." Fitted values are constructed using the observed points lying within a fixed-size window or bandwidth around each point to be fitted. A weighted-linear regression is used to determine each fitted point, with the weights chosen according to their distance from the fitted point relative to the density function of a normal-distributed random variable centered at the fitted point.[8] For each trend curve estimated, the window includes two-thirds of the total number of points. Seventeen equally spaced points were fitted for each county, with in-between points being interpolated from the fitted values. The residuals from these fitted trend curves constitute the estimated shocks that

[8] In general, a higher order polynomial regression may also be used, but a locally linear trend seemed appropriate for the present analysis.

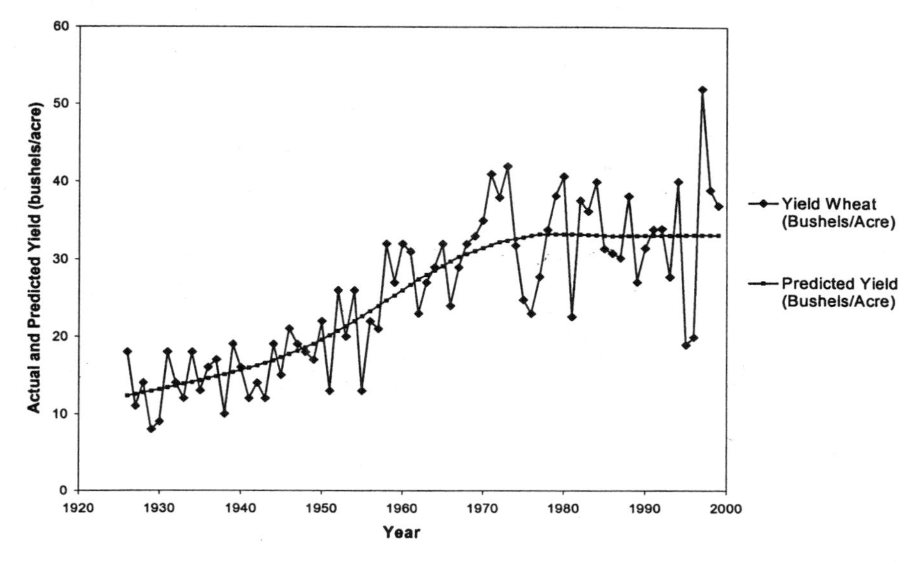

Figure 2. Actual and Predicted Wheat Yield: Cowley County, Kansas

are used throughout the rest of the chapter.[9] For Cowley County, these residuals are plotted against the lagged residuals in Figure 3, which shows no apparent relationship ($\rho = -0.013$).[10]

The transitory nature of yield shocks is formally demonstrated using a unit root test. The unit root test reported in Table 1a strongly rejects the null hypothesis of a unit root in favor of a trend-stationary alternative (Dickey and Fuller 1979). This simple test assumes a linear time trend rather than the non-parametric curve used for the ensuing analysis. The same test performed on the log-log relationship produced similar results (Table 1b). These results are especially convincing given the fact that unit root tests are not very powerful – the unit root null hypothesis is difficult to reject. In contrast, Table 1c reports unit root tests for prices in the state of Kansas. Despite a clear downward trend in prices (see Figures 4a and 4b), the test only just rejects (at 5 percent) a pure random walk without a drift, and fails to reject (at 10 percent

---

[9] Note that the trend curves were estimated for the log of yields against the log of time. The model is $\log(y_{it}) = f_i(\log(t)) + \varepsilon_{it}$. The fitted values in Figure 2 are determined using the following second-order approximation derived using the propagation of error method: $E[y_{it}] = \exp(E[\log(y_{it})])(1 + 0.5s^2)$. $E[y_{it}]$ is the fitted value on the plot, $E[\log(y_{it})]$ is the loess-fitted value, and s^2 is the estimated variance of ε_{it}.

[10] The degree of autocorrelation is different for different counties and generally not significant. If one pools the shocks over all counties, however, there is a slight but statistically very significant correlation equal to about 0.13 (R^2 equals about 0.01). Within the structural model, this slight yield autocorrelation could arise from autocorrelation in planting decisions coupled with heterogeneous quality of land within each county. It also arises from autocorrelation in capital investment and other input choices that follow rationally from autocorrelation in prices, regardless of whether or not liquidity constraints matter.

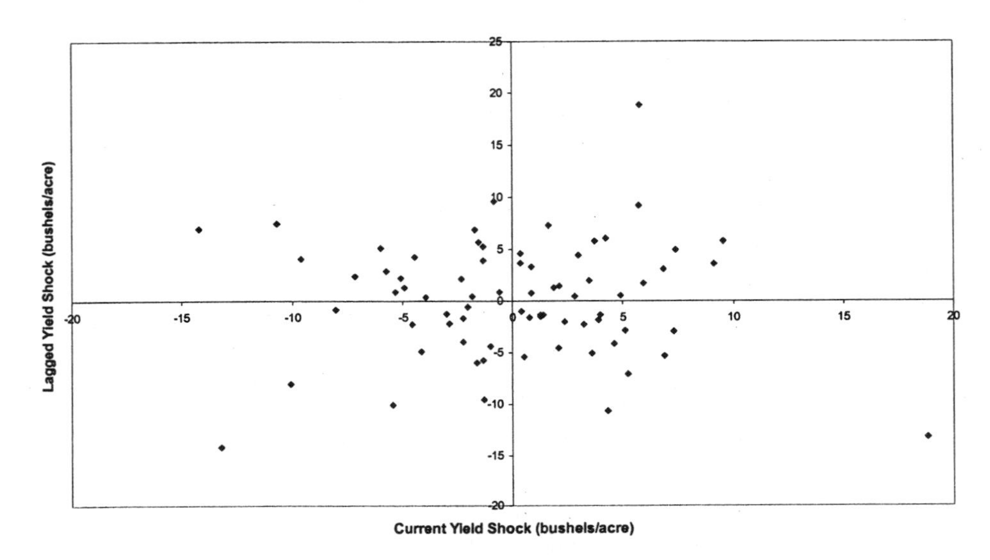

Figure 3. Estimated Yield Shocks and Lagged Yield Shocks: Cowley County

a pure random walk with a drift. Contrary to yield shocks, price shocks have a large degree of persistence.

The large relative variance of the yield shocks (which provides power to the liquidity tests in the following sections) also is evident in Figure 2. The coefficient of variation (the standard deviation of shocks divided by the loess-predicted yield) is equal to about 0.23. There is also a large degree of variation of shocks across space, which means that shock effects are unlikely to be confounded by spatial heterogeneity. Although shocks of bordering counties are correlated, they are uncorrelated for counties separated by less than one state. See, for example, the scatter plots of close-together counties Sherman and Cheyenne in Figure 5a (r = 0.85) as compared to the farther-apart counties Cowley and Cheyenne in Figure 5b (r = 0.04). The relative distances between these counties can be seen in the map of Kansas in Figure 6. Both time-series and cross-sectional yield variation of nearly 2,000 counties therefore serve as a powerful source of identification.

Planted Acreage Comparisons

Tables 2 and 3 present results from the planted-acres regressions. Table 2 presents the estimated coefficients, standard errors, t-statistics, and p-values from regressing fixed effects of six mutually exclusive groups on the log of

Table 1a. Wheat Yields Unit Root Test: Cowley County, Kansas

Variable	Coefficient	Std. Error	t value	Pr(>|t|)*
(Intercept)	6.21814	1.82356	3.410	0.00109
Y_{t-1}	-0.71905	0.11718	-6.136	4.70e-08
Time	0.35993	0.06797	5.295	1.34e-06

Dependent variable: $\Delta y_t = y_t - y_{t-1}$
R-squared: 0.3533 Adjusted R-squared: 0.3345
F-statistic: 18.85 on 2 and 69 degrees of freedom p-value: 2.948e-007*

Table 1b. Log Wheat Yields Unit Root Test: Cowley County, Kansas

Variable	Coefficient	Std. Error	t value	Pr(>|t|)*
(Intercept)	1.700782	0.271441	6.266	2.77e-08
$\log(y_{t-1})$	-0.694815	0.110778	-6.272	2.70e-08
Time	0.013874	0.002822	4.916	5.75e-06

Dependent variable: $\Delta\log(y_t) = \log(y_t) - \log(y_{t-1})$
R-squared: 0.364 Adjusted R-squared: 0.3456
F-statistic: 19.75 on 2 and 69 degrees of freedom p-value: 1.653e-007*

Table 1c. Wheat Prices Unit Root Test: Kansas

Variable	Coefficient	Std. Error	t value	Pr(>|t|)*
(Intercept)	0.495	0.250	1.982	0.0515
$\log(p_{t-1})$	-0.107	0.054	-1.981	0.0515
Time	-0.007	0.003	-2.365	0.0208

Dependent variable: $\Delta\log(p_t) = \log(p_t) - \log(p_{t-1})$
R-squared: 0.078 Adjusted R-squared: 0.051
F-statistic: 2.91 on 2 and 69 degrees of freedom p-value: 0.061*

* Note that reported p-values are incorrect for unit root tests (see Dickey and Fuller 1979). The p-values for the standard Dickey-Fuller tests are much larger than those reported. The 10 percent critical t value for the coefficient of $\log(P_{t-1})$ against the null hypothesis of a random walk with drift is about –2.58. The 10 percent critical value of the F-type test is about 3.94.

planted acres across all counties and years that include five non-missing lagged values. The first of the six groups includes all county-years in which all five of the lagged estimated yield shocks are positive, the second includes all county-years in which one of the five shocks is negative, the third includes all counties in which two of the five are negative, and so on. An intercept captures the average value of the first group (with five positive or "good" shocks), and the remaining coefficients estimate the difference between the first group and

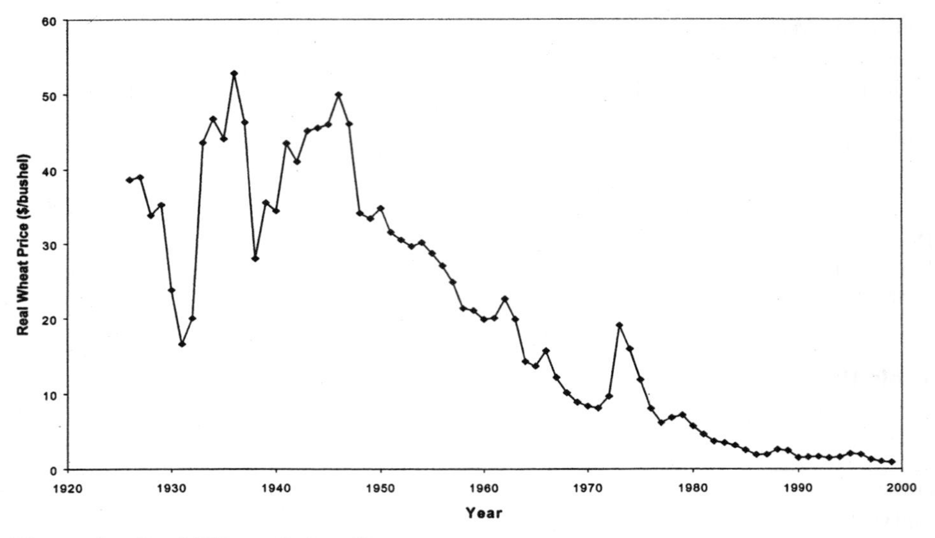

Figure 4a. Real Wheat Price: Kansas

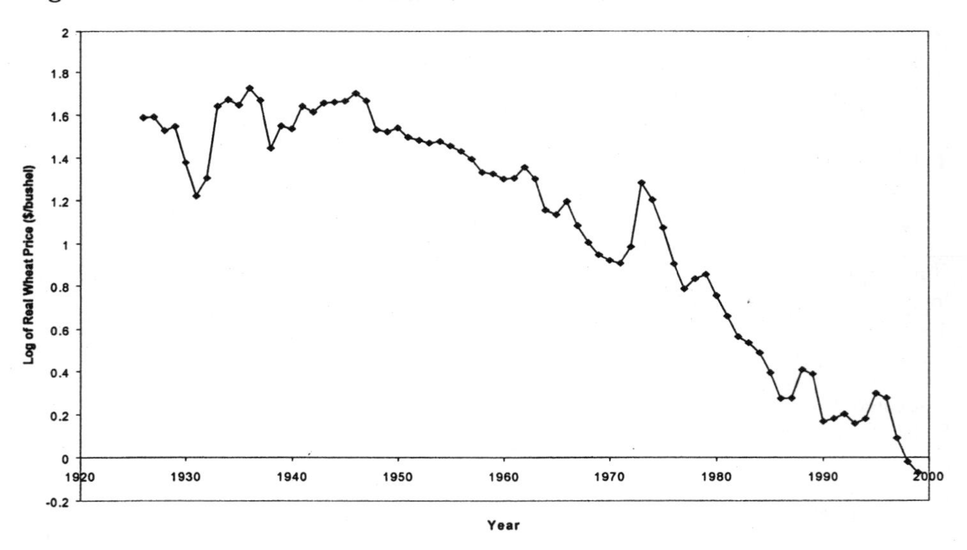

Figure 4b. Log of Real Wheat Price: Kansas

each of the other five groups, which include different numbers of "bad" shocks. Every coefficient is highly significant, with the estimated level of planted acres diminishing as the number of "bad" shocks grows.

This simple regression provides clear evidence that past yield shocks matter for subsequent plantings. The question then arises as to whether the coefficients represent a subtle form of bias and therefore provide misleading results. If so, it is difficult to determine by what means such a bias could enter into the regression. Lagged yield shocks arise mostly from weather shocks and hold little if any correlation with subsequent fundamentals, such

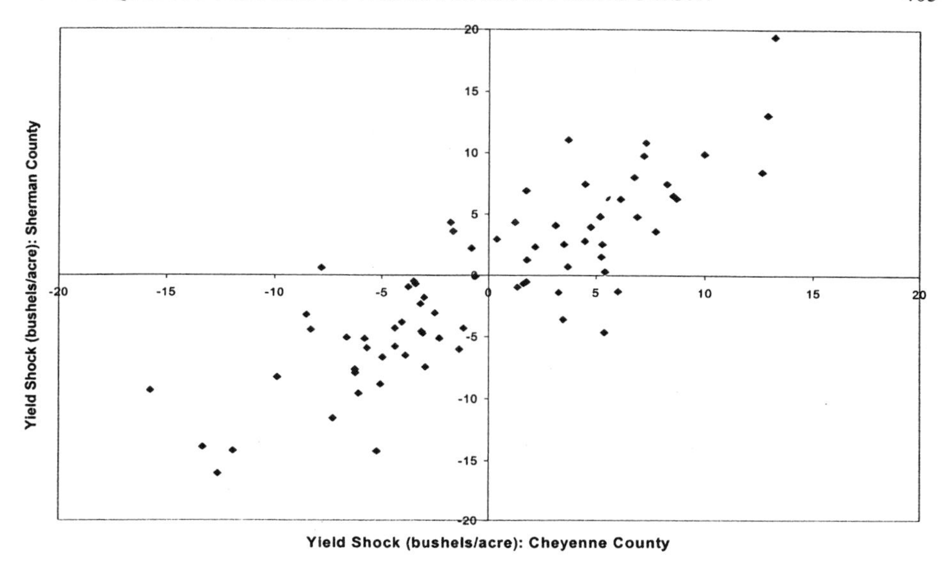

Figure 5a. Estimated Wheat Yield Shocks in Close-Together Counties

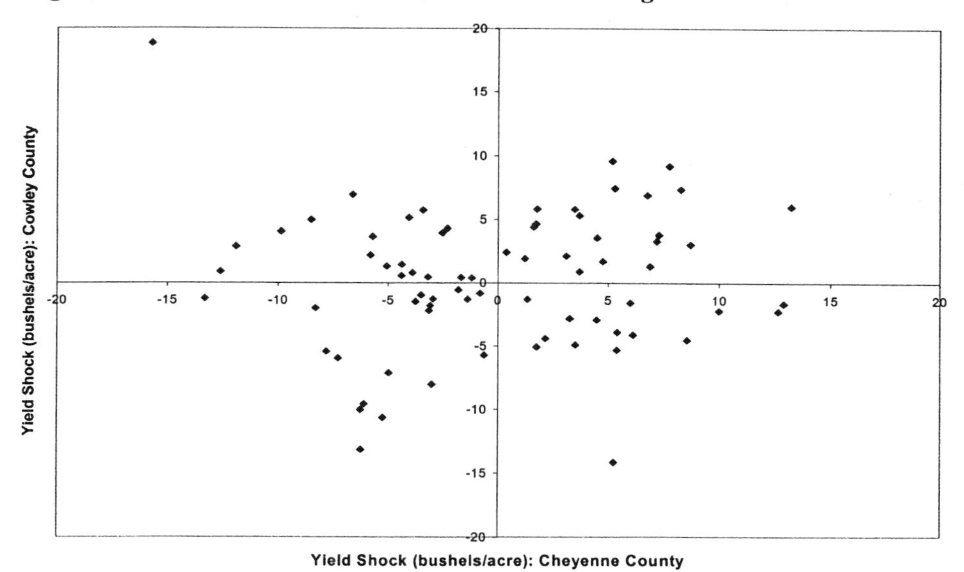

Figure 5b. Estimated Yield Shocks in Farther-Apart Counties

as prices and subsequent yields. So, even though there are many missing variables that explain the large standard error of the regression, these missing variables are unlikely to be correlated with the indicator variables, and therefore should not bias the coefficients. Although it is conceivable that yield shocks could be correlated with cost factors in the subsequent periods, such as residual soil moisture content, it seems inconceivable that this

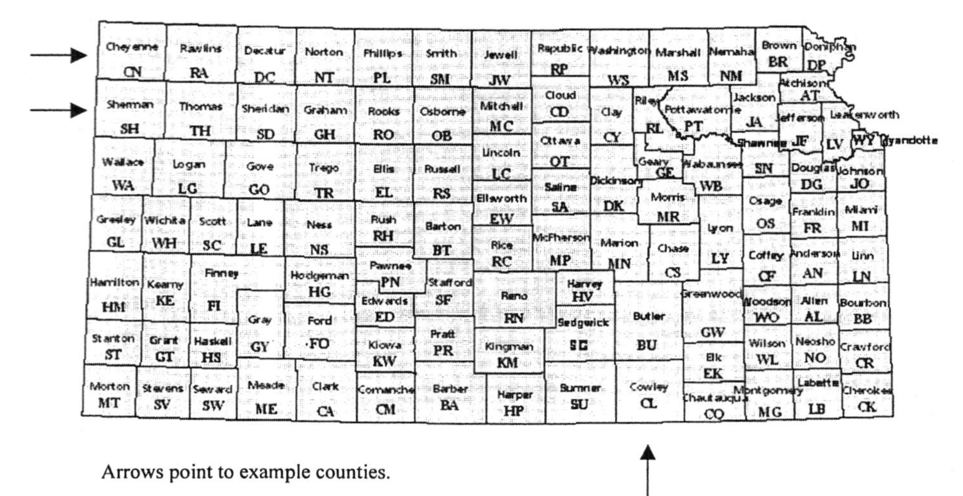

Figure 6. County Map of Kansas

relationship could extend for more than a single year. The estimated planted-acreage responses over the last five shocks are therefore difficult to reconcile using a cost-based explanation.

To be more confident of the results reported in Table 2, we would like to control for the most obvious variation in planted acres. Much of the planted acres variation comes from differences between counties and within-county time trends. Clearly, some counties plant more wheat acreage than others; and some counties have increased wheat acreage while others have moved out of wheat and into higher-value crops such as corn and soybeans. It could be that these trends hold a subtle correlation with lagged yield shocks. Although there are enough degrees of freedom to estimate separate fixed-effects and

Table 2. Fixed Effects of Lagged Shocks on Plantings

Variable	Parameter Estimate	Standard Error	t Value	Pr > |t|
Intercept (Five Good Shocks)	9.989	0.030	337.17	<.0001
One Bad Shock	-0.443	0.034	-13.14	<.0001
Two Bad Shocks	-0.818	0.032	-25.44	<.0001
Three Bad Shocks	-0.966	0.032	-29.76	<.0001
Four Bad Shocks	-1.135	0.0350	-32.45	<.0001

Dependent variable: log (planted wheat acres) for all years and counties with five non-missing previous observations
Sample size: 85,292 observations. Includes about 2,200 counties
R-squared: 0.0321
Adj. R-squared: 0.0320

trends for every county, this is difficult in practice, as it requires inversion of an extremely large data matrix (about 85,000 by about 4,000). Such a regression also would fail to capture non-linearity in planted-acreage trends. Alternatively, we estimated planted-acreage trends for each county separately using loess regressions – the same way yield trends were estimated as described above in the section on estimation of yield shocks. The six yield-shock indicator variables were then regressed against the residuals of the loess-estimated planted acreages.

Results from this regression are provided in Table 3. This somewhat more complicated regression mainly confirms the findings reported in Table 2. It shows that the residuals are on average decreasing with respect to the number of prior bad shocks, with the exception of the difference between three and four bad shocks. The difference between these two coefficients, however, is less than half of one standard error. Because the dependent variable is the residual of a log-log regression of plantings on time, the coefficients imply that the average yield shock causes approximately a 2 percent change in subsequent acreage plantings. We feel these estimates should be viewed as conservative, because the county-level trends were not estimated simultaneously with the shock effects, so possible over-fitting of the non-parametric loess estimators could bias downward the magnitude of the fixed-effects coefficients.

Table 3. Fixed Effects of Lagged Shocks on Residual of Loess-Estimated Plantings

Variable	Parameter Estimate	Standard Error	t Value	Pr > |t|
Intercept (Five Good Shocks)	0.0384	0.00505	7.59	<.0001
One Bad Shock	-0.0122	0.00575	-2.12	0.0337
Two Bad Shocks	-0.0318	0.00548	-5.79	<.0001
Three Bad Shocks	-0.0452	0.00553	-8.17	<.0001
Four Bad Shocks	-0.0435	0.00597	-7.29	<.0001
Five Bad Shocks	-0.0964	0.00691	-13.95	<.0001

Dependent variable: residual from county-level loess-estimated plantings:

$$y_{it} = \log(l_{it}) - f_i(\log(t))$$

Sample size: 85,255 observations. Includes about 2,200 counties (i)
R-squared: 0.0037
Adj. R-squared: 0.0036

Note: The sample size is somewhat smaller than that reported in Table 2 because loess estimates of plantings were made only for suitably long county-level time series.

The results summarized in Tables 2 and 3 demonstrate that total planted acres and changes in planted acres are significantly impacted by past yield shocks. In addition, as the number of negative yield shocks received in the past five years increases, the more land is taken out of wheat production. This result confirms that farmers' decisions are impacted by unexpected income shocks and that farmers are unable to perfectly cope with risk. In the next section we add structure to our test of the importance of liquidity in farm production decisions.

Structural Estimation of the Planting Decision

Methods. Consider the following parameterization of equation (6), which was derived in the section on theory and is replicated below for convenience.

$$E_{it}\left[p_{it}(s_{it})y_{it}(l_{it}, x_{it}', s_{it}) + l_{it}p_{it}(s_{it})\frac{\partial y_{it}}{\partial l_{it}} - c'(l_{it})m(x_{it}', s_{it}) \right.$$

$$\left. + \frac{a_i}{1+r}(l_{i(t+1)} - l_t) - a_i(l_{it} - l_{i(t-1)}) \mid \Theta_{it} \right] = \lambda_{it}[c'(l_{it})m(x_{it}', s_{it})] \geq 0 . \tag{6}$$

$$\frac{\partial y_{it}}{\partial l_{it}} = \alpha_{0i} + \alpha_{li}t \tag{7}$$

$$c'(l_{it}) = \beta_{0i} + \beta_{li}l_{it} \tag{8}$$

$$m(x_{it}', s_{it}) = \gamma_{0i} + \gamma_{li}t + s_{it} . \tag{9}$$

The variable s_{it} is a scalar random variable with zero mean that captures variation in the state vector s_{it}. The parameters α_{0i}, α_{li}, β_{0i}, β_{li}, γ_{0i}, and γ_{li} can be collectively accounted for empirically but cannot be estimated separately from the reduced form. In the estimating equation, the multiplier term should be inseparable from these parameters as well. To test whether or not the multiplier is non-zero, we need to somehow divide the sample into two more groups which conceivably have different average multipliers and test for a difference between these two groups.

If the model is "correct" (that is, it encompasses all expected acreage choices), then it is not important how these two groups are delineated, except that one group should have a greater likelihood of being contemporaneously constrained. If, on the other hand, the model is not *exactly* correct, but provides crude approximation of behavior, then the test is more believable if the two groups have different chances of being constrained but are otherwise similar. This way, the test is relatively robust to misspecification of the model. So long as missing features of the model are not correlated with the way in which the sample is divided, the test should remain valid. We therefore "divide" our sample by including two indicator variables based on past yield shocks: one indicates whether the two immediately preceding shocks were negative and the other indicates whether they were both positive. The group with the negative shocks should have a slightly higher chance of being constrained, but should otherwise be similar to the positive-shock group. This construction of the problem allows us to estimate the average difference in the multiplier between these two groups and the group in which one of the last two shocks was positive and the other negative (captured by the constant). These indicator variables are denoted as D_{it}^G and D_{it}^B, the good and bad shocks respectively. Finally, suppose that expectations errors can be captured by the random variable η_{it}, which leads to the following reduced form:

$$\begin{aligned} p_{it}y_{it} &= \alpha_{0i}l_{it}p_{it} + \alpha_{li}tl_{it}p_{it} + \beta_{0i} + \beta_{li}l_{it} + \beta_{2i}l_{it}t + \beta_{3t}t - a_i'(l_{i(t+1)} - l_{it}) \\ &+ a_i(l_{it} - l_{i(t-1)}) + \phi_i^G D_{it}^G + \phi_i^B D_{it}^B + s_{it}(\beta_{0i} + \beta_{li}) + \eta_{it}, \end{aligned} \tag{10}$$

where the coefficients β_{ji}' are reduced-form scalars derived from the interaction β and γ coefficients above and a_i' equals $a/(1+r_t)$, which assumes that the real discount rate is constant over time.

Estimation of equation (10) using ordinary least squares will lead to bias resulting from missing variables bias, specification bias, measurement error, and dependency bias. Dependency bias is caused by the confounding of the expectation error η_{it} with changes in the state variable s_{it}, which is linked to the explanatory variable l_{it}. The expectation error also is sure to be correlated with current prices, and also may be correlated with planted acreage, because some planting decisions may respond to weather information learned early in the season. We correct these biases, in addition to biases caused by possible measurement error, by using lagged values of prices, yields, and planted acreages to instrument $l_{it}p_{it}$ and l_{it}. For reasons described earlier, missing variable and specification biases should be minimal with respect to the coefficients of interest, ϕ_i^G and ϕ_i^B.

We are interested in testing whether ϕ_i^G and ϕ_i^B differ from zero. Because each estimate results from only between 12 and about 65 degrees of freedom, there is not enough power to test any individual estimate against a suitable null hypothesis (for example, $\phi_i^G = \phi_i^B = 0$). Given 1,952 such estimates, however, there is enough power to ask more general and more interesting questions about the average values of ϕ_i^G and ϕ_i^B, taken across counties. These two parameters, and corresponding estimates of them, can be written as follows:

$$\phi^G = \sum_{i=1}^{N} w_i\phi_i^G, \phi^B = \sum_{i=1}^{N} w_i\phi_i^B, \tag{11}$$

$$\hat{\phi}^G = \sum_{i=1}^{N} w_i \hat{\phi}_i^G, \hat{\phi}^B = \sum_{i=1}^{N} w_i \hat{\phi}_i^B, \tag{12}$$

where the weights w_i are chosen to be the inverse of the estimated standard errors from the estimates ϕ_i^G and ϕ_i^B.[11] Assuming that these estimates are independent, the standard error for $\hat{\phi}^G$ approximately equals the square root of $\sum_{i=1}^{N} w_i [SE(\hat{\phi}_i^G)]^2$.[12]

Intuitively, identification of the parameters comes from predictable variation in county-level expected marginal profit-per-acre caused by county-level lagged yield shocks. Expectations about marginal profit-per-acre (i.e., the *predictors*) were made using the instruments (lagged yields, prices, and planted acres). The structural estimates therefore attempt to control for lingering changes in investment opportunities that may be correlated with past yield shocks, despite their apparent transitory nature. For example, yields are at least partly a function of average land quality, which is likely correlated with planted acres, which are serially correlated. Indeed, pooling shocks from all counties together, one finds a slight but statistically very significant autocorrelation in yields equal to about 0.13. Although the simple regressions in the last section do not control for such autocorrelation, the structural estimates do – by comparing only the variation in marginal profits that is predictable using past observable values.[13]

Because yield shocks are caused mainly by weather variability, are at least mostly transitory, and therefore should be uncorrelated with missing variables that affect the fundamental profitability of subsequent decisions, the estimates should be robust to specification and missing-variables biases. In other words, the source of identification is likely independent of missing

[11] This choice of weights minimizes the approximate standard errors of $\hat{\phi}^G$ and $\hat{\phi}^B$.

[12] Although $\hat{\phi}_i^G$'s and $\hat{\phi}_i^B$'s are not strictly independent across counties, we do not feel this independence assumption to be an especially distorting one. Dependence, which will not bias the parameter estimates $\hat{\phi}^G$ and $\hat{\phi}^B$ but will bias standard error estimates, could arise from spatial correlation of both expectation error and lagged yield shocks, which probably does exist. Examination of the data shows that spatial correlation of yield shocks is modest, however, and diminishes to zero for counties separated by as little as one state. The average level of dependence therefore is likely to be small. Moreover, the spatial dependence is likely overwhelmed by expectation error that is not spatially correlated (such as aggregate price shocks). The reported standard errors therefore underestimate the true values, but probably by very little. Bootstrapped confidence intervals for $\hat{\phi}_i^G$ and $\hat{\phi}_i^B$ verify this conjecture.

[13] Biased estimates could still result, however, if missing variables in the marginal profit equation are in some way correlated with past yield shocks. For example, severe weather in the current year could influence both current yields and residual moisture in the subsequent year, and thereby increase subsequent costs.

variables and functional form errors, so these will add noise to the estimates of $\hat{\phi}^G$ and $\hat{\phi}^G$, but not bias.

Results. A summary of the structural estimates is reported in Tables 4 and 5. Table 4 summarizes the results from the instrument equations. The relatively high R^2 values imply that the instruments are good ones. Table 5 provides two estimates for the change in multiplier λ_{it} between three sub-samples of the data. Similar in spirit to the quasi-experimental approach taken in the last section, the sample was broken into three mutually exclusive sub-samples: one that includes all observations where the two immediately preceding estimated yield shocks are positive ("Two Good Shocks"), a second in which the last two shocks are negative ("Two Bad Shocks"), and a third

Table 4. Summary of First-Stage Regressions

Endogenous Variable	Mean R^2	Mean Adjusted R^2	Mean Error D.F.	Min. Error D.F.
$l_{it}p_{it}$	0.424	0.268	31.98	10
l_{it}	0.735	0.653	31.98	10
$l_t - l_{it-1}$	0.685	0.586	31.98	10
$l_{it+1} - l_{it}$	0.193	0.079	34.98	13

Notes:
1. One linear regression was estimated for each county (i) with 20 or more observations.
2. The covariates (instruments) are two years of lagged prices, two years of lagged planted acreage, two years of lagged yields, time (year), and an intercept. For the lead acreage change, however, only one lag of each of these variables was used.

group in which one of the last two shocks is positive and the other negative. The third group is captured by an intercept. Although we are unable to obtain an estimate for the level of λ_{it}, because it cannot be separated from the constant value that measures a component of marginal cost, the dummy variables measure changes in the average value of λ_{it} between the three groups. Following two bad shocks, the average value of λ_{it} is significantly greater than following one bad shock and one good shock. Following two good shocks, the average value of λ_{it} is estimated to be less than one bad and one good shock, but not significantly so. These results are consistent with the picture presented in Figure 1a: as the liquidity constraint tightens (following two bad shocks), marginal profit increases; as it relaxes (following two good shocks), marginal profit decreases. The coefficients can be interpreted as the change in the average value (in 1982-84 dollars) of having a credit line extended to finance planting of an additional acre. The change is with respect to the average value of λ_{it} in the third group, with one bad and one good shock. In other words, the average yield shock changes the marginal illiquidity cost by about \$1.50 to

Table 5. Summary of Group Mean-ϕ_{it} Estimates from Second-Stage Regressions

Group	Simple Average Over Counties	Simple Standard Error	Simple t-statistic	Weighted Average	Weighted Standard Error	Weighted t-statistic
Two Good Shocks (ϕ_i^G)	-1.3827	0.4280	-3.231	-0.5605	0.3423	-1.637
Two Bad Shocks (ϕ_i^B)	2.8304	0.4015	7.050	2.2098	0.3594	6.149

Notes:
1. This table summarizes structural estimates of equation (10).
2. The coefficients can be interpreted as the change in the average value (in 1982-84 dollars) of having a credit line extended to finance planting of an additional acre. The change is with respect to the average value of $c'()m()\lambda_{it}$ in the missing group: those observations for which one of the last two yield shocks is positive and the other one negative.
3. One linear regression was estimated for each county (i) with 20 or more observations. There are 1,952 counties in the sample and a total of about 85,000 observations with non-missing observations.
4. The covariates include the estimated values from the first-stage regressions in Table 4, these estimated values interacted with time, and two 0-1 indicator variables, one for two consecutively positive lagged shocks ("Two Good Shocks") and one for two consecutively negative lagged shocks ("Two Bad Shocks").
5. The other coefficients, which serve mainly as controls, vary widely across counties.
6. Bootstrapped confidence intervals are consistent with the simple-average estimates; there appears to be no substantive correlation in estimates across counties.

$2.50 (in 1982-84 dollars) per acre of wheat acres not planted due to the credit constraint. This value also could be interpreted as an estimated lower bound on the average value of λ_{it}. For comparison, the USDA-ERS nationwide average cost estimates for 1999 were $69.58 for total costs and $33.02 for operation costs (in comparable 1982-84 dollars).

CONCLUSIONS

This chapter shows how lagged yield shocks, which are both large and transitory, have persistent economic effects that result from farmers' inability to cope perfectly well with risk. We have estimated these effects using a model that subsumes risk-neutral and forward-looking farmers who make rational land use decisions but are constrained in the amount that they can borrow. Put succinctly, liquidity constraints and risk matter to farm production decisions.

At least implicitly, this work responds to important questions raised by Richard Just in last year's meetings of the Northeastern Agricultural and Resource Economics Association (Just 2000). With regard to long-run versus short-run risks, he wrote:

> "...the consequences of risk may not be great in the short-run unless serial correlation is high (which makes it a longer-run problem). Farmers can easily shift major purchases of machinery and equipment and even consumer-durable consumption decisions from one time period to another so that short-run fluctuations in revenue need not cause drastic consequences. Furthermore, many sources of agricultural credit are structured to allow considerable variability in debt repayment as long as solvency is not in question (which is a longer-run problem). So why is almost all risk research in agriculture carried out in a short-run context that focuses on year-to-year variability? Why does risk research not focus on longer-term swings in agricultural production conditions? Certainly, some longer-term swings have been observed and have had serious consequences. For example, the commodity boom of the 1970's with the resulting high investment in land and machinery followed by the high interest rates of the 1980s caused the highest rate of farm failures since the Great Depression" (p. 142).

Just points out that farmers have many risk-coping tools at their disposal. Because these tools mainly involve tradeoffs over time, modeling and measurement of risk effects should be dynamic. He also suggests that solvency is a binding constraint, which implies that collaterizable wealth or internal net worth, in addition to fundamental profitability, matters for production choices. We incorporate these key features into our analysis.

We depart from Just by focusing on transitory rather than persistent shocks. We feel that to answer certain questions there are good reasons to examine year-to-year variability rather than longer-run swings. Updated information about fundamental expected profitability, in addition to solvency constraints or risk aversion, provides motives for farmers to adjust production decisions. Although persistent shocks, due to their longevity, may ultimately induce greater risk-coping responses than transitory shocks, it is more difficult to empirically disentangle these risk effects from the information effects that confound them. If one wishes to estimate the risk effects of persistent shocks, one must identify their degree of persistence (prices, for example, have frustratingly variable autocorrelation functions) as well as hold deep faith in the specified model. Unless the empirical model correctly controls for information effects, the skeptic portrayed in the introductory section will suspect that estimated risk effects are actually spurious findings driven by lingering information effects.

We feel that this chapter makes several contributions to the study of agricultural risk. The approach and the results bridge issues of risk management and natural resource management in a way that differs markedly from the existing agricultural literature. Contrary to some assertions, we show that risk matters to land use decisions, a production choice with environmental consequences (Sumner and Lee 2000, LaFrance, Shimshack, and Wu 2000, Antle and Just 1990, Soule, Nimon, and Mullarkey 2000). In addition, contrary to static models with risk-averse farmers, we propose a model that captures the fundamentally dynamic nature of risk effects that result from a credit constraint rather than preferences. In our empirical estimates of these effects we take care to disentangle information effects from risk effects by using transitory yield shocks as a source of identification. By combining both structural and quasi-experimental empirical techniques, the approach makes the results more robust to model misspecification. Finally, we feel that this empirical approach can be used to examine how other inputs are influenced by risk and, by comparing the magnitude of shock responses across policy regimes, to evaluate how past changes in government agricultural programs have affected risk coping.

REFERENCES

Alderman, H., and C.H. Paxson. 1994. "Do the Poor Insure? A Synthesis of the Literature on Risk and Consumption in Developing Countries." *Economics in a Changing World: Proceedings of the Tenth World Congress of the International Economic Association, Moscow* (Vol. 4: Development, Trade and the Environment). IEA Conference Volume, No. 110. New York: St. Martin's Press.

Antle, J.M., and S.M. Capalbo. 2001. "Agriculture as a Managed Ecosystem: Implications for Econometric Analysis of Production Risk." In R.E. Just and R.D. Pope, eds., *A Comprehensive Assessment of the Role of Risk in U.S. Agriculture*. Boston, MA: Kluwer Academic Publishers.

Antle, J., and R. Just. 1990. "Interactions between Agricultural and Environmental Policies: A Conceptual Framework." *American Economic Review* 80: 197-202.

Carroll, C. 1997. "Buffer Stock Saving and the Life-Cycle /Permanent Income Hypothesis." *Quarterly Journal of Economics* 112: 1-55.

Chavas, J., and M.T. Holt. 1990. "Acreage Decisions Under Risk: The case of Corn and Soybeans." *American Journal of Agricultural Economics* 24: 529-538.

___. 1996. "Economic Behavior Under Uncertainty: A Joint Analysis of Risk Preferences and Technology." *Review of Economics and Statistics* 78: 329-335.

Deaton, A. 1992. *Understanding Consumption*. Clarendon Lectures in Economics. Oxford University Press.

Dickey, D.A., and W.A. Fuller. 1979. "Distribution of the Estimators for Autoregressive Time Series with a Unit Root." *Journal of the American Statistical Association* 74: 427-431.

Dixit, A.K., and R.S. Pindyck. 1994. *Investment Under Uncertainty*. Princeton, NJ: Princeton University Press.

Freund, R.J. 1956. "The Introduction of Risk into a Programming Model." *Econometrica* 24: 253-264.

Hubbard, R.G. 1998. "Capital-Market Imperfections and Investment." *Journal of Economic Literature* 36: 193-225.

Hubbard, R.G., and A.K. Kashyap. 1992. "Internal Net Worth and the Investment Process: An Application to U.S. Agriculture." *Journal of Political Economy* 100: 506-534.

Hubbard, R.G., A.K. Kashyap, and T.M. Whited. 1995. "International Finance and Firm Investment." *Journal of Money, Credit, and Banking* 27: 683-701.

Jaffee, D.M., and T. Russell. 1976. "Imperfect Information, Uncertainty, and Credit Rationing." *Quarterly Journal of Economics* 90: 651-666.

Just, R.E. 2000. "Some Guiding Principles for Empirical Production Research in Agriculture." *Agricultural and Resource Economics Review* 29: 138-158.

Just, R.E., and R.D. Pope. 1991. "Stochastic Specification of Production Functions and Economic Implications." *Journal of Econometrics* 7: 67-86.

Kochar, A. 1999. "Smoothing Consumption by Smoothing Income: Hours-of-Work Responses to Idiosyncratic Agricultural Shocks in Rural India." *The Review of Economics and Statistics* 81: 50-61.

LaFrance, J.T., J. Shimshack, and S. Wu. 2000. "Crop Insurance and the Environment." Working Paper. Department of Agricultural and Resource Economics, University of California, September.

Leathers, H.D., and J.C. Quiggin. 1991. "Interactions Between Agricultural and Resource Policy: The Importance of Attitudes Toward Risk." *American Journal of Agricultural Economics* 73: 757-764.

Loehman, E., and C. Nelson. 1992. "Optimal Risk Management, Risk Aversion, and Production Function Properties." *Journal of Agricultural and Resource Economics* 17: 219-231.

Marra, M.C., and G.A. Carlson. 2001. "Agricultural Technology and Risk." In R.E. Just and R.D. Pope, eds., *A Comprehensive Assessment of the Role of Risk in U.S. Agriculture.* Boston, MA: Kluwer Academic Publishers.

Modigliani, F., and M.H. Miller. 1958. "The Cost of Capital, Corporation Finance and the Theory of Investment." *American Economic Review* 48: 261-297.

Morduch, J. 1995. "Income Smoothing and Consumption Smoothing." *Journal of Economic Perspectives* 9: 103-114.

Paxson, C.H. 1992. "Using Weather Variability to Estimate the Response of Savings to Transitory Income in Thailand." *American Economic Review* 82: 15-33.

Pope, R.D., and R.E. Just. 1991. "On Testing the Structure of Risk Preferences in Agricultural Supply Analysis." *American Journal of Agricultural Economics* 73: 743-748.

Pope, R.D., and R.A. Kramer. 1979. "Production Uncertainty and Factor Demands for the Competitive Firm." *Southern Economic Journal* 46: 489-501.

Rambaldi, A.N., and P. Simmons. 2000. "Response to Price and Production Risk: The Case of Australian Wheat." *The Journal of Futures Markets* 20: 345-359.

Sandmo, A. 1971. "On the Theory of the Competitive Firm under Price Uncertainty." *American Economic Review* 61: 65-73.

Schwartz, E.S. 1997. "The Stochastic Behavior of Commodity Prices: Implications for Valuation and Hedging." *The Journal of Finance* 52: 924-973.

Soule, M., W. Nimon, and D. Mullarkey. 2000. "Risk Management and Environmental Outcomes: Framing the Issues." Paper presented at the workshop, "Crop Insurance, Land Use, and the Environment," September 20-21, 2000, Economic Research Service, Washington, D.C.

Stiglitz, J.E., and A. Weiss. 1981. "Credit Rationing in Markets with Imperfect Information." *American Economic Review* 71: 393-410.

Sumner, D.A., and H. Lee. 2000. "Discussion of the Effects of Federal Risk Management Subsidies on Crop Production Patterns." Prepared for workshop on risk management research, Washington D.C., September 20, 2000.

Townsend, R.M. 1995. "Consumption Insurance: An Evaluation of Risk-Bearing Systems in Low-Income Economies." *Journal of Economic Perspectives* 9: 83-102.

Zeldes, S.P. 1989. "Consumption and Liquidity Constraints: An Empirical Investigation." *Journal of Political Economy* 97: 305-346.

Chapter 19

PRECISION FARMING TECHNOLOGY AND RISK MANAGEMENT

James A. Larson, Burton C. English, and Roland K. Roberts
The University of Tennessee, Knoxville

INTRODUCTION

Farm fields have numerous areas that differ from one another with respect to soil type, topography, microclimate, and other factors that influence crop yields. Crop response to a particular input may differ across these areas within a field. Furthermore, differences in inherent soil properties across a field may result in non-uniform temporal variation in yield response to stochastic events such as weather, disease, insect pressure, and weed pressure. Thus, spatial variation in temporal production risk could occur as stochastic temporal yield-limiting factors interact with inherent yield-limiting soil properties (Nielsen 2001). With the availability of so-called "precision farming" technologies (also referred to as variable-rate, prescription, site-specific, or soil-specific farming), the cost of gathering information about within-field variability has decreased, making it possible for farmers to consider using site-specific management practices in crop production (Lowenberg-DeBoer and Swinton 1997). Farmers can now gather information about variation in yield-limiting soil properties within a farm field to make management choices about applying crop inputs using variable rate technology. Thus, the potential for precision farming as a risk management tool is in allowing farmers to gather more and better information and gaining increased control over production (Lowenberg-DeBoer 1999).

The purpose of this chapter is to examine the potential for precision farming technologies in managing agricultural risk. The first section describes what precision farming is in greater detail and summarizes the current status of its adoption by farmers. Next is a discussion of the potential uses and implications of precision farming technology in risk management. Then we present an empirical evaluation of variable rate technology as a risk management tool. The chapter concludes with a discussion of future research needs for evaluating the risk management potential for alternative precision farming technologies.

BACKGROUND

All precision farming technologies involve some form of collecting and analyzing site-specific information for management decision making (Swinton and Lowenberg-DeBoer 1998). Precision farming includes a wide array of site-specific technologies, from farmers using flags to partition a field into management zones to on-board computers interfacing with satellites pinpointing precise coordinates within a field. Some of the more basic types of site-specific technologies include soil testing for nutrients and pests, plant tissue sampling for nutrients and pests, field scouting, plant mapping, aerial photography, and using soil survey maps. These and other basic technologies provide a producer with site-specific information to aid in management decisions.

In addition, several more sophisticated computer-based technologies allow a farmer to gather and analyze field data and apply inputs in response to that information. Optical sensing equipment allows on-the-go information to be collected and processed as equipment moves through the field. Current examples of in-field sensing technologies include yield monitors and soil organic matter sensors. Remote sensing technologies include aerial photography and satellite imagery.

Geographic Information Systems (GIS) are databases that provide electronic information about site-specific field attributes. Global Positioning Systems (GPS) allow site-specific information to be collected through interface with satellites. GPS technology can allow a computer to know the exact location (to within a few meters or less) of farm equipment that it is monitoring and controlling. A GPS receiver can be attached to machinery such as tractors, combines, ground and aerial sprayers, or all-terrain vehicles. For example, GPS technology can be used with a yield monitor on a combine or cotton picker to measure within-field yield variability. This information can then be converted from raw data into a yield map using precision farming computer applications. A database of spatial and temporal yield variability over time can be used with other information, such as a topographical map of the field, to make crop management decisions for the specified field. GPS guidance systems on sprayers and planters can guide equipment to help avoid skips and overlaps and may reduce labor requirements.

Grid or management-zone soil sampling is based on GPS technology. Grid soil sampling partitions the field into grids of a specified size and pulls soil samples from the grids. Management-zone soil sampling involves sampling from each of several management zones, which are identified by characteristics such as soil type or topography. Information gathered from soil sampling, as well as other information such as soil electro-conductivity, may then be used to generate variable rate fertilizer recommendations for different grids or management zones.

Variable rate technology (VRT) application of crop-related inputs such as seed, lime, fertilizer, herbicide, and pesticide is another important aspect of precision farming that uses GPS. VRT involves applying crop inputs in a non-uniform manner based on varying needs throughout a field. With VRT, input application maps for management zones in the field are created from information analyzed from yield monitor, soil sample, or other field data. Inputs are then applied using a GPS-guided single- or multi-product fertilizer spreader or spray application equipment. Real-time, on-the-go field sensors are also being developed to diagnose crop problems and apply inputs.

Advantages of variable rate application may include higher average yields, lower farm input costs, improved farm profits, and environmental benefits from applying fewer inputs (Kitchen et al. 1996, Koo and Williams 1996, National Research Council 1997, Sawyer 1994, Watkins, Lu, and Huang 1999). Several studies (Babcock and Pautsch 1998, Bongiovanni and Lowenberg-DeBoer 1999, Bullock et al. 1998, English, Roberts, and Mahajanashetti 1999, Lowenberg-DeBoer 1996, Lowenberg-DeBoer and Aghib 1999, Lowenberg-DeBoer and Swinton 1997, Roberts, English, and Mahajanashetti 2000, Thrikawala et al. 1999, Watkins, Lu, and Huang 1999), along with several reviewed by Lambert and Lowenberg-DeBoer (2001), have assessed the economic potential of VRT. Lambert and Lowenberg-DeBoer indicated that about 60 percent of the studies they reviewed reported positive economic benefits from using VRT for fertilizer application. Principal factors influencing the profitability of VRT are the crop, the input, their prices, the cost of VRT versus URT (uniform rate technology) application, the distribution across a field of management zones with different yield responses (spatial variability), and the magnitudes of the differences in yield response across management zones (English, Roberts, and Mahajanashetti 1999, Forcella 1993).

Evidence about the degree to which farmers have adopted precision farming technologies is also scattered and incomplete. Nevertheless, a few studies provide estimates of adoption rates for a limited number of precision farming technologies. In a 1996 USDA personal interview survey of 1,673 corn farmers in 16 states, 5.5 percent of producers reported using yield monitors (Daberkow and McBride 1998). A 1997 mail survey of 1,000 farmers in four North-Central states – Illinois, Indiana, Iowa, and Wisconsin – found that 9.8 percent of farmers used yield monitors (Khanna, Epouhe, and Hornbaker 1999). Novartis (1999) surveyed 2,366 corn producers and found that 16 percent had yield monitors in 1998, but only 4.5 percent had those monitors linked to GPS. Of large farms with more than a thousand acres, 47.5 percent had yield monitors, with 26.2 percent having the monitor attached to GPS (Novartis 1999). Most farmers who had adopted yield monitors were using them as a stand-alone technology for making decisions (Khanna, Epouhe, and Hornbaker 1999). Khanna, Epouhe, and Hornbaker (1999) found about half the farmers surveyed were using yield monitors to make input decisions but two-thirds of yield-monitor users were not using variable rate technology to apply those inputs. Farmers appeared to be establishing a history of field variability before making variable input decisions (Swinton, Harsh, and Ahmad 1996). Yield monitor adopters also appeared to be creating databases for drainage decisions and land rental negotiations (Khanna, Epouhe, and Hornbaker 1999).

VRT has become more widely available to producers through off-farm service providers. In 2000, more than 30 percent of fertilizer dealers nationwide offered controller-driven variable rate application services, up from 13 percent in 1996 (Akridge and Whipker 1996, 2000). Farmer adoption of VRT has been relatively high for some high-value crops such as sugar beets. Nitrogen was applied using VRT on an estimated 40 percent of sugar beet acreage in the Red River Valley of North Dakota and Minnesota in 1999 (Franzen 2000). Adoption of site-specific practices for lower-value crops such as wheat, soybeans, and corn has been more limited. Daberkow and McBride (1998) determined that 3.7 percent of corn farmers used variable rate applications in 1996. Khanna, Epouhe, and Hornbaker (1999) found that farmer usage of VRT in Illinois, Iowa, Indiana, and Wisconsin was 11.9 percent for fertilizer, 2.5 percent for pesticides, and 2.4 percent for seed in 1997. Adopters of these advanced technologies were younger, had more education and farming experience, used computers, and ran larger farm operations than non-adopters (Khanna, Epouhe, and Hornbaker 1999).

PRECISION FARMING TECHNOLOGY AND RISK MANAGEMENT

The electronic information technologies that are an integral part of precision farming allow farmers to apply electronic monitoring and control techniques widely used in other industries for data collection, data processing, decision analysis, and precision application of inputs in agricultural production (Lowenberg-DeBoer and Swinton 1997, Lowenberg-DeBoer and Boehlje 1996). This section describes some of the potential uses and implications for these information technologies in risk management.

Variable rate technology may be a helpful risk management tool for reducing year-to-year variability in yields and net returns. Many yield-limiting soil physical properties change little from year to year – for example, topography, depth to fragipan (a loamy, brittle subsurface soil horizon that restricts crop root growth), tilth, organic matter content, and so forth – while some soil chemical properties, such as plant-available nitrogen, can change markedly from season to season. Crop yield response to an applied input for a given combination of soil properties and their interactions with one another could be predicted reasonably well if weather were not variable across years. Temporal variation in weather interacts with the soil environment in a number of ways. First, weather affects several other stochastic yield-limiting factors such as weed and insect pressures, which in turn affect yield. Second, weather affects soil chemical properties through leaching, erosion, plant uptake, and so forth. Third, weather interacts with soil properties and exacerbates interactions among soil chemical and physical properties. For example, ridges within a field may provide high yields during wet years and low yields during dry years, while just the opposite would be true for low-lying areas where yield might be reduced by drowning in wet years and increased in dry years. In addition, different amounts of rainfall would affect yields through temporal variation in the amounts of nitrogen leached from these two topographical areas.

Temporal variation in yield caused by weather and its interactions with soil properties was discussed by Nielsen (2001), who showed markedly different yield patterns for the same 30-acre corn field in 1997 and 1998. In addition, Lamb et al. (1997) were unable to reliably predict corn yield for the same spot in a field in a 5-year study. Huggins and Alderfer (1997) found that 67 percent of temporal yield variability was attributable to climatic variability, while only 15 percent was attributable to the nitrogen treatments under farmer control. Eghball and Varvel (1997) found that spatial variability was overshadowed by temporal variability.

One potential for VRT to have risk management benefits is based on the concept that creating a more spatially homogeneous growing environment across the field may reduce the number of problem areas in the field that drive down overall yields in some crop years (Lowenberg-DeBoer and Swinton 1997). Another potential source of risk reduction emanates from the possibility that temporal yield variability may be different across soil types, causing field average yield variability to be different under VRT than under whole field management. Thus far, the literature evaluating VRT as a risk management tool is limited. Braga, Jones, and Basso (1999) found that VRT application of nitrogen did not reduce production risk associated with weather. Nevertheless, Lowenberg-DeBoer (1999) and Lowenberg-DeBoer and Aghib (1999) found that VRT application of phosphorus and potassium can reduce net return variability when compared with using URT. In addition, Dillon, Mueller, and Shearer (2001) found that a variable rate seeding and planting date decision rule can be an effective risk management strategy for corn farmers. Also, farmers who apply larger than optimal amounts of an input as insurance against uncertainty may reduce input use under VRT because of greater knowledge about spatial variability (Hennessy and Babcock 1998). Finally, Isik, Khanna, and Winter-Nelson (1999) concluded that adoption of VRT under uncertainty would increase with higher soil fertility and with greater variation in soil quality.

Besides changes in yields and input costs with VRT, site-specific management may provide risk benefits with respect to environmental quality, food safety, management of differentiated products, and increased span of managerial control (Pierce and Nowak 1999). A few studies have addressed limited aspects of the potential impact on environmental quality (Babcock and Pautsch 1998, English, Roberts, and Mahajanashetti 1999, Thrikawala et al. 1999, Watkins, Lu, and Huang 1999). These studies showed the potential to improve net returns, reduce nitrogen usage, and positively impact groundwater quality with variable rate technology.

Precision farming technology may have other risk management benefits for farmers and agribusinesses. Documentation provided through electronic monitoring and control of agricultural production may provide an effective way to reduce environmental and food contamination risk (Lowenberg-DeBoer and Boehlje 1996). Monitoring and control technologies could be used by producers, input suppliers, and processors as a way to trace input usage in production for the purpose of reducing insurance premiums and liability claims and as documentation in environmental and food safety court cases. For example, computer-generated input application maps used to make variable rate applications of fertilizers and chemicals can be used by farmers to document their practices. In addition, Akridge and Whipker (2000) reported that 12 percent of agribusiness input suppliers in 2000 were using field mapping with GIS to document their work for billing, insurance, and legal purposes.

Better information about crop progress through sensing and remote sensing data may also help producers in their marketing decisions (Taylor 1998, Lowenberg-DeBoer 1999). Farmers may be more willing to use forward contracting or hedging strategies if they have more accurate information about yields. Finally, insurance providers may be willing to provide lower crop insurance premiums if it can be shown that VRT application of inputs can reduce yield risk (Lowenberg-DeBoer 1999).

Lowenberg-DeBoer and Swinton (1997) suggest that site-specific management technologies have the potential to increase the average acreage of crop farms and dramatically change linkages in the agribusiness sector. Electronic monitoring and control techniques may allow farmers to increase their span of control (Lowenberg-DeBoer and Boehlje 1996). Development of cost-effective and reliable on-the-go and remote plant-sensing technology may allow farmers to devote fewer specialized resources to labor-intensive crop-monitoring activities. Development of labor-saving sensing technologies suggests the potential for a larger scale of operation in crop production. Precision farming also has the potential to facilitate much closer ties among producers, input suppliers, and processors. Incentives to strengthen agribusiness linkages come from economies of scale in the analysis of spatial information, the potential development of proprietary crop production practices (e.g., specialized genetically modified crops), and the links between spatial management of inputs and crop product quality. Consequently, processors may be more willing to enter into production contracts with farmers who are skilled users of information technologies to manage crop production. Increased availability of production contracts could potentially stabilize farm income (Lowenberg-DeBoer 1999).

As with the adoption of other agricultural technologies, precision farming may increase some types of risk. Cochrane's (1958) treadmill model of agricultural technology adoption indicates that early adopters of technology are more likely to survive than late adopters. Farmers unable to effectively use advanced information technology to produce commodities to specification may have a harder time surviving in the future if contract production becomes much more prevalent in crop agriculture. Application of inputs with VRT does not necessarily preclude the possibility of a crop failure, so yield risk under some weather scenarios may actually increase with precision farming (Lowenberg-DeBoer 1999). Increased investment in grid soil sampling, VRT application of inputs, and other precision farming services may actually exacerbate revenue losses in a low yield year and cause increased financial risk for farmers. Producers who adopt precision technologies also face technological and human capital risk (Lowenberg-DeBoer 1999). Many precision farming technologies are subject to rapid obsolescence as new and better technologies are developed. Farmers complain that equipment and software are often incompatible and some technologies are not reliable (Roberts and English 1999). Moreover, farmers have indicated that some technologies have a steep learning curve and require a significant amount of time, knowledge, and skill to operate the equipment and interpret the data (Roberts and English 1999). Specialized training and skills are required for many information technologies, so the loss of an employee with these skills may be a serious problem for a farm operation.

Finally, Byerlee and de Polanco (1986) hypothesize that technology packages usually can be divided into subsets or clusters of components that allow critical interactions to be exploited and enable adoption to follow a stepwise pattern. In their scenario, elements initially adopted will be those that provide the highest rate of return on capital expenditures. Because precision farming is not a single technology but a group of technologies, farmers may adopt precision farming technologies in a piecemeal manner (Khanna, Epouhe, and Hornbaker 1999, Khanna 2001). Farmers with fields with larger spatial variability may be able to obtain greater potential benefits from the adoption of precision farming technologies. However, spatial variability in fields is not known with certainty and may be discovered over time by an agricultural producer (Jaenicke and Cohen-Vogel 2000). Leathers and Smale (1991) indicate that when farmers are uncertain about the impact of a new technology, it is rational for them to sequentially adopt components rather than adopt the complete package at one time. For example, farmers may enroll a field in a variable rate application program sponsored by their local fertilizer dealer (Lowenberg-DeBoer 1999). This can be a low-cost method for a farmer to learn about spatial variability. Another example is farmers purchasing yield-monitoring equipment but not using variable rate application of inputs until they have built up a yield history on the field (Swinton, Harsh, and Ahmad 1996). Risk perceptions and psychological processes may enter into the sequential adoption decision for precision farming technology (Feder 1982). The psychology literature indicates that cognitive biases affect the use of information on probabilistic judgments (Tversky and Kahneman 1974). Decision maker perceptions potentially influence the choice of adoption of more advanced technology (Batte, Jones, and Schnitkey 1990). Thus, farmer perceptions about potential impacts of site-specific management on profitability, environmental quality, and other factors can influence the adoption decision.

VRT EMPIRICAL ANALYSIS

The purpose of this section is to provide an empirical analysis to evaluate the potential of VRT as a risk management tool. The production setting is a set of simulated corn fields located in western Tennessee. The hypothesis addressed by this research is that using VRT to apply nitrogen across management zones within a corn field reduces yield risk compared with whole field management using uniform rate technology (URT). The specific objectives of the empirical analysis were to 1) evaluate the risk effects of using VRT instead of URT for nitrogen fertilizer application to corn fields under alternative field spatial and temporal variability scenarios, 2) determine what a risk-averse farmer would be willing to pay for the information required to make the VRT nitrogen decision, and 3) analyze the potential effects of risk-aversion behavior on total nitrogen fertilizer usage and environmental losses with VRT compared with URT. The conceptual framework, empirical framework, and results and analysis are presented in the next three sections.

Conceptual Framework

Farmers interested in adopting VRT are assumed to employ management zones for deciding whether or not to make variable or uniform rate nitrogen fertilizer decisions. Management zones within a field are identified using characteristics such as soil type and topography, which might affect yield response to nitrogen.

Assume a field has m soil management zones. Let δ_i be the proportion of the field in the ith management zone such that $\sum_{i=1}^{m} \delta_i = 1$. With the VRT nitrogen application decision, a crop production function is estimated for each management zone such that

$$y_{i,t} = f_i(n_{i,t}) + h_i^{1/2}(n_{i,t})\varepsilon_{i,t}\,, \tag{1}$$

where i is the management zone subscript, t is a year subscript, y is per acre yield, n is the amount of nitrogen applied per acre, and ε is a random variable with mean zero. Following Just and Pope (1978), the production function is separated into deterministic and stochastic components:

$$E(y_{i,t}) = f_i(n_{i,t})\text{ , and}$$

$$\sigma^2_{y_{i,t}} = h_i(n_{i,t}),$$

where $E(y)$ is expected yield, σ^2_y is variance of yield, and $\sigma^2_{i,t} = E(\varepsilon^2_{i,t})$.

Assuming output price is random and input prices are known with certainty, expected per acre profit for management zone i is

$$E(\pi_{i,t}) = E[\bar{p}f_i(n_{i,t}) - rn_{i,t} - vrtc - oc], \tag{2}$$

where $\bar{p}$ is expected crop price (\$/bu), r is nitrogen price (\$/acre), *vrtc* is the cost of variable rate application of nitrogen (\$/acre), and *oc* is variable costs other than for nitrogen and variable rate application (\$/acre). For this analysis, the costs of gathering information and using it to estimate the management zone yield response functions are treated as sunk costs and thus are not explicitly accounted for in equation (2). Variance of per acre profit for management zone i when both p and y are random and independent is

$$Var(\pi_{i,t}) = [f(n_{i,t})]^2\sigma^2_p + \bar{p}^2[h(n_{i,t})] + \sigma^2_p[h(n_{i,t})] \tag{3}$$

(Bohrnstedt and Goldberger 1969).

With the VRT decision rule, expected per acre profit is a weighted average of the proportions of each soil management zone in the field:

$$E(\pi_{vrt}) = \sum_{i=1}^{m} \delta_i E(\pi_{i,t}). \tag{4}$$

Variance of per acre profit VRT is also a weighted average of the proportions of each soil management zone such that

$$Var(\pi_{vrt}) = \sum_{i=1}^{m} \delta_i^2 Var_i(\pi_{i,t}) + 2\sum_{i=j}^{m}\sum^{m} Cov_{ij}(\pi_{i,t}, \pi_{j,t}), \tag{5}$$

where $Cov_{ij}(\pi_{i,t}, \pi_{j,t})$ is the covariance between management zones i and j.

Equations (4) and (5) can be used to estimate the certainty equivalent (CE) for VRT and URT application of nitrogen. The CE for a risky decision is the return on a risk-free investment that makes the decision maker indifferent between the payoff from the uncertain decision and the payoff from the risk-free investment. The CE has a monotonically increasing relationship with expected utility, i.e., the CE from decision a is greater than the CE of decision b, if and only if decision a yields greater expected utility than decision b. This relationship allows inferences to be made about the expected utilities of the VRT and URT decision criteria. Finally, the CE is denominated in monetary units and thus can be used as an indicator of the economic significance of the differences between VRT and URT decisions. The CE of per acre profit for VRT can be approximated by

$$CE(\pi_{vrt}) = E(\pi_{vrt}) - \lambda / 2Var(\pi_{vrt}), \tag{6}$$

(Robison and Barry 1987), where λ is the value of the Pratt-Arrow absolute risk aversion coefficient. Freund (1956) has shown the linear mean-variance objective function is consistent with the negative exponential utility function (assumes constant absolute risk aversion) and normally distributed profits.

With URT, field average yields rather than management zone yields are used to make optimal nitrogen fertilization decisions. Thus, the same nitrogen rate is applied across the field with the URT decision criterion. Expected per acre profit for URT is

$$E(\pi_{urt}) = E\left[\sum_{i=1}^{m} \delta_i \bar{p} f_i(n_t) - rn_t - urtc - oc\right], \tag{7}$$

where *urtc* is the cost of uniform rate application of nitrogen. The same nitrogen rate is applied to all management zones in the field, so variance of profit in management zone *i* is

$$Var_i(\pi_{i,t}) = [f_i(n_t)]^2 \sigma_p^2 + \bar{p}^2 [h_i(n_t)] + \sigma_p^2 [h_i(n_t)], \tag{8}$$

and field-level variance of profit for URT is

$$Var(\pi_{urt}) = \sum_{i=1}^{m} \delta_i^2 Var_i(\pi_{i,t}) + 2 \sum_{i=}^{m} \sum_{j}^{m} Cov_{ij}(\pi_{i,t}, \pi_{j,t}). \tag{9}$$

The CE income for URT is calculated using

$$CE(\pi_{urt}) = E(\pi_{urt}) - \lambda / 2Var(\pi_{urt}). \tag{10}$$

Thus, if $CE(\pi_{vrt}) > CE(\pi_{urt})$, then VRT application of nitrogen is a risk-efficient strategy for a risk-averse farmer because it provides a higher expected utility. The difference in the CE income, $CED = CE(\pi_{vrt}) - CE(\pi_{urt})$, provides an estimated monetary value that a risk-averse farmer could pay to acquire the information required to make the VRT decision, e.g., the cost of collecting and analyzing yield monitor or soil sample data to make the VRT nitrogen decision.

Empirical Framework

Analysis of VRT versus URT requires estimates of yield distributions for alternative soil management zones within a field and for the whole field. A problem with evaluating spatial versus temporal yield variability within a field and their effects on yield risk is lack of sufficient years of yield observations for the management zones and for the field (Lamb et al. 1997). Until data from long-term experimental and on-farm field trials are available, one way to overcome this lack of data is to use a crop-growth simulation model, such as the Environmental Policy Integrated Climate (EPIC) model (Sharpley and Williams 1990, Williams et al. 1990), to develop meta-response functions for a number of management zones that have homogeneous soil properties occurring in fields within a particular geographic area (English, Roberts, and Mahajanashetti 1999). For the case of nitrogen application to corn, yield distributions can be developed for several soil management zones and nitrogen levels through simulation over a large number of years, and meta-response functions could be estimated for each management zone. The analysis could be done for fields with different proportions of their area in each management zone to determine if the degree of spatial variability affects the relative risk of VRT and URT.

Yields for three representative management zones were obtained by using the EPIC crop model for three western Tennessee soil types suited to corn production. The modeled soils were Collins (0 percent slope with no fragipan), Memphis (1 percent slope with 42-inch depth to fragipan), and Loring (3 percent slope with 30-inch depth to fragipan). These three soil types are frequently found in many western Tennessee farm fields. In general, the Collins soil is the most productive and the Loring soil is the least productive. Memphis is intermediate in productivity. Reduced tillage practices were assumed for all three soils. These practices included chisel plowing and a single disking, leaving more than 30 percent residue cover after planting (Uri 1999). Average monthly rainfall and temperature data recorded at the Covington Weather Station in western Tennessee (U.S. Department of Commerce) were used to generate daily stochastic weather data for the EPIC yield simulations.

The following procedures were used to simulate corn yield distributions with EPIC and estimate yield response for VRT and URT. The first step was to estimate yield response functions for each soil type. The data for response model estimation were created by simulating 100 years of weather-generated yields for each soil type assuming 11 nitrogen application rates ranging from 0 to 223 lbs/acre in 22-pound increments. Because starting soil parameter values in EPIC were not available, the first two years of simulated yields for each application rate were dropped before estimating the yield response functions.

Preliminary analysis of the data suggested that a quadratic-plus-plateau response model would best represent the mean yields generated by EPIC. Furthermore, in several field experiments, the quadratic-plus-plateau model better explained corn yield response to applied nitrogen than other models considered (Bullock and Bullock 1994, Cerrato and Blackmer 1993, Decker et al. 1994). The second step was to estimate quadratic-plus-plateau mean yield response functions for each soil type as expressed in equation (11):

$$\begin{aligned} Y &= \beta_0 + \beta_1 N_t + \beta_2 N_t^2 + e_t && \text{if } N < N^c, \\ Y &= Y^p && \text{if } N \geq N^c, \end{aligned} \tag{11}$$

where Y is corn yield (bu/acre), N is the nitrogen fertilizer rate (lbs/acre), β_0, β_1, and β_2 are parameters to be estimated by regression, and N^c and Y^p are the critical nitrogen rate and plateau yield, respectively.

The third step was to estimate the variance of yield function for each soil type using residuals obtained from the mean yield models. Variance of corn yield was specified as a function of nitrogen such that

$$\ln \hat{e}_t^2 = \alpha_0 + \alpha_1 N_t + \mu_t , \tag{12}$$

where $\ln \hat{e}_t^2$ is the natural log of the squared residuals, α_0 and α_1 are parameters to be estimated by regression, and μ is a random error term.

Many of the prior nitrogen fertilizer studies have indicated that applied nitrogen is moderately risk-increasing (Roumasset et al. 1989). However, some studies have found that nitrogen fertilizer reduces risk (e.g., Antle and Crissman 1990, Lambert 1990). The potential impact of nitrogen fertilizer on yield variability is influenced by soil-specific factors (e.g., soil water-holding capacity, organic matter, soil pH, etc.), the crop production system (e.g., dryland production versus irrigated production), and other management factors. Therefore, the hypothesized sign for N was uncertain.

The estimated mean yield function corrected for heteroscedasticity and the yield variance function for each management zone soil were substituted into equations (6) and (10) and used to predict CE-maximizing applied nitrogen rates, yields, and net revenues. A corn price of \$2.94/bu and a nitrogen fertilizer cost of \$0.33/lb of pure nitrogen were used to calculate CE-maximizing nitrogen rates. These prices are the means of 1984 through 1999 annual Tennessee prices inflated to 1999 dollars using the Implicit Gross Domestic Product Price Deflator index (Tennessee Department of Agriculture, various 1985 through 2000 annual issues, Council of Economic Advisers 2000). For the purpose of this study, custom rates for VRT and URT services were used for calculating costs. Results were generated assuming a cost difference of \$3/acre between VRT and URT for the application of nitrogen. This additional cost of VRT versus URT was close to the mean of \$3.08/acre found by Roberts, English, and Sleigh (2000) in a survey of firms that provided precision farming services to Tennessee farmers. Other costs of production that did not vary in this analysis were from an extension service enterprise budget for corn (Gerloff 2000).

Optimal input decisions for the maximization of CE in equations (6) and (10) were solved for risk neutrality ($\lambda = 0$) and several levels of risk aversion, $\lambda = 0.004/\$$, $\lambda = 0.008/\$$, $\lambda = 0.010/\$$, $\lambda = 0.012/\$$, and $\lambda = 0.014/\$$, consistent with the range of risk aversion for per acre profit used by Lambert (1990) and Larson et al. (1998). Certainty equivalent differences (CEDs) for risk-neutral and risk-averse VRT and URT decision criteria were evaluated for 18 field scenarios, each having a different mix of soils. The proportions of each soil type in the field were varied from 0 to 80 percent in 20 percent increments such that the sum of the percentages in the three soils equaled 100 percent and at least two soils existed in each field. For example, one field examined was assumed to be 0 percent Collins, 20 percent Memphis, and 80 percent Loring soils (0-20-80), while another field was assumed to be 60, 20, and 20 percent Collins, Memphis, and Loring soils (60-20-20), respectively.

The final study objective was accomplished through input of the estimated optimal VRT and URT nitrogen fertilization rates back into EPIC to estimate the amount of nitrogen lost per acre to the environment (N_{loss}). Nitrogen loss was calculated by summing nitrate leaching, surface runoff, and sub-surface flow for each soil series as indicated by output from EPIC. The nitrogen loss difference (NLD), defined as N_{loss} with VRT minus N_{loss} with URT, was calculated for each field.

Results and Discussion

Table 1 presents the estimated corn yield response functions for Collins, Memphis, and Loring management zone soils. The linear and quadratic coefficients for all equations have the expected signs, and the asymptotic standard errors are low relative to the magnitudes of the coefficients. The estimated nitrogen coefficient in each yield variance equation is significant with a positive sign. Regression results indicate that higher nitrogen levels increased yield risk for all three soils. However, the degree of yield variability is markedly different in each soil management zone. Evaluation of the variance equation coefficients with respect to nitrogen rate indicates that the marginal Loring soil has the largest yield variance and the highly productive Collins soil has the smallest yield variance. The standard deviation of corn yields on the Collins soil is 6.20 bu/acre compared with 25.24 bu/acre on the Loring soil when nitrogen is applied at the profit-maximizing level on each soil using VRT (Table 2). Memphis is intermediate in yield variability among the three soil types, but in relative terms its yield variability is closer to Loring than to Collins.

Optimal nitrogen application rates, net revenues, and certainty equivalent incomes for simulated fields with 100 percent Collins, Memphis, or Loring soils and for alternative levels of risk aversion are presented in Table 2.

Table 1. Estimated Corn Yield Response Functions for Applied Nitrogen (N) for Collins, Memphis, and Loring Soils

Soil	Equation
Collins	
Mean Yields[a]	$Y = 15.012 + 1.634N - 0.0039N^2$ if $N < 208.00$
	(0.486) (0.011) (0.000)
	$Y = 184.92$ if $N \geq 208.00$
Variance of Yields[b]	$ln\hat{e}_t^2 = 2.392 + 0.006N$
	(0.114) (0.001)
Memphis	
Mean Yields[a]	$Y= 10.512 + 1.429N - 0.0036N^2$ if $N < 198.41$
	(1.109) (0.036) (0.000)
	$Y = 152.30$ if $N \geq 198.41$
Variance of Yields[b]	$ln\hat{e}_t^2 = 1.573 + 0.024N$
	(0.109) (0.001)
Loring	
Mean Yields[a]	$Y = 6.793 + 1.232N - 0.0034N^2$ if $N < 181.92$
	(1.381) (0.041) (0.000)
	$Y = 118.87$ if $N \geq 181.92$
Variance of Yields[b]	$ln\hat{e}_t^2 = 1.708 + 0.029N$
	(0.145) (0.001)

[a] Weighted least square results. Numbers in parentheses are asymptotic standard errors.
[b] Numbers in parentheses are standard errors.

Optimal nitrogen rates are largest for the Collins soil and smallest for the Loring soil. The profit-maximizing ($\lambda = 0$) nitrogen rate for Collins is 194 lbs/acre compared with 183 lbs/acre for Memphis and 165 lbs/acre for Loring. Because nitrogen is a risk-increasing input for each soil type, the optimal nitrogen rate declines with increasing risk aversion. At the slight risk aversion level ($\lambda = 0.004$), nitrogen usage declines by less than 1 percent (2 lbs/acre) on the highly productive Collins soil. By comparison, the nitrogen application rate drops by 8 percent (13 lbs/acre) on the marginal Loring soil. At the extreme risk aversion level ($\lambda = 0.014$), nitrogen usage on the Collins soil declines by only 4 percent (8 lbs/acre), but drops by 20 percent (33 lbs/acre) on the Loring soil.

The predicted impact that risk-aversion behavior has on the nitrogen application rates for VRT and URT for the 18 field scenarios is presented in Table 3. Several important findings can be obtained from this table.

First, results indicate that the field average amount of nitrogen applied using the VRT decision criterion is similar to the URT field average nitrogen rate when assuming risk-neutral behavior ($\lambda = 0$). The difference between

Table 2. Optimal Nitrogen Application Rates, Yields, Net Revenues, Certainty Equivalents, and Risk Premiums for Alternative Risk Aversion Criteria

Absolute Risk Aversion λ	Applied Nitrogen Rate (lb/acre)	Corn Yield Mean (bu/acre)	Corn Yield Standard Deviation (bu/acre)	Net Revenue Mean ($/acre)	Net Revenue Standard Deviation ($/acre)	Certainty Equivalent ($/acre)	Risk Premium[a] ($/acre)
100 Percent Collins Soil Field							
0.000	193.71	184.12	6.20	330.63	115.02	330.63	0.00
0.004	191.99	183.92	6.17	330.60	114.88	304.20	26.39
0.008	189.89	183.63	6.13	330.46	114.69	277.85	52.61
0.010	188.65	183.45	6.10	330.34	114.56	264.71	65.62
0.012	187.25	183.23	6.08	330.15	114.42	251.60	78.55
0.014	185.66	182.96	6.04	329.88	114.24	238.53	91.35
100 Percent Memphis Soil Field							
0.000	182.83	151.42	20.13	238.09	111.22	238.09	0.00
0.004	174.16	150.18	18.12	237.30	107.40	214.23	23.07
0.008	166.64	148.66	16.54	235.32	104.25	191.84	43.47
0.010	163.11	147.81	15.85	233.98	102.81	181.13	52.85
0.012	159.68	146.89	15.21	232.42	101.42	170.70	61.72
0.014	156.31	145.91	14.60	230.65	100.07	160.55	70.09
100 Percent Loring Soil Field							
0.000	165.35	117.94	25.24	145.40	105.04	145.40	0.00
0.004	152.20	115.87	20.90	143.68	95.09	125.60	18.08
0.008	143.23	113.80	18.37	140.53	89.25	108.67	31.86
0.010	139.46	112.76	17.40	138.73	86.97	100.91	37.82
0.012	135.99	111.72	16.55	136.82	84.95	93.53	43.30
0.014	132.75	110.68	15.80	134.82	83.12	86.47	48.36

[a] The risk premiums were calculated as one-half the standard deviation of net revenue multiplied by the absolute risk aversion coefficient λ.

VRT and URT (NAD) was less than 1 lb/acre across the 18 field scenarios, with VRT nitrogen rates being slightly lower than URT nitrogen rates. The primary impact of the VRT decision criterion on nitrogen usage is the redistribution of nitrogen among the soil management zones in the field as suggested in Table 2 by the different nitrogen rates for each soil.

Second, less nitrogen was lost to the environment (NLD) with VRT under profit maximization, indicating that the corn crop used fertilizer nitrogen more efficiently with VRT. Results suggest that the amount of nitrogen lost to the environment may be modestly reduced from between 0.1 and 3.5 pounds per

Table 3. Uniform Nitrogen Application Rate (URT), Variable Nitrogen Application Rate (VRT), Nitrogen Application Rate Difference (NAD), and Nitrogen Loss Difference (NLD) for Alternative Risk-Aversion Decision Criteria and Field Scenarios

		$\lambda = 0.000$[a]				$\lambda = 0.008$				$\lambda = 0.014$			
Field No. and		N Rate				N Rate				N Rate			
Soil Mix[b]		URT	VRT	NAD	NLD	URT	VRT	NAD	NLD	URT	VRT	NAD	NLD
		lb/acre											
1	0-20-80	169.02	168.85	-0.18	-0.54	147.38	148.11	0.73	0.00	136.81	137.81	1.00	0.17
2	0-40-60	172.60	172.34	-0.26	-0.80	151.72	152.89	1.18	0.05	141.07	142.70	1.63	0.33
3	0-60-40	176.09	175.84	-0.26	-0.79	156.30	157.57	1.27	0.10	145.62	147.41	1.79	0.42
4	0-80-20	179.50	179.33	-0.17	-0.52	161.24	162.15	0.91	0.10	150.63	151.95	1.32	0.35
5	20-0-80	171.72	171.02	-0.70	-2.41	150.89	153.03	2.14	-1.51	140.61	144.08	3.47	-1.07
6	20-20-60	175.21	174.52	-0.69	-2.32	155.29	157.80	2.50	-1.20	144.95	148.96	4.01	-0.63
7	20-40-40	178.60	178.01	-0.59	-1.98	160.01	162.46	2.46	-0.91	149.67	153.67	3.99	-0.31
8	20-60-20	181.92	181.51	-0.41	-1.40	165.13	167.03	1.89	-0.69	154.97	158.19	3.22	-0.19
9	20-80-0	185.16	185.00	-0.15	-0.57	170.80	171.49	0.69	-0.58	161.10	162.53	1.43	-0.42
10	40-0-60	177.72	176.69	-1.02	-3.50	158.81	162.61	3.80	-2.08	148.69	155.09	6.39	-1.14
11	40-20-40	181.02	180.19	-0.83	-2.85	163.70	167.26	3.55	-1.57	153.67	159.77	6.11	-0.62
12	40-40-20	184.25	183.68	-0.57	-1.96	169.08	171.80	2.72	-1.14	159.36	164.28	4.92	-0.36
13	40-60-0	187.41	187.18	-0.23	-0.84	175.10	176.25	1.15	-0.84	166.12	168.59	2.47	-0.52
14	60-0-40	183.36	182.36	-0.99	-3.40	167.45	171.96	4.50	-1.90	157.70	165.72	8.02	-0.59
15	60-20-20	186.50	185.86	-0.64	-2.22	173.17	176.48	3.31	-1.29	163.94	170.19	6.26	-0.24
16	60-40-0	189.58	189.36	-0.22	-0.83	179.62	180.90	1.28	-0.81	171.60	174.48	2.88	-0.40
17	80-0-20	188.68	188.04	-0.64	-2.20	177.48	181.05	3.58	-1.18	168.91	175.93	7.02	0.10
18	80-20-0	191.68	191.53	-0.15	-0.54	184.50	185.45	0.95	-0.53	177.89	180.17	2.29	-0.17

[a] λ is the value for the coefficient of absolute risk aversion.

[b] Percentages of the field in Collins, Memphis, and Loring soils, respectively.

acre by profit-maximizing farmers who adopt VRT. The greatest benefit from reduced nitrogen loss is on fields that have the largest spatial variability, i.e., fields that have both the marginal Loring soil and the productive Collins soil (fields 5, 6, 7, 8, 10, 11, 12, 14, 15, and 17). The relative proportions of marginal and productive soils in the field also influence nitrogen losses.

Third, in contrast to the profit-maximizing nitrogen rate results, the field average nitrogen rates for VRT are higher than the nitrogen rates for URT under the assumption of risk aversion. Even though nitrogen is a risk-increasing input for each of the soil management zones, nitrogen usage with VRT does not decline as much as under the URT decision criterion. Reallocation of nitrogen among the management zones with the VRT decision criterion may allow a favorable tradeoff in increased yield variance with more nitrogen use on the Collins soil, and decreased yield variance with reduced nitrogen use on the Loring soil. Results suggest that when nitrogen is risk-increasing, fields with large spatial and temporal yield variability among soil management zones may present risk-averse farmers who adopt VRT with an opportunity to exploit tradeoffs in temporal yield variability among soil types. This finding is consistent with Isik, Khanna, and Winter-Nelson (1999), who hypothesize that the adoption of VRT under uncertainty will be greater for farmers with fields that have large spatial variation in soil quality.

Finally, while nitrogen lost to the environment for both VRT and URT declines with increasing risk-aversion behavior because of reduced nitrogen usage, the NLD advantage of VRT over URT diminishes as the level of risk aversion increases. Under extreme risk aversion ($\lambda = 0.014$), increased nitrogen usage with VRT caused larger nitrogen losses to the environment relative to the URT decision criterion for 5 of the 18 field scenarios (fields 1, 2, 3, 4, and 17). However, the difference in nitrogen losses to the environment was small (1 lb/acre or less) for all 18 field scenarios. This finding suggests that the potential environmental benefits of VRT may be diminished or not realized if the adopters of the technology are risk-averse.

Estimated differences in the mean and standard deviation of net revenues for VRT and URT are presented in Table 4. Results indicate that the VRT decision criterion produced an across-the-board reduction in net revenue variability for both risk-neutral and risk-averse behavior. For profit maximization, the standard deviation of net revenue is reduced between 0.57 percent to 5.5 percent depending on the field scenario. Similar percentage reductions in revenue variability with VRT are found under the five risk-aversion categories. In general, the relative advantage of VRT in lowering variability declines as the level of risk aversion increases. Lower revenue variability with VRT is consistent with the results for VRT application of phosphorus and potassium reported by Lowenberg-DeBoer (1999) but is not consistent with the results for VRT application of nitrogen reported by Braga, Jones, and Basso (1999).

Notwithstanding the positive benefits on net revenue variance, the VRT decision criterion generally has a negative impact on mean net revenues as indicated by the net revenue difference (NRD) values in Table 4. Under profit

Table 4. Net Revenue Difference (NRD) and Percentage Change in the Standard Deviation of Net Revenue (STDD) for Alternative Variable Rate and Uniform Rate Risk-Aversion Decision Criteria and Field Scenarios

		λ=0.000[a]		λ=0.004		λ=0.008		λ=0.010		λ=0.012		λ=0.014	
Field No. and	Soil Mix[b]	NRD ($/acre)	STDD (%)	NRD ($/acre)	STDD (%)	NRD ($/acre)	STDD (%)	NRD ($/acre)	STDD (%)	NRD ($/acre)	STDD (%)	NRD ($/acre)	STDD (%)
1	0-20-80	-2.49	-1.28	-2.38	-0.91	-2.19	-1.29	-2.09	-1.20	-1.98	-1.13	-1.88	-1.06
2	0-40-60	-2.24	-1.90	-2.09	-1.36	-1.78	-1.23	-1.60	-1.04	-1.43	-0.87	-1.25	-0.72
3	0-60-40	-2.25	-1.88	-2.10	-1.36	-1.77	-0.36	-1.58	-0.14	-1.38	0.06	-1.18	0.24
4	0-80-20	-2.51	-1.23	-2.41	-0.91	-2.18	0.29	-2.04	0.44	-1.89	0.57	-1.73	0.69
5	20-0-80	-1.56	-3.73	-1.36	-3.25	-0.94	-3.34	-0.68	-3.17	-0.41	-3.00	-0.13	-2.84
6	20-20-60	-1.51	-3.89	-1.29	-3.33	-0.79	-3.54	-0.48	-3.28	-0.15	-3.02	0.20	-2.78
7	20-40-40	-1.70	-3.39	-1.52	-2.94	-1.05	-2.48	-0.75	-2.20	-0.41	-1.93	-0.05	-1.66
8	20-60-20	-2.13	-2.29	-2.03	-2.10	-1.72	-1.26	-1.50	-1.08	-1.25	-0.90	-0.97	-0.73
9	20-80-0	-2.79	-0.63	-2.80	-0.78	-2.78	-1.01	-2.75	-1.03	-2.69	-1.03	-2.62	-1.03
10	40-0-60	-0.91	-5.46	-0.65	-4.87	0.04	-4.31	0.48	-3.88	0.99	-3.45	1.54	-3.03
11	40-20-40	-1.27	-4.51	-1.08	-4.11	-0.49	-3.41	-0.08	-2.99	0.40	-2.57	0.93	-2.14
12	40-40-20	-1.87	-2.97	-1.79	-2.89	-1.42	-1.98	-1.13	-1.71	-0.78	-1.42	-0.37	-1.13
13	40-60-0	-2.68	-0.93	-2.72	-1.18	-2.70	-1.33	-2.65	-1.31	-2.55	-1.26	-2.41	-1.20
14	60-0-40	-0.97	-5.17	-0.78	-4.83	-0.11	-3.29	0.41	-2.77	1.02	-2.24	1.73	-1.70
15	60-20-20	-1.72	-3.26	-1.66	-3.25	-1.28	-1.91	-0.94	-1.58	-0.51	-1.21	0.02	-0.82
16	60-40-0	-2.69	-0.90	-2.74	-1.17	-2.74	-1.01	-2.69	-0.95	-2.59	-0.86	-2.43	-0.74
17	80-0-20	-1.68	-3.16	-1.64	-3.20	-1.28	-1.55	-0.93	-1.23	-0.46	-0.87	0.14	-0.48
18	80-20-0	-2.80	-0.57	-2.84	-0.77	-2.87	-0.43	-2.84	-0.38	-2.78	-0.30	-2.67	-0.21

[a] λ is the value for the coefficient of absolute risk aversion.
[b] Percentages of the field in Collins, Memphis, and Loring soils, respectively.

maximization, VRT did not produce a positive return for any of the 18 field scenarios. Results indicate that risk-neutral farmers would not adopt VRT. However, farmers might adopt VRT if the cost of VRT application were reduced enough through a subsidy. The minimum amount of the subsidy required to induce a farmer to adopt VRT is indicated by the NRD. The level of the NRD depends on spatial variability, differences in the yield response functions among soil types, and input and output prices. The amount of the subsidy varies in this study from field to field because of the difference in spatial variability across fields. Some fields produce a positive tradeoff between the mean and standard deviation of net revenues with VRT when risk-averse preferences are assumed. When $\lambda = 0.014$, five of the eighteen fields generate a positive increase in mean return along with a reduction in net revenue variance (fields 6, 10, 11, 14, and 17).

Differences in certainty equivalent (CED) for VRT and URT are presented in Table 5. The CED provides an indication of whether VRT is a risk-efficient strategy by providing a higher expected utility. Results indicate that relatively large differences in spatial and temporal yield variability are required for VRT to be utility-maximizing for risk-averse decision makers. At the slight risk-aversion level ($\lambda = 0.004$), the VRT decision criterion produces a larger expected utility on four of the eighteen fields (fields 6, 10, 11, and 14). These four fields tend to have the largest spatial and temporal variability. The number of field scenarios on which VRT is risk efficient rises to 6 when $\lambda = 0.008$ (fields 5, 6, 7, 10, 11, and 14). VRT is risk inefficient on fields that do not combine the marginal Loring soil with the high quality Collins soil (fields 1, 2, 3, 4, 9, 13, 17, and 18). The number of fields on which VRT provides higher expected utility rises to 9 in the $\lambda = 0.01$, $\lambda = 0.012$, and $\lambda = 0.014$ risk-aversion categories. At these three levels of risk aversion, VRT is risk efficient on fields with smaller spatial and temporal variability than under lower absolute risk aversion.

The CEDs in Table 5 are denominated in monetary units and provide an indication of the amount of money farmers can spend to obtain the information required to make the VRT decision. The largest CEDs tend to come from fields exhibiting the largest spatial and temporal variability. In the slight risk aversion category ($\lambda = 0.004$), farmers would be willing to spend from $0.02/acre (field 6) to $1.31/acre (field 4) to obtain the information required to make the VRT decision. The monetary value that an extremely risk averse decision maker ($\lambda = 0.014$) could spend for VRT information to make the VRT decision ranges from $0.72/acre to $4.31/acre.

Because the VRT profit-maximizing nitrogen rates in Table 3 reduce temporal yield variability, the risk-neutral VRT nitrogen application rates may also be of benefit for risk-averse decision makers. CEDs for the risk-neutral nitrogen rates for different levels of risk aversion are presented in Table 6. CEDs are the largest in fields that have the greatest spatial and temporal variability. For example, field 10, which has large spatial and temporal variability, has CED values that range from $1.55/acre for $\lambda = 0.004$ to

Table 5. Differences in Certainty Equivalent Income for Variable Rate and Uniform Rate Application of Nitrogen Fertilizer on Corn for Risk-Averse Decision Criteria

Field No. and Soil Mix[a]		λ=0.004[b]	λ=0.008	λ=0.010	λ=0.012	λ=0.014
		$/acre				
1	0-20-80	-2.04	-1.46	-1.28	-1.11	-0.96
2	0-40-60	-1.55	-1.10	-0.92	-0.77	-0.64
3	0-60-40	-1.53	-1.56	-1.48	-1.43	-1.40
4	0-80-20	-2.01	-2.38	-2.40	-2.44	-2.50
5	20-0-80	-0.15	0.99	1.49	1.96	2.39
6	20-20-60	0.02	1.07	1.58	2.04	2.47
7	20-40-40	-0.30	0.27	0.66	1.02	1.35
8	20-60-20	-1.11	-0.96	-0.71	-0.48	-0.26
9	20-80-0	-2.44	-2.03	-1.81	-1.59	-1.37
10	40-0-60	1.29	2.53	3.19	3.79	4.31
11	40-20-40	0.65	1.39	1.93	2.41	2.84
12	40-40-20	-0.50	-0.25	0.10	0.43	0.72
13	40-60-0	-2.17	-1.77	-1.52	-1.27	-1.03
14	60-0-40	1.31	2.06	2.63	3.13	3.55
15	60-20-20	-0.17	-0.01	0.34	0.65	0.92
16	60-40-0	-2.17	-2.01	-1.84	-1.67	-1.52
17	80-0-20	-0.11	-0.02	0.31	0.58	0.80
18	80-20-0	-2.45	-2.50	-2.44	-2.40	-2.37

[a] Percentages of the field in Collins, Memphis, and Loring soils, respectively.
[b] λ is the value for the coefficient of absolute risk aversion.

$6.45/acre for $\lambda = 0.014$. Results indicate that risk-averse farmers for fields with sufficiently large spatial and temporal variability would be willing to pay for information required to make the VRT nitrogen application decision using the profit-maximizing nitrogen application rates.

Empirical Analysis Summary and Limitations

Results of the empirical analysis indicate that VRT may be risk efficient for risk-increasing inputs in fields that have large differences in spatial and temporal variability among soil management zones. Reallocation of nitrogen among the management zones with the VRT decision criterion may present risk-averse farmers with an opportunity to exploit favorable risk-return tradeoffs in fields with large spatial and temporal yield variability among soil management zones. Even though nitrogen is a risk-increasing input for each of the soil management zones, nitrogen usage with VRT does not decline as much as under the URT decision criterion because of the favorable mean-

Table 6. Differences in Certainty Equivalent Income for Variable Rate and Uniform Rate Application of Nitrogen Fertilizer on Corn for Risk-Neutral Decision Criteria

Field No. and Soil Mix[a]		λ=0.004 [b]	λ=0.008	λ=0.010	λ=0.012	λ=0.014
		$/acre				
1	0-20-80	-2.49	-1.91	-1.33	-1.04	-0.75
2	0-40-60	-2.24	-1.36	-0.49	-0.05	0.39
3	0-60-40	-2.25	-1.36	-0.48	-0.03	0.41
4	0-80-20	-2.51	-1.91	-1.32	-1.02	-0.72
5	20-0-80	-1.56	0.08	1.73	2.55	3.37
6	20-20-60	-1.51	0.25	2.01	2.88	3.76
7	20-40-40	-1.70	-0.13	1.44	2.23	3.02
8	20-60-20	-2.13	-1.04	0.05	0.59	1.13
9	20-80-0	-2.79	-2.48	-2.17	-2.01	-1.86
10	40-0-60	-0.91	1.55	4.00	5.22	6.45
11	40-20-40	-1.27	0.81	2.88	3.92	4.96
12	40-40-20	-1.87	-0.46	0.95	1.65	2.36
13	40-60-0	-2.68	-2.23	-1.77	-1.55	-1.32
14	60-0-40	-0.97	1.44	3.85	5.06	6.26
15	60-20-20	-1.72	-0.16	1.40	2.18	2.96
16	60-40-0	-2.69	-2.24	-1.80	-1.57	-1.35
17	80-0-20	-1.68	-0.13	1.43	2.21	2.99
18	80-20-0	-2.80	-2.51	-2.21	-2.07	-1.92

[a] Percentages of the field in Collins, Memphis, and Loring soils, respectively.
[b] λ is the value for the coefficient of absolute risk aversion.

variance tradeoffs. Consequently, nitrogen losses to the environment for some field spatial variability situations may be greater with VRT relative to URT under the assumption of risk aversion.

There are several important limitations to this analysis. First, the empirical VRT risk analysis accounted for spatial variation and correlation in yields among soil types but does not account for some other potentially important correlations among soil types. For example, subsurface soil water flows between different soil management zones may influence temporal yield variability. Most crop simulation models are designed to simulate yields for a single soil type and do not have algorithms to account for interactions among soil types in a field. Second, the mean-variance approach used in this analysis may place an overly simplistic representation on risk. While mean-variance may provide a useful framework for deriving VRT and URT decision rules, it may not provide a clear picture on downside risk that farmers are most often worried about, i.e., the impact of VRT on the lower tail of the distribution of outcomes. This may be especially important for risk-reducing inputs such as pesticides that truncate bad outcomes.

Modeling the VRT decision in a mean-semivariance framework is one possible extension of the model presented here.

FUTURE RESEARCH NEEDS

Studies have indicated that various aspects of precision farming are being adopted by farmers. Other research has demonstrated that these information technologies have the potential to improve net revenues, reduce input usage, and mitigate the environmental consequences of crop production. However, economic research on the potential for information technology in managing risk in crop production in agriculture is limited. Better information about spatial and temporal variability in crop yields may help farmers apply inputs in a more risk-efficient manner.

An important impediment to the analysis of the risk management benefits of precision farming is a lack of long-term experimental plot and farm-level data for developing empirical risk models. A concerted effort to document yield and environmental response for a variety of crops, management choices, soil series, and weather conditions would be beneficial to farmers, agribusiness firms, and policymakers who are evaluating the economic and environmental impacts of precision farming. As an alternative, crop simulation models are well suited to evaluate the consequences of site-specific management because they can account for the daily spatial and temporal interactions affecting plant growth and yield. An appropriately adapted and validated crop model can provide useful "what if" information that would be difficult and costly to derive from long-term large plot field experiments. However, additional work is needed to develop crop models that account for certain types of interactions among soil types in a field, e.g., subsurface soil water flows between different soil management zones that may influence temporal yield variability.

Besides additional data requirements, research is needed to develop a more general risk framework linking spatial variability with temporal variability in farm fields. In addition, the question of whether risk-averse farmers who adopt precision technology would behave in the manner predicted by expected utility models should be investigated further. For example, are farmer attitudes towards risk identical for different field spatial and temporal variability scenarios or different crop situations? Direct elicitation of farmer risk preferences, subjective probabilities, or other risk measures in the context of precision farming would provide valuable insight into the risk implications of site-specific management.

Finally, precision farming technology may have other risk management benefits for farmers and agribusinesses. Electronic monitoring and control of agricultural production may provide an effective way to reduce environmental and food contamination risk. Better information about crop progress through sensing and remote sensing data may help farmers reduce labor costs, assist in forward contracting and hedging of crop production decisions, and facilitate contract production relationships between farmers and processors. The risk implications of these potential uses of precision farming technology need to be explored.

REFERENCES

Akridge, J.T., and L.D. Whipker. 1996. "Precision Ag Services Dealership Survey Results." Center for Agricultural Business Staff Paper No. 96-11. West Lafayette, IN: Purdue University.

___. 2000. "2000 Precision Agriculture and Enhanced Seed Dealership Survey Results." Center for Agricultural Business Staff Paper No. 00-4. West Lafayette, IN: Purdue University, 2000. Available online at http://mollisol.agry.purdue.edu/SSMU/.

Antle, J.M., and C.C. Crissman. 1990. "Risk, Efficiency, and the Adoption of Modern Crop Varieties: Evidence from the Philippines." *Economic Development and Cultural Change* 38: 517-530.

Babcock, B.A., and G.R. Pautsch. 1998. "Moving from Uniform to Variable Fertilizer Rates on Iowa Corn: Effects on Rates and Returns." *Journal of Agricultural and Resource Economics* 23: 385-400.

Batte, M.T., E. Jones, and G.D. Schnitkey. 1990. "Computer Use by Ohio Commercial Farmers." *American Journal of Agricultural Economics* 72: 935-945.

Bohrnstedt, G.W., and A.S. Goldberger. 1969. "On the Exact Covariance of Products of Random Variables." *Journal of the American Statistical Association* 64: 1439-1442.

Bongiovanni, R, and J. Lowenberg-DeBoer. 1999. "Economics of Variable Rate Lime in Indiana." In P.C. Robert, R.H. Rust, and W.E. Larson, eds., *Proceedings of the Fourth International Conference on Precision Agriculture*. Madison, WI: ASA, CSSA, and SSSA.

Braga, R.P., J.W. Jones, and B. Basso. 1999. "Weather Induced Variability in Site-Specific Management Profitability: A Case Study." In P.C. Robert, R.H. Rust, and W.E. Larson, eds., *Proceedings of the Fourth International Conference on Precision Agriculture*. Madison, WI: ASA, CSSA, and SSSA.

Bullock, D.G., and D.S. Bullock. 1994. "Quadratic and Quadratic-plus-plateau Models for Predicting Optimal Nitrogen Rate of Corn: A Comparison." *Agronomy Journal* 86: 191-195.

Bullock, D.G., D.S. Bullock, E.D. Nafziger, T.A. Doerge, S.R. Paszkiewicz, P.R. Carter, and T.A. Peterson. 1998. "Does Variable Rate Corn Seeding Pay?" *Agronomy Journal* 90: 830-836.

Byerlee, D., and E.H de Polanco. 1986. "Farmers' Stepwise Adoption of Technological Packages: Evidence from the Mexican Altiplano." *American Journal of Agricultural Economics* 68: 519-527.

Cerrato, M.E., and A.M. Blackmer. 1993. "Comparison of Models for Describing Corn Yield Response to Nitrogen Fertilizer." *Agronomy Journal* 85: 138-143.

Cochrane, W.W. 1958. *Farm Prices: Myth and Reality*. Minneapolis, MN: University of Minnesota Press.

Council of Economic Advisers. 2000. *Economic Report of the President*. Washington, D.C.: U.S. Government Printing Office.

Daberkow, S.G., and W.D. McBride. 1998. "Adoption of Precision Agriculture Technologies by U.S. Corn Producers." *Journal of Agribusiness* 16: 151-168.

Decker, A.M., A.J. Clark, J.J. Meisinger, F.R. Mulford, and M.S. McIntosh. 1994. "Legume Cover Crop Contributions to No-Tillage Corn Production." *Agronomy Journal* 86: 126-135.

Dillon, C., T. Mueller, and S. Shearer. 2001. "An Assessment of the Profitability and Risk Management Potential of Variable Rate Seeding and Planting Date." Paper presented at the annual meeting of the Southern Agricultural Economics Association, January 27-31, 2001, Fort Worth, TX.

Eghball, B., and G.E. Varvel. 1997. "Fractal Analysis of Temporal Yield Variability of Crop Sequences: Implication for Site-Specific Management." *Agronomy Journal* 89: 851-855.

English, B.C., S.B. Mahajanashetti, and R.K. Roberts. 1999. "Economic and Environmental Benefits of Variable Rate Application of Nitrogen to Corn Fields: Role of Variability and Weather." Department of Agricultural Economics and Rural Sociology, Staff Paper 99-09. The University of Tennessee, Knoxville.

English, B.C., R.K. Roberts, and S.B. Mahajanashetti. 1999. "Spatial Break-Even Variability for Variable Rate Technology Adoption." In P.C. Robert, R.H. Rust, and W.E. Larson, eds., *Proceedings of the Fourth International Conference on Precision Agriculture*. Madison, WI: ASA, CSSA, and SSSA.

Feder, G. 1982. "Adoption of Interrelated Agricultural Innovations: Complementarity and the Impacts of Risk, Scale, and Credit." *American Journal of Agricultural Economics* 64: 95-101.

Forcella, F. 1993. "Value of Managing Within-Field Variability." In F.J. Pierce and E.J. Sadler, eds., *Proceedings of the First Workshop Soil Specific Crop Management: A Workshop on Research and Development Issues*. Madison, WI: ASA, CSSA, and SSSA.

Franzen, D. 2000. "North Dakota Report." North Central Regional Project 180 – Site Specific Management. 2000 Annual Meeting, January 2000, Bozeman, MT.

Freund, R.J. 1956. "Introduction of Risk into a Programming Model." *Econometrica* 24: 257-263.

Gerloff, D. 2000. "Corn Budgets for 2000." AE&RD No 42. The University of Tennessee Agricultural Extension Service, Knoxville.

Hennessy, D., and B. Babcock. 1998. "Information, Flexibility, and Value Added." *Information Economics and Policy* 10: 431-449.

Huggins, D.R., and R.D. Alderfer. 1997. "Yield Variability within a Long-Term Corn Management Study: Implications for Precision Farming." In F.J. Pierce and E.J. Sadler, eds., *The State of Site Specific Management for Agriculture*. Madison, WI: ASA, CSSA, and SSSA.

Isik, M., M. Khanna, and A. Winter-Nelson. 1999. "Investment in Site-Specific Crop Management Under Uncertainty." Paper presented at the Annual Meeting of the American Agricultural Economics Association, August 8-11, 1999, Nashville, TN.

Jaenicke, E.C., and D.R. Cohen-Vogel. 2000. "Sequential Adoption of Precision Farming Technology Under Uncertainty." Paper presented at the Annual Meeting of the Southern Agricultural Economics Association, January 29 – February 2, 2000, Lexington, KY.

Just, R., and R.D. Pope. 1978. "Stochastic Specification of Production Functions and Econometric Implications." *Journal of Econometrics* 7: 67-86.

Khanna, M. 2001. "Sequential Adoption of Site-Specific Technologies and Its Implications for Nitrogen Productivity: A Double Selectivity Model." *American Journal of Agricultural Economics* 83: 35-51.

Khanna, M., O.F. Epouhe, and R. Hornbaker. 1999. "Site-Specific Crop Management: Adoption Patterns and Incentives." *Review of Agricultural Economics* 21: 455-472.

Kitchen, N.R., K.A. Sudduth, S.J. Birrell, and S.C. Borgelt. 1996. "Missouri Precision Agriculture Research and Education." In P.C. Robert, R.H. Rust, and W.E. Larson, eds., *Proceedings of the Third International Conference on Precision Agriculture*. Madison, WI: ASA, CSSA, and SSSA.

Koo, S., and J.R. Williams. 1996. "Soil-Specific Production Strategies and Agricultural Contamination Levels in Northeast Kansas." In P.C. Robert, R.H. Rust, and W.E. Larson, eds., *Proceedings of the Fourth International Conference on Precision Agriculture*. Madison, WI: ASA, CSSA, and SSSA.

Lamb, J.A., R.H. Dowdy, J.L. Anderson, and G.W. Rehm. 1997. "Spatial Stability of Corn Grain Yields." *Journal of Production Agriculture* 10: 410-414.

Lambert, D. 1990. "Risk Considerations in the Reduction of Nitrogen Fertilizer Use in Agricultural Production." *Western Journal of Agricultural Economics* 15: 234-244.

Lambert, D., and J. Lowenberg-DeBoer. 2001. "Precision Farming Profitability Review." Site-Specific Management Center. West Lafayette, IN: Purdue University. Available online at http://mollisol.agry.purdue.edu/SSMU/ (accessed February 2, 2001).

Larson, J.A., R.K. Roberts, D.D. Tyler, B.N. Duck, and S.P. Slinsky. 1998. "Nitrogen Fixing Winter Cover Crops and Production Risk: A Case Study for No Tillage Corn." *Journal of Agricultural and Applied Economics* 30: 163-174.

Leathers, H.D., and M. Smale. 1991. "Bayesian Approach to Explaining Sequential Adoption of Components of a Technological Package." *American Journal of Agricultural Economics* 73: 734-742.

Lowenberg-DeBoer, J. 1996. "Precision Farming and the New Information Technology: Implications for Farm Management, Policy, and Research: Discussion." *American Journal of Agricultural Economics* 78: 1281-1284.

___. 1997. "Taking a Broader View of Precision Farming Benefits." *Modern Agriculture* 1: 32-33.

___. 1999. "Risk Management Potential of Precision Farming Technologies." *Journal of Agricultural and Applied Economics* 31: 275-285.

Lowenberg-DeBoer, J., and A. Aghib. 1999. "Average Returns and Risk Characteristics of Site Specific P and K Management: Eastern Corn Belt On-Farm Trial Results." *Journal of Production Agriculture* 12: 276-282.

Lowenberg-DeBoer, J., and M. Boehlje. 1996. "Revolution, Evolution or Dead-End: Economic Perspectives on Precision Agriculture." In P.C. Robert, R.H. Rust, and W.E. Larson, eds., *Proceedings of the Fourth International Conference on Precision Agriculture*. Madison, WI: ASA, CSSA, and SSSA.

Lowenberg-DeBoer, J., and S.M. Swinton. 1997. "Economics of Site-Specific Management in Agronomic Crops." In F.J. Pierce and E.J. Sadler, eds., *The State of Site Specific Management for Agriculture*. Madison, WI: ASA, CSSA, and SSSA.

National Research Council. 1997. *Precision Agriculture in the 21st Century: Geospatial and Information Technologies in Crop Production*. Washington, D.C.: National Academy Press.

Nielsen, R.L. 2001. "Site-Specific Management of Corn: Perennial Versus Sporadic Yield Limiting Factors." Site-Specific Management Center. West Lafayette, IN: Purdue University. Available online at http://mollisol.agry.purdue.edu/SSMU/ (accessed February 2, 2001).

Novartis Seeds. 1999. "On the Farm." *@gInnovator* (online newsletter), February 1999, p. 2. Available online at http://www.agriculture.com/technology/aginnovator.

Pierce, F.J., and P. Nowak. 1999. "Aspects of Precision Agriculture." In D.L. Sparks, ed., *Advances in Agronomy*, D.L Sparks ed., 67: 1-85. New York: Academic Press.

Roberts, R.K., and B.C. English. 1999. "Use of Precision Farming Technologies by Crop Producers in Tennessee: Results from Surveys of Agricultural Extension Agents, Off-Farm Service Providers, and Producers." Contract report to Cotton Incorporated. Department of Agricultural Economics and Rural Sociology, The University of Tennessee, Knoxville.

Roberts, R.K., B.C. English, and S.B. Mahajanashetti. 2000. "Evaluating the Returns to Variable Rate Nitrogen Application." *Journal of Agricultural and Applied Economics* 32: 133-143.

Roberts, R.K., B.C. English, and D.E. Sleigh. 2000. "Precision Farming Services in Tennessee: Results of a 1999 Survey of Precision Farming Service Providers." Tennessee Agricultural Experiment Station, Research Report 00-06. The University of Tennessee, Knoxville.

Robison, L.J., and P.J. Barry. 1987. "The Competitive Firm's Response to Risk." New York: Macmillan.

Roumasset, J.A., M.W. Rosengrant, U.N. Chakravorty, and J.R. Anderson. 1989. "Fertilizer and Crop Yield Variability: A Review." In J.R. Anderson and P.B.R. Hazell, eds., *Variability in Grain Yields*. Baltimore, MD: John Hopkins University Press.

Sawyer, J.E. 1994. "Concepts of Variable Rate Technology with Considerations for Fertilizer Application." *Journal of Production Agriculture* 7: 195-201.

Sharpley, A.N., and J.R. Williams. 1990. "EPIC – Erosion Productivity Impact Calculator (Vol. I): Model Documentation." USDA Technical Bulletin No. 1768.

Swinton, S.M., S.B. Harsh, M. Ahmad. 1996. "Whether and How to Invest in Site-Specific Crop Management: Results of Focus Group Interviews in Michigan, 1996." Department of Agricultural Economics Staff Paper 96-37, Michigan State University, East Lansing.

Swinton, S.M., and J. Lowenberg-DeBoer. 1998. "Evaluating the Profitability of Site-Specific Farming." *Journal of Production Agriculture* 11: 439-446.

Taylor, O. 1998. Yield Monitors Are Marketing Tools." *Soybean Digest*, October, p. 6.

Tennessee Department of Agriculture. *Tennessee Agriculture.* Tennessee Agricultural Statistics Service, Nashville, TN. (Various annual issues, 1985 through 2000.)

Thrikawala, S., A. Weersink, G. Kachanoski, and G. Fox. 1999. "Economic Feasibility of Variable-Rate Technology for Nitrogen on Corn." *American Journal of Agricultural Economics* 81: 914-927.

Tversky, A., and D. Kahneman. 1974. "Judgement Under Uncertainty: Heuristic and Biases." *Science* 185: 1124-1131.

Uri, N.D. 1999. *Conservation Tillage in U.S. Agriculture: Environmental, Economic, and Policy Issues.* New York: Food Products Press.

U.S. Department of Commerce. *Climatological Data, Tennessee.* National Oceanic and Atmospheric Administration (NOAA), National Climatic Data Center, Nashville, TN. (Various issues.)

Watkins, K.B., Y.C. Lu, and W.Y. Huang. 1999. "Economic Returns and Environmental Impacts of Variable Nitrogen Fertilizer and Water Applications." In P.C. Robert, R.H. Rust, and W.E. Larson, eds., *Proceedings of the Fourth International Conference on Precision Agriculture.* Madison, WI: ASA-CSSA-SSSA.

Williams, J.R., P.T. Dyke, W.W. Fuchs, V.W. Benson, O.W. Rice, and E.D. Taylor. 1990. In A.N. Sharpley and J.R. Williams, eds., "EPIC – Erosion Productivity Impact Calculator (Vol. II): User Manual." USDA Technical Bulletin No. 1768.

Part 5

POLICY ISSUES RELATING TO RISK IN AGRICULTURE

Chapter 20

CROP INSURANCE AS A TOOL FOR PRICE AND YIELD RISK MANAGEMENT

Keith H. Coble and Thomas O. Knight
Mississippi State University and Texas A&M University

INTRODUCTION

Agricultural production uncertainty due to events such as drought, frost, insect infestation, and disease is widely recognized. Concern about the effects of this uncertainty on the financial stability of farm firms has generated long-standing interest in providing risk management instruments to protect producers from such production risks. In the United States, crop insurance stands as the most prominent risk-sharing mechanism for agricultural yield risk, while price risk protection has traditionally been afforded through forward pricing instruments such as futures contracts and through price-oriented government programs such as marketing loan programs and deficiency payments. Recent introduction of crop revenue insurance has, however, blurred the separation between mechanisms for price and yield risk protection.

As an area of research, the general topic of crop insurance has seen a dramatic increase in activity during the past decade. A search of the *Uncover* database indicates that during 1995-2000 the number of crop insurance related articles published in refereed agricultural economics journals was double that of the previous six-year period. We believe that a number of forces have attracted researchers to this seemingly narrow research topic. Primary among these is the fact that issues that arise in crop insurance are at the interface of a number of subdisciplines within the agricultural economics profession. Not surprisingly, crop insurance has attracted the interest of many who are interested in the modeling of risk and risk behavior. It also fits within the broader framework of farm management decision making. Further, it has appealed to those interested in information economics because the apparent market failure in agricultural insurance has been attributed to adverse selection and moral hazard. Finally, crop insurance, as it exists in the United States and most of the world, is a government-subsidized program. Thus, there is a significant farm policy dimension to research in this area.

When attempting to characterize the current literature on crop insurance we believe that one can predicate much of the existing research upon the following stylized characterization of the crop insurance market:

- Yield and revenue risk is pervasive in agricultural production, yet there appears to be little demand for an actuarially fair crop insurance product.
- There is a perceived private market failure to provide multiple-peril insurance products, which has led to government intervention.

For example, a desire to understand low participation led to numerous studies investigating the elasticity of crop insurance demand. More recently this concern has led to numerous attempts to design alternative insurance plans that are more attractive to producers. The second issue may be characterized as a problem of supply. These issues led to at least two veins of literature. First to emerge was a set of papers investigating the hypothesis that asymmetric information exists in the available programs. Second, a body of literature developed that took asymmetric information as a given and investigated alternative insurance designs that might reduce the problem.

In this chapter, we attempt to provide a synthesis of the fundamental economic context of crop insurance and the existing crop insurance literature. In particular, we lay out a conceptual framework for the existing literature and discuss the major topics in this literature from within this unified framework. This also serves as a springboard for a discussion of future directions that we expect crop insurance research to take.

Historical Overview

A brief summary of the legislative history of the modern U.S. multi-peril crop insurance program is provided in Table 1. The Crop Insurance Improvement Act of 1980 is the foundation of the modern era of U.S. crop insurance for a number of reasons. Most important, it introduced premium subsidies, allowed private sector delivery of federal crop insurance, and set the stage for a shift from a common yield guarantee for all farms in an area to coverage based on an individual farm yield. In the two decades since the 1980 Act, both the role of the private sector in the delivery of crop insurance and premium subsidies have increased enormously. The 1980 Act also started a trend toward program expansion, increasing the number of insurable crops and regions from the fairly limited set that existed prior to 1980. Figure 1 summarizes some major attributes of the U.S. crop insurance program since 1980. The aggregate total liability by year charts the program's sevenfold growth over the past twenty years. Figure 1 also reports the total government subsidy and loss ratio (indemnities/total premiums) over the 1981-2000 period.

By the late 1980s there was significant concern about the actuarial soundness of the expanding program. During the 1980s and up through 1993 the crop insurance program failed to break even in terms of its loss ratio in any year. As

Table 1. Recent Legislative History of U.S. Federal Crop Insurance

Crop Insurance Improvement Act (1980)

- Shifted the policy focus from free disaster assistance to federal crop insurance.
- Introduced a premium subsidy for federal crop insurance.
- Allowed the private sector to deliver federal crop insurance.
- Greatly expanded insurable crops and areas.
- Allowed implementation of insurance guarantees based on individual versus area expected yield.

Federal Crop Insurance Commission Act (1988)

- Mandated "the thorough review of the federal crop insurance program and the development of recommendations...to improve the program."

Food, Agriculture, Conservation, and Trade Act (1990)

- Included a special title for crop insurance and disaster assistance that emphasized correcting the problems with federal crop insurance.
- Federal Crop Insurance Corporation mandated to test-market new products.
- Private insurance companies authorized to develop supplemental products that could be packaged together with the federal crop insurance product.
- Mandated a premium rate increase for federal crop insurance to reduce excess losses.
- Federal Crop Insurance Corporation mandated to take actions to control fraud.

Crop Insurance Reform Act (1994)

- Developed more restrictive procedures for passage of future free disaster assistance.
- Required farmers to sign up for catastrophic federal crop insurance (CAT) in order to be eligible for price and income support programs.
- Increased premium subsidies.

Federal Agricultural Improvement and Reform Act (1996)

- Severed the cross-compliance linkage between CAT and farm program benefits. Farmers could opt out of CAT by waiving rights to future disaster payments.

Agricultural Risk Protection Act (2000)

- Provided for an estimated $8 billion additional crop insurance spending over a five-year period. This additional spending will largely support further increases in the subsidies for crop insurance.
- Equalize the percent subsidy between revenue and yield insurance.
- Mandate efforts to insure livestock, forage, cost of production, and other new insurance designs.
- The act also mandated that USDA become more of a regulator of privately developed products rather than carry out its own development program.
- Required USDA to reimburse development costs of privately developed insurance products approved for federal reinsurance, regardless of whether the development activity is commissioned by USDA or privately initiated.

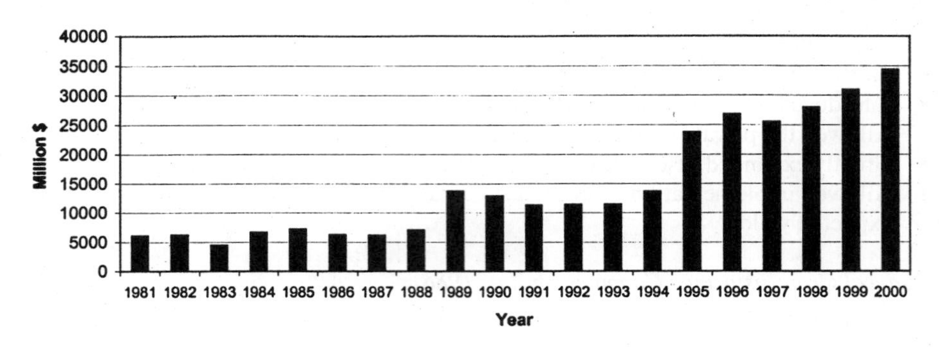

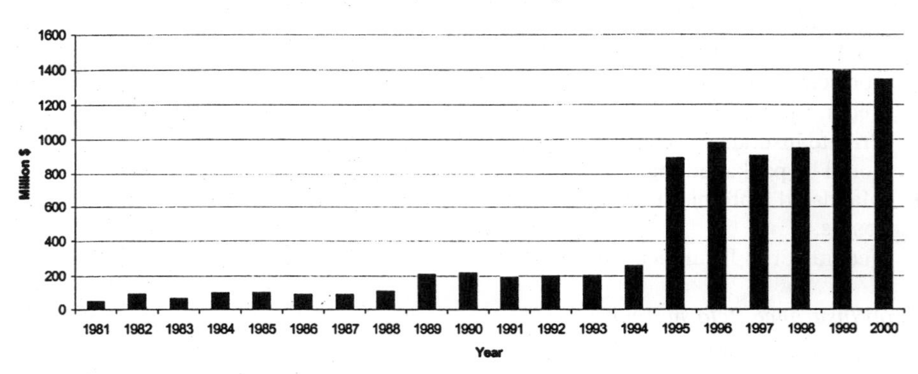

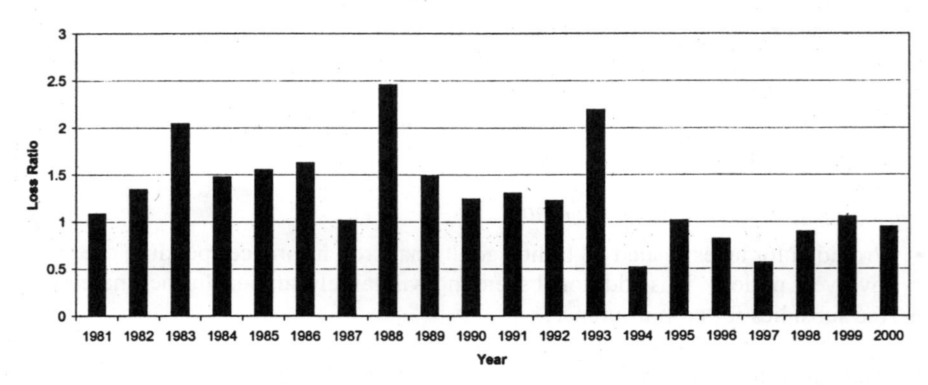

Figure 1. Summary of U.S. Crop Insurance Liability, Subsidy, and Loss Ratio, 1981-2000

Source: USDA Risk Management Agency Actuarial Summary

Figure 1 shows, the loss ratio approached 2.5 in the drought year of 1988. The crop insurance program also failed to preclude ad hoc disaster legislation from being enacted in several years.

The 1990 Farm Bill was another watershed in the evolution of the crop insurance program. At one point in the deliberations the administration recommended elimination of the program and returning to a standing disaster program. Ultimately, the Food, Agriculture, Conservation, and Trade Act of 1990 mandated a premium subsidy increase to attempt to bring more people into the program, included provisions to improve actuarial soundness, and allowed for private insurance companies to develop supplemental products that could be packaged together with the federal crop insurance programs. This was a step in the direction of allowing private companies to develop new products and deliver the federally developed products while being protected by a federal reinsurance agreement.

The next significant policy change was the Crop Insurance Reform Act of 1994. This legislation, enacted during a period of fiscal austerity, was largely precipitated by a desire to eliminate continued ad hoc disaster payments. It instituted catastrophic coverage, a low-coverage, effectively free insurance. Catastrophic coverage was meant to provide government support roughly equivalent to ad hoc disaster programs, and insurance was perceived to be less subject to abuse than disaster programs. Again, premium subsidies were increased, as reflected in Figure 1. Interestingly, catastrophic coverage was tied to other farm program benefits in that producers were required to sign up for catastrophic coverage in order to be eligible for price and income support programs. This linkage did not last long. It was repealed in the 1996 Federal Agricultural Improvement and Reform Act.

Most recently, crop insurance reform was undertaken in the Agricultural Risk Protection Act of 2000 (ARPA), which was projected to add $8 billion of additional spending on crop insurance over a five-year period. This additional spending is largely designated to increase subsidies on crop insurance from the level set in the 1994 Reform Act. It also takes a new step in making USDA's Risk Management Agency (RMA) more of a regulator than a program developer, while giving the private sector an increased role in designing new crop insurance programs. This legislation also mandates attempts to expand crop insurance programs into areas such as livestock, specialty crops, and forage, and the piloting of a cost-of-production policy.

CONCEPTUAL FRAMEWORK

In this section we develop a model of actual production history (APH) yield insurance coverage to provide a framework for much of the discussion that follows.[1] The model we propose extends Just, Calvin, and Quiggin's (JCQ)

---

[1] A model of yield rather than revenue insurance is chosen for several reasons: (a) APH yield coverage was given a dominant role in the U.S. crop insurance program prior to 1996 and, as a result, a preponderance of prior research relates to this program; (b) catastrophic and higher-level APH coverage continues to account for 61 percent of total acres insured in 2000; (c) many elements of the APH structure are preserved in revenue insurance designs such as crop revenue

(1999) model of crop insurance participation incentives by incorporating moral hazard in addition to the risk aversion, subsidy, and adverse selection incentives considered by JCQ. This decomposition of incentives facilitates a unified treatment of issues such as crop insurance demand, adverse selection, and moral hazard, which have dominated the crop insurance policy debate and the prior literature.

Assume a risk-averse crop producer confronting yield risk only. The producer's yield is denoted y, with distribution $F_{ij}(y)$, where the subscripts i and j respectively identify producer risk type and input vector. To incorporate adverse selection we define two risk types, 0 and 1, such that $F_{0j}(y) \neq F_{1j}(y)$. Three input levels are considered. Input vector x_0 is that chosen by an uninsured producer. Input vector x_1 is the input vector of an insured producer, given that actions are fully observable (i.e., that there is no moral hazard). Moral hazard is introduced into the model by allowing the producer to shift to an input vector x_2, without government observation of the change. In this notation, we represent the government's perception of an insured farmer's yield distribution as $F_{01}(y)$. That is, the government assumes the producer is of type 0 and applies inputs x_1 when insured.[2]

A farmer participating in APH insurance selects a coverage level C, between 50 percent and 85 percent of the APH yield for the insured unit, and a price guarantee p_g, between 50 percent and 100 percent of the RMA's estimated harvest period price. Defining the producer's APH or insurable yield as μ, then a producer whose yield falls below $C\mu$ receives an indemnity of p_g $(C\mu - y)$ per acre. Premiums, γ, are a function of C, μ, and p_g, as well as the assumed producer type and input level implicit in the distribution F_{01}. Given the government subsidy, producer premiums are derived by multiplying the premium by $(1-S)$, where $S(0<S<1)$ is the proportionate subsidy offered by the government. The farm's per acre revenue with insurance may be written as

---

coverage (CRC) and revenue assurance (RA); and (d) much of the current crop insurance debate focuses on perceived shortcomings of the APH program and related revenue insurance designs.

[2] Clarification is in order regarding the input vectors x_0, x_1, and x_2. We allow the optimal input vector with insurance and full observability of inputs, vector $\boldsymbol{x}_1$, to deviate from the uninsured vector x_0 in order to recognize that, in general, risk sharing is expected to result in an input response to insurance coverage even if the response is appropriately incorporated into the premium. Introduction of moral hazard may appear inconsistent with this observability assumption. A clear alternative would be to assume $x_1 = x_0$ when moral hazard is introduced, such that any deviation from uninsured behavior is defined as moral hazard. This assumption could be made without changing the nature of the model implications discussed later. However, as will be discussed later, APH rate calculation procedures, which are based on historical loss experience, presumably capture the effect of an average historical input response. Thus, we have chosen to simultaneously represent the three input vectors, x_1, x_2, and x_3, reflecting insurance rates that are not based on uninsured behavior even when inputs are not observable and moral hazard behavior occurs. It is recognized that models of signaling could more formally represent this phenomenon, but such models are not utilized here because they would add complexity while offering little additional insight.

$$\Phi(y) = \begin{Bmatrix} py - (1-S)\gamma(C, p_g, \mu, x_1, F_{01}), \textbf{if } y \geq C\mu \\ py + p_g(C\mu - y) - (1-S)\gamma(C, p_g, \mu, x_1, F_{01}), \textbf{if } y < C\mu \end{Bmatrix}, \quad (1)$$

where p is the price received for output produced.

Assuming a choice criterion of expected utility maximization, the incentive to participate in or purchase insurance is given by

$$CE[\Phi(y) \mid F_{ij} - wx_j] - CE[\Phi(y) \mid F_{i0} - wx_0] = \Delta_1 + \Delta_2 + \Delta_3 + \Delta_4, \quad (2)$$

where CE represents the certainty equivalent operator, which for notational convenience is assumed to take into account initial wealth and acreage; w is a nonstochastic input price vector; and Δ_k, $k = 1\text{-}4$, represent partitioned effects of risk aversion (Δ_1), and expectation incentives (i.e., incentives deriving from differences in expected returns with and without insurance) of the government subsidy (Δ_2), adverse selection (Δ_3), and moral hazard (Δ_4). Following JCQ, the first three of these incentives are defined as

$$\Delta_1 = R[\Phi(y) \mid F_{ij} - wx_j] - R[py \mid F_{i0} - wx_0], \quad (3)$$

$$\Delta_2 = E[\Phi(y) \mid F_{01}] - E[py \mid F_{01}], \text{ and} \quad (4)$$

$$\Delta_3 = \{E[\Phi(y) \mid F_{11}] - E[py \mid F_{11}]\} - \{E[\Phi(y) \mid F_{01}] - E[py \mid F_{01}]\}, \quad (5)$$

while the moral hazard incentive is

$$\Delta_4 = E[\Phi(y) \mid F_{i2} - wx_2] - E[\Phi(y) \mid F_{i1} - wx_1], \quad (6)$$

where R and E are, respectively, the risk premium and expectation operators; like the CE, the arguments of these operators implicitly include initial wealth and acreage. It is clear that with no subsidy, moral hazard, or adverse selection the risk-aversion effect is given by

$$\Delta_1 = R[\Phi(y) \mid F_{01}(y)dy - wx_1] - R[py \mid F_{00}(y)dy - wx_0]. \quad (3a)$$

The subsidy effect on expected return is given by

$$\Delta_2 = S\gamma(C, p_g, \mu, x_1, F_{01})A, \quad (4a)$$

where A represents acres. Thus, in isolation, the insurance participation incentive provided through government subsidization of a risk-neutral decision maker is the dollar amount of the subsidy. The adverse selection incentive is

$$\Delta_3 = A\left[\int_0^{C\mu} p_g(C\mu - y)F_{11}(y)dy - \int_0^{C\mu} p_g(C\mu - y)F_{01}(y)dy\right],$$

or

$$\Delta_3 = A\left[\int_0^{C\mu} p_g(C\mu - y)F_{11}(y)dy - \gamma(C, p_g, \mu, x_1, F_{01})\right]. \tag{5a}$$

This is the difference in expected indemnity for farm type 1 versus type 0, or equivalently the difference in expected indemnity for farm type 1 and the actuarially fair premium for farm type 0. In isolation the moral hazard effect on expected return is given by

$$\begin{aligned}\Delta_4 = A\Big\{&\left[\int_0^{\infty} py \mid F_{02}(y)dy - \int_0^{\infty} py \mid F_{01}(y)dy\right] \\ &+\left[\int_0^{C\mu} p_g(C\mu - y)F_{02}(y)dy - \int_0^{C\mu} p_g(C\mu - y)F_{01}(y)dy\right] \\ &+ w(x_1 - x_2)\Big\},\end{aligned} \tag{6a}$$

which is the sum of moral hazard effects on expected market revenue, the expected indemnity, and input cost. The combined effect of risk aversion, subsidy, adverse selection, and moral hazard is given by

$$\begin{aligned}&R[\Phi(y) \mid (F_{12} - wx_2] - R[py \mid F_{10} - wx_0] \\ &+ A\Big\{\left[\int_0^{C\mu} p_g(C\mu - y)F_{12}(y)dy - (1-S)\gamma(C, p_g, \mu, x_1, F_{01})\right] \\ &+\left[\int_0^{\infty} py \mid F_{12}(y)dy - \int_0^{\infty} py \mid F_{11}(y)dy\right] \\ &+ \ w(x_1 - w_2)\Big\}.\end{aligned} \tag{7}$$

The first term in (7) represents the difference in risk premium for a farm of type 1 which, when insured, engages in morally hazardous behavior through use of inputs x_2, and the risk premium for the same farm when uninsured. It should be emphasized that this difference is, in general, affected by the subsidy, adverse selection, and moral hazard. Remaining terms in (7) reflect combined effects of subsidy, adverse selection, and moral hazard on expected return. The second term is the difference in expected indemnity for a farm of type 1, applying inputs x_2, and the producer premium based on assumptions of farm type 0 and input use x_1. The third and fourth terms, respectively, are differences in expected market revenue and input cost resulting from morally hazardous behavior by a farm of type 1.

The four components of the incentive to purchase insurance are illustrated graphically in Figure 2. This graph is a modification of one used by Hirshleifer

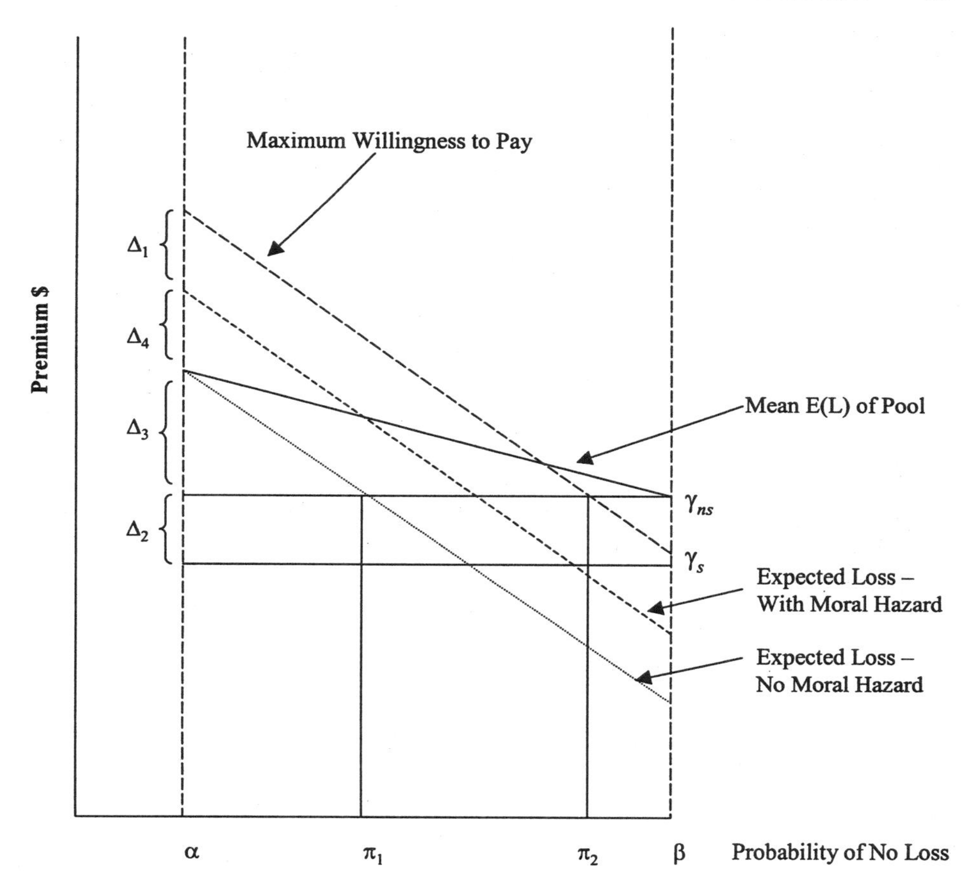

Figure 2. Illustration of the Incentives to Purchase Crop Insurance

and Riley (1992) in examining adverse selection processes, with generalization to facilitate inclusion of moral hazard and subsidy incentives. The decision setting illustrated is that of a group of potential insureds, each facing a Bernoulli risk, with differing probabilities of not incurring a loss π ranging on a continuum $\alpha \leq \pi \leq \beta$, where $\alpha > 0$ and $\beta < 1$. Premium, expected losses, and willingness to pay for insurance are shown on the vertical axis. The solid line labeled *expected loss with no moral hazard* depicts the expected loss for individuals with different probabilities of no loss under the assumption that there is no moral hazard behavior. The second solid line labeled *mean expected loss of pool* shows average losses for pools of insureds with probabilities of loss ranging from α to each risk level π. The dashed line labeled *expected loss with moral hazard* incorporates the effect of moral hazard on expected losses, while the dashed line labeled *maximum willingness to pay* adds in the risk aversion incentive.[3]

[3] In depicting the moral hazard effect on this graph we are compromising technical accuracy in the interest of simplicity of presentation. The probabilities π ranging from α to β can be thought of as

Two possible premium rates are shown in Figure 2 to illustrate salient points about government subsidized insurance. The first, γ_{ns}, is the breakeven premium rate for the pool, assuming that all individuals in the pool from risk type α to β purchase insurance and that there is no moral hazard. With forced participation and no moral hazard, this premium rate would be actuarially sound for the insurer (expected losses equal to total premiums) but would be actuarially fair only for insured individuals of risk type π_1, whose expected loss is equal to γ_{ns}. Individuals with lower probability of no loss (to the left of π_1) would have a positive net return from the insurance (i.e., a positive adverse selection incentive), while those with higher probability of no loss (to the right of π_1) would have a negative expected net return to the insurance (a negative adverse selection effect). When moral hazard and risk aversion incentives are taken into account, individuals of risk types up to π_2 would purchase the insurance. Those with probability of no loss greater than π_2 would not purchase because they realize a negative adverse selection incentive that more than offsets their positive moral hazard and risk aversion incentives. Use of a government subsidy to induce full participation is illustrated through a shift to a subsidized premium rate γ_s. At this rate negative adverse selection effects for insureds of risk types to the right of π_1 are offset by combined effects of positive subsidy, moral hazard, and risk aversion incentives. To further illustrate, the four incentives for an individual of type α are shown as Δ_1, Δ_2, Δ_3, and Δ_4. As riskiness decreases moving to the right, the adverse selection incentive Δ_3 decreases, becoming negative for those less risky than π_1 . For the least risky type β, the negative adverse selection incentive is exactly offset by positive subsidy, risk aversion, and moral hazard incentives such that participation in the subsidized insurance program is induced.

PREVIOUS LITERATURE

Having defined the insurance participation incentives, we now examine the crop insurance literature and relate it to our conceptual framework. The summary provided here is intentionally terse. We suggest Knight and Coble (1997) or Goodwin and Smith (1995) for a more complete treatment. Table 2 lists several of the most relevant refereed journal articles. This table groups past literature into sub-categories and in chronological order to give a sense of the major veins of this literature and the progression of research through time.

---

being associated with input level x_1 – the input level that would be applied by an insured farmer under full information (with no moral hazard). Moral hazard behavior changes these probabilities for producers of all risk types. This could be thought of as a leftward shift in probabilities for producers of all risk types to lower levels of probability of no loss, say π'. Accordingly we would have $\alpha' < \alpha$ and $\beta' < \beta$. It would be possible to show this shift through use of a second graph. However, for sake of simplicity it is adequate to consider the horizontal axis as measuring both π and $\pi' < \pi$, with the lines labeled *expected loss with moral hazard* and *maximum willingness to pay* being drawn with reference to π' and all other lines referring to probabilities π.

Table 2. Selected Works Representing the Major Veins of Crop Insurance Literature

Crop Insurance Demand	
Simulation of Insurance Demand	
▪ Kramer and Pope (1982) ▪ Lemieux, Richardson, and Nixon (1982) ▪ King and Oamek (1983)	▪ Lee and Djogo (1984) ▪ Zering, McCorkle and Moore (1987) ▪ Mapp and Jeter (1988)
Aggregate Econometric Insurance Demand	
▪ Niewoudt, Johnson, Womack, and Bullock (1985) ▪ Gardner and Kramer (1986) ▪ Hojjati and Bockstael (1988)	▪ Goodwin (1993) ▪ Barnett and Skees (1995) ▪ Richards (2000)
Disaggregate Econometric Insurance Demand	
▪ Calvin (1992) ▪ Vandeveer and Loehman (1994)	▪ Smith and Baquet (1996) ▪ Coble, Knight, Pope, and Williams (1996)
Asymmetric Information	
Moral Hazard	
▪ Nelson and Loehman (1987) ▪ Chambers (1989) ▪ Horowitz and Lichtenberg (1993) ▪ Ramaswami (1993) ▪ Quiggin, Karagiannis, and Stanton (1993)	▪ Just and Calvin (1993b) ▪ Vercammen and van Kooten (1994) ▪ Smith and Goodwin (1996) ▪ Babcock and Hennessy (1996) ▪ Coble, Knight, Pope, and Williams (1997) ▪ Wu (1999)
Adverse Selection	
▪ Ahsan, Ali, and Kurian (1982) ▪ Skees and Reed (1986) ▪ Nelson and Loehman (1987) ▪ Goodwin (1994)	▪ Luo, Skees, and Marchant (1994) ▪ Ker and McGowan (2000) ▪ Knight and Coble (1999) ▪ Just, Calvin, and Quiggin (1999)
Alternative Designs	
Area Yield Insurance	
▪ Halcrow (1949) ▪ Miranda (1991) ▪ Skees, Barnett, and Black (1997)	▪ Goodwin and Ker (1998) ▪ Mahul (1999) ▪ Ker and Goodwin (2000)
Revenue Insurance	
▪ Gray, Richardson, and McClaskey (1995) ▪ Hennessy, Babcock and Hayes (1997) ▪ Stokes (2000)	▪ Wang, Hanson, Myers, and Black (1998) ▪ Coble, Heifner, and Zuniga (2000)

The Demand for Crop Insurance

Crop insurance participation has been a major policy issue throughout the period since 1980. A desire to induce increased participation has been the primary motivation for many of the program modifications discussed earlier. Clearly, the demand for crop insurance is driven by all of the incentives in equation (7). Although the primary motivation for reforms aimed at reducing adverse selection and moral hazard has been a desire to reduce perceived program abuses, a secondary objective of these reforms has been to encourage program participation by preventing these information asymmetries from driving low-risk producers out of the program. However, the primary mechanism that has been used to encourage participation has been increased government premium subsidies (see the pattern of subsidy increases in Table 1). Referring back to Figure 2, this can be viewed conceptually as shifting the producer premium rate downward from γ_{ns}, representing the actuarially fair premium rate with the entire pool participating and no moral hazard, toward γ_s, the subsidized rate that provides sufficient subsidy to attract full participation.

The literature investigating crop insurance demand can generally be described as evolving from simulation studies, to econometric studies with aggregate data, and then to econometric studies utilizing firm-level data. The econometric studies have generally included as explanatory variables some measure of expected return to insurance, constructed to attempt to subsume all of the expectation incentives in equation (7). Results of such studies uniformly indicate that farmers with higher expected returns to insurance are more likely to insure. In most studies, farm and farm operator characteristics are included as proxies for risk and risk aversion, which underlie the risk aversion incentive. We believe it is clear that from a policy perspective the demand or participation elasticities produced by these studies are their most important outputs. Despite considerable variation in crops and regions examined, elasticities reported from the firm-level studies (Calvin 1992, Just and Calvin 1990, Coble et al. (1996), Smith and Baquet 1996) are consistent in suggesting that crop insurance demand is inelastic. Interestingly, it appears that these elasticity estimates are fairly robust to recent significant program modifications. Provisions of the 1998 ad hoc disaster relief bill provided an additional 30 percent premium subsidy for the 1999 crop year. Coble and Barnett (1999) examined the participation effects of this change in insurance cost. When measuring participation as acres insured, the point elasticity was 0.65, a value consistent with the range of previous estimates.

Moral Hazard

Theoretical models of moral hazard in multiple peril crop insurance (MPCI) are well developed and consistently support the conclusion that the moral hazard effect on input use, output, and expected indemnities is ambiguous unless strong assumptions are made about risk preferences and the risk properties of inputs. Thus, the literature has turned to empirical studies where results have been less uniform and generally less compelling than those on program participation.

Chemical inputs have been a focus of several empirical investigations of moral hazard. Horowitz and Lichtenberg (1993) estimated chemical use and insurance choices of corn producers in ten Midwestern states. Results indicated that MPCI participation induced increased use of nitrogen fertilizer, insecticides, herbicides, and total pesticides. Smith and Goodwin (1996) followed with an econometric system model for a set of Kansas dryland wheat farms in the 1991 production year. Results of Smith and Goodwin's study indicated that MPCI participation had a significant negative effect on total chemical input expenditures of the study farms. Babcock and Hennessy (1996) also addressed these issues using Iowa corn test plot data to estimate yield distributions conditional upon nitrogen fertilizer use. They concluded that, over most of the range of nitrogen fertilizer levels commonly applied, fertilizer use decreased the probability of low yields and that optimal nitrogen fertilizer levels decreased with insurance for all levels of risk aversion examined.

Quiggin, Karagiannis, and Stanton (QKS) (1993) examined the relationship between MPCI participation, input use, and output through the estimation of revenue and input share equations. Although they were unable to separate adverse selection and moral hazard effects, QKS obtained results indicating that MPCI participants had significantly lower output (revenue) than nonparticipants and lower expenditures on variable inputs. However, the input use results were not statistically significant. Coble et al. (1997) emphasized the effects of moral hazard on output and MPCI indemnities. Their analysis of moral hazard in Kansas wheat data found significant evidence of moral hazard increasing the yield shortfalls of insured producers relative to uninsured producers. Finally, Vercammen and van Kooten (1994) examined the dynamics of repeated (annual) contracts when coverage is based on a historical moving average of yields. They illustrated the potential for an optimal strategy of cyclical moral hazard behavior.

Adverse Selection

Adverse selection has been widely discussed, but research in this area has largely been indirect. Given that the adverse selection effect (Δ_3 in equations 5 and 5a) can take either positive or negative values for different producers in a heterogeneous pool, adverse selection can potentially have a pronounced effect on insurance participation. Past participation studies have implicitly addressed adverse selection through inclusion of measures of returns to insurance as discussed earlier. While results of these participation studies suggest adverse selection occurs, they do not illuminate the underlying causes.

Skees and Reed (1986) examined MPCI premium rate setting assumptions that could give rise to adverse selection. Specifically, they examined the relationship between the mean and standard deviation of farm-level yields. Hypothesis tests failed to reject independence of standard deviation and mean yield, suggesting a flaw in rating practices at the time which, as the authors observed, implicitly assumed a constant coefficient of variation for farms within a county. Goodwin (1994) continued this line of investigation to examine the validity of uniform premiums for farms with equal mean yields. Regressions of yield standard deviation on mean yield produced no consistent results, with generally low explanatory power. Goodwin concluded that APH rating methods based on assumed relationships between the mean yield and yield variability will introduce adverse selection.

Just, Calvin, and Quiggin (1999) developed the previously mentioned model of MPCI participation incentives in which the benefits from MPCI were decomposed into a risk aversion incentive and an expected revenue incentive (i.e., $\Delta_2 + \Delta_3$ in equation 2). Quantitatively, JCQ found insured farmers realized a risk reduction 65 percent larger than would have been realized by farmers who did not insure. In general, farmers who insured realized a positive effect on expected revenue, whereas this effect would have been negative for farmers who did not insure. Decomposition of the expected revenue effect showed a positive actuarial (subsidy) incentive for both insured and uninsured farmers. However, the estimated asymmetric information (adverse selection) incentive was negative for both insured and uninsured farmers. This somewhat surprising result was attributed to Federal Crop Insurance Corporation's (FCIC) APH yield expectations failing to account for trends such that both insured and uninsured producers had expected yields that were higher than those used by the FCIC.

Luo, Skees, and Marchant (1994) examined the potential for intertemporal adverse selection in crop insurance due to the insured utilizing a superior forecast of weather conditions prior to insurance sign-up. They concluded that a significant potential for intertemporal adverse selection existed in Midwest corn insurance decisions. Ker and McGowan (2000) followed in this same vein by examining the potential for a crop insurance company to use weather information to optimize the allocation of liability within the reinsurance agreement provided by USDA. They found that using El Niño/La Niña information could generate significant rents for the insurance company.

Alternative Insurance Plans

Area yield insurance has been advocated as an alternative to MPCI at least since the mid-1900s (see Halcrow 1949). Support for an area yield plan strengthened in the early 1990s, largely as a result of the poor actuarial performance and associated high government cost of MPCI. The essential difference between MPCI and area yield insurance is that under area yield coverage both insurable yield and indemnities are based on yields for a geographic area such as a county. Suggested advantages of area yield coverage are virtual elimination of adverse selection and moral hazard due to removal of the direct link between farm-level yields and indemnities, and reductions in administrative costs associated with elimination of the need to verify farm yield histories and adjust losses at the farm level. The primary disadvantage of area yield coverage is that loss protection is imperfect because of less-than-perfect correlation between area- and farm-level yields (in essence a type of basis risk).

The literature relating to area yield insurance has largely centered around the "optimal" design and implementation of such a policy given insureds with varying correlation to the area yield (Miranda 1991, Smith, Chouinard, and Baquet 1994, Carriker et al. 1991, Williams et al. 1993, Mahul 1999). It has also proven an interesting case study, in that a program very much along the lines proposed has been widely piloted (Skees, Barnett, and Black 1997).

Crop revenue insurance provides a guarantee based on revenue for a given crop or for all insurable crops grown on a farm. Like yield insurance, revenue insurance coverage can be based on either farm- or area-level measures of outcome. An advantage of crop revenue insurance over yield insurance is that unfavorable outcomes that trigger losses are defined in terms of a superior measure of farm business performance; hence, indemnity payments should be better correlated with need. A disadvantage of revenue insurance is the added difficulty of developing premium rates when the joint distribution of two or more random variables must be assessed.

The U.S. crop insurance program introduced the first of several revenue insurance designs in 1996 and these insurance plans have now captured a more than 50 percent share of the corn and soybean crop insurance markets. However, the literature on these designs has lagged behind implementation. Hennessy, Babcock, and Hayes (1997), as well as Gray, Richardson, and McClaskey (1995), conducted analyses comparing revenue insurance programs with alternative insurance and farm policy scenarios. Stokes proposed an options pricing approach to rating revenue insurance. Because revenue insurance subsumes both price and yield risk, questions have arisen regarding the correlation of yield and price at the farm level. The aforementioned articles, as well as Coble, Heifner, and Zuniga (2000) and Wang et al. (1998), have investigated these relationships and found evidence that in some crops and regions there is enough correlation between farm and national yield that assumed independence of price and farm yield is rejected.

FUTURE DIRECTIONS FOR CROP INSURANCE RESEARCH

In this section we offer our vision of future priorities in crop insurance research. At least two caveats apply. First, by nature it is difficult to predict the path of progress in any area of scientific endeavor. A single important theoretical development can open new avenues of investigation and significantly redirect subsequent theoretical and empirical research. Second, crop insurance research is often policy-driven and, as the summary of legislation in Table 1 suggests, is in a particularly fluid, if not volatile, policy area. Therefore, we will focus our comments on four areas of investigation that we believe would make a significant contribution in the current policy context. However, we will not attempt to predict the path of future policy developments.

How Effective Are Federally Subsidized Crop Insurance Programs in Reducing Farm Risk?

Crop insurance is one of many risk mitigation measures available to producers. To the extent that insurance substitutes for savings, diversification, and other risk management tools, slippage occurs. Here our use of the term "slippage" is essentially analogous to its traditional use in referring to failure of acreage diversion programs to achieve proportionate effects on domestic supply. In an insurance context, it means that the insurance is substituted for other risk management measures so that the marginal effect of the insurance protection on farm income or financial risk is less than anticipated. This issue has implications for the value of the crop insurance as a policy instrument and in understanding producer decision making. It relates to Wright and Hewitt's (1990) hypothesis that we fail to accurately characterize crop insurance demand because of a failure to account for diversification and other effects that influence the purchase decision.

A small body of prior research has broached the issue of how insurance interacts with other risk management tools. Work such as that of Wang et al. (1998) and Coble, Heifner, and Zuniga (2000) examines the relationship between alternative insurance designs and forward pricing. In general, these studies indicate that these risk management tools are not separable: some degree of complementarity or substitutability exists. However, our understanding of the nature and extent of these interrelationships is very limited. Additional research could enhance our understanding of these issues and contribute to better policy decisions.

How Does Crop Insurance Affect Acreage, Output, Input Use, and Environmental Quality?

The enactment of ARPA continues a pattern of increasing crop insurance subsidies and allowable coverage levels. For example, the subsidy percentage for a 75 percent coverage policy has increased from 17 percent to 55 percent since 1993. Maximum coverage levels of 85 percent are now widely available. Thus, under ARPA, producers will be allowed lower deductibles and greater subsidies. We believe there is an inherent set of research issues that will become more prominent in this new context. Specifically, the collateral effects of insurance on crop acreage, output, and input use have implications for aggregate supply and demand balances, trade relationships, and agriculture's effect on the environment.

The acreage effect of crop insurance would clearly appear to be a question of magnitude. That is, a policy with clear risk-reducing attributes, augmented with subsidies, would tend to induce increased production. However, quantifying this effect econometrically has proven challenging. Studies by Wu (1999) and Goodwin, Smith, and Hammond (2000) have attempted to isolate the effect of insurance from other acreage-inducing policies such as deficiency payments, loan programs, and disaster programs. The issue of the effect of insurance on input use appears to be even more complex. As indicated in our earlier discussion of prior research on moral hazard, the effect of insurance on input use appears both theoretically and empirically ambiguous. Clearly, the aggregate output effect would combine the induced acreage effect with the per acre input intensity.

We believe that these will be important policy issues as an expanding crop insurance program attracts greater scrutiny under trade agreements and from environmental groups. Further, we believe that expansion of crop insurance programs to a wider range of specialty crops will intensify the debate on acreage effects because the potential for significant market effects is much greater for such crops than for traditionally insured major commodities such as corn, wheat, cotton, and soybeans. In the policy debate prior to the enactment of ARPA, there appeared to be little attention given to the implications of providing government support to producers through a crop insurance mechanism (Coble 1999). Given that crop insurance subsidies are distributed much differently and more heterogeneously than benefits of price-triggered programs such as market loan programs, the nature of farm program distortions may be significantly altered.

Can Progress Be Made in the Rating and Economic Evaluation of Existing and New Insurance Designs?

We have suggested there is a large social cost associated with government provision of insurance in an adversely selected market. The upshot of that argument is that reductions in adverse selection would have significant value. One may ask whether we can reduce moral hazard and adverse selection and the need for subsidy by designing and rating products more effectively. The rating problem centers largely on statistical issues arising from use of limited data when losses are driven by infrequent but catastrophic events. This problem is made increasingly difficult by the rapid expansion of insurance products. For example, the RMA now reinsures five forms of revenue insurance, all introduced since 1995. Further, ARPA legislation strongly promotes continued development of alternative plans of insurance.

Rigorous comparisons of alternative rate estimators are needed. To date, most analysis of rating issues has focused on various aspects of the experience-based APH yield insurance design. We are not aware of previous research examining the predictive capability of alternative simulation procedures currently in use. This failure to test rating model calibration should not be excused on grounds of data limitations. At a minimum, the data available from the APH program along with readily available commodity futures and options price series provide a substantial basis for such analysis.

Can We Credibly Quantify and Forecast Insurability?

The question of insurability of agricultural risks in general has been central to the debate regarding government intervention in provision of crop insurance. In a concise but insightful discussion of private market insurability, Chambers (1989) observed that "Insurability hinges on whether individually rational insurance policies are available, i.e., policies which make both farmers and insurance companies better off than in the absence of insurance" (p. 604). Over time, insurance experts have identified at least six *ideal* conditions for a risk to be considered insurable (Rejda 1995, pp. 23, 24):

(1) Determinable and measurable loss
(2) Large number of roughly homogeneous, independent exposure units
(3) Accidental and unintentional loss
(4) No risk of catastrophic losses for the insurer
(5) Calculable chance of loss
(6) Economically feasible premium

In reality, most insurance products do not perfectly satisfy these criteria, with the question of insurability hinging on the degree of deviation. In the present context, federal reinsurance greatly reduces the importance of conditions (2) and (4), which relate to the effects of credit rationing or a steeply sloping supply of funds schedule on the ability of private insurance companies to finance an insurance program (Chambers 1989). Miranda and Glauber (1997) examined the risk exposure of an insurance firm holding a portfolio of U.S. crop insurance liability and compared it to other lines of insurance. Also Duncan and Myers (2000) examine specifically the role of reinsurance in insurance markets where condition (2) is not met. They suggest that reinsurance may reduce premiums and increase coverage when an equilibrium exists, but that public subsidies may be required when there is complete market failure. Large federal subsidies undoubtedly go a long way in addressing condition (6) – that the insurance is affordable for farmers. However, the remaining conditions (1, 3, and 5) apply in order for government-reinsured insurance programs to meet Congressional mandates for actuarial soundness and avoidance of excessive fraud and abuse.

As previously mentioned, the ARPA legislation strongly encourages the development of new insurance designs. This has created an environment where producer groups and policymakers are proposing a variety of new insurance designs. Economists will be asked to evaluate these designs in terms of their actuarial soundness, producer cost/ benefits, and implications for the agricultural sector. At this point, when asked, economists have little ability to characterize the outcomes with anything other than our subjective opinion. We are especially confounded by the disaggregate nature of crop insurance effects while policymakers want and often need aggregate analysis.

The development of credible models to evaluate alternative insurance designs would make a significant contribution. At least two issues come to mind. First, greater effort should be devoted to modeling producer behavior when confronted with alternative insurance designs. In particular, these models should include moral hazard and adverse selection behavior on the part of producers so that the relative implication of alternative insurance designs can be assessed. This would necessarily require a movement away from naive simulation approaches and toward models that optimize producer behavior. Second, crop insurance as a policy tool significantly complicates the assessment of farm policy as compared to traditional price-oriented domestic support programs. Given that price risk is more homogeneous than either yield or revenue risk, modeling farm policy with highly aggregated models is much less appropriate when farm-level yield or revenue risk is an integral part of the triggering mechanism. For example, calculating government cost of corn deficiency payments involved one random variable – national season average price. The aggregate cost of an insurance-based program necessarily involves the random yield (or revenue) for every insured unit. This has implications for producer welfare, but also greatly complicates the estimation of government program costs. To date we have seen little progress toward models capable of quantifying a multitude of firm-level risks and appropriately aggregating the outcomes, though both types of analysis are needed by policymakers.

CONCLUDING COMMENTS

During the past 20 years, U.S. crop insurance programs have grown sevenfold to a liability of $35 billion per year. This has come about largely due to farm policy initiatives that have increased insurance subsidies, crops covered, and the menu of insurance products offered. During the same time period, agricultural economists have addressed a number of theoretical and empirical issues associated with crop insurance. In particular, given the combined problems of low participation and poor actuarial performance, research has emphasized the interrelated problems of demand and asymmetric information. Early articles such as Ahsan, Ali, and Kurian (1982), Chambers (1989), and Nelson and Loehman (1987) adapted seminal asymmetric information articles such as Rothschild and Stiglitz (1976) to the specific crop insurance context. However, none of these articles provided empirical evidence of the asymmetric information phenomena they discussed. Much of the subsequent literature has asserted asymmetric information as a motivating factor or attempted to empirically quantify its magnitude or effects.

Our assessment of the progress made in empirically quantifying the effect of asymmetric information is that it has been hampered by a lack of appropriate firm-level data that measures the relevant economic variables. Thus, the pace of further progress, at least in some areas of investigation, is likely to depend upon data availability. In particular, farm-level panel data specifically designed to capture relevant risk and firm-level attributes would greatly enhance our ability to contribute to a better understanding of risk management generally and crop insurance specifically. Cross-sectional information generally precludes objective risk measures and any ability to characterize dynamic behavior such as saving and borrowing. We echo Just's comments at the SERA-IEG-31 meetings (see the Preface) in 2000 (see Just forthcoming):

> "To make better progress and to open broad legitimate debate on many issues, the profession needs broad access to a common micro-level data set. These data would ideally include data on wealth, asset vintages, debt structure, mid-season information affecting input usage, local attributes of farms and farmers, etc., and reflect the joint distribution of these characteristics among farms/farmers."

A case in point relating to crop insurance is the Quiggin, Karagiannis, and Stanton (1993) paper. They recognized that with a cross-sectional analysis there is no opportunity to distinguish individual-specific from time-varying effects and thus to separate adverse selection and moral hazard. The study of moral hazard by Coble et al. (1997) was possible due to the availability of panel data for a particular location, but our experience strongly indicates that results from one crop or region where data happen to be available cannot be generalized to other crops and regions.

Turning from the empirical to more theoretical issues, we previously noted that Ahsan, Ali, and Kurian (1982), Chambers (1989), and Nelson and Loehman (1987) were the first to apply the seminal asymmetric information articles such as Rothschild and Stiglitz (1976), Raviv (1979), and Holmstrom (1979) to the crop insurance problem. However, we would note that the game theoretic literature cited in the more recent crop insurance research has generally not advanced beyond those same seminal moral hazard and adverse selection articles. For example, little of the optimal incentive contract literature, such as Crocker and Morgan (1998), has been drawn upon. To our knowledge, the principal-agent modeling of crop insurance adverse selection has not yet made the distinction between the objectives of a private and public sector principal. Many of the equilibria discussed in the adverse selection literature require competition and assume profit-maximizing behavior of the insurer. That probably does not characterize the social objective function of USDA. Further, the incentives and design of repeated contracts is now a well-developed literature, but Vercammen and van Kooten (1994) stand out as one of the few attempts to consider any dynamic behavior in crop insurance contracts. We would suggest that attempts to apply or adapt more sophisticated game theoretic models merit serious consideration.

In conclusion, we have suggested many opportunities to advance crop insurance research both empirically and theoretically. And at least in the short run, political attention will also create demand for further crop insurance research. However, the pace of progress depends on many factors, including efforts to collect appropriate data, advances in statistical procedures to quantify risk, and advances in the theoretical base on which we examine crop insurance issues.

REFERENCES

Ahsan, S.M., A. Ali, and N. Kurian. 1982. "Toward a Theory of Agricultural Insurance." *American Journal of Agricultural Economics* 64: 520-529.

Babcock, B.A., and D. Hennessy. 1996. "Input Demand under Yield and Revenue Insurance." *American Journal of Agricultural Economics* 78: 416-427.

Barnett, B.J., and J.R. Skees. 1995. "Region and Crop Specific Models of the Demand for Federal Multiple Peril Crop Insurance." *Journal of Insurance Issues* 19: 47-65.

Calvin, L. 1992. "Participation in the U.S. Federal Crop Insurance Program." Washington, D.C.: U.S. Department of Agriculture, ERS Technical Bulletin No. 1800 (June).

Carriker, G.L., J.R. Williams, G.A. Barnaby, and J.R. Black. 1991. "Yield and Income Risk Reduction Under Alternative Crop Insurance and Disaster Assistance Designs." *Western Journal of Agricultural Economics* 16: 238-250.

Chambers, R.G. 1989. "Insurability and Moral Hazard in Agricultural Insurance Markets." *American Journal of Agricultural Economics* 71: 604-616.

Coble, K.H. 1999. Written testimony before the U.S. Senate Agriculture, Nutrition, and Forestry Committee (hearing conducted March 10, 1999).

Coble, K.H., and B.J. Barnett. 1999. "Preliminary Evidence of Crop Insurance Demand Elasticity from the 1999 Response to Additional Subsidies." Department of Agricultural Economics, Mississippi State University (October), http://www.agecon.msstate.edu/risk/documents.

Coble, K.H., R. Heifner, and M. Zuniga. 2000. "Implications of Crop Yield and Revenue Insurance for Producer Hedging." *Journal of Agricultural and Resource Economics* 25: 432-452.

Coble, K.H., T.O. Knight, R.D. Pope, and J.R. Williams. 1996. "Modeling Farm-Level Crop Insurance Demand with Panel Data." *American Journal of Agricultural Economics* 78: 439-447.

___. 1997. "An Expected Indemnity Approach to the Measurement of Moral Hazard in Crop Insurance." *American Journal of Agricultural Economics* 79: 216-226.

Crocker, K.J., and J. Morgan. 1998. "Is Honesty the Best Policy? Curtailing Insurance Fraud Through Optimal Incentive Contracts." *Journal of Political Economy* 106: 355-375.

Duncan, J., and R.J. Myers. 2000. "Crop Insurance Under Catastrophic Risk." *American Journal of Agricultural Economics* 82: 842-855.

Gardner, B.L., and R.A. Kramer. 1986. "Experience With Crop Insurance Programs in the United States. In P. Hazell, C. Pomareda, and A. Valdes, eds., *Crop Insurance for Agricultural Development, Issues and Experience.* Baltimore: The John Hopkins University Press.

Goodwin, B.K. 1993. "An Empirical Analysis of the Demand for Crop Insurance." *American Journal of Agricultural Economics* 75: 425-434.

___. 1994. "Premium Rate Determination in the Federal Crop Insurance Program: What Do Averages Have to Say About Risk?" *Journal of Agricultural and Resource Economics* 19: 382-395.

Goodwin, B.K., and A.P. Ker. 1998. "Nonparametric Estimation of Crop Yield Distributions: Implications for Rating Group-Risk Crop Insurance Contracts." *American Journal of Agricultural Economics* 80: 139-153.

Goodwin, B.K., and V.H. Smith. 1995. *The Economics of Crop Insurance and Disaster Aid.* Washington, D.C.: The American Enterprise Institute Press.

Goodwin, B.K., V.H. Smith, and C. Hammond. 2000. "An Ex-Post Evaluation of Conservation Reserve, Federal Crop Insurance and Other Government Programs: Program Participation and Soil Erosion." Paper presented at USDA's "Crop Insurance, Land Use, and the Environment Conference," September 20-21, Washington, D.C.

Goodwin, B.K., and M. Vandeveer. 2000. "An Empirical Analysis of Acreage Distortions and Participation in the Federal Crop Insurance Program." Paper presented at USDA's "Crop Insurance, Land Use, and the Environment Conference," September 20-21, Washington, D.C.

Gray, A.W., J.W. Richardson, and J. McClaskey. 1995. "Farm-Level Impacts of Revenue Assurance." *Review of Agricultural Economics* 17: 171-183.

Halcrow, H.G. 1949. "Actuarial Structures for Crop Insurance." *Journal of Farm Economics* 31: 418-443.

Hennessy, D.A., B.A. Babcock, and D.J. Hayes. 1997. "The Budgetary and Producer Welfare Effects of Revenue Assurance." *American Journal of Agricultural Economics* 79: 1024-1034.

Hirshleifer J., and J.G. Riley. 1992. *The Analytics of Uncertainty and Information.* New York: Cambridge University Press.

Hojjati, B., and N.E. Bockstael. 1988. "Modeling the Demand for Crop Insurance." Article No. A-4704, Maryland Agricultural Experiment Station.

Holmstrom, B. 1979. "Moral Hazard and Observability." *Bell Journal of Economics* 10: 74-91.

Horowitz, J.K., and E. Lichtenberg. 1993. "Insurance, Moral Hazard, and Chemical Use in Agriculture." *American Journal of Agricultural Economics* 75: 926-935.

Just, R.E. 2002. "Risk Research in Agricultural Economics: Opportunities and Challenges for the Next Twenty-Five Years." *Agricultural Systems* (special issue: "Advances in Risk Impacting Agriculture and the Environment") (forthcoming).

Just, R.E., and L. Calvin. 1990. "An Empirical Analysis of U.S. Participation in Crop Insurance." Unpublished report to the Federal Crop Insurance Corporation.

___. 1993. "A Moral Hazard in U.S. Crop Insurance: An Empirical Investigation." Unpublished manuscript, University of Maryland (April).

Just, R.E., L. Calvin, and J. Quiggin. 1999. "Adverse Selection in U.S. Crop Insurance: Actuarial and Asymmetric Information Incentives." *American Journal of Agricultural Economics* 80: 834-849.

Ker, A.P., and B.K. Goodwin. 2000. "A Nonparametric Estimation of Crop Insurance Rates Revisited." *American Journal of Agricultural Economics* 83: 463-478.

Ker, A.P., and P. McGowan. 2000. "Weather-Based Adverse Selection and the U.S. Crop Insurance Program: The Private Insurance Company Perspective." *Journal of Agricultural and Resource Economics* 25: 386-410.

King, R.P., and G.E. Oamek. 1983. "Risk Management by Colorado Dryland Wheat Farmers and the Elimination of Disaster Assistance Programs." *American Journal of Agricultural Economics* 65: 247-255.

Knight, T.O., and K.H. Coble. 1997. "A Survey of Literature on U.S. Multiple Peril Crop Insurance Since 1980." *Review of Agricultural. Economics* 19: 128-156.

Knight, T.O., and K.H. Coble. 1999. "Actuarial Effects of Unit Structure in the U.S. Actual Production History Crop Insurance Program." *Journal of Agricultural and Applied Economic* 31: 519-535.

Kramer, R.A., and R.D. Pope. 1982. "Crop Insurance for Managing Risk." *Journal of American Society of Farm Managers and Rural Appraisers* 46: 34-40.

LaFrance, J. 2000. "Subsidized Crop Insurance and the Environment." Paper presented at USDA's "Crop Insurance, Land Use, and the Environment Conference," September 20-21, Washington, D.C.

Lee, W.F., and A. Djogo. 1984. "The Role of Federal Crop Insurance in Farm Risk Management." *Agricultural Finance Review* 44: 15-24.

Lemieux, C.M., J.W. Richardson, and C.J. Nixon. 1982. "Federal Crop Insurance vs. ASCS Disaster Assistance for Texas High Plains Cotton Producers: An Application of Whole Farm Simulation." *Western Journal of Agricultural Economics* 7: 141-153.

Luo, H., J.R. Skees, and M.A. Marchant. 1994. "Weather Information and the Potential for Intertemporal Adverse Selection." *Review of Agricultural Economics* 16: 441-451.

Mahul, O. 1999. "Optimum Area Yield Crop Insurance." *American Journal of Agricultural Economics* 81: 75-82.

Mapp, H.P., and K.L. Jeter. 1988. "Potential Impact of Participation in Commodity Programs and Multiple Peril Crop Insurance on a Southwest Oklahoma Farm." *Multiple Peril Crop Insurance: A Collection of Empirical Studies.* Southern Cooperative Series Bulletin No. 34, Washington, D.C.

Miranda, M.J. 1991. "Area-Yield Crop Insurance Reconsidered." *American Journal of Agricultural Economics* 73: 233-242.

Miranda, M.J., and J.W. Glauber. 1997. "Failure of Crop Insurance Markets." *American Journal of Agricultural Economics* 79: 206-215.

Nelson, C.H., and E.T. Loehman. 1987. "Further Toward a Theory of Agricultural Insurance." *American Journal of Agricultural Economics* 69: 523-531.

Niewoudt, W.L., S.R. Johnson, A.W. Womack, and J.B. Bullock. 1985. "The Demand for Crop Insurance." Department of Agricultural Economics, University of Missouri-Columbia, Agricultural Economics Report No. 1985-16 (December).

Patrick, G.F. 1988. "Mallee Wheat Farmers' Demand for Crop and Rainfall Insurance." *Australian Journal of Agricultural Economics* 32: 37-49.

Quiggin, J., G. Karagiannis, and J. Stanton. 1993. "Crop Insurance and Crop Production: Empirical Study of Moral Hazard and Adverse Selection." *Australian Journal of Agricultural Economics* 37: 95-113.

Ramaswami, B. 1993. "Supply Response to Agricultural Insurance: Risk Reduction and Moral Hazard Effects." *American Journal of Agricultural Economics* 75: 914-925.

Raviv, A. 1979. "The Design of an Optimal Insurance Policy." *American Economics Review* 69: 84-96.

Rejda, G.E. 1995. *Principles of Risk Management and Insurance.* New York: Harper Collins College Publishers.

Richards, T.J. 2000. "A Two-Stage Model of the Demand for Specialty Crop Insurance." *Journal of Agricultural and Resource Economics* 25: 177-194.

Rothschild, M., and J. Stiglitz. 1976. "Equilibrium in Competitive Insurance Markets: An Essay on the Economics of Imperfect Information." *Quarterly Journal of Economics* 90: 629-649.

Skees, J.R., B.J. Barnett, and J.R. Black. 1997. "Designing and Rating an Area Yield Crop Insurance Contract." *American Journal of Agricultural Economics* 79: 430-438.

Skees, J.R., and M.R. Reed. 1986. "Rate Making for Farm-Level Crop Insurance: Implications for Adverse Selection." *American Journal of Agricultural Economics* 68: 653-659.

Smith, V.H., and A. Baquet. 1996. "The Demand for Multiple Peril Crop Insurance: Evidence from Montana Wheat Farms." *American Journal of Agricultural Economics* 78: 189-210.

Smith, V.H., H.H. Chouinard, and A.E. Baquet. 1994. "Almost Ideal Area Yield Crop Insurance Contracts." *Agricultural and Resource Economics Review* 23: 75-83.

Smith, V.H., and B.K. Goodwin. 1996. "Crop Insurance, Moral Hazard, and Agricultural Chemical Use." *American Journal of Agricultural Economics* 78: 428-438.

Stokes, J.R. 2000. "A Derivative Security Approach to Setting Crop Revenue Coverage Insurance Premiums." *Journal of Agricultural and Resource Economics* 25: 159-176.

Vandeveer, M.L., and E.T. Loehman. 1994. "Farmer Response to Modified Crop Insurance: A Case of Corn in Indiana." *American Journal of Agricultural Economics* 76: 128-140.

Vercammen, J., and G.C. van Kooten. 1994. "Moral Hazard Cycles in Individual-Coverage Crop Insurance." *American Journal of Agricultural Economics* 76: 250-261.

Wang, H.H., S.D. Hanson, R.J. Myers, and J.R. Black. 1998. "The Effects of Crop Yield Insurance Designs on Farmer Participation and Welfare." *American Journal of Agricultural Economics* 80: 806-820.

Williams, J.R., G.R. Carriker, G.A. Barnaby, and J.K. Harper. 1993. "Crop Insurance and Disaster Assistance Design for Wheat and Grain Sorghum." *American Journal of Agricultural Economics* 75: 435-447.

Wright, B.D., and J.D. Hewitt. 1990. "All Risk Crop Insurance: Lessons From Theory and Experience." Giannini Foundation, California Agricultural Experiment Station, Berkeley, CA (April).

Wu, J. 1999. "Crop Insurance, Acreage Decisions, and Nonpoint-Source Pollution." *American Journal of Agricultural Economics* 81: 305-320.

Zering, K.D., C.O. McCorkle, and C.V. Moore. 1987. "The Utility of Multiple Peril Crop Insurance for Irrigated, Multiple-Crop Agriculture." *Western Journal of Agricultural Economics* 12: 50-59.

Chapter 21

RISK MANAGEMENT AND THE ROLE OF THE FEDERAL GOVERNMENT

Joseph W. Glauber and Keith J. Collins*
U.S. Department of Agriculture

INTRODUCTION

One of the key characteristics of agriculture is the inherent production risks facing producers from adverse weather, pests, and diseases. These risks have been used to justify government intervention in the form of disaster assistance payments, emergency loans, livestock feed assistance programs, crop insurance, and other subsidized assistance schemes. Yet while government intervention to provide assistance has been widely supported in the United States, the form of assistance has been much debated.

Since 1980, the principal form of crop loss assistance in the United States has been provided through the federal crop insurance program. The Federal Crop Insurance Act of 1980 was intended to replace disaster programs with a subsidized insurance program that farmers could depend on in the event of crop losses. Crop insurance was seen as preferable to disaster assistance because it was less costly and hence could be provided to more producers, less likely to encourage moral hazard, and less likely to encourage producers to plant crops on marginal lands (U.S. GAO 1989).

Over the past twenty years, the program has grown from a pilot program insuring 30 crops in 4,683 county crop programs in 1980 to over 100 crops in 36,262 county crop programs in 2000. In 2000, over 200 million acres were enrolled in the program, compared with only 26 million in 1980. The enrolled acres accounted for about 75 percent of eligible acreage in 2000. Total liability of the program in 2000 was $34.1 billion, almost 10 times the level of liability insured in 1980.

Despite this growth, the crop insurance program has not replaced other disaster programs as the sole form of assistance. Over the past 20 years, producers received an estimated $15 billion in supplemental disaster payments in

---

* The authors are Deputy Chief Economist and Chief Economist, respectively. The views expressed here are theirs and do not reflect those of USDA.

addition to $22 billion in crop insurance indemnities. Citing failures of the crop insurance program to attract sufficient participation at sufficiently high coverage levels, Congress has passed two crop insurance reform bills since 1980, in 1994 and 2000, that have increased the scope of the program and the size of government costs. The Agricultural Risk Protection Act of 2000 provides $8.2 billion in subsidies over five years to encourage the purchase of federal crop insurance. Projected annual costs of the program under this legislation are estimated at $3 billion, almost double the annual costs under the previous program and a tenfold increase over spending levels of the early 1980s.

As the costs of the program have grown, criticisms have arisen that the high level of subsidies may affect producers' planting decisions and input use. To the degree that these subsidies increase crop production, their benefit to producers may be offset by lower market revenues.

In this chapter we examine the history of the federal crop insurance program. We first consider whether there is an inherent market failure that justifies government intervention. We then review the experience of the federal crop insurance program, in particular, over the past 20 years since passage of the Federal Crop Insurance Act of 1980. We examine the underlying goals of the 1980 Act and assess how the program has met these goals. We then consider the costs of the program and whether the program has had significant effects on production and crop prices.

THE INSURABILITY OF CROP YIELDS AND THE RATIONALE FOR GOVERNMENT INTERVENTION

A primary justification for government intervention has been the failure of private agricultural insurance markets (see, for example, Appel, Lord, and Harrington 1999, Hazell, Pomareda, and Valdez 1986, Goodwin and Smith 1995). Valgren (1922) describes the disastrous experiences of fire insurance companies that offered crop insurance in the Dakotas and Montana in 1917 and the early 1920s. Severe droughts caused widespread crop losses in those states. The insurance companies had not protected themselves from such large losses and were unable to indemnify the insured farmers. As Valgren concluded, "the outcome of this first attempt to provide a general crop coverage is much to be regretted" (p. 17). Other private ventures to establish multiple peril crop insurance prior to 1938 met with similar results (Kramer 1983).

Arguably, private crop insurance markets today are crowded out by subsidized crop insurance and other agricultural support programs. However, whether a viable market for agricultural insurance could exist today in the absence of government programs is not clear. There has been substantial development in financing catastrophic risks, particularly over the past 10 years (see, for example, Froot 1999, Cutler and Zeckhauser 1999, Kleindorfer and Kunreuther 1999), and there has been much interest in developing private crop insurance products outside of the United States (Skees, Hazell, and Miranda 1999, European Commission 2001, Meuwissen 2000). Yet apart from similarly subsidized crop insurance programs in other countries (e.g., Canada, Japan), no large-scale private crop insurance markets have emerged to date (Goodwin and Smith 1995, Wright and Hewitt 1994).

One of the reasons why private crop insurance markets have not developed is the relatively low demand for crop insurance. Despite large subsidies in the United States, crop insurance participation has been relatively low. Farmers and ranchers use a variety of risk management strategies to mitigate the risks that they face (Harwood et al. 1999, U.S. GAO 1999), many of which compete with crop insurance. These include futures and options markets, contracting, cultural practices that reduce crop loss (e.g., irrigation, pesticide use), crop and livestock diversification, non-farm income, savings and borrowing, leasing, federal price and income support programs, and federal disaster assistance payments.

A number of studies have estimated the demand for crop insurance (for a survey of this literature, see Knight and Coble 1997, Goodwin and Smith 1995). Most have concluded that the demand for crop insurance is inelastic, ranging from -0.2 to -0.92 (Goodwin and Smith 1995). A recent study by Just, Calvin, and Quiggin (1999) found that for producers participating in the federal crop insurance program, risk aversion was a minor part of their incentive to participate. Rather, their decision to participate was driven by the size of the expected benefit (due to premium subsidies).

On the supply side, researchers have questioned the viability of private crop insurance markets because of the presence of moral hazard and adverse selection problems (Ahsan, Ali, and Kurian 1982, Chambers 1989, Nelson and Loehman 1987, Goodwin and Smith 1995). Moral hazard occurs when an insured producer can increase his or her expected indemnity by actions taken after buying insurance. To combat moral hazard, insurance contracts typically include deductibles, co-payment provisions, or other mechanisms where losses are shared between the insurer and the insured. However, because of the high costs of monitoring agricultural production, private crop insurance would require relatively high deductibles or high premium costs. Either of these reduce producer demand for insurance (Goodwin and Smith 1995).

Adverse selection occurs when a producer has more information about the risk of loss than the insurer does, and is better able to determine the fairness of premium rates (Harwood et al. 1999). As a result, those who are overcharged are less likely to purchase insurance, while those who are undercharged are more likely to overpurchase insurance. Over time, indemnities will exceed premiums in such markets, and raising premium rates for all insureds will potentially create an even more adversely selected market as the less risky participants drop out of the program. More accurate risk classification reduces adverse selection problems, but risk classification, like monitoring for moral hazard, is potentially costly. Compulsory insurance coverage can mitigate adverse selection by forcing lower-risk buyers to buy coverage, but as pointed out by Appel, Lord, and Harrington (1999), mandatory coverage generally reduces the welfare of these buyers and therefore can be politically unpopular.

Another factor often cited for why there is no significant private market for crop insurance is the fact that yield losses tend to be positively correlated across farmers (Bardsley, Abey, and Davenport 1984, Miranda and Glauber 1997, Duncan and Myers 2000). Because of this, insurers can not easily diversify their risks across space and, in the absence of reinsurance, would have to hold large reserves in the event of a large crop loss. As a result, a higher premium loading would be necessary to cover the insurer's opportunity cost of capital (Appel, Lord, and Harrington 1999). In practice, however, insurance companies can diversify their risks through the use of reinsurance. Crop liabilities, while large, are small relative to the size of the global reinsurance market. Nonetheless, reinsurance comes at some cost to the insurance company, which will be reflected in higher premium costs for producers.

The problems of adverse selection, moral hazard, and correlated risks are certainly not unique to crop insurance. Other lines of insurance face similar problems, yet private markets exist. The costs of addressing these problems for crop insurance are possibly high enough to make the costs of crop insurance too high for most producers to support a viable market, except perhaps in limited markets and regions. Crop insurance would likely be unaffordable for most producers in high-risk areas.

This potential disparity in availability of private insurance between regions and crops is sometimes cited as a reason for government intervention (U.S. GAO 1980, Appel, Lord, and Harrington 1999), but here again crop insurance is not unique. Many risk management tools used by farmers are available only in certain regions. For example, cash forward contracting is widely available for corn and soybean producers in the Midwest, but the same is not necessarily true for producers in regions where basis risk is high. But there is little impetus for government intervention in those markets.

While the conclusions drawn from the above studies would argue that the case for government intervention in crop insurance markets is weak on economic efficiency and equity grounds, Congress has come to a different conclusion. For the past 70 years, they have provided assistance to farmers and ranchers for crop and livestock losses (Dyson 1988). The debate over the past 70 years has focused not on whether Congress should provide assistance, but rather on the form that assistance should take.

CROP INSURANCE AND FEDERAL DISASTER ASSISTANCE POLICY

Prior to the 1930s, there was little federal role in providing disaster assistance to farmers and ranchers. In 1886, Congress appropriated $10,000 for the Department of Agriculture to purchase seed for drought-stricken farmers in Texas, but President Grover Cleveland vetoed the act with the message, "Federal aid in such cases encourages the expectation of paternal care on the part of government and weakens the sturdiness of our national character" (Porter 1988, p. 13).

With the New Deal legislation in the 1930s, this sentiment changed considerably as Congress and the Roosevelt Administration came to the aid of Dust Bowl farmers. Since the 1930s, federal disaster assistance policy to farmers has been provided through three programs: crop insurance, emergency loans, and direct disaster payments.

Federal crop insurance was first authorized in title V of the Agricultural Adjustment Act of 1938 (Benedict 1953, Kramer 1983, Goodwin and Smith 1995). The program was offered on a pilot basis and initially covered wheat only. In 1939, about 165,000 wheat policies were issued on approximately 7 million acres in 31 states (Rowe and Smith 1940). As first envisioned as part of Secretary Wallace's concept of an "ever-normal granary," crop insurance premiums and indemnities were to be made in-kind, but by 1940, these payments were largely made in cash. Premiums were established to equal indemnities over a period of years, although the government absorbed all delivery and operating costs.

For its first 40 years, the federal crop insurance was run as a pilot program, offered for a limited number of crops and in a limited number of counties. County crop programs were often withdrawn if heavy losses were experienced and coverage levels were adjusted to limit loss exposure. By 1980, only about half of the nation's counties and 26 crops were eligible for insurance coverage (Chite 1988).

Established in 1949, the Farmers Home Administration's (FmHA) emergency loan program provided emergency loans at subsidized interest rates to eligible producers who had sustained actual losses as a result of natural disasters. In the mid-1970s, the program was expanded to include loans for purposes other than actual losses, such as expanding farm operations. By 1980, the costs of the FmHA emergency loan program exceeded $245 million (U.S. GAO 1989).

The disaster payments program was authorized by the Agriculture and Consumer Protection Act of 1973 and the Rice Production Act of 1975. The program paid producers of program crops (corn, barley, oats, sorghum, wheat, cotton, and rice) who had been prevented from planting a crop or who experienced lower yields because of natural disasters. Producers received payments for crop losses in excess of one-third of their program yields. Payment rates were equal to the higher of the deficiency payment rate or one-third of the target price. The program offered essentially free insurance to those producers who complied with production adjustment requirements of the price and income support programs. Effective coverage levels were increased in the Food and Agriculture Act of 1977. Under the 1977 Act, wheat and feed grain producers received yield loss disaster payments if their yields fell below 60 percent of the farm program yield. Payment rates were set equal to 50 percent of the target price. Rice and cotton producers received payments when yields fell below 75 percent of their program yield, but their payment rate was set equal to only one-third of the target price (Johnson 1980).

Between 1974 and 1980, the government paid an average of $436 million per year in disaster payments (Chite 1988). The disaster program was popular with program crop producers because it provided disaster protection with no premium costs, and coverage in high-risk areas where crop insurance was not available (Gardner and Kramer 1986). However, by the late 1970s the program had come under heavy criticism for its cost and for encouraging production in high-risk areas. Critics maintained that the disaster program encouraged moral hazard and that the "prevented planting" provisions provided incentives to expand production in arid areas for the sole purpose of collecting payments (Miller and Walter 1977). In 1978, the Carter Administration proposed replacing the disaster payments program with a greatly expanded crop insurance program. Two years later, Congress passed the Federal Crop Insurance Act of 1980.

The 1980 Act made crop insurance the primary form of disaster protection. Disaster assistance remained available only for producers of program crops in counties where crop insurance was not available. Producers could purchase yield coverage at 50, 65, and 75 percent of their normal yields. To encourage participation, crop insurance premiums were subsidized 30 percent for 50 and 65 percent coverage. Producers who insured at 75 percent received the same subsidy as for 65 percent coverage. Crop insurance coverage for program crops was rapidly expanded to all counties where program crops were grown and to major producing areas for many other crops.

Proponents of crop insurance recognized that participation in the program was key to eliminating disaster assistance. Testifying before the Senate Committee on Agriculture, Nutrition, and Forestry in 1978, Secretary Bergland put it thus: "If less than 60 to 70 percent of the farmers are protected, it is likely that a sense of sympathy will prompt the system to provide protection for those who did not participate. Reaching the target level of participation would require both a well developed program and the termination of other programs that would provide protection and thereby competed with the new system" (Bergland, quoted in Wright and Hewitt 1994). When the 1980 Act was passed, Congress envisioned a participation rate approaching 50 percent of eligible acres by the end of the decade (Chite 1988).

Despite the premium subsidies and expanded coverage, participation in the program grew slowly during the 1980s. By 1986, fewer than 56 million acres were insured, almost double the 1980 level, but only 20 percent of eligible acres (Table 1). A drought in the Southeast prompted Congress to pass supplemental disaster payment legislation (PL 99-500 and PL 99-591), which applied only to crops of that year. Producers who were unable to plant their crops or who had low yields could apply for disaster payments. Unlike the disaster payments program of the 1970s, which had been limited to program crops, the 1986 disaster legislation opened payments to producers of non-

Table 1. Federal Crop Insurance Program

Year	Policies (thous)	Acres (mil)	Participation Rate[a]	Total Premium (mil $)	Producer Premium (mil $)	Liability (mil $)	Indemnity (mil $)	Loss Ratio (total)	Loss Ratio (producer)
1981	416.8	45.0	16	376.8	329.8	5,981.2	407.3	1.08	1.23
1982	386.0	42.7	15	396.1	304.8	6,124.9	529.1	1.34	1.74
1983	310.0	27.9	12	285.8	222.1	4,369.9	583.7	2.04	2.63
1984	389.8	42.7	16	433.9	335.6	6,619.6	638.4	1.47	1.90
1985	414.6	48.6	18	439.8	339.7	7,159.9	683.1	1.55	2.01
1986	406.9	48.7	20	379.7	291.6	6,230.0	615.7	1.62	2.11
1987	433.9	49.1	22	365.1	277.5	6,094.9	369.8	1.01	1.33
1988	461.0	55.6	25	436.4	328.4	6,964.7	1,067.6	2.45	3.25
1989	949.7	101.7	40	819.4	613.1	13,620.7	1,215.3	1.48	1.98
1990	893.7	101.3	40	835.5	620.5	12,818.2	1,033.6	1.24	1.67
1991	706.2	82.3	33	736.4	546.5	11,209.2	958.5	1.30	1.75
1992	663.1	83.1	31	758.7	562.0	11,333.6	922.5	1.22	1.64
1993	678.8	83.7	32	755.6	555.6	11,352.6	1,653.4	2.19	2.98
1994	800.4	99.6	38	948.9	694.1	13,600.9	600.9	0.63	0.87
1995	2,039.2	220.6	41/85	1,543.0	653.8	23,724.5	1,566.5	1.02	2.40
1996	1,623.3	205.0	44/76	1,838.4	856.4	26,882.0	1,491.0	0.81	1.74
1997	1,319.6	181.9	43/67	1,773.8	872.3	25,457.6	993.5	0.56	1.14
1998	1,242.4	181.7	46/69	1,874.2	928.4	27,910.2	1,659.3	0.89	1.79
1999	1,288.1	196.3	53/73	2,304.4	916.2	30,869.4	2,421.5	1.05	2.64
2000[b]	1,313.1	204.1	56/76	2,502.59	1,174.3	34,107.6	2,408.8	0.96	2.05

[a] Acres insured as percent of eligible acres. For 1995-2000, first number is for buy-up enrollment, second number is for buy-up plus CAT.
[b] Projected.

Source: U.S. Department of Agriculture (2001a)

program crops. The 1986 program cost $634 million.

In 1988, a major drought struck the Midwest. Crop insurance participation had grown to only 25 percent of eligible acreage, and was even lower still in many of the states adversely affected by the drought. For example, less than 13 percent of eligible corn acreage and 10 percent of eligible soybean acreage was insured in Illinois and Indiana in 1988 (Chite 1988). In response to the drought, Congress again passed disaster legislation. Under the Disaster Assistance Act of 1988, about $3.5 billion was provided to producers who suffered crop losses. Widespread crop losses in the Southern Plains prompted Congress to pass disaster legislation in 1989 as well, which totaled almost than $1.5 billion.

To promote crop insurance, Congress required those who received disaster payments in 1988 and 1989 to purchase crop insurance in the following year. However, this requirement could be waived for a broad number of reasons. In the 1989 program, producers who had purchased crop insurance received higher coverage (65 percent of yields compared with 60 percent) under the disaster program than those who had not purchased insurance. Participation grew to 40 percent of eligible area in 1990 and 1991, but fell back again in 1992.

With the large expenses for disaster aid, coupled with a perceived failure of the crop insurance program, debate between disaster assistance and crop insurance began anew. In 1989, the General Accounting Office (GAO) examined disaster payments, crop insurance, and emergency loans and concluded that crop insurance was the preferable form of providing protection because it was less likely to encourage moral hazard and encourage production on marginal lands. At the same time, GAO recognized the problems with the crop insurance program, including poor actuarial performance and low participation (U.S. GAO 1989).

In late 1988, Congress enacted the Federal Crop Insurance Commission Act of 1988 (Chite 1989). The Act authorized the formation of a 25-member Commission to identify the major problems with the crop insurance program and to make recommendations for "such changes as needed to improve the program so as to lessen, if not eliminate, the need for additional disaster payment programs..." (PL 100-546). The Commission's report was released in July 1989 and included both administrative recommendations and proposed legislative changes to improve participation rates. The Commission proposed that subsidy levels for 75 percent coverage be increased from 13 percent to 30 percent and additional subsidies for 65 percent coverage be temporarily increased from 30 percent to 50 percent. For areas where participation was low, they recommended that coverage levels be increased to 80 percent of yield. The Commission contended that higher coverage levels and greater premium subsidies would increase participation and hence greatly lessen, if not preclude, the need for supplemental disaster assistance.

Congress and the Bush Administration rejected these recommendations as too costly. In its 1990 farm bill and fiscal year (FY) 1991 budget proposals, the Bush Administration proposed eliminating the federal crop insurance program altogether and replacing it with a standing disaster program that would be triggered by area losses (Gardner 1994). However, the proposal received little interest in Congress. Likewise other crop insurance reform bills fared poorly. Congressman Glenn English proposed offering producers free 50 percent yield coverage, but his proposal was rejected as too costly. In the end, the 1990 farm bill included a number of minor changes to the crop insurance program, but no major reforms.

There was no disaster legislation in 1990, but supplemental appropriation legislation enacted in November 1991 authorized a total of $995 million for crop losses in either 1990 or 1991. Because claims exceeded funding, a factor of 50 percent was used to prorate payments. The supplemental appropriation legislation also authorized $775 million for crop losses in any one of the three years 1990, 1991, or 1992 for which producers had not already requested assistance. GAO reported that of the payments that went to producers growing insurable crops, about half went to those who had not purchased insurance (U.S. GAO 1994).

In Spring 1993, floods throughout most of the Midwest caused extensive crop damage. In response, Congress enacted legislation that provided $2.5 billion in disaster payments in FY 1994. Pressure mounted again for crop insurance reform. In August 1993, Department of Agriculture Secretary Espy convened a forum to discuss possible changes to the crop insurance program, and in its FY 1995 budget, the Clinton Administration proposed additional subsidies to increase participation. During previous years, such proposals were rejected because of the tight budget constraints on agricultural spending. The Clinton Administration proposed that the budget baseline be increased to reflect the reality that disaster payments had been provided through supplemental appropriations for every crop year since 1988. The Senate Budget Committee agreed. The FY 1995 budget baseline, in the absence of crop insurance reform, assumed that disaster payments averaging $1 billion per year would continue for the 1994 and subsequent crops. With an additional $1 billion to fund reforms, Congress debated and passed the Crop Insurance Reform Act of 1994.

Under the 1994 Act, producers of insurable crops were eligible to receive a basic level of coverage, catastrophic risk protection (CAT), which initially covered 50 percent of a producer's approved yield at 60 percent of the expected market price. Producers who elected coverage levels of at least 65 percent of the approved yield at 100 percent of the expected market price were eligible for a subsidy equal to the premium rate for a policy guaranteeing 50 percent of the approved yield at 75 percent of the expected market price. CAT coverage was required for producers who participated in the commodity price support and production adjustment programs, farm credit, or certain other farm programs (so-called linkage). While the premium cost of CAT coverage was fully subsidized by the government, producers were required to pay a sign-up fee equal to $50 per crop per county.

At the time of enactment, about 80 percent of the eligible acreage was expected to be enrolled in the program (USDA 1995). While most of the initial increase in participation was expected to be at the CAT level, more purchases of higher coverage levels were anticipated as producers became familiar with the program and realized that the government intended to rely on crop insurance as a substitute for the ad hoc assistance provided for free in the past.

A record 220 million acres were enrolled in the program in 1995, over 80 percent of eligible acres, with over half of these at the CAT level. Over 105 million acres were enrolled at the buy-up coverage levels, also a record high. However, over time, the CAT program proved unpopular among many producers who believed that the value of the CAT coverage was not worth the nominal administrative fee. As a result, Congress modified the linkage requirement in the Federal Agriculture Improvement and Reform Act of 1996. For the 1996 and subsequent crop years, producers could forego CAT coverage by waiving their eligibility for any other emergency crop loss assistance. Participation in the CAT program dropped sharply. In 1996, there were 87 million acres enrolled in CAT coverage, a drop from the previous year of almost 25 percent. By 1998, acreage enrolled in CAT had fallen to less than 60 million acres, a decline of 49 percent from 1995. Acres insured at the buy-up levels increased to 120 million acres, but not enough to offset the decline in CAT acres.

Under pilot authorities of the 1994 Act, revenue insurance products were first introduced for the 1996 crop year. By 1998, three revenue insurance programs (Crop Revenue Coverage, Income Protection, and Revenue Assurance) were offered to producers in selected locations for selected crops. Two more (Gross Revenue Income Protection and Adjusted Gross Revenue) were approved for sale in 1999. In 1998, almost 12 million acres of corn, 10 million acres of soybeans, and 5 million acres of wheat were enrolled in revenue insurance programs. Of total buy-up enrollment, revenue insurance accounted for over 32 percent of the corn acreage, 35 percent of the soybean acreage, and about 16 percent of the wheat acreage. In Iowa, over 50 percent of the total corn acreage enrolled in buy-up was enrolled under a revenue insurance plan.

The 1994 Act was temporarily successful in preventing disaster legislation. No disaster aid was provided for the 1994-97 crops. Market prices were high and crop conditions generally favorable over the period, with the exception of quality problems experienced by Northern Plains wheat producers and a drought in 1995 affecting Southern Plains winter wheat producers.

But the dam burst again in 1998, following a wet spring in California and a drought in Texas and much of the Southeast. In late 1997, prices for many field crops declined significantly, causing a drop in farm income for many producers. Crop insurance was criticized by many as providing an inadequate safety net (Chite 2000). Supplemental legislation enacted in late 1998 provided $5.8 billion in emergency spending for producer assistance, including $2.4 billion for emergency financial assistance to farmers who suffered crop losses due to natural disasters in 1998 or who suffered crop losses in three of the previous five years. Despite signing waivers indicating that they would be ineligible for disaster payments, producers who had declined insurance coverage were made eligible for disaster assistance.

Of the total disaster payments, $400 million was provided as incentive payments to farmers who purchased buy-up coverage for their 1999 crops. The subsidy reduced farmer premiums by an estimated 30 percent. Reduced costs for buy-up insurance led to widespread increases in participation in 1999. Buy-up acreage increased in nearly every state, and climbed nationwide from 120 million acres in 1998 to 143 million in 1999 (Dismukes and Glauber 2000). For the first time since the 1980 Act, participation in higher coverage levels exceeded 50 percent. Almost three-quarters of all eligible acreage was enrolled in either CAT or buy-up coverage.

The increased participation did not stop Congress from passing supplemental disaster legislation again in 1999. The Agriculture Appropriations Act of October 22, 1999, included $1.2 billion for disaster relief and $400 million for additional crop insurance subsidies for the 2000 crop year. One month later, Congress passed the Consolidated Appropriations Act that added another $576 million for assistance to repair and replace crops, buildings, and land damaged by Hurricane Floyd.

After having spent nearly $4 billion in disaster assistance over the previous two years, Congress and the Administration again decided that the crop insurance program was in need of further reform and larger subsidies. CAT polices were criticized as providing inadequate coverage. By 1999, the maximum CAT indemnity that would be paid out in the event of a total crop failure was only 28 percent of the producer's expected revenue. Because the premium subsidies were fixed dollar amounts, the subsidy share of total premiums declined as coverage increased beyond 65 percent. For example, under the 1994 Act, premiums for 65 percent coverage were subsidized at 42 percent; for 75 percent coverage, 24 percent; and for 85 percent coverage, 13 percent. Because the increase in out-of-pocket costs for higher coverage levels was so large, most producers who purchased buy-up coverage chose 65 percent coverage. In 1997, fewer than 10 percent of the acres enrolled in buy-up coverage were enrolled at coverage levels higher than 65 percent. With the higher subsidies following the 1998 disaster legislation, producers increased coverage levels. In 1999, almost 25 percent of the area was at coverage levels higher than 65 percent.

To encourage participation at higher levels of coverage, several reform proposals emerged in the Senate and House that would continue the increased subsidies at higher coverage levels. Some critics voiced concern over new crop insurance spending, since the 1994 reforms had not prevented ad hoc disaster assistance (Chite 2000). Despite these concerns, Congress adopted in its FY 2000 budget resolution a reserve fund of $6 billion over a multi-year period to fund the added costs of reforms. The FY 2001 budget resolution increased the amount available for new crop insurance spending from $6 billion over 4 years to $8.2 billion over 5 years. The Agricultural Risk Protection Act was signed by the President on June 20, 2000.

The Agricultural Risk Protection Act increased the size of premium subsidies for most buy-up levels (Table 2). For example, premium subsidies for 75 percent coverage increased from 24 percent under the 1994 Act to 55 percent under the 2000 Act. Subsidies for 85 percent coverage rose from 13 percent to 38 percent. In addition to increased subsidy rates, the new legislation authorized a pilot livestock insurance program, allowed producers to improve their yield guarantees despite multi-year losses, and encouraged private-sector development of new types of insurance products.

Table 2. Premium Subsidy Rate – Multiple Peril Crop Insurance

Coverage Level[a]	Federal Crop Insurance Act of 1980	Federal Crop Insurance Reform Act of 1994	Agricultural Risk Protection Act of 2000
55/100	30.0	46.1	64.0
65/100	30.0	41.7	59.0
75/100	16.9	23.5	55.0
85/100	---	13.0	38.0

[a] Percent of expected yield/percent of expected price.

Early indications suggest that producers of 2001 crops are purchasing higher levels of coverage and moving to the more expensive revenue products (USDA 2001b). It remains problematic whether the additional subsidies will lead to higher participation and prevent Congress from passing future ad hoc disaster legislation. Four months after Congress passed the Agricultural Risk Protection Act, the FY 2001 Agricultural Appropriations Act provided about $1.7 billion for crop disaster losses during crop year 2000.

THE COSTS OF CROP INSURANCE

From 1981-99, the costs of the federal crop insurance program totaled more than $15 billion (Table 3). Total disaster payments were $16.5 billion over the same period. Crop insurance costs have grown proportionately with increased premium subsidies and higher program participation. Over the period 1990-94, crop insurance costs averaged $711 million per year. The 1994 Act roughly doubled the costs of the program to $1.5 billion, on average, over 1995-99. With passage of the Agricultural Risk Protection Act of 2000, the annual program costs are expected to double again to $3.1 billion in 2001 (USDA 2001a).

In addition to premium subsidies, the costs of crop insurance include total excess losses (indemnities minus total premium costs, i.e., including premium

Table 3. Federal Costs of the Crop Insurance Program, 1981-99 (million dollars)

Crop Year	Premium Subsidy	Excess Losses	Delivery Costs	Underwriting Gains	Total
1981	47.0	30.5	0.0	0.3	77.8
1982	90.7	133.5	18.5	2.6	245.3
1983	64.0	296.8	26.2	(2.4)	384.6
1984	98.6	203.4	75.7	(0.4)	377.3
1985	100.3	242.9	107.3	3.4	453.9
1986	88.4	234.7	101.3	8.0	432.4
1987	87.9	3.4	107.0	16.7	215.0
1988	108.3	630.9	154.7	(8.0)	885.9
1989	206.6	402.9	265.9	28.1	903.5
1990	215.6	133.6	271.6	52.2	673.0
1991	190.3	216.4	245.2	42.0	693.9
1992	197.0	158.0	246.0	22.6	623.6
1993	200.3	898.0	249.8	(82.5)	1,265.6
1994	255.0	(349.6)	291.7	104.4	301.5
1995	889.6	29.7	373.1	130.9	1,423.3
1996	983.6	(321.7)	490.4	245.8	1,398.1
1997	903.3	(785.8)	450.3	352.5	920.3
1998	948.6	(199.3)	426.9	283.2	1,459.4
1999	1,352.6	109.5	500.0	280.4	2,242.5

Source: U.S. Department of Agriculture (2001a)

subsidies) and delivery expenses (reimbursement to the reinsured companies plus net underwriting gains through the Standard Reinsurance Agreement). During the 1980s, rapid program expansion and widespread droughts in the Midwest led to an annual average of almost $500 million in excess losses. During the 1990s, the actuarial performance of the program improved considerably. Total premiums (including premium subsidies) exceeded total indemnities over the period 1990-99. Delivery expenses accounted for less than 25 percent of total program costs during the 1980s. Companies received cost reimbursements for delivering crop insurance but shared in very little of the underwriting risks with the government. At the prodding of Congress and GAO, USDA increased the risk sharing in the Standard Reinsurance Agreement (SRA). Favorable actuarial performance led to large underwriting gains through the SRA. By the late 1990s, companies were earning on average over $250 million annually in underwriting gains. These gains plus expense reimbursements averaged almost 50 percent of total program costs from 1990-99.

Because the demand for crop insurance is generally inelastic with respect to premium (Goodwin 1993, Knight and Coble 1997), the marginal per-acre costs of enrolling additional acres into the program is high. Consider the current market for crop insurance. In 1999, roughly 143 million acres out of 270 million eligible acres were enrolled at buy-up levels of coverage for a participation rate of about 53 percent. The average premium for buy-up acres in

1999 was $14.01 per acre, with an average producer payment of $6.43 per acre and an average premium subsidy of $7.58.

To increase the buy-up participation rate to 65 percent, 31.5 million additional acres would need to be enrolled, an increase of 22 percent. Assuming a demand elasticity of -0.6, the premium paid by producers would have to decline by roughly 37 percent, to $4.08 per acre. The per-acre subsidy would increase to $9.93 per acre, an increase of $2.35 per acre. However, since the additional premium subsidies would apply to all buy-up acres, total additional premium subsidy costs would increase by $650 million ($2.35 x 143.4 million acres + $9.93 x 31.5 million acres). The marginal cost per additional acre would be $20.65. Marginal cost would increase the more inelastic the demand or the higher the target level of participation. For example, assuming a demand elasticity of -0.3 and assuming an increase in buy-up participation to 75 percent, additional program costs would likely exceed $2.2 billion, with a marginal per-acre cost of $37.74. While some of these costs would be offset by decreases in CAT participation, delivery costs (expense reimbursements and underwriting gains) would increase the marginal costs to $45 per acre.

THE EFFECTS OF CROP INSURANCE ON PRODUCTION

Proponents of federal crop insurance argued that crop insurance was a less costly means of providing crop loss protection than disaster assistance. As has been shown in previous sections, however, it is clear that crop insurance has been costly, both from the standpoint of the high marginal costs to encourage additional participation in the program and from the fact that, despite large subsidies and participation rates above 50 percent, Congress has continued to pass supplemental disaster assistance.

However, the growth in the level of crop insurance subsidies raises additional concerns. At time of passage of the Federal Crop Insurance Act of 1980, proponents of crop insurance argued that crop insurance was less likely than disaster assistance to encourage moral hazard problems and to encourage production of riskier crops on marginal lands – criticisms of the disaster payment program of the 1970s. Per-acre crop insurance premium subsidies have grown from an average $2 per acre in the mid-1980s to over $7 per acre in 1999. At the same time, deductibles have been lowered. In 1999, USDA introduced 85 percent yield coverage for many crops and regions. Under the 2000 Act, policies at 85 percent yield or revenue coverage will receive a 38 percent premium subsidy. The sharp increase in total subsidy costs has raised concerns that crop insurance may be distorting production decisions.

Over the past few years, there has been some anecdotal evidence that suggests that the crop insurance program has increased production of certain crops. For example, in 1999, North Dakota durum wheat producers were offered a revenue insurance policy that was based on a biased formula for establishing the revenue guarantee. As a result, sales of the contract increased significantly and it is estimated that durum wheat area increased by as much as 1 million acres (25 percent) over what would have been planted in the absence of the policy (Glauber 1999). After the 1999 crop year, watermelon producers asked USDA to terminate a pilot insurance program for watermelons after a large increase in watermelon area (USDA 1999). Concerns about the effects of revenue insurance on plantings prompted the potato growers to lobby successfully to have potatoes excluded from revenue coverage under the 2000 Act (Thompson 2000). Cotton growers have raised concerns that the added land provisions of the crop insurance program combined with increased subsidies have caused cotton producers in the mid-South to increase 2001 crop cotton plantings (Robinson 2001).

Soule, Nimon, and Mullarkey (2000) provide a review of the empirical literature on the effects of crop insurance on production. The literature suggests that for a risk-averse producer, insurance causes lower use of risk-reducing production inputs. Although this result is supported in a number of studies, there is not a consensus, as results depend on the extent to which farmers are risk averse and whether inputs are risk-reducing. However, many inputs that reduce risk, such as pesticides and fertilizer, are probably output-increasing over a wide range of likely conditions. Thus, by reducing such inputs, crop insurance would usually be output-reducing. Many of the studies that examined input use were theoretical, or if empirical, applied to narrow geographic regions and specific crops.

Other studies, mostly empirical, have examined the effects insurance has on the mix and level of crops planted. The general conclusion is that more, higher-risk acreage is brought into production, although studies differ on the magnitude of the expansion. For example, Wu (1999) concluded that crop insurance for corn causes increased corn-planted area as producers shift from other crops into corn. Young et al. (1999) modeled crop insurance subsidies paid during 1995-98 as price enhancements in a national multiple-crop simulation model. They concluded that acreage of 6 of 8 major insured crops rises in response to insurance subsidies. Raising subsidies from zero to about $1.4 billion results in a net increase of 600,000 acres in planted area, about 0.2 percent, and results in market price reductions of less than one percent for most commodities. Young et al. offer reasons why their results may be understated, such as using national data instead of individual producer data, although their inclusion of administrative and other non-premium subsidies may overstate the price enhancement a producer sees when making the planting decision.

Keaton, Skees, and Long (1999) estimated a much larger effect on planted area. They modeled six crops at the crop-reporting district level and compared the effects on planted acreage of the disaster payment regime of 1978-82 with crop insurance during 1988-92. They concluded that a 10 percentage point increase in the crop insurance participation rate is associated with increased planted area of the six crops of 5.9 million acres, and that 50 percent of insurable acres participating in insurance would increase planted area by some 50 million acres, a 20 percent increase from 1978-82 planted area. The

Young et al. (1999) and Keaton, Skees, and Long (1999) results are substantially different in magnitude. The Keaton, Skees, and Long results may be overstated by including Conservation Reserve Program acreage and land idled under annual programs, which increased sharply between the two time periods the study compared.

An econometric study by Goodwin and Vandeveer (2000) used a multi-equation structural model of corn and soybeans in the Midwest. They concluded that a 50 percent decrease in premiums, with a concomitant increase in loss ratios, would raise participation sharply – about 30 percent for corn and about 20 percent for soybeans – but that planted acreage rises much less – 2.2 to 3.3 percent. Orden (2001) summarized several studies, concluding that the effect on overall crop production of crop insurance subsidies during 1998-2000 was to increase output by 0.28 to 4.1 percent.

The upshot of these and a number of other studies is that whether to use crop insurance premium subsidies as a key part of the farm safety net is much like evaluating price and income support subsidies that are tied to production. Insurance subsidies distort farmer planting decisions, causing more production of the subsidized crops. The result is a reduction in market prices that offsets the income benefits of insurance subsidies, reducing the efficiency of insurance subsidies as a means of supporting income.

The extent to which subsidy premiums are offset by lower market prices is an empirical issue. Table 4 provides a general indication of the price effects derived using Orden's (2001) range of the general crop production effects due to insurance, and the acreage effects of Goodwin and Vandeveer (2000) under their scenario of a 50 percent decrease in premium cost for corn and soybeans and a consequent increase in the loss ratio. Two alternative price elasticities of demand are used for corn, -0.3 and -0.6. The market price declines range from $0.01 to $0.25 per bushel, depending on elasticity assumptions. This is roughly similar to an example of Babcock and Hart (2000) that concluded that a range of price changes due to elimination of crop insurance subsidies for corn is $0.02 to $0.16 per bushel.

Table 4. Change in Corn Market Price Due to Subsidization[a]
(dollars per bushel)

Assumed Production Increase (Percent)	**Assumed Elasticity of Demand** -0.3	-0.6
0.28 (Orden 2001 – low end)	-0.02	-0.01
2.20 (Goodwin and Vandeveer 2000 – low end)	-0.13	-0.07
3.30 (Goodwin and Vandeveer 2000 – high end)	-0.20	-0.10
4.10 (Orden 2001 – high end)	-0.25	-0.12

[a] The change is calculated using the 1999/2000 average corn farm price of $1.82 per bushel.

For the 1999 corn crop for participants in crop insurance, the premium subsidy for all types of policies averaged $0.05 per bushel (assuming a yield of 125 bushels per acre on insured acres). This was 54 percent of the total premium. Hence, one can conclude that the price effect may well fully offset the income benefit of the premium subsidy for insured producers. In addition, 77.4 million acres of corn were planted in 1999 but only 52.4 million were insured, so production on 25 million acres likely faced a reduced price due to insurance, without receiving any benefit of premium subsidies.

CONCLUSIONS

Many have heralded the U.S. crop insurance program as a model for other countries (e.g., Smith 2001). In recent years, a number of countries and the European Union have considered implementing similar programs (European Commission 2001, Meuwissen 2000). Yet before embarking on such a path, countries would be wise to examine the U.S. experience over the past 20 years.

Twenty years after passage of the Federal Crop Insurance Act of 1980, the debate continues over whether crop insurance will ever become the sole means of providing crop loss protection to U.S. producers. Participation in the crop insurance program has grown to a point where more than half of eligible acres are enrolled at buy-up levels and another 25 percent are enrolled at catastrophic coverage levels. Yet despite the steady increase in participation, Congress continues to pass supplemental disaster legislation to aid producers suffering from crop losses. Moreover, the marginal costs of enrolling more acreage in the program are high and increasing, suggesting that the costs of increasing buy-up participation rates from 50 to 75 percent will be far higher than increasing participation from 25 to 50 percent.

As subsidies increase and deductibles are lowered, the effects of crop insurance on acreage planting decisions become a greater concern. During the debates in the late 1970s and 1980s, proponents of crop insurance, such as the GAO, argued that crop insurance was preferable to disaster assistance because disaster assistance provided essentially free coverage to producers. It is an empirical issue whether a fully subsidized policy at 60 percent yield coverage causes more or fewer distortions than a partially subsidized policy that provides higher levels of protection. There is concern on the part of some farmers that crop insurance subsidies are too high and are resulting in increased acreage and lower crop prices. That these concerns will result in future legislation to reverse the level of these subsidies seems unlikely; nonetheless, the effects of these subsidies on acreage and program costs will likely draw increased scrutiny from Congress and farmers.

REFERENCES

Ahsan, S.M., A.A.G. Ali, and N.J. Kurian. 1982. "Toward a Theory of Agricultural Insurance." *American Journal of Agricultural Economics* 64: 520-529.

Appel, D., R.B. Lord, and S. Harrington. 1999. "The Agricultural Research, Extension and Education Reform Act of 1998, Section 535 Crop Insurance Study." Prepared by Milliman and Robertson Inc, July 23.

Babcock, B., and C. Hart. 2000. "A Second Look at Subsidies and Supply." *Iowa Ag Review* 6: 3.

Bardsley, P., A. Abey, and S. Davenport. 1984. "The Economics of Insuring Crops Against Drought." *Australian Journal of Agricultural Economics* 28:1-14.

Benedict, M.R. 1953. *Farm Policies of the United States, 1790-1950.* New York: The Twentieth Century Fund.

Chambers, R.G. 1989. "Insurability and Moral Hazard in Agricultural Insurance Markets." *American Journal of Agricultural Economics* 71: 604-616.

Chambers, R.G., and J. Quiggin. 2001. "Decomposing Input Adjustments Under Price and Production Uncertainty." *American Journal of Agricultural Economics* 83: 20-34.

Chite, R.M. 1988. "Federal Crop Insurance: Background and Current Issues." Congressional Research Service Report No. 88-739 ENR (December).

___. 1989. "Crop Insurance Reform: A Review of the Commission Recommendations." Congressional Research Service Report No. 89-624 ENR (November).

___. 2000. "Federal Crop Insurance: Issues in the 106th Congress." Congressional Research Service Issue Brief IB10033 (June).

Coble, K.H., T.O. Knight, R.D. Pope, and J.R. Williams. 1997. "An Expected-Indemnity Approach to the Measurement of Moral Hazard in Crop Insurance." *American Journal of Agricultural Economics* 79: 216-226.

Cutler, D.M., and R.J. Zeckhauser. 1999. "Reinsurance for Catastrophes and Cataclysms." In K.A. Froot, ed., *The Financing of Catastrophe* Risk. Chicago: University of Chicago Press.

Dismukes, R., and J. Glauber. 2000. "Crop and Revenue Insurance: Premium Discounts Attractive to Producers." *Agricultural Outlook.* AGO-269 (March), pp. 4-6.

Duncan, J., and R.J. Myers. 2000. "Crop Insurance Under Catastrophic Risk." *American Journal of Agricultural Economics* 82: 842-855.

Dyson, L.K. 1988. "History of Federal Drought Relief Programs." USDA, Economic Research Service, Staff Report No. AGESS880914 (October).

European Commission. 2001. "Risk Management Tools for EU Agriculture: With a Special Focus on Insurance." Agriculture Directorate-General, January.

Froot, K.A. 1999. "Introduction." In K.A. Froot, ed., *The Financing of Catastrophe Risk.* Chicago: University of Chicago Press.

Gardner, B.L. 1994. "Crop Insurance in U.S. Farm Policy." In D.L. Hueth and W.H. Furtan, eds., *Economics of Agricultural Crop Insurance: Theory and Evidence.* Boston: Kluwer Academic Publishers.

Gardner, B.L., and R.A. Kramer. 1986. "Experience with Crop Insurance Programs in the United States." In P. Hazell, C. Pomerada, and A. Valdez, eds., *Crop Insurance for Agricultural Development: Issues and Experience.* Baltimore: Johns Hopkins University Press.

Glauber, J.W. 1999. "Declaration in the U.S. District Court for the District of North Dakota, Southeastern Division." Testimony given in *Paul Wiley et al. v. Daniel Glickman*, Civil No. A3-99-32 (May 18).

Goodwin, B.K. 1993. "An Empirical Analysis of the Demand for Multiple-Peril Crop Insurance." *American Journal of Agricultural Economics* 75: 425-434.

Goodwin, B.K., and V.H. Smith. 1995. *The Economics of Crop Insurance and Disaster Aid.* Washington, D.C.: The AEI Press.

Goodwin, B.K., and M. Vandeveer. 2000. "An Empirical Analysis of Acreage Distortions and Participation in the Federal Crop Insurance Program." Paper presented at the workshop, "Crop Insurance, Land Use, and the Environment," USDA, Economic Research Service, Washington, D.C., September 20-12.

Harwood, J., R. Heifner, K. Coble, J. Perry, and A. Somwaru. 1999. "Managing Risk in Farming: Concepts, Research, and Analysis." USDA, Economic Research Service, Agricultural Economic Report No. 774 (March).

Hazell, P., C. Pomareda, and A. Valdez. 1986. "Introduction." In P. Hazell, C. Pomareda, and A. Valdez, eds., *Crop Insurance for Agricultural Development: Issues and Experience.* Baltimore, MD: Johns Hopkins University Press.

Innes, R., and S. Ardila. 1994. "Agricultural Insurance, Production and the Environment." In D.L. Hueth and W.H. Furtan, eds., *Economics of Agricultural Crop Insurance: Theory and Evidence.* Boston, MA: Kluwer Academic Publishers.

Johnson, J. 1980. "Alternatives and Congressional Action in Developing the Food and Agriculture Act of 1977." *Agricultural-Food Policy Review*, ESCS-AFPR-3. USDA, Economic, Statistics, and Cooperatives Service (February).

Just, R.E., L. Calvin, and J. Quiggin. 1999. "Adverse Selection in Crop Insurance." *American Journal of Agricultural Economics* 81: 834-849.

Keeton, K., J. Skees, and J. Long. 1999. "The Potential Influence of Risk Management Programs on Cropping Decisions." Selected paper presented at the annual meeting of the American Agricultural Economics Association, August 8-11, Nashville, TN.

Kleindorfer, P.R., and H.C. Kunreuther. 1999. "Challenges Facing the Industry in Managing Catastrophic Risks." In K.A. Froot, ed., *The Financing of Catastrophe Risk.* Chicago: University of Chicago Press.

Knight, T.O., and K.H. Coble. 1997. "Survey of Multiple Peril Crop Insurance Literature Since 1980." *Review of Agricultural Economics* 19: 128-156.

Kramer, R.A. 1983. "Federal Crop Insurance." *Agricultural History* 97: 181-200.

Meuwissen, M.P.M. 2000. "Insurance as a Risk Management Tool for European Agriculture." Ph.D. thesis, Wageningen University, The Netherlands.

Miller, T.A., and A.S. Walter. 1977. "Options for Improving Government Programs that Cover Crop Losses Caused by Natural Hazards." USDA, Economic Research Service. ERS No. 654 (March).

Miranda, M.J., and J.W. Glauber. 1997. "Systemic Risk, Reinsurance, and the Failure of Crop Insurance Markets." *American Journal of Agricultural Economics* 79: 206-215.

Nelson, C.H., and E.T. Loehman. 1987. "Further Toward a Theory of Agricultural Insurance." *American Journal of Agricultural Economics* 69: 523-531.

Orden, D. 2001. "Should There Be a Federal Income Safety Net?" Paper presented at the Agricultural Outlook Forum 2001, Washington, D.C., February 22.

Porter, J.M. 1988. "Drought in the United States: A Short History." USDA, Economic Research Service, Staff Report No. AGES881020 (December).

Robinson, E. 2001. "Cotton Insurance: Loophole for Abuse?" *Delta Farm Press*, March 16. Available at http://industryclick.com.

Rowe, W.H., and L.K. Smith. 1940. "Crop Insurance." *1940 Yearbook of Agriculture: Farmers in a Changing World.* U.S. Department of Agriculture.

Skees, J.R., and B.J. Barnett. 1999. "Conceptual and Practical Considerations for Sharing Catastrophic/Systemic Risks." *Review of Agricultural Economics* 21: 424-441.

Skees, J.R., P. Hazell, and M. Miranda. 1999. "New Approaches to Crop Yield Insurance in Developing Countries." International Food Policy Research Institute. EPTD Discussion Paper No. 55 (November).

Smith, D.R. 2001. "The Public/Private Risk Management Partnership." Paper presented at the Agricultural Outlook Forum 2001, Washington, D.C., February 22.

Soule, M., W. Nimon, and D. Mullarkey. 2000. "Risk Management and Environmental Outcomes: Framing the Issues." Paper presented at the workshop, "Crop Insurance, Land Use, and the Environment," Economic Research Service, Washington, D.C., September 20-12.

Thompson, J. 2000. "Revenue Insurance Distorting Market Prices." *Potato Grower* (December). Available at http://www.potato.grower.com.

U.S. Department of Agriculture. 1995. *FY 1996 Budget Summary* (February).

___. 1999. "USDA to Suspend Pilot Watermelon Crop Insurance Program." Risk Management Agency. Press release, September 12. Available at http://www.act.fcic.usda.gov/news/pr.

___. 2001a. *FY 2002 Budget Summary* (April).

___. 2001b. "Revenue Insurance, 70/75 Sales Skyrocket." Risk Management Agency. Available at http://www.act.fcic.usda.gov/news on April 24.

U.S. General Accounting Office. 1980. "Federal Disaster Assistance: What Should the Policy Be?" GAO/PAD-80-39 (June).

___. 1989. "Disaster Assistance: Crop Insurance Can Provide Assistance More Effectively than Other Programs." GAO/RCED-89-211 (September).

___. 1993. "Crop Insurance: Federal Program Has Been Unable to Meet Objectives of 1980 Act." Testimony of J.W. Harman, Director, Food and Agriculture Issues, Resources, Community and Economic Development Division, before the Subcommittee on Rural Development, Agriculture and Related Agencies, Committee on Appropriations, U.S. House of Representatives. GAO/T-RCED-93-12 (March 3).

___. 1994. "Disaster Assistance: Problems in Administering Agricultural Payments." Testimony of William E. Gahr, Associate Director, Food and Agriculture Issues, Resources, Community and Economic Development Division, before the Committee on Agriculture, Nutrition and Forestry, U.S. Senate. GAO/T-RCED-94-187 (April 13).

___. 1999. "Agriculture in Transition: Farmers' Use of Risk Management Strategies." GAO/RCED-99-90 (April).

Valgren, V.N. 1922. "Crop Insurance: Risks, Losses, and Principles of Protection." USDA, Bulletin No. 1043 (January 23).

Westcott, P.C., and C.E. Young. 2000. "U.S. Farm Program Benefits: Links to Planting Decisions and Agricultural Markets." *Agricultural Outlook*, AGO-275 (October).

Wright, B.D., and J.A. Hewitt. 1994. "All-Risk Crop Insurance: Lessons From Theory and Experience." In D.L. Hueth and W.H. Furtan, eds., *Economics of Agricultural Crop Insurance: Theory and Evidence*. Boston, MA: Kluwer Academic Publishers.

Wu, J. 1999. "Crop Insurance, Acreage Decisions, and Nonpoint-Source Pollution." *American Journal of Agricultural Economics* 81: 305-320.

Young, C.E., R.D. Schenpf, J.R. Skees, and W.W. Lin. 1999. "Production and Price Impacts of U.S. Crop Insurance Subsidies: Some Preliminary Results." Proceedings of the 1999 Federal Forecasters Conference.

Chapter 22

RISK CREATED BY POLICY IN AGRICULTURE

Bruce L. Gardner
University of Maryland, College Park

INTRODUCTION

The "Freedom to Farm" program of the 1996 FAIR Act has been criticized on the grounds that it provides inadequate measures to protect farmers from market price risks. The underlying assumption is that farm policy can and should reduce the risks farmers face by stabilizing prices and incomes. At the same time, as the 1995-96 debate leading up to the FAIR Act and the current debate about its continuation after 2002 show, policies themselves can prove unsustainable, can change unexpectedly, or can have unanticipated consequences that destabilize the farm economy. This chapter considers the issue of risks stemming from uncertainties about policies. The topic has been much less investigated than the stabilization role of policies, and the discussion to follow will accordingly be preliminary and tentative. Nonetheless, I will try to bring some empirical evidence as well as conceptual treatment of the subject to bear on the issues.

BACKGROUND: SELECTED HISTORICAL EPISODES

The most famous cases of risks created by policies have involved those governments that have attempted to exert the broadest command over agriculture. The collectivization of Russian agriculture in the 1920s, Khrushchev's virgin lands program of the 1950s, China's Great Leap Forward of the 1960s, have all been widely acknowledged as disastrous errors that destabilized farming to the extent of causing serious food shortages and even the demise of thousands of farmers.

Democratic governments have tended to be removed from power before such drastic results could be achieved. Olson (2000) cites this as the primary reason why democracies have better records of economic growth. But even in countries with democratic institutions, there are many instances of governments being responsible for dramatic policy changes that have alternately led agriculture in developing countries to major spurts of capital investment, land clearing, and infrastructure development, only to see the returns fizzle and often the entire investment wiped out by policy reversals or inability to implement the policy announced. For detailed descriptions of notable examples in developing countries, see the five volumes of Krueger, Schiff, and Valdés (1991).

Even in cases where appropriate stabilization policies seem in principle straightforward to implement, such as stabilizing returns for a commodity sold into well-functioning international markets, politics can undo stabilizing ideas. A notable historical case is Alfred Marshall's analysis of Britain's mismanagement of the Corn Laws, under which a policy "designed to keep prices steady at a high level, made them very unsteady" (Marshall 1920, p. 749). A more recent set of examples involves marketing boards in many countries. The World Bank concluded that "the history of marketing board operations in Africa and elsewhere suggests that the stabilization objective can gradually give way to the objective of raising revenues at the expense of the producers" (World Bank, 1986, p. 90); and at the behest of many observers and interested parties, marketing boards have been abandoned or greatly reduced in power in most countries that have had them.

In international finance, the term "sovereign risk" refers to the prospect that a government will interfere with the repayment of debt to a foreign creditor. More broadly the term covers a wide range of defaults by governments.[1] Recent literature has treated this formally as the inability of a sovereign, by virtue of its sovereign powers, to precommit to actions. Baumol notes the emphasis given by Hicks (1969) to the lack of creditworthiness of powerful monarchs in history, and their consequent difficulty in obtaining loans, as an illustration that "the ability to undertake an enforceable commitment is itself a very valuable asset. But the king had no way he could commit himself. If he defaulted there was no higher court in which the debtor could sue and, as a consequence, kings were required to pay very high rates of interest" (Baumol 1990, p. 1712).

One way of avoiding sovereign risk is to carry out policies through international institutions where possible. However, the commodity policy area in which this approach has been most thoroughly developed, stabilization through International Commodity Agreements, has provided chastening experiences. International Commodity Agreements have utilized storage and export controls in attempts to stabilize prices of almost all the major internationally traded commodities during the post-World War II period. Despite backing and financing through the United Nations and many national govern-

[1] Firms offering insurance against such risks cover legal risks (e.g., confiscation of property, discriminatory legislation, failure to perform on contracts), market risks (e.g., devaluation of currency, trade embargoes), and credit risks. See for example the websites www.sovereignbermuda.com or www.intlfinanciallaw.com.

ments, by the mid-1990s all the agreements having storage or other controls intended to stabilize prices had been abandoned. The general assessment is that the agreements when operating left price variability "more or less unchanged" (Gilbert 1996, p. 16), but the stabilization mechanisms periodically collapsed in crises; and in cases where the agreements suddenly collapsed, as happened with the tin agreement in 1985, prices fell drastically. In the other three agreements that Gilbert considered (cocoa, sugar, and coffee), the average price fell by about 40 percent in the year following the agreement's lapse. This suggests that from producers' viewpoint the agreements not only were ineffective at stabilization, but in the long term increased risk in the sense that grower investments made under the assumption that the agreements would provide a stable market over the long term turned out to incur major losses.

Destabilizing policies have made unwelcome appearances within the United States, too. Following are details on notable instances.

Grain Price Supports

The Federal Farm Board was introduced to support and stabilize the grain markets in 1929. Following generally weak markets in the 1920s, U.S. production in 1929 was down from the preceding year, and projected ending stocks were lower by 100 million tons (Benedict 1953, p. 261). Yet prices were only slightly higher in July 1929 than they had been a year earlier.[2] This led the Board to believe wheat was undervalued and that a stabilization effort was warranted. The Board announced intentions to support the price in August 1929, and by October was offering loans to farmers to keep their crops off the market. This held prices steady until the end of 1929 (despite the stock market crash), but early in 1930 wheat prices began to fall sharply, mainly because export demand weakened. The \$1.30 price of July 1929 had declined to \$0.83 in July 1930 – a 41 percent drop in one year. Prices continued to decline and at the end of June 1931, when the six-market price was \$0.67, the Board decided it could no longer fight the market and "adopted a policy of liquidating its holdings with as little adverse effect on the market as possible" (Benedict 1953, p. 262). Whereupon, in July 1931, the price fell a further 30 percent to \$0.465 per bushel, a level at which prices remained for most of the following two years until the Board was liquidated in May 1933.

A similar story could be told about the Board's actions in cotton.

The Federal Farm Board has been cited as a policy failure in that it lost about \$300 million in storage operations while failing to stem the tide of falling prices. For our purposes the salient point is that the program was actively destabilizing. It held prices higher than they otherwise would have

[2] The six-market average price (Chicago, Minneapolis, Kansas City, St. Louis, Omaha, and Duluth) quoted by USDA was \$1.30 per bushel in July 1929 and \$1.26 in 1928. This and other data cited are from USDA 1933, p. 418.

been during a relatively high price period (1929-30) and depressed prices during a lower-price period (1931-32).

One could say, that was then, this is now, and today government agencies are better prepared to make appropriate choices in stabilizing markets. From the New Deal through the 1960s the system of acreage diversion and Commodity Credit Corporation (CCC) storage programs evolved and arguably provided improved stabilization services. Assessments such as Cochrane and Ryan (1976) have argued that reduced farmer risks under these programs were a significant stimulus to investment and productivity growth in U.S. agriculture after 1940. Even studies that have been critical of the benefits as compared to the costs of particular programs, such as the Farmer Owned Reserve Program, found a net stabilization effect (e.g., U.S. GAO 1981).

However, the record is mixed even with respect to more recent commodity policy. The big test for stabilization policy in the post-World War II era was the commodity price boom of the 1970s, initially triggered by the Soviet wheat purchases of the 1970s. Without prior expectation of significant purchases the Soviets bought a quantity of wheat equal to one-sixth of the U.S. crop during July and August 1972. This shock caused wheat prices to rise from $1.51 per bushel (Kansas City cash) in June to $2.09 in September. Futures prices for distant delivery dates rose correspondingly as Soviet purchases were considered likely into the next crop year and beyond. This situation was a good one for the stabilization capabilities of the U.S. wheat program to take action. As of the end of 1971, the Commodity Credit Corporation owned 371 million bushels of wheat, one-fifth of a year's output and the highest level since the mid-1960s. In 1972, 20 million acres of wheat base were idled in acreage set-asides. The stabilization tools available were to draw down CCC stocks and eliminate set-asides. Some stocks were sold, but in the fall of 1972 a set-aside program was again implemented for the 1973 crop. And even after Soviet purchases were known in August, USDA continued its wheat export subsidy program, further exacerbating the scarcity of wheat in the U.S. There were reasonable political arguments for continuing both export subsidies and set-asides, in that it was not certain that the Soviets would keep buying in 1973, and USDA judged it politically more costly to have prices fall after a transient high than to have prices rise even higher if, as expected, the Soviets continued to buy. In fact, world demand generally was high in 1973 and the Kansas City price that was $2.09 per bushel in September 1972 rose to $5.00 a bushel in September 1973. Thus U.S. wheat policy was destabilizing – it continued to boost prices when a true stabilization policy would have taken steps to reduce them. What is most notable here is that the government knew what it ought to do for stabilizing purposes, but for political reasons chose not to. (For further details see Schnittker 1973 or Sanderson 1975.)

The preceding destabilizing policies were not of course the primary cause of the commodity price boom of the 1970s. It is nonetheless arguable that commodity policy added to rather than buffered market shocks, even though the policies most criticized by producer interests were temporary embargoes on export sales to particular countries that were thought to have price-depressing effects.[3] The overall results affected buyers adversely in the first instance, but the long-term effect was quite likely destabilizing for farmers too. When set-asides were finally ended in 1974, and USDA was urging farmers to plant "fencerow to fencerow" (as Secretary of Agriculture Butz put it), farmers could reasonably suppose that conditions were appropriate for output expansion and investment. U.S. wheat area planted increased from 55 million acres in 1972 to 80 million in 1976. Prices duly plummeted.

The political element of stabilization policy also was apparent in the last implementation of an Acreage Reduction Program (ARP) for corn, a 7.5 percent reduction for the 1995 crop. The Secretary of Agriculture had authority under provisions of the 1990 Farm Act to set the ARP between zero and 7.5 percent given the projected stocks/use ratio. Administration economists as well as independent outside analysts tended to believe that demand prospects and projected stock levels called for a zero ARP as being best for stabilization purposes; but based on industry arguments that large supplies could be price depressing, the Secretary chose the largest ARP allowable. However, as expected, 1995 turned out to be a high-price, excess-demand year, and the ostensibly stabilizing ARP rules were in fact destabilizing. Again the issue turned not on information or expertise, but rather on interest-group politics.

Dairy Program

Dairy policy provides an example where a policy regime that has been generally accepted as price and income stabilizing has at the same time created considerable policy risk. Figure 1 shows the time series of prices received by farmers for milk from 1915 to 2000, put in real terms using the GDP deflator to obtain a constant-dollar measure over this long period. For purposes of comparison, the real price received by farmers for hogs is also shown.

The dairy price support program began in 1948. It is apparent that after that time milk prices varied less around their long-term trend (a 0.5 percent annual rate of decline), until after 1990, when price supports were reduced sufficiently to become largely ineffectual and prices began to fluctuate more. Note also that prices on average have been lower after 1990. Partly this

[3] In the aftermath of these events, the Economic Research Service of USDA carried out a study, at the mandate of Congress, "to determine the losses suffered by U.S. farm producers during the last decade as a result of embargoes..." (USDA 1986, p. ii). The study, drawing on many academic contributors as well as USDA Economic Research Service analysts, concluded that the embargoes had quite small effects, which were more than offset by other policies. This is a case however where opinions of the academic and grassroots farming communities remained far apart, and indeed the Secretary of Agriculture at the time (Richard Lyng) repudiated the findings of the study.

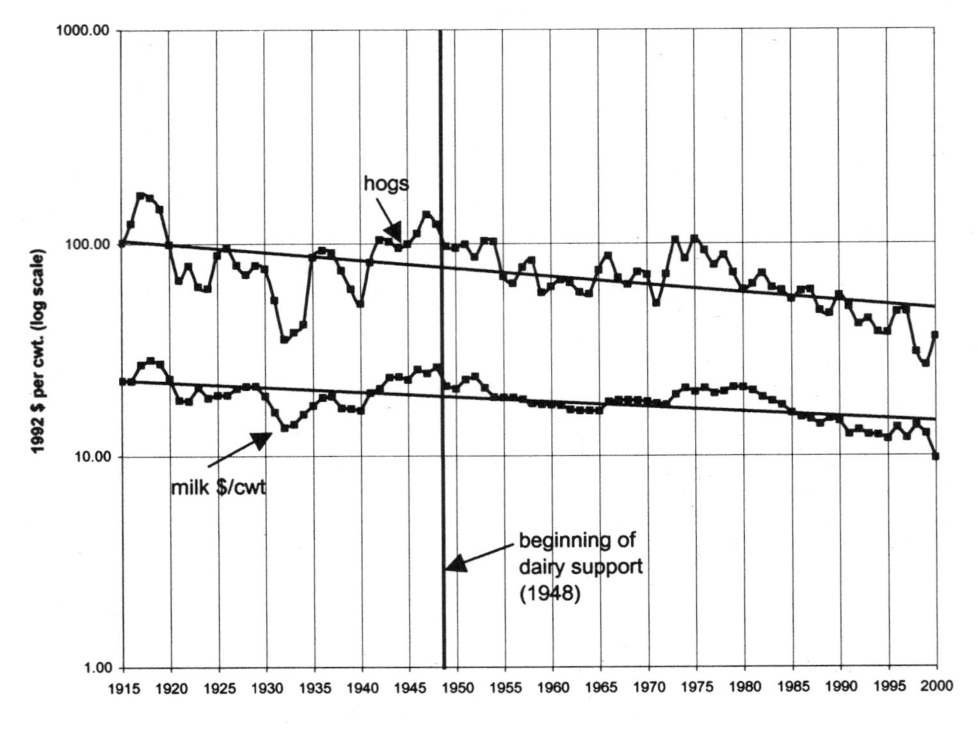

Figure 1. Farm Prices Received (Real), Milk and Hogs
Source of data: USDA and Historical Statistics of the United States

reflects lower real feed costs and reduced costs in large-scale operations – factors also at work in the even greater declines in hog prices. But it also suggests that dairy farmers, having placed their economic fate in the hands of a government program, became vulnerable to changes in that program even though the program itself, when functioning, was an effective stabilization measure.

The story is amplified by the simultaneous existence of the milk marketing order system, which regulates spatial price differences and differentials between fluid milk for drinking and milk used in manufactured dairy products. Both the 1990 and 1996 Farm Acts mandated reforms of the marketing order system, at the behest of producers in the upper Midwest, the demand for whose milk is reduced by regulations and spatial pricing that stimulate production elsewhere. The spatial element of pricing policy has been further intensified by the Northeast Dairy Compact, which gives the New England states a still further price advantage. Now other Northeastern states see the extension of this concept to cover their producers as a key requirement for their economic survival. This is just to say that the U.S. dairy industry did obtain stabilizing market policies, but at the price of substantial policy risk. In the end, it is not possible to say that dairying has proven economically more rewarding than hog production despite the much greater policy support that dairying has had.[4]

Emergency Loans, Crop Insurance, and Disaster Payments

Disaster and insurance programs are paradigms of risk-reducing policies, in that they provide benefits to farmers just when bad states of nature occur. Even here, however, policy risk can be important. One element is that farmers can never be sure whether, when, and with what conditionalities such benefits will be provided. Moreover, it is important to take into account the response of farmers to the existence of these programs. Glauber and Collins (this volume) review a number of studies that find significant effects of the current subsidized crop insurance program on farmers' choices of crops and acreage planted. In a case study that illustrates the resulting policy risk, Gardner and Kramer (1986) document the expansion of wheat acreage under protection provided by the Disaster Payments Program of the late 1970s in counties in Colorado and West Texas where production was so risky that the Federal Crop Insurance Corporation would not write crop insurance policies. The Disaster Payments Program regularized emergency assistance by essentially giving away crop insurance on terms specified in advance in legislation. This removed the ad hoc nature of typical emergency relief and thus could be seen as reducing policy risk. But the risk of an end to the program remained, and in fact Congress in the 1980 Farm Act chose not to reauthorize the program. Thus, those who adopted wheat production in the riskiest areas found themselves, so to speak, up a dried-up creek without a paddle.

Value of Production Quotas

Tobacco quotas provide a case study that permits measurement of the extent of policy risk as farmers perceive it. At various times and places, there have existed both rental and sale markets for tobacco quota, conveying the right to sell a pound of tobacco. These have operated under various restrictions, such as the buyer having to operate within the same county as the seller, but there were few such restrictions for flue-cured tobacco between 1982 and 1987. Sumner and Alston (1985) estimated the sale value of tobacco quota in North Carolina at that time to be 3 to 5 times its rental value (giving the right to sell tobacco in the current year only). Earlier evidence is consistent with

[4] Solid comparable evidence on the returns earned from production of each product are not available, but McBride (1995) for hogs and El-Osta and Johnson (1998) for dairy provide the most detailed assessments on a national scale, using data of the early 1990s. With free entry into production of both hogs and dairy it is hard to see how either enterprise could long sustain net advantages over the other.

these ratios, finding for example that in an area where quota rented for 30 cents a pound, quota sold for $1.20.

These data permit the measurement of policy risk, as indicated by the expected rate of depreciation of the tobacco program's benefits. The rental value of quota can be estimated under the assumption that willingness to pay for admission to the program measures the program's economic value. The sale value of quota is determined by current and expected future rental values as

$$V_1 = \sum_{t=1}^{\infty} R_1[(1-\delta)/(1+i)]^t \,, \tag{1}$$

where R_i is the rental value of quota in the current year. R is assumed to remain constant in future years except for depreciation at annual rate δ, and i is the constant riskless rate of interest, t is an index for years, and V_1 is the current value of the future stream of R's, i.e., the value of quota. With the expression inside square brackets less than one, the sum of the infinite series converges, and dividing by R_1 obtains[5]

$$V_1 / R_1 = (1+i)/(i+\delta) \,. \tag{2}$$

Using data relevant to tobacco quota in the mid 1980s, we have $V/R = 4$ and $i = 0.07$. Solving (2) for δ then gives $\delta = 0.1975$. That is, market data are consistent with an expectation that benefits from the tobacco program will depreciate at a rate of almost 20 percent annually. This result could also be expressed by saying that the reason an investor could buy tobacco quota and rent it out for a current return of 25 percent is that the market provides a risk premium (reflecting both the probability of future program decline and any risk aversion that exists) equal to (25 - 7 =) 18 percent. These data suggest substantial perceived policy risk, at a time when the tobacco program had been in place for almost half a century with no imminent threats to its future. Travena and Keller (1970), albeit with less direct market data, estimated a similar V/R ratio for cotton allotments in the 1960s.

Note that risk as measured here refers to expectation of loss rather than uncertainty of that expectation. However, δ could alternatively measure the probability that each year's rental value will be 0 rather than R. Under that interpretation the year-ahead expected value of R is $(1-\delta)R$, and risk as measured by the coefficient of variation of R is $[\delta(1-\delta)]^{.5}$.

As still another alternative, we could let $\delta = 0$ and ask, for $V = 4R$, what value of t preserves the equality in equation (1). The answer for $i = 0.07$ is just

[5] This assumes that the rental return in the first year is received at the same time the quota is purchased, so the discounted series has the form: $1 + x + x^2 + x^3 + \ldots$, which converges to $1/(1-x)$. Using $x = (1-\delta)/(1+i)$ gives equation (2).

less than 5 years. That is, the data are consistent with an expected future length of life of the program of less than 5 years, after which no support is expected.

Each of these approaches indicates that policy risk is real and important in the minds of market participants. Indeed, the premium on current returns to tobacco quota has a direct analogy to the high interest rates mentioned earlier that kings have had to pay because of sovereign risk. The sizes of both premia are indicators of the likelihood of default.

Macroeconomic Policy

Monetary and fiscal policies are typically treated as stabilizing rather than destabilizing, but as with farm policy mistakes and/or political expediency, can create risks too. The agricultural sector has long been seen as particularly susceptible to being whipsawed by effects of macro policies that have economy-wide targets but sometimes insufficiently appreciated sectoral consequences. A well-known example in U.S. history is the concern, classically expressed in the "Cross of Gold" speech of William Jennings Bryan, that monetary policies intended to reign in inflation keep real interest rates high and, it is alleged, farm prices low relative to nonfarm prices. At least through the 1940s, macroeconomic fluctuations were seen as a prime generator of economic instability in agriculture, a view developed in Schultz (1945). A further development of this reasoning was the argument of Schuh (1974) that monetary and exchange rate policies caused an overvalued U.S. dollar, which weakened export demand and hence U.S. farm incomes in the 1950s and 1960s.

The most recent and serious economic swings attributed to macroeconomic factors occurred in the late 1970s and 1980s. The inflation and low (even negative) real rates of interest of the 1970s encouraged highly leveraged buying of farm real estate that drove up asset prices unsustainably, the crash then triggered by the Federal Reserve's attack on inflation beginning in 1979. The Fed's tightening in this view generated the extraordinarily high real interest rates of the early 1980s, which coupled with declines in commodity prices made it impossible for many farmers to cover their debt service at a time when land values had declined well below prices paid to purchase the land, resulting in financial failures at a rate not seen since the Great Depression. Policy risk does not get much more serious than that.

Regulation

Since environmental concerns began to get embodied in legislation with the Clean Water Act of 1972 and the Endangered Species Act of 1973, farmers have expressed worry about the risks posed by mandatory regulation of their production practices or bans on practices or uses of particular inputs. It is not so much that draconian measures have actually been imposed, but rather the prospect that with the legal framework in place for doing so, such regulations might be imposed in the future. The impact of this policy risk is not primarily a matter of current production decisions, but rather occurs through farmers' decisions to invest (or not invest) in their operations.

An example is the regulation of poultry litter under a Maryland nutrient management law intended to reduce nitrogen and phosphorus entering ground and surface water from farmers' fields fertilized with poultry litter. The proposed regulations implementing this law require each farm to have a nutrient management plan and to apply poultry litter and other fertilizers in accordance with that plan. The regulations would require "co-permitting," with the broiler grower and the processor (or other "integrator" who writes a production contract with the grower) being jointly responsible for the nutrient management plan being carried out. Processors have argued that such legal liability makes them responsible for actions they cannot adequately monitor, and the potential costs of which would make Maryland non-competitive with other locations for contracted broiler growing. This in turn limits the industry's current undertaking of new broiler contracts, and farmers' opportunities to undertake investment in chicken houses and other fixed equipment.

CONCEPTUAL BASIS FOR ANALYSIS OF POLICY RISK

Analytically, the broiler grower's situation under policy uncertainty is an application of the "real options" approach to the theory of investment. In that theory, a decision to invest is the exercise of an option, the date of exercise being the time of financial commitment to the fixed capital stock involved. The greater the uncertainty about future market conditions over the lifetime of the capital stock, the greater the time value of the option, which is lost when the option is exercised. Therefore, for any given expected value of expected returns, greater long-term uncertainty causes less investment at any given time, *even if investors are risk neutral* (the classic reference is Dixit and Pindyck 1994). Policy uncertainty enters the picture by increasing the uncertainty of returns to investment. Thus policy uncertainty is expected to reduce investment, other things equal.[6]

We can also consider the decision to regulate as a public investment, with the relevant uncertainty being the payoff to the investment, for example uncertainty about the increase in water quality caused by a regulation of farmers' practices. If this uncertainty is sufficient, it is optimal to postpone a regulation for which expected benefits exceed expected costs, even for a risk-neutral government. Consider also the farmer as counterparty of the govern-

---

[6] An exceptional case occurs, however, if policy risk is correlated with pre-existing market risks in such a way that unfavorable policy outcomes tend to coincide with favorable market outlooks, and vice versa. Then uncertainty in policy could actually reduce the overall uncertainty in returns to investment.

ment's option to invest in a regulatory program. Possible future regulation under existing laws gives government agencies an option to impose costs. Farmers or others subject to regulation from this perspective are writers of an option, which they have to provide to the regulators (often without the premium that induces commodity market participants, for example, to sell options on futures; although agricultural regulations have typically provided some offsetting compensation). The expected loss on writing this option is another way of looking at what reduces the return to investment by farmers, and hence the level of investment in agriculture.

Consider a situation in which farmers make annual decisions about current input commitments and about longer-term investments. Their short-term managerial decision is the allocation of fixed land and labor among commodities and quantities of purchased inputs (which may include additional hired land and labor), and their long-term decisions are about what investment in equipment and improvements in their capital stock to make. They look at expected prices for the current year and longer-term expectations about market conditions, future technological change, and policies that may affect the profitability of both short-term production choices and longer-term investments.

Following are two polar extremes one might use in modeling policy changes in this context. First, farmers might take the policies as given, except that when new laws are passed or regulations are promulgated, farmers assume they will be implemented as described for an indefinite period. If Congress in 2002 passes a law that preserves the $17 billion Freedom to Farm and loan program payments of the last two crop years (as the Farm Bureau and most other farm groups recommend), and converts the fixed payments to deficiency payments with acreage restrictions, using annual acreage controls as necessary to keep surplus production under control and stabilize prices, farmers will assume this policy will continue indefinitely. Second, an alternative approach is to model farmers as disbelieving the permanence of new laws or regulations, and assuming they will go away after a few years or depreciate in value as discussed earlier with respect to the tobacco program.

What analytical difference does it make which view a farmer takes? For current production decisions, for most kinds of policies, probably little. The exception would be current production decisions that somehow generate fixed costs in the making of future decisions. But for longer-term investment decisions, policy uncertainty can more easily make a big difference, as the real options approach formalizes. Some farmers would choose not to replace aging equipment if they thought the probability was substantial that future regulation would make farming uneconomical. In Maryland, it has been argued that investment in new poultry operations has been substantially reduced because processors are not willing to write new contracts that would require expanding processing capacity under the potential costs of liability for growers' violations of chicken-litter handling requirements under mandated nutrient management plans. But empirical work to quantify such effects is lacking.

Welfare Economics

The preceding discussion of reduced investment suggests the possibility of social losses under policy uncertainty. The normative analysis of random policy instruments has been carried furthest in the literature on optimal taxation. It has been noted that random auditing implies randomness in tax rates facing individual filers who choose to underreport their incomes (a low tax rate if undetected but a quite high rate if detected). Stiglitz (1982) showed that random taxation can be optimal, by reducing the deadweight loss from labor market distortion per dollar of revenue raised under some circumstances. Pestieau, Possen, and Slutsky (1998) derive optimal random tax rates when some citizens are risk preferers, giving taxpayers a choice of relatively low rate *t* or a random process that generates on average a higher rate.[7] The gains from randomness result from convexity of taxpayer utility in income (or concavity of deadweight losses in taxes assessed), as occurs for risk preferers. However, for commodity taxes or subsidies we have convexity of deadweight losses in the policy instrument, as in the basic case of linear supply and demand where doubling the subsidy per unit output increases the deadweight loss triangle four times, so a mean-preserving increase in the variability of the subsidy increases deadweight losses. Nonetheless, in the second-best situations that arise from taxation to counter environmental damage from production, with an appropriate functional form of the damage function (sufficiently concave in production), random taxation could be optimal.

Considering the distortion of farmers' investment decisions under policy risk, it may be thought that a similar possibility of welfare gain arises. In the case of disaster assistance – for example, payment of indemnities to crops lost in floods – social costs are generated by the incentives the program provides to convert land to crop production in flood-prone areas. The moral hazard and attendant costs are reduced if farmers who plant in flood-prone areas are uncertain whether they will be bailed out, and similar reductions in deadweight losses due to incentive effects could be generated by uncertainty in any distorting program's benefits. However, this uncertainty reduces the farmer's expected gain from the policy. Confining the analysis to policy risk implies comparing, for example, policy (1), which generates farmer benefits B with certainty, to a policy (2) that generates 0 for the farmer with 50 percent probability and 2B with 50 percent probability. (In the flood indemnity example, B in policy (1) could mean 50 percent coverage of losses, and situation (2) is an equal chance of a policy that covers all losses or no losses.) Again the results will depend on the functional form of farmer utility in B.

[7] An approximation to this is what some state governments actually do: confront everyone with a basic tax schedule and in addition sell lottery tickets at prices that exceed the expected payout and raise net revenue for the state.

Political Economy

So far we have discussed policy, whether random or not, as imposed exogenously upon farmers by government. How does the situation change analytically when we consider that policy is influenced by the actions of farmers and other interest groups? When policy is the result of lobbying, policy risk arises from randomness in the process through which lobbying generates policy. Risks arise not only because lobbying groups are uncertain how policymakers will respond to their arguments, campaign contributions, and get-out-the-vote efforts, but in addition the efforts of farmers are likely to stimulate lobbying by others (agribusiness, environmental groups), the results of which cannot be predicted with precision.

Some literature relevant to this situation analyzes the lobbying equilibrium, and improves social benefit-cost accounting by adding the costs of lobbying to the deadweight losses resulting from policies that are enacted. As shown by Coggins (1995), it is quite possible in lobbying for a price support program that the gains to farmers are more than dissipated by lobbying costs.[8] A common finding in this literature is that interest groups that are evenly balanced in wealth and initial political influence dissipate more in the lobbying game than occurs when the groups are unevenly matched (so the losers know from the start that lobbying against farmers, say, is hopeless). In this situation, policy risk is uncertainty about lobbying effectiveness, and this will make it less clear whether competing groups are unevenly matched or not. This would induce competing groups to act as if they were more equally matched, and thus dissipation of policy benefits through lobbying costs would tend to be larger, the greater the policy risk.

EVIDENCE ON RISKS AND POLICY

The earlier historical examples indicate that government policies can be destabilizing as well as stabilizing. The relevance for the economic behavior of farmers is how past policies have actually worked, how that experience translates to expectations, and how the resulting expectations about policies influence production and investment decisions. I will consider some empirical evidence on two aspects of this situation, both in partial and preliminary investigations using U.S. data. The first is the extent to which ostensibly price stabilizing policies actually do stabilize prices, i.e., whether policy risk possi-

[8] The extreme case occurs if proponents and opponents reach a standoff and no policy is enacted. Of course, if the interest groups foresee this outcome they would not lobby in the first place, or at least would modify their actions, perhaps using some random process themselves. The usual noncooperative Nash equilibrium in the lobbying game assumes that lobbying begins if an interest group sees gains given the level of opponents' lobbying efforts, which is initially zero. But interest groups sometimes can think further ahead than that.

bly dominates the market risks reduced by policy; and the second is investment in agriculture.

Corn and Soybean Prices Under the FAIR Act

One of the most persistent criticisms of the Freedom to Farm provisions of the Federal Agricultural Improvement and Reform (FAIR) Act of 1996, as mentioned earlier, is that price stability achieved by pre-1996 programs has been abandoned. Secretary of Agriculture Glickman stated the Clinton Administration's belief that "the 1996 Freedom to Farm bill fails to provide an effective safety net for American farmers" (Glickman 2000). On the Republican side, House Agriculture Committee Chairman Combest argues for modifications to make FAIR Act payments "countercyclical" (*New York Times*, May 14, 2001, p. 1).

Pre-1996 programs placed floors under market prices through CCC loan support and stockholding, at least until the late 1980s. The programs stabilized farmers' receipts through deficiency payments that rose as market prices fell, and reduced acreage through annual set-asides when markets were weak. Historical episodes discussed earlier indicate that the stabilization possibilities have not been fully realized because of questionable implementation of these provisions of law, but this is far from showing that prices were less stable under these programs than would have been the case without them.

Empirical investigation of this topic is difficult because the programs were ubiquitous between 1933 and 1995, except for exceptional periods of price boom in the World War II period and in the mid-1970s. It is hard to test the counterfactual situation of no programs econometrically because despite many efforts we do not have models of commodity markets sufficiently precise that we can confidently say what prices would have been in the absence of program constraints and incentives. Now we are accumulating post-1995 evidence. That has been seen as an unstable period so far, but the question is how different would recent history have been if pre-1996 programs had been maintained?

Figure 2 shows price data for corn and soybeans before and after the FAIR Act. It is difficult to know when to place the date at which the FAIR Act regime replaced the pre-1996 program as the policy environment for commodity prices. The 1996 crops were the first ones produced and marketed under the FAIR Act, but the influence of the Act (and the absence of the pre-1996 programs) should have been felt before the 1996 crop year began, in September 1996. On the other hand, the failure of the 1995 version of Freedom to Farm to be enacted suggests that the old regime should be regarded as having been still in place before January 1996. For purposes of the following analysis, I take the inception of the FAIR Act, in farmers' and traders' expectations and actions, to have occurred at the time of the key votes in Congress that ensured passage of the FAIR Act; accordingly I take March 1996 and after as the FAIR Act period, and February 1996 and earlier as belonging to the prior regime.

March 1996 was three months before the mid-1990s peak in corn prices and almost a year before the peak in soybean prices. So that peak, and the subsequent price declines, are part of the FAIR Act experience, the basis for claims that the FAIR Act has been insufficient as stabilization policy. However, as Figure 2 indicates, there was plenty of price variability before 1996 (even omitting the early 1970s price run-ups).

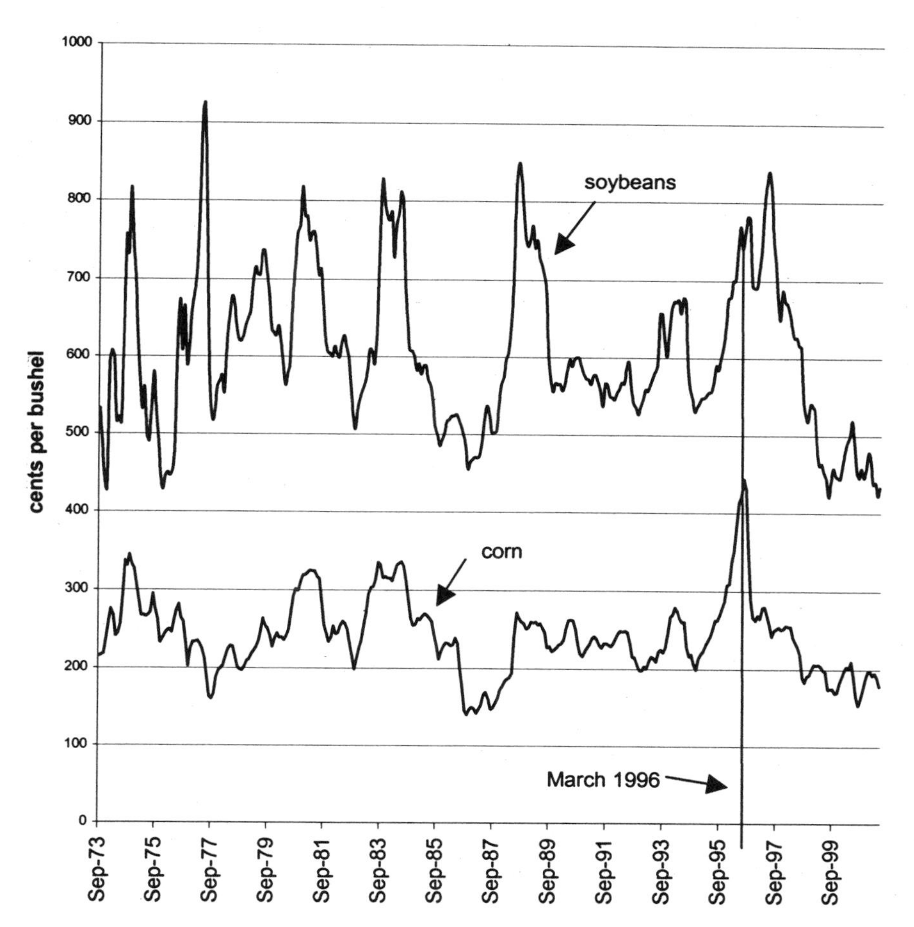

Figure 2. Corn and Soybean Prices Received by Farmers, Monthly, September 1973 - September 1999

Source of data: USDA, National Agricultural Statistics Service, Agricultural Prices

Some evidence for corn and soybeans is as follows:

	pre-FAIR (270 months)		FAIR (63 months)	
	---------- cents per bu. ----------			
	corn	beans	corn	beans
mean price:				
nominal	244	611	235	583
real (1996$)	349	872	228	566
standard deviation of price:				
nominal	43.4	95.2	68.9	131.8
real	62.0	136	66.8	128.0
volatility (standard deviation of price *changes*)				
cents/bu.	9.1	24.8	10.8	19.8
relative (% change)	3.8	4.0	4.4	3.4

The average farm-level price was slightly lower under the first five years of the FAIR Act than during the preceding 20 years in nominal terms, but real (deflated by the GDP deflator) prices averaged much lower in 1996-2001 than in 1975-1995. Several measures of price variability are shown. The standard deviation of price within each sub-period around the mean price for that sub-period was much greater in nominal terms under the FAIR Act but in real terms increased only from 62 cents (1996$) before FAIR to 66.8 cents after, and the standard deviation of real soybean prices was lower under FAIR. A closely related but distinct measure of instability is volatility (using the jargon of finance), the standard deviation of month-to-month prices. Volatility is much lower than the standard deviation of prices themselves because high and low prices tend to be grouped together, as Figure 2 shows. Volatility, measured in terms of either changes in cents or in percentage changes, increased during the FAIR Act period for corn but decreased for soybeans.

Time-series regressions were carried out on several measures of volatility as related to commodity stock levels and changes in USDA's supply-demand estimates as well as a dummy for months before and after March 1996. In addition to monthly changes in farm prices received, the analysis also considered prices on the day just after USDA's crop production estimates were announced, for the same set of USDA reports covered in USDA (2001). The results were similar to those of the raw data above in that the coefficients for the FAIR Act period showed slightly higher volatility for corn and lower volatility for soybeans. But in neither case is the effect statistically significant. I conclude therefore that despite the assertions of uncertainty created by the FAIR Act as compared to prior "countercyclical" policies, in fact there has been no significant increase in price instability in the corn and soybean markets.

Similar results are found in a recent study using a different methodology of simulating a supply-demand model with optimal private-sector stockholding under FAIR Act and pre-FAIR Act parameters and constraints (Lence and Hayes 2000). Lence and Hayes find support for the hypothesis that "the changes made when FAIR was enacted did not lead to a permanent significant increase in the volatility of farm prices or revenues" (p. 23). The economic reason for the results of these simulations, which arguably are the underlying reasons for the findings above looking at actual price behavior, is that farmers' profit-seeking behavior in a free-market environment generates stockholding and acreage planting decisions that adjust as well to market conditions as USDA regulatory regimes were previously able to accomplish.

If the FAIR Act has had little effect on market prices and farmers' behavior, that suggests that the corresponding policy risk – uncertainty about continuation of the FAIR Act – is also unlikely to be a major economic factor. But there are other channels of influence besides short-term price stabilization and farmers' decisions about acreage allocation and holding commodity stocks. It is possible that AMTA payments have provided capital that kept some farms in business that otherwise would have folded, or kept them growing program crops rather than alternative crops.[9] And, it has been argued that AMTA payments have been used to finance investments, for example in new equipment, that otherwise would not have been made. If true this effect implies the farmer faces capital rationing or else is basing investment decisions on something other than expected profitability, since AMTA payments do not change the rate of return to investment in such equipment (unless they do serve to keep the farm in business, but then the investment effect is not an *additional* one).

Evidence on Investment

The idea that farmers' investment decisions are influenced by long-term confidence instilled by a commodity policy regime was strongly argued by Sally Clarke (1994). Focusing on investment in tractors in the Midwest in the 1930s and 1940s, using primarily farmers' diaries and other non-quantitative information, she concluded that "farmers' willingness to invest turned in large part on the long-term changes initiated by the New Deal farm policy" (p. 200). Her position elaborates that expressed earlier by Willard Cochrane, who in his book with Mary Ryan connects commodity support programs with productivity growth as follows:

[9] Notwithstanding AMTA payments being associated with production flexibility and treated as lump-sum payments in some economic analyses, producers who are known to have planted fruit or vegetable crops on acreage base on which AMTA payments have been contracted become ineligible for the payments.

> "What did the price and income support programs have to do with these gains in agricultural productivity? They had a lot to do with it. They provided the stable prices, hence price insurance, to induce the alert and aggressive farmers to invest in new and improved technologies and capital items, and the reasonably acceptable farm incomes and asset positions to induce lenders to assume the risk of making farm production loans (Cochrane and Ryan 1976, p. 373).

If this is true, uncertainties associated with the policies themselves, such as current uncertainty about what farm programs will follow the FAIR Act's expiration in 2002, would be expected to retard investment in agriculture.

Figure 3 shows an estimate of the capital stock on farms since 1950, and of net investment (expenditures on capital equipment minus depreciation and obsolescence). Capital is defined to include durable machinery and equipment, and excludes real estate and irrigation equipment, purchased inputs normally used up in a year, and inventories of commodities and storable inputs such as fertilizer.[10] U.S. agriculture experienced two periods of sustained investment boom, 1945-55 and 1964-80, and two periods of disinvestment, 1930-35 and 1980-present.

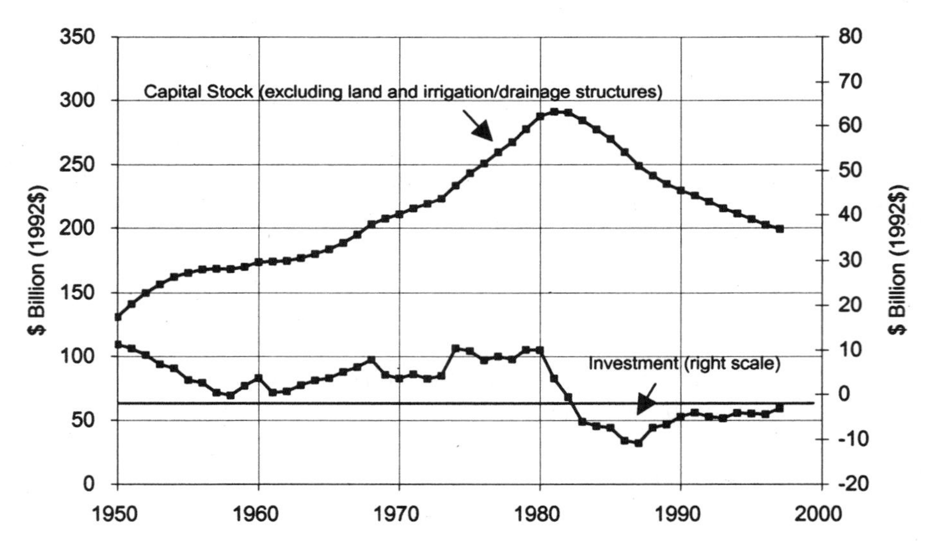

Figure 3. Investment and Capital Stock in U.S. Agriculture
Source: USDA, Economic Research Service

[10] The information base for capital measurement is questionable for many items, and some important conceptual issues in capital measurement are not totally resolved, notably the appropriate measurement of depreciation. Two alternative measures of U.S. farm capital are available. One is published by the Bureau of Economic Analysis of the U.S. Commerce Department (1997) as Fixed Reproducible Tangible Wealth. The second, which is shown in Figure 3, is constructed by the Economic Research Service of USDA as part of its output, input, and productivity measurement program (Ball et al. 1997).

The fewness and length of the periods of investment and disinvestment suggest that farmers are not reacting to short-term changes, but rather focusing on their expectations about the longer-term outlook. This is consistent with attributing sea-changes in investment to changes in policies that are expected to be permanent. A problem with the Cochrane-Clarke hypothesis about New Deal policies and investment, is that the boom did not get under way until a decade after the policy regime change. Possibly it took time to convince farmers of the permanence of the regime change, and that coupled with World War II impediments caused the delay. It is also possible, however, that farmers like many others saw World War II rather than the New Deal as the bringer of prosperity, and only when the widely feared post-War resumption of general economic depression failed to occur did investment pick up strongly. This would suggest that the key factors in the acceleration of investment were the improved general economic outlook plus the fact that by the end of the 1940s the remaining capital stock in agriculture was largely depreciated and/or obsolete; confidence in government as guarantor of profitable investment in farming was not a key factor.

The causes of the post-1970 investment surge are more readily apparent in the commodity boom of that period. Likewise, the huge decline in capital after 1980 – with about 40 percent of the 1980 capital stock disappearing by 1995 as compared to a 25 percent decline in 1930-35 – can be seen as a reaction to overexpansion during the 1970s.[11] Tying investment to long-term optimism or pessimism about commodity markets is too confining anyway. As mentioned earlier, a long period of disinvestment almost inevitably means that a subsequent period of re-investment will not be brief. And, capital investment and disinvestment occur because of factor market trends. It is likely that the initial surge of investment in the 1960s was fostered by rising real wage rates inducing substitution of capital for labor, especially capital embodying new labor-saving technology. The reduction in capital since 1980 is partly a response to the development of reduced-tillage crop production, which required less tractor and mechanical cultivation service.

Factor markets and technological change are also influenced by policies, and this is another avenue through which policy uncertainty can make a difference. For example, Sunding and Zilberman (2001) cite evidence that irrigation subsidies have retarded the adoption of new water-saving technology. Uncertainty about the size and administration of such subsidies in the future is likely to affect farmers' investment decisions not only about specific fixed capital equipment associated with particular crops that are more or less dependent on irrigation, but even about whether to continue farming as opposed to selling land for development or other purposes.

---

[11] Note that the measured increase in capital of the 1970s is not directly attributable to the land price increases of those years. Both the Commerce-BEA and USDA-ERS measures of capital are attempts to measure physical quantities of capital, not capital gains due to price increases, and in any case neither the Commerce nor the USDA measure includes real estate in the capital measured.

POLICY AND RESEARCH IMPLICATIONS

The welfare economics of agricultural stabilization policies has sought for estimates of the gains to be had when social costs of missing markets for risk are remedied through policies that reduce risk or the financial consequences of risk. This chapter has focused on the much less investigated topic of the consequences of risk in the policy process itself. The spirit of this chapter is analogous to that of the literature on political economy, which attempts to treat policies as endogenous, as part of the economic situation to be explained rather than as an exogenous cause of that situation. In the analysis of stabilization policies, the approach is to consider the consequences of uncertainties in lobbying as well as in autonomous governmental action.

An important related subject that this chapter has not considered is the design of governmental institutions in the light of policy risk. How do these risks alter the pros and cons of transparency in governmental decision making, legislative requirements that benefit-cost analysis of proposed policies be carried out, legislative rules that allow minorities to slow or stop policy innovations that majorities prefer, restrictions on lobbying by interest groups, and related matters?

With respect to the underlying factual situation, further work is needed to assess how important policy risk is, who gains and loses from uncertainty about policies, the extent to which such uncertainty stems from imperfections of knowledge about the economy, policymakers' uncertainties about the electoral consequences of their policy choices, or the existence of rational waiting before exercising the real option that a policy commitment entails.

These are all difficult questions to address productively in terms of an appropriate conceptual framework, mobilization of relevant facts, and analyses of alternative hypotheses (in positive economics) and alternative policies and institutions (in normative economics). The methodology for answering them will have to vary depending on the issue. With respect to the consequences of policy risk for investment in agriculture, arguably the most important area to investigate, a promising framework is the theory of investment under uncertainty using the real option model of Dixit and Pindyck (1994), as discussed earlier. Implementing the theory empirically is not straightforward, however. One of the few agricultural examples is Plato (2000). He explains the decision to utilize soybean-crushing capacity as a function of the spread between the prices of soybeans and their products, and the volatility of that spread.

The application to commodity policy risk would be to explain investment in fixed capital as a function of expected market returns, the uncertainty of market returns, policy-generated returns, the uncertainty of those returns, and the covariance of market and policy-generated returns. Observations for econometric investigation could be a pooled time series and cross section of commodities or geographic areas such as states. A problem with a commodity approach is that one needs commodity-specific investment, which is hard to obtain. But there are data for some specialized machinery and equipment. A problem with using states as observations is that while state-specific investment data are available, state-specific policy variables are harder to get. We do have government payments by state, but that does not cover all program benefits. Data on expectations are always difficult. For market returns we have to use lagged prices, or futures prices for some commodities. For the key variable of policy uncertainty, one has to search hard for appropriate measures. Data on implied depreciation rates or the default probability in policy benefits is a feasible source for some commodities, as discussed earlier for tobacco quota. For other commodities one can estimate program rent-to-value ratios under the assumption that expected program values are capitalized into real estate prices, as is plausible for most programs (which are tied to land in the sense that the benefits convey when land is sold rather than being tied to the farmer as a person). One can also resort to subjective measures of program risk, from appropriately conducted surveys, although this is a big and difficult job and not feasible in work with historical data.

REFERENCES

Ball, V.E., J.-C. Bureau, R. Nehring, and A. Somwaru. 1997. "Agricultural Productivity Revisited." *American Journal of Agricultural Economics* 79: 1045-1063.

Baumol, W.J. 1990. "Sir John Versus the Hicksians." *Journal of Economic Literature* 28: 1708-1715.

Benedict, M. 1953. *Farm Policies of the United States, 1790-1950.* New York: The Twentieth Century Fund.

Clarke, S. 1994. *Regulation and the Revolution in United States Farm Productivity.* New York: Cambridge University Press.

Cochrane, W.W., and M.E. Ryan. 1976. *American Farm Policy, 1948-1973.* Minneapolis: University of Minnesota Press.

Coggins, J. 1995. "Rent Dissipation and the Social Cost of Price Policy." *Economics and Politics* 7: 147-166.

Dixit, A., and R. Pindyck. 1994. *Investment Under Uncertainty.* Princeton: Princeton University Press.

El-Osta, H., and J. Johnson. 1998. "Determinants of Financial Performance of Commercial Dairy Farms." USDA, Economic Research Service, Technical Bulletin No. 1859 (July).

Gardner, B., and R. Kramer. 1986. "Experience with Crop Insurance Programs in the United States." In Hazell, Pomerada, and Valdés, eds., *Crop Insurance for Agricultural Development.* Baltimore: Johns Hopkins Press.

Gilbert, C.L. 1996. "International Commodity Agreements: An Obituary Notice." *World Development* 24: 1-19.

Glauber, J.W., and K.J. Collins. 2001. "Risk Management and the Role of the Federal Government." In R.E. Just and R.D. Pope, eds., *A Comprehensive Assessment of the Role of Risk in U.S. Agriculture.* Boston, MA: Kluwer Academic Publishers.

Glickman, D. 2000. "Statement on Signing of Emergency Farm Assistance and Crop Insurance Reform Bill." USDA News Release 0201.00, June 20, 2000 (available at http://usda.gov/news//releases/).

Hicks, J.R. 1969. *A Theory of Economic History.* Oxford: Oxford University Press.

Krueger, A.O., M. Schiff, and A. Valdés. 1991. *The Political Economy of Agricultural Pricing Policy* (Vols. 1-5). Baltimore: Johns Hopkins University Press.

Marshall, A. 1920. *Industry and Trade.* London: Macmillan.

Lence, S.H., and D.J. Hayes. 2000. "U.S. Farm Policy and the Variability of Commodity Prices and Farm Revenues." Working Paper No. 00-WP 239, Center for Agricultural and Rural Development (CARD), Iowa State University.

McBride, W.D. 1995. "U.S. Hog Production Costs and Returns, 1992: An Economic Basebook." Economic Research Service, USDA, Agricultural Economic Report No. 724 (November).

Olson, M. 2000. *Power and Prosperity.* New York: Basic Books.

Pestieau, P., U.M. Possen, and S. Slutsky. 1998. "The Value of Explicit Randomization in the Tax Code." *Journal of Public Economics* 67: 87-103.

Plato, G. 2000. "The Soybean Processing Decision." Mimeo, Economic Research Service, USDA (November).

Sanderson, F. 1975. "The Great Food Fumble." *Science* (May): 503-509.

Schnittker, J. 1973. "The 1972-73 Food Price Spiral." *Brookings Papers on Economic Activity*, pp. 498-507.

Schuh, G.E. 1974. "The Exchange Rate and U.S. Agriculture." *American Journal of Agricultural Economics* 56: 1-13.

Schultz, T.W. 1945. *Agriculture in an Unstable Economy.* New York: McGraw Hill.

Stiglitz, J.E. 1982. "Self-selection and Pareto Efficient Taxation." *Journal of Public Economics* 17: 213-240.

Sumner, D., and J. Alston. 1985. "Removal of Price Supports and Supply Controls for Tobacco." NPA Report No. 220, National Planning Association, Food and Agriculture Committee, Washington, D.C. (December).

Sunding, D., and D. Zilberman. 2001. "The Agricultural Innovation Process: Research and Technology Adoption in a Changing Agricultural Sector." In B. Gardner and G. Rausser, eds., *Handbook of Agricultural Economics* (Vol. 1). Amsterdam: North-Holland.

Travena, B.J., and L.H. Keller. 1970. "Lease and Sale Transfers of Cotton Allotments in Tennessee." Tennessee Agricultural Experiment Station, Knoxville.

U.S. Department of Agriculture. 1933. *Yearbook of Agriculture, 1933.* Washington, D.C.: U.S. Government Printing Office.

___. 1986. "Embargoes, Surplus Disposal, and U.S. Agriculture." Economic Research Service, Agricultural Economic Report No. 564 (December).

___. 2001. "Price Reactions After USDA Crop Reports." National Agricultural Statistics Service, No. Pr Rc 1 (01), May 2001.

U.S. Department of Commerce. 1997. "Fixed Reproducible Tangible Wealth in the United States." Bureau of Economic Analysis.

U.S. General Accounting Office. 1981. "Farmer-Owned Grain Reserve Program Needs Modification." Report to Congress, CED-81-70 (June).

World Bank. 1986. *World Development Report 1986.* Oxford: Oxford University Press.

Chapter 23

RISK MANAGEMENT AND THE ENVIRONMENT

Mark Metcalfe, David Sunding, and David Zilberman
University of California, Berkeley

INTRODUCTION

Environmental and resource management issues have played a major role in agriculture. On the one hand, there is increased awareness of the environmental side effects resulting from agriculture and the problems associated with pollution, depletion of natural resources, and the effects on human and wildlife health caused by farming activities. On the other hand, it has also been realized that farmers are stewards of the land, and their actions contribute to the enhancement of environmental quality and also generate value that is not compensated directly by markets. Both the control of negative externalities as well as the compensation for positive ones are subject to public sector involvement through government policies, and therefore much of the research on the environment and agriculture is policy-driven and provides input for both regulatory action and farmers' response (Lichtenberg, forthcoming).

Agricultural activities rely on and manipulate natural systems. The randomness inherent in these systems is a major source of risk for farmers, and the subsequent farming activities undertaken to manage this risk have the potential to significantly impact environmental quality. We argue that, when it comes to environmental issues, the notion of risk has alternative definitions and that decision making under uncertainty can be applied creatively to better define risk. In particular, most of the economic literature is concerned with the economic risk associated with randomness of income or another monetary variable. Of course this randomness may be related to some natural phenomena such as weather variations or pest infestation, but economists rarely deal explicitly with the nature of the random phenomenon itself.

In this chapter, we examine two interpretations of risk management. First, we discuss policy strategies that aim to manage environmental and health risks; and second, we examine the implications of strategies that aim to manage the economic and production risks facing farmers. Thus, one major part of this chapter presents methodologies and results examining environmental regulatory policies enacted to protect public health by limiting public exposure to high concentrations of agricultural contamination. Examples of these types of regulations are pesticide use restrictions limiting pesticide applications and water quality policies protecting public health from the nutrients generated by agricultural wastes.

The other major part of this chapter presents a methodological approach and some results on how the management of production, market risks, and agriculture influences environmental quality. Specifically, we examine the influence of crop insurance on the choices of cultivated acreage, variable inputs, and the technology used in crop production, and how these choices influence environmental quality. Before we proceed with these two main sections of the chapter, we first identify some principles of methodology and modeling that apply to both of these types of risk-related issues.

MAJOR ELEMENTS OF ANALYSIS OF RISK AND THE ENVIRONMENT

Studies that examine the management of risks in agriculture and the subsequent consequences for public health and environmental quality have several major features in common. These features suggest a variety of opportunities for synergies in terms of data collection, development of methodologies, and improved coordination between the literature examining the environmental implications of managing economic risks and the literature on the management of environmental and health-related risks. These features include the following:

(1) *The specification of production technologies, including the understanding of how human activities affect agricultural production as well as environmental quality.* Much of the emphasis in agricultural economics has been in quantifying production functions, namely, the relationship between the level of input use by farmers and the overall agricultural output. In the last 40 years, and especially since the publication by Just and Pope (1978), there has been much emphasis on risk and the role of inputs in determining agricultural production. It is understood that changes in inputs can affect not only mean output but also the variability, or the distribution, of that output. Because of this important understanding, studies that examine the impact of agriculture on the environment require a more developed understanding of how economic activities relate to environmental quality. Quantifying these relationships is a major challenge because of the lack of adequate measures of environmental quality that account for the impact of agricultural activities both on the average level of performance as well as on the variance of outcomes.

(2) *The need for multidisciplinary knowledge.* A complete understanding of the influence that production and management activities have on agriculture and the environment requires incorporating the knowledge and understanding of other disciplines into agricultural production and environmental management models. For example, the analysis of pest risks requires understanding of the phenomenon of resistance, of how pest population dynamics affect overall productivity, and of the processes that influence the rate of transfer of chemicals in the environment. This implies the need for incorporation of biological knowledge into economic models in order to generate new hybrids that have an economic decision making framework augmented by biological and physical relationships.

(3) *The need to incorporate heterogeneity.* The majority of economic theory has been developed under the assumption of a representative firm and the notion of identical firms within an industry. These assumptions are especially problematic when it comes to environmental problems, as the realization of environmental effects may well depend on when and where pollution occurs. Incorporation of differences in agro-ecological conditions is crucial in adequately explaining the evolution and the structure of agricultural systems and their impact on the environment. Thus incorporating heterogeneity is a major challenge for the analysis of risk management and the environment.

(4) *Understanding the multidimensionality of impacts.* Economic and production activities may affect several dimensions of environmental quality. In many cases, a response to one sort of environmental quality concern may result in environmental problems in another medium. For example, concern about soil erosion may lead to transition to a low-tillage system that relies heavily on herbicides and thus generates a new source of environmental and health risks. The trade-offs that one may see are not only between productivity and environmental quality but also between the different dimensions of environmental performance.

(5) *Distinguishing between various sources of variability.* Key performance measures such as changes in yield per acre, pollution per acre, or the number of sick days per worker due to exposure to environmental risks all vary to some degree. Sources of these variations can be decomposed into (i) heterogeneity representing differences in conditions among individuals and locations; (ii) a lack of knowledge that results in the use of estimates that have a high degree of uncertainty; and (iii) natural randomness. Each of these sources of variability is treated differently in policy making. Heterogeneity is addressed through the development of discriminatory policies that encourage responses appropriate for particular situations, and a lack of knowledge is addressed through continuous learning and improved estimation of key parameters. There are a wide array of policies that are used to address natural randomness, including insurance, storage, and the requirement of protective

clothing, to name a few. Below we will deal with problems that originate with random events, but their analysis must recognize the existence of heterogeneity and incomplete information.

ENVIRONMENTAL HEALTH RISK MANAGEMENT

Agriculture is the source of many regulations that are imposed to limit environmental health risk. The Food Quality Protection Act (FQPA) of 1996 was designed to protect public health by reducing the risks associated with over-exposure to pesticides (Byrd 1997, Rawlins 1998, Schierow 2000), and water quality regulations are imposed at both the federal and state levels to protect health from agricultural waste generated in livestock production (Innes 2000, Metcalfe 2000, Hochman and Zilberman 1979, U.S. EPA 1999).

Designing and implementing policies to reduce health risks requires a great deal of information in order to determine the acceptable levels of public exposure and also to decide how to best assure that these levels are not exceeded. Given this large information requirement, policies directed at controlling environmental risks in agricultural production inevitably face a tremendous amount of uncertainty when attempts are made to relate potential environmental risks to the processes that generate these risks.

For example, the health risks to an individual exposed to pesticides are dependent not only on the total amount of pesticides that are physically applied to a crop, but also on the amount of this total to which the individual is actually exposed and the extent to which an individual is biologically affected by the chemicals. Given variability in environments, randomness in nature, and heterogeneity in populations, the true extent of health risks is actually unknown. Incorporation of this uncertainty into economic modeling ultimately leads to the design of policies better able to limit health risk.

The health risk resulting from overexposure to agricultural contaminants is the probability that an individual selected randomly from a population consequently will suffer adverse health effects. Effective modeling of this health risk requires a conceptual and qualitative understanding of the relationships that exist in the risk-generation model (Spear 1978, Bogen 1990). The concept of the risk-generation (or risk-assessment) model has been used in past studies to examine policy options for health risk issues with respect to agricultural externalities (Lichtenberg, Zilberman, and Bogen 1989, Lichtenberg and Zilberman 1989, Hanemann, Lichtenberg, and Zilberman 1989, Harper and Zilberman 1992, Sunding and Zivin 2000, Zivin and Zilberman, forthcoming).

The risk-generation model is comprised of four separate processes: contamination, transport, exposure, and dose-response. Contamination is the physical introduction of a substance into the environment. In terms of pesticides, this is the process of application when chemicals are applied to the desired crop. Transport (or the fate of a contaminant) is the movement of a contaminant away from the location of its intended use. The spread of pesticides through the air or the run-off of animal waste from fields into nearby surface and ground waters are examples of the transport process. Exposure is the actual level of contact an individual has with a contaminant. Again with respect to pesticides, this can occur through consumption of food, through breathing or touching the chemical directly, or through contact with water that has been contaminated. The dose-response process is the translation of exposure levels into adverse health effects. These effects can be acute, such as immediate poisoning, or more chronic, such as an illness that develops over time. The risk-generation process is illustrated in Figure 1.

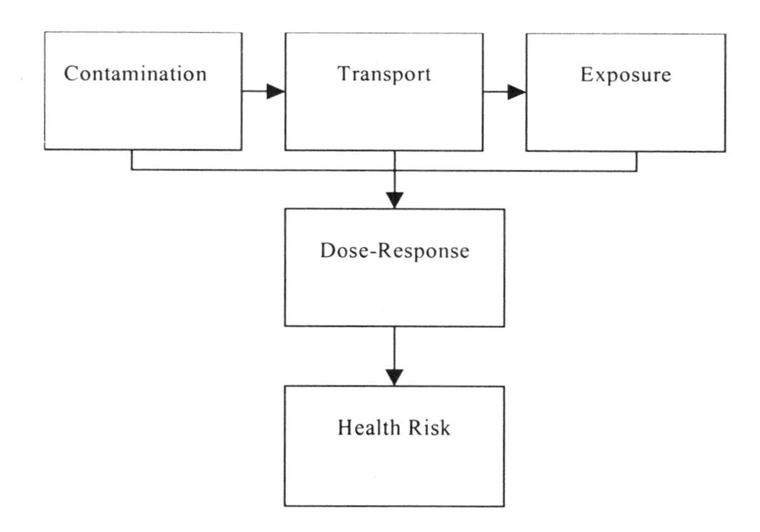

Figure 1. The Risk-Generation Process

Mathematically, let r represent individual health risk such that

$$r = f_1(B_1, B_2) f_2(B_3) f_3(B_4) f_4(B_5), \tag{1}$$

where $f_1(B_1, B_2)$ is a function relating the level of contamination (B_1) and the level of contamination control activities undertaken (B_2). The function $f_2(B_3)$ is the transport coefficient, which is dependent on environmental factors (B_3) such as rainfall, wind velocity, and soil characteristics. The exposure of an individual to a contaminant is determined as $f_1(B_1, B_2) \cdot f_2(B_3) \cdot f_3(B_4)$, where $f_3(B_4)$ is the exposure coefficient and is dependent on activities like consumer education and the use of worker safety equipment to help limit exposure (B_4). The functional relationship defined by $f_4(B_5)$ is the dose-response function, which relates the level of exposure to the probability of health risk and is dependent on activities such as medical treatment and preventative vaccinations (B_5).

A common functional form that is used in empirical work is a log transformation of a multiplicative combination of the variables in equation (1). For simplicity, assume that there are five activities (represented by B_1 through B_5 above) undertaken to reduce risk. Therefore, the log of risk is written as

$$r = \sum_{i=1}^{5} b_i , \tag{2}$$

where b_i is the log level of B_i. This functional form is desirable because of its simplicity and its ability to describe risk in a wide variety of health risk problems (Crouch and Wilson 1981).

There is uncertainty involved in determining each of the variables in the model. The contamination and contamination control variables (b_1 and b_2) require information on the use of the contaminant, which may not be known with certainty or may not be directly observable. The measure of the environmental factor (b_3) is dependent on the ability of the local environment to absorb contaminants. This value is difficult to determine as there is randomness and great variability from region to region. The measure of the exposure risk-reducing factor (b_4) is dependent on the level of public awareness and on the education and training of agricultural workers, both of which are often unknown. Accurately defining the dose-response variable (b_5) is dependent on having scientific knowledge as to the human health effects of various levels of exposure to hazardous substances, and this information is often lacking when it is time to develop risk-reducing policies.

This uncertainty is incorporated into many empirical applications by assuming that the parameters of the model are random variables with a joint lognormal distribution, which allows each of the log parameters to be distributed joint normal. Making this assumption for the log risk model in equation (2), the mean and variance of log risk are equal to

$$E(r) = \sum_{i=1}^{5} E(b_i) \tag{3}$$

$$V(r) = \sum_{i=1}^{5} V(b_i) + 2\sum_{i \neq j} COV(b_i, b_j) , \tag{4}$$

where $E(\cdot)$ is the expected value function, $V(\cdot)$ is the variance function, and $COV(\cdot)$ is the covariance function.

A common methodology used to incorporate the existence of this uncertainty is the safety rule approach (Kataoka 1963, Lichtenberg and Zilberman 1988). The safety rule methodology minimizes the social cost of regulation subject to a constraint allowing the risk of adverse health effects to exceed a certain threshold level, R, no more than some acceptable percentage of time, α. The value of α can be thought of as a measure of the level of aversion to uncertainty in society, where lower values of α represent lower tolerances for violations and a greater aversion to uncertainty. The point being that a society adverse to uncertainty is willing to pay higher regulatory costs to maintain a low α and reduce the uncertainty involved in health risks.

To mathematically determine this safety rule, let $r(\alpha)$ denote the value of log risk exceeded with probability α, and let $F(\alpha)$ denote the value of the standard normal distribution exceeded with probability α. Knowing that log risk is distributed normally and, therefore, $F(\alpha)=[r(\alpha)-E(r)]/V(r)^{1/2}$, a safety rule can be defined as

$$r(\alpha)=\sum_{i=1}^{5}E(b_i)+F(\alpha)\left[\sum_{i=1}^{5}V(b_i)+2\sum_{i\neq j}COV(b_i,b_j)\right]^{1/2}\leq R. \qquad (5)$$

This safety rule relationship expresses log risk as a function of the sum of mean risk and uncertainty. Uncertainty is weighted by the value of $F(\alpha)$, which can be thought of as the level of regulators' aversion to uncertainty. Therefore, when regulators are risk neutral, that is to say they are indifferent between meeting or exceeding the maximum contamination level, α = ½ and the uncertainty term is removed from the model. Values of α greater than ½ represent policymakers' desire to achieve smaller probabilities of exceeding the maximum allowable contamination level, and therefore the weight of uncertainty in the decision making process, $F(\alpha)$, increases accordingly.

The safety rule approach is based on determination of the maximum allowable level of contamination and the desired margin of safety. This is appealing to policymakers since both of these concepts are easily understandable in a policy framework. Scientific experimentation with contaminants can be undertaken to provide guidance as to the determination of appropriate levels of these parameters.

The ultimate objective of policymakers is to minimize the total social cost of regulatory policy subject to the safety rule. Assume that the policymaker has n policies at his disposal to affect the five variables from the risk-generation model. For example, regulations could be imposed that require workers to undergo training on the correct procedures for pesticide application or to wear protective clothing when applying pesticides. Both of these policies would help to reduce the risk generated by the exposure process (b_4) in the risk-generation model. Assuming that each risk-reducing policy imposes a cost equal to $M_k : k = 1,\ldots,n$ allows the policymakers' objective to be written mathematically as

$$\max_{M_k} - \sum_{k=1}^{n} M_k , \tag{6}$$

subject to equation (5). First-order conditions of this optimization with respect to each policy *k* are calculated as

$$-\sum_{i=1}^{5} \frac{\partial E(b_i)}{\partial M_k} + \left[\frac{F(\alpha) V(r)^{1/2}}{M_k} \right] t_i e_{i,k} \leq \frac{1}{\lambda}, \tag{7}$$

where $t_i = [V(b_i) + \sum_{i \neq j} COV(b_i, b_j)] / V(r)$ is the share of uncertainty provided by the ith variable, $e_{i,k} = [M_k / V(b_i)^{1/2}] \cdot [\partial V(b_i)^{1/2} / \partial M_k]$ is the elasticity of the standard deviation of the ith variable with respect to the cost of the kth policy, and λ is the shadow price of the safety rule constraint. Equation (7) demonstrates that each policy enacted has an effect on mean risk and an effect on the uncertainty surrounding that risk.

The benefits of using the risk-generation model and safety rule approach are seen in the variety of policies that can be examined. The optimal portfolio of policies that satisfies equation (7) can be used to balance mean risk reduction and the reductions in uncertainty. Lichtenberg and Zilberman (1988) demonstrate that increasing aversion of society to uncertainty (a lowering of the α standard) leads to the use of policies that reduce the uncertainty of risk.

Policymakers evaluate the impact of intervention on expected risk reduction and also on the expected costs that will be incurred. Sometimes, it may be most effective to impose a direct control on the source of contamination only; however, in cases where the marginal cost of pollution reduction is high, a combination of other types of intervention, such as worker safety regulations or medical treatment requirements, may be optimal. This result is immediately appealing because it provides more policy options to encourage target risk reduction to be achieved at the least cost.

It is important to recognize that risk assessment studies may produce estimates of risk that are derived under varying degrees of reliability. Using similar methodology and the same data set, two studies could calculate very different dollar values for risk-reduction policies because they are not consistent in their measurement of risk. For example, in the safety rule approach, choosing to examine a higher value of α would provide a higher cost for regulatory compliance, while a lower α would yield lower dollar values. Another possible source of inconsistency is found in the measurement of risk itself. The multiplication of many worst-case estimates in the risk-generation model would lead to wildly unrealistic risk estimates. This is known as the creeping safety problem and it is important to address this issue in order to maintain consistent policy analysis.

The consistency of environmental policies can be observed through the implied shadow value of risk. This shadow value can be interpreted as the regulatory cost of saving a statistical life. Environmental policies are consistent if they imply that the value of a statistical life is similar across policies. A study of environmental policies by Cropper et al. (1992) found a high degree of variability associated with this value. It is probably the case that this inconsistency of policies is due to an inconsistent policy process as well as to the fact that risk estimators were derived under different degrees of statistical reliability. Lichtenberg and Zilberman (1988) have argued that since it is an objective of policy analysis to compare policies that affect risk, then analysis must be consistent and estimates of risk with the same statistical reliability must be used.

A study by Lichtenberg, Zilberman, and Bogen (1989) examines cancer risk arising from dibromochloropropane (DBCP) pesticide contamination of groundwater drinking supplies in the Central Valley of California. The cancer risk of an individual is assumed to be determined by the factors of the risk-generation model, and the policy measures available to reduce this risk are concerned with controlling the contamination process of risk generation. The cost-minimizing safety rule approach is utilized to calculate cost efficiency for cost/risk trade-offs when risk remains below an expected safety threshold 95 percent of the time. The policies found to be least-cost are different in urban as compared to rural regions, which demonstrates that examining the differences in regional impacts is important in conducting health risk analysis.

Zivin and Zilberman (forthcoming) use the health risk generation methodology to examine the risks associated with drinking water contamination by the parasite cryptosporidiosis. They incorporate heterogeneous populations in order to examine optimal policies for the general public and also for the segment of the population that is at greater risk. Their results show that it is important to account for differences in populations as it allows for tagging policies, such as installation of end-point filtration and provision of safe bottled drinking water to the most affected, in situations where they can provide a least-cost policy alternative.

Sunding and Zivin (2000) develop a health risk generation model to examine policy options to protect agricultural workers exposed to pesticides. Their modeling of the incentives that influence pesticide use incorporates the uncertainty inherent in pest population dynamics. Results demonstrate that when examining the health effects associated with pesticide regulations, it is important to incorporate this uncertainty of insect population dynamics because the probability of infestation, and the consequent damage, will drive farmers' incentives to apply or not apply pesticides.

This section examined the issue of environmental health risk regulation. It emphasized the importance of the risk-generation model and the cost-minimizing safety rule approach to policy evaluation. It is important that the evaluation of environmental health risk is consistent across policies. To ensure that our policy analysis is consistent, we as economists must take care to consider the issues of the measurement and consistency of risk when we obtain policy-relevant information from other disciplines. Obtaining such an understanding of these issues in health risk will provide us with a much richer set of policy options when we must determine the best combination of risk and economic well-being.

IMPACTS OF RISK MANAGEMENT STRATEGIES

Agricultural production is subject to random variables that affect productivity, and a number of mechanisms can be used to reduce the economic risks to farmers. These mechanisms can alter resource allocation in farming and, thus, influence environmental quality. Agricultural activities impact natural resources and the environment in three important ways:

(1) *Negative externalities are generated.* Pesticide use can cause food safety, worker safety, and environmental health problems, e.g., contaminate bodies of water, harm fish and wildlife, etc. (National Research Council 2000). In addition, fertilizers and pesticides can cause water quality problems.

(2) *Resources are depleted.* Tillage activities may create soil erosion problems, thereby reducing the availability of soil as a resource. Also, overapplication of certain pesticides can increase pest resistance, thereby reducing the common property resource of pest vulnerability to treatment (Hueth and Regev 1974).

(3) *Environmental amenities are reduced.* Farming activities can divert water that supports fish and wildlife, and tilling land can destroy the habitat of valuable species of flora and fauna.

Environmental policies in agriculture are designed to address these types of problems. The pollution from agricultural practices is controlled through regulations such as those limiting pesticide use and setting water quality standards. Depletion of resources is partly addressed through subsidization of practices such as low tillage (as in the Environmental Quality Incentives Program, or EQIP) and through the establishment of the Conservation Resource Program (CRP) and other purchasing fund programs (Wu and Babcock 1996) that fund environmental and resource conservation activities.

We present two modeling frameworks examining the impact of agricultural risk management on the environment. These models include a presentation of the agricultural economy and the relationship between agricultural activities and environmental quality. Production activities affect the environment through variable inputs, application technologies, and land allocation. In the first model, we rely on the results of Sandmo (1971) to analyze the effects of crop insurance on the environment and to determine the impacts on the choice of variable inputs in the production of a crop. The second model considers the impacts of risk management strategies on the adoption of available "green technologies."

The Environmental Impacts of Crop Insurance

Suppose an industry produces a crop using variable inputs on production units (e.g., fields) of the same size. Assume that each producer is risk averse and these production units are heterogeneous in productivity, risk, and environmental amenities. (Note that for the moment we omit the influence of size differences on risk aversion and instead will examine that issue in the next model.) The utility of producers is a function of profit π, so that $u = u(\pi)$, and we assume a decreasing marginal utility of profit $u'>0, u''<0$.

We utilize a Just-Pope production function that varies across production units. Output per unit is denoted by y and is dependent on the variable input per unit and two heterogeneity parameters: α, which is a productivity parameter, and β, which is a risk parameter. Specifically,

$$y = af(x) + \beta g(x)\varepsilon ,$$

where ε is random with $E(\varepsilon) = 0$. Thus, expected output is $E(y) = af(x)$, and the variance of output is $\beta^2[g(x)]^2\sigma^2$, where $\sigma^2 = E(\varepsilon^2)$. We consider the situations where the input is risk-reducing $(g'<0)$ and also the situations where it is risk-increasing $(g'>0)$. We assume that the elements of production functions are well behaved to ensure internal solutions; thus, we expect that $f''<0$. We also assume that production requires a fixed per-acre cost, c, and initially assume a given output price, p, and a given price of variable input, w.

Let z_1 and z_2 be the measures of the environmental side effects of a production unit. Pollution per unit, z_1, is a function of input use and the pollution parameter γ, so that

$$z_1 = \gamma h(x) ,$$

with $h'>0$, $h''>0$. The parameter γ captures differences in the impacts of pollution. That is, a farm geographically located closer to a city may generate more realized pollution and is therefore associated with a higher γ. The loss of environmental amenities per unit occurring because of production $z_2 = \delta$ varies across production units. This variable can be interpreted as the value of native plants and wildlife that would have utilized the resources of a production unit, if it had not gone into operation.

Let us assume that β is the only source of heterogeneity in production across production units, and thus $\alpha = 1$. The total number of production units in the industry is N_0, and β has a density function $\psi(\beta)$ defined on $\underline{\beta} < \beta < \overline{\beta}$, so that $\int_{\underline{\beta}}^{\overline{\beta}} \psi(\beta)\, d\beta = 1$. The number of units in the interval $\beta_1 - \Delta\beta/2 \leq \beta \leq \beta_1 + \Delta\beta/2$ is represented as $N_0 \psi(\beta) \Delta\beta$. When each unit maximizes the expected utility of profits, the optimal input level is determined by solving $\max_{x \geq 0} E\{U(p[f(x) + \beta g(x)\varepsilon] - wx - c)\}$, and a necessary condition holding at the optimal input use level x^*, when $x^* \geq 0$, is

$$E\{U'(\pi) \cdot [p[f'(x)] + \beta g'(x)\varepsilon] - w]\} = 0. \tag{8}$$

Following Feder's (1977) generalization of Sandmo, the first-order condition (8) can be rewritten as

$$pf'(x) - w + RAV = 0, \tag{9}$$

where

$$RAV = \frac{\operatorname{cov}[U'(\pi), \beta pg'(x) \cdot \varepsilon]}{E\{U'(\pi)\}}.$$

At the optimal, $x = x^*(> 0)$, the average value of the marginal product of the variable input $pf'(x^*)$ is equal to the price of the variable input amended by a risk adjustment variable (RAV). This variable is the covariance of the marginal utility of profit, $U'(\pi)$, and the value of marginal risk of input use, $\beta pg'(x) \cdot \varepsilon$, divided for monetizing purposes by average marginal utility, $E\{U'(\pi)\}$. This variable can be rewritten as

$$RAV = \frac{\beta pg' \operatorname{cov}[U'(\pi), \varepsilon)}{E\{U'(\pi)\}}. \tag{10}$$

We assume a decreasing marginal utility of profit, and since a high π is associated with a high value of ε, the covariance between $U'(\pi)$ and ε is negative. Therefore, the risk adjustment coefficient is negative when the input is risk-increasing, $g'(x) > 0$, but negative when the variable input is risk-reducing, $g'(x) < 0$. The optimal input use level when there is no risk, x_c, is determined when $pf'(x_c) = w$. Figure 2 relates the optimal input use under risk aversion, x^* to x_c.

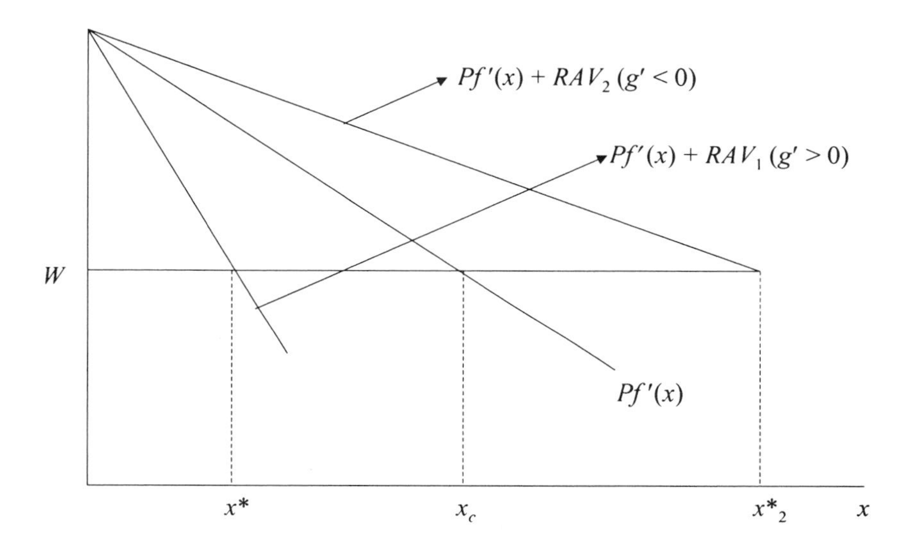

Figure 2. Optimal Input Under Risk Aversion

As Figure 2 indicates, risk aversion leads to an increase in input use when the input is risk-reducing (at $x = x_2^*$), and to a reduction in input use when the input is risk-increasing (at $x = x_1^*$). The level of input use is also influenced by the level of risk as represented by the risk parameter β. Metcalfe and Zilberman (2001), using the methodology of Sandmo, demonstrate that an increase in the value of β results in an increase in x^* when x is risk-reducing, and leads to a decrease in x^* when x is risk-increasing.

For a firm to operate with $x > 0$, expected utility has to be positive, namely, $E\{U(p[f(x) + \beta g(x)\varepsilon] - wx - c)\} \geq 0$. For a given solution, an increase in β reduces expected utility since it increases risk without increasing mean profit. Adjustment of x reduces the cost of risk bearing but also tends to reduce average profit. Thus, there is a critical β value above which production units will not operate. Let this critical value of β be denoted by $\hat{\beta}$. Thus, the expected supply of the industry at a given period will be

$$N_0 \int_0^{\hat{\beta}} \left(f[x^*(\beta)] + \beta g[x^*(\beta)]\right) \psi(\beta) d\beta. \tag{11}$$

The level of pollution, z_1, produced by a unit with a given β is $h[x^*(\beta)]$. If x is risk-increasing, firms with larger β's produce less pollution because they use less input. However, if x is risk-reducing (as Babcock claims is the case with fertilizers and pesticides), firms with a larger β will have a larger input use and thus will be more polluting. In our case, expected aggregate pollution denoted by Z_1 is determined by

$$Z_1 = \int_0^{\hat{\beta}} (\bar{\gamma} \mid \beta) h[x^*(\beta)] \psi(\beta)\, d\beta, \tag{12}$$

where $(\bar{\gamma} \mid \beta) = E(\gamma \mid \beta)$ is the average damage per unit of externality for units with a given β.

Agricultural production uses resources (land) that is able to provide environmental amenities. The aggregate expected loss of amenities, Z_2, through production is thus

$$Z_2 = \int_0^{\hat{\beta}} (\bar{\Gamma} \mid \beta) \psi(\beta) d\beta, \tag{13}$$

where $E(\delta \mid \beta) = \bar{\Gamma} / \beta$ is the expected loss of environmental amenities per unit of production with a given β.

Metcalfe and Zilberman (2001) demonstrate that an increase in the riskiness of ε (measured, by a change in its variance, as σ_ε^2) will lead to an increase in RAV and a deviation from the risk-neutral solution. An increase in the variance of ε will increase x^* for a given β when the input is risk-reducing, and will reduce x^* when x is risk-increasing. Higher riskiness will also reduce the critical value of β. Insurance can be viewed as an arrangement that either reduces the variance of ε or reduces the value of β for all producers.

Suppose that after the introduction of insurance, the risk random variable ε is transformed to $\varepsilon_1 = \eta\varepsilon$ with $0 \leq \eta < 1$. The analysis can be pursued in this case if one uses the same ε but every β is modified to $\beta = \eta\beta$. In this case, the range of values of β with $x^* > 0$ will increase to $\hat{\beta}/\eta$. Let $x_\eta^*(\beta)$ denote optimal input use when cost-free insurance reduces the production risk to $\eta\varepsilon$. The change in input use due to insurance is then represented by

$$\Delta x_\eta^*(\beta) = x_\eta^*(\beta) - x^*(\beta),$$

where $x^*(\beta)$ is optimal use without insurance.

For simplicity, we ignore moral hazard and asymmetric information and consider insurance when there is full information (except, of course, regarding ε). When comparing the impact of insurance on aggregate pollution, note that the change in aggregate pollution after insurance is

$$\Delta Z_{1\eta} = \int_{\hat{\beta}}^{\hat{\beta}/\eta} (\bar{\gamma} \mid \beta) h[x_\eta^*(\beta)]\psi(\beta) d\beta + \int_0^{\hat{\beta}} (\bar{\gamma} \mid \beta) \{h[x_\eta^*(\beta)] - h[x^*(\beta)]\}\psi(\beta) d\beta \,. \tag{14}$$

where the first right-hand term is the extensive margin effect and the second right-hand term is the intensive margin effect. This change in aggregate pollution, due to insurance, is decomposed in equation (14) into the *extensive margin effect,* which is always positive, since the number of utilized production units increases with insurance, and into the *intensive margin effect*, which is negative when the variable input is risk-reducing (and thus x^* declines with insurance as less risk-reducing input is required) and positive when the input is risk-increasing (and thus x^* can be used more since there is now insurance to cover the increased risk). This suggests that introduction of insurance will unambiguously increase pollution if x is risk-increasing, and that the introduction of insurance will reduce pollution if x is risk-reducing and the reduction in the intensive margin effect dominates the increase in pollution due to the extensive margin effect.

The magnitude of the extensive vs. the intensive margin effect depends on the correlation between γ and β. When units with more riskiness (a higher β) generate more pollution (a higher γ), then there is a positive correlation between γ and β, therefore $\partial(\bar{\gamma}/\beta)/\partial\beta > 0$. This positive correlation increases the extensive margin vs. the intensive margin because the extensive margin is comprised of those production units with higher risk (a higher β). This positive correlation increases the likelihood that the introduction of yield insurance will lead to an overall increase in pollution. When there exists a negative correlation between γ and β, then more risky units generate less pollution damage per unit and the introduction of insurance may actually reduce the pollution level, as the intensive margin effect is likely to increase relative to the extensive margin effect when $\partial(\bar{\gamma} \mid \beta)/\partial\beta < 0$.

The additional loss in environmental amenities occurring from increased use of the production units in agriculture, as a result of insurance, is

$$\Delta Z_{2\eta} = N_0 \int_{\hat{\beta}}^{\hat{\beta}|\eta} (\bar{\Gamma} / \beta)\psi(\beta) d\beta \,. \tag{15}$$

Introduction of insurance will increase land use and, thus, reduce the resources available for ecological activities.

It should be noted that this analysis demonstrates that various measures of environmental quality may respond differently to risk management strategies. When a variable input is risk-increasing, both the measure of pollution levels and the measure of loss in amenities increase as a result of the introduction of insurance. Conversely, if a variable input is risk-reducing and the intensive marginal effect dominates the extensive marginal effect, we know aggregate pollution levels will decline, but the aggregate loss in environmental amenities will still increase. Therefore, it is important that appropriate measures be used to capture these different types of environmental quality effects.

Thus far, we have analyzed outcomes for a price-taking industry, but in actuality we are very concerned with the impact of insurance policies on producers who have significant market power. In these cases, our previous analysis can provide a starting point to assess the impact of insurance policies on supply, and one has to add a demand equation for each existing state of nature. In this case, the exact outcome depends on whether or not producers are able to use price insurance tools (futures markets, price supports, etc.) and then separate the tools for crop insurance. If they use only one of these tools, none of them, or even another option such as revenue insurance, then land allocation and input choice under each of these policies will vary substantially. That will result in both increased pollution and a loss of environmental amenities.

The analysis thus far ignores accumulation of inventories. We assume that the indicators of environmental quality are dependent on input use and land use, and therefore they are not deterministic in our analysis. However, if one allows inventory spillovers, then output prices, variable input, and land use will change over time in response to modifications in economic conditions and stock situations.

Just and Zilberman (1988) examine the relationship between management of production stock under uncertainty and the environment. They suggest that, in some cases, there is joint management of commodity stocks as well as of natural resources (say, water). They show that a treatment of commodity stocks to control risk and increase economic surplus that ignores the dynamics of the resource stock may be suboptimal. Similarly, management of resource stocks has to consider economic conditions and the dynamics of the resource stock.

The old theory developed by Tinbergen (1952) on targets and tools should not be ignored as we consider the relationship between risk management and environmental quality. If one is interested in the management of environmental quality, then it is useful to develop a tool to attain this specific objective. The development of these tools should, of course, take into account the use of other policy tools, for example, risk management strategies. Furthermore, analysis that leads to the design of risk management strategies must take into account the implications of preexisting environmental regulations. As Lichtenberg and Zilberman 1986 have shown, ignoring agricultural policies when considering environmental regulation may be misleading, and the opposite is true when one considers the design of agricultural and risk management strategies. Ignoring management policies in this case may lead to the wrong conclusions.

We formally examine the effect of insurance on the variance of production, but it should be noted that there is also a well-known effect on the production mean. Crop insurance reduces the downside risk of producers but does not impose any upper limit on higher yields and therefore the activity is not mean-preserving. The implications of an increase in the production mean are increases in profitability and also in the expected value of production. These increases in profitability and wealth imply increased input use and therefore an increase in overall environmental damage.

Risk Management, Technology Adoption, and the Environment

Environmental quality is affected by farmers' choices of cultivated acreage, variable input use, land use, and technology. The previous section presented a model to assess the impact of risk on the environment and its effect on cultivated acreage and variable input use. This section presents a conceptual approach and discussion of issues concerning the relationship between the introduction of new technologies under risk and the implications for the environment.

Some of the most important implications that agricultural practices impose on the environment have resulted from the introduction of new technologies. Mechanical cultivation, for example, drastically increased soil erosion and contributed to the intensification of agricultural production. Also, mechanization drastically altered agriculture by contributing to the migration of population from rural areas into the cities. Technology has thus changed the design of the rural scale by expanding modern farms and replacing the traditional farm communities, which provided valuable environmental and cultural amenities. Similarly, some of the most important environmental issues in agriculture relate to the use of synthetic chemicals. While these modern technologies have imposed negative environmental side effects, they have also provided significant benefits to agriculture.

Some of the most important challenges in agriculture today involve refining and modifying modern technologies while reducing environmental disruption and maintaining and improving production efficiency. New innovations such as low- and no-tillage, modern irrigation technologies (low-pressure sprinkler and drip irrigation, computerized irrigation, etc.), technologies known as precision farming, and precision application of inputs such as pesticides are examples of technologies that may significantly improve both environmental quality and production.

Caswell, Lichtenberg, and Zilberman (1990) presented a model that analyzed the impact of modern irrigation technologies on productivity and the environment. These new technologies were developed to increase the efficiency of variable inputs, be they water or chemicals. The input use efficiency is a measure of the fraction of applied inputs that are actually used in production. In many cases, pollution is a residue that is unutilized in the production process. Thus, an increase in input use efficiency can reduce the pollution per unit of applied input. These types of technologies may improve environmental quality by reducing applied input and, in particular, the residue associated with these inputs.

Khanna and Zilberman (1996) expanded this model to examine a wide array of technologies in agriculture and natural resources (also see National Research Council 2000). Adoption of these new technologies requires investment in new equipment, which may sometimes increase the cost of equipment and may actually increase the cost per unit of output. Sometimes it may also increase the cost of labor (for example, integrated pest management requires extra monitoring effort). However, the extra cost per unit of production may be compensated for by the increase in productivity, as in the case of modern irrigation technologies. In addition, introduction of a modern technology may require significant investments in learning, training, and improvement of farm infrastructure (purchase of new machinery and replacement of pipes and pumps). These fixed costs are not scale-dependent and, therefore, may give larger farms an advantage. That is, the fixed costs related to modern technologies may require extra credit; therefore, credit constraints from risks to both farmers and lenders may impede adoption and improvements in environmental quality.

Another risk-related feature of a new technology is the significant uncertainty about its performance. While the main element of the variances of traditional technologies reflects the impact of randomness to climate and prices, when it comes to new technologies, the risk perceived by farmers may also reflect the lack of knowledge about the feature, the capability of the new technology, and the fit of the technology to the particular needs and circumstances of the farm and the farmer. In addition, there are differences in the results obtained by a new technology in an experimental plot and in actual field performance. In many cases, technologies need time before they can reach their potential, and that contributes to the uncertainty about new technologies. Marra and Carlson (1987) present a wide array of modeling approaches for adoption under conditions of risk and uncertainty. We choose a relatively simple portfolio framework introduced by Just and Zilberman to further analyze the relationship between risk management adoption and the environment.

Just and Zilberman's (1988) model on adoption under uncertainty is used to analyze the impacts of risk management on the environment. Suppose a farm has $\overline{L}$ total acres, and we consider technologies (or crops) with constant returns to scale. Let π_1 and π_2 be profits per acre of technologies 1 and 2, respectively. These profits are assumed to be distributed normally. Let mean profits per acre of technology i be denoted by $\mu_i = E(\pi_i)$, let the variance of profits per acre of technology i be σ_i^2, and let the covariance between the per-acre profits of the technologies be $\sigma_{12} = \rho\sigma_1\sigma_2$, where ρ is a correlation coefficient.

Assume that technology 1 is a modern technology that results in less pollution. Let z_1 be pollution (or the externality) per acre of this technology, and let z_2 be pollution per acre of the traditional technology – therefore, $z_1 < z_2$. We also assume that the new technology requires an annualized fixed cost investment of K dollars. Assuming risk-averse behavior, the farmer has to choose (i) whether to adopt or not adopt, and also (ii) if adopting, how much land to allocate to the modern technology. The expected utility of profits under adoption is denoted by $U_A = \max_{L_1} EU(\pi_1 L_1 + \pi_2 L_2 - K)$, and the decision of the farmer is therefore

$$U_A = \max_{L_1} EU[L_1(\pi_1 - \pi_2) + \pi_2 \bar{L} - K],$$

where $L_1 > 0$ and $L_2 \geq 0$. More trivially, expected utility when the new technology is not adopted, $(L_1 = 0)$, is

$$U_N = EU(\pi_2 \bar{L}).$$

Adoption occurs when $U_A > U_N$ and is partial when $0 < L_1 < \bar{L}$. Full adoption may occur if the new technology is, on average, more profitable $(\mu_1 > \mu_2)$ and risk considerations are not sufficient to lead to diversification. When the farm is relatively small and gains from adoption are insufficient to cover the fixed cost, adoption will not occur at all.

Land diversification, the amount of land in the modern technology, is shown by Just and Zilberman (1988) to be equal to

$$L_1^* = \frac{\mu_1 \mu_2}{v(\pi_1 - \pi_2)\phi(w)} + \frac{\sigma_2^2 - \sigma_{12}}{v(\pi_1 - \pi_2)},$$

where $v(\pi_1 - \pi_2) = \sigma_1^2 + \sigma_2^2 - 2\sigma_{12}$, and $\phi(w)$ is a measure of absolute risk aversion dependent on the level of wealth.

To better interpret the right-hand side of the equation, note that expected profits increase linearly with the land area allocated, but the variance of profits is a quadratic function of the area allocated to various crops. The first element of the right-hand side of the equation represents the trade-off between changes in expected profits and increases in variance due to the marginal transition of land from the traditional to the modern technology. The second element reflects the correlation effect. It suggests that an increase in farm size will increase the acreage allocated to technology 1 if the profits of the two technologies are not highly correlated.

Just and Zilberman (1988) argue that, under most plausible conditions, the modern technology will be considered if it has either a higher profit per acre or a lower variance relative to the traditional technology. The less correlated the risks of the two technologies, the higher the adoption rate becomes. From the perspective of a grower, the variance of profit per acre for a technology can be decomposed into variances that reflect randomness of yields, prices, etc., as well as variances that reflect uncertainty concerning the properties of the technologies. New technologies are shrouded with much higher uncertainty and, therefore, it is quite plausible that the overall risk is perceived to be higher than with traditional technologies. Thus, it is quite plausible to assume that $\sigma_1^2 > \sigma_2^2$.

When the variance associated with the modern technology is relatively higher than the variance of the traditional technology, the modern technology must provide a higher expected profit per acre (e.g., $\mu_1 > \mu_2$) in order to be adopted. Therefore, the adoption rates of modern technologies can be enhanced by environmental regulations or resource policies that on average increase their profitability relative to traditional technologies. Indeed, adoption of water-saving modern technologies increases when water prices are higher (Caswell and Zilberman 1986). It should be noted that the expected profit we consider is not necessarily the true expected profit but instead the expected profit as perceived by the farmer. As Feder and O'Mara (1982) suggest, the expected profit from a new technology may be underestimated by farmers who consider adoption. Indeed, policymakers have been quite frustrated with low adoption rates of technologies such as integrated pest management, low tillage, and modern irrigation technologies, especially since in some cases there seems to be significantly higher expected profits available.

When the cleaner modern technologies have a higher expected profit, Just and Zilberman (1988) show that, under plausible conditions, their adoption is likely to increase by reducing their risk per acre. Profit per acre is equal to revenue minus cost, and risk reduction has to reduce the risk of some, if not all, of the components. Not many policies reduce the variance of profit without affecting the mean. Forward contracting may eliminate price risks, but still growers will be exposed to yield risk. Crop insurance increases the downside yield risk and also increases expected profit. Similarly, revenue insurance reduces downside yield risk and also increases expected profit. Thus, introduction of policies such as price supports or revenue insurance that favor the modern, cleaner technology will lead to its further adoption. Examples of such policies are programs that condition entitlement to government insurance on adoption of desirable practices (e.g., programs such as EQIP).

Many risk-reducing strategies are introduced by the private sector in its attempts to encourage adoption of new technologies. Dealers of new technologies promote these technologies through demonstrations, which aim to reduce uncertainty about performance of the technology, and also through money-back guarantees that reduce risk, increase yield, and sometimes even provide yield guarantees.

Contracting is another form of policy that aims to reduce the risk associated with adoption of new technologies. If the new technology is embodied in a new product, the supplier of inputs may also provide a contract to buy the product at a certain price. In principle, this contract eliminates most of the yield and price risks and its mechanisms assure fast adoption. The fast growth of the broiler industry in the United States is credited to some extent to the marketing contracts that were provided by seed companies. Prices for broilers were rather unstable, and not many farmers were ready to risk adopting this new form of farming. Thus, major seed producers assured farmers a market for their final product, and that led to the adoption of broilers. Of course, broilers are not necessarily an example of a green technology, but there is evidence of growers who switch to organic agriculture or pesticide-free products because of a contract with a potential buyer.

The magnitude of the effect on the adoption of green technologies occurring from traditional risk management strategies, such as price supports, and crop and revenue insurance, depends on the relative impact on the mean and variance of profits per acre for the new technologies relative to the traditional ones. If the new technology is embodied in a new product that is not covered by crop insurance or price supports, then these risk processes actually discourage adoption and do not contribute to environmental quality and improvement. This indeed may be the situation when entitlements to government programs are tied to traditional crops – the adoption of alternative crops being thus discouraged. Increased flexibility of commodity programs, particularly insurance programs, is thus desirable and may provide a better incentive for the adoption of greener and more nontraditional crops.

CONCLUSIONS

This chapter examines some of the issues related to risk and environmental policy in agriculture and demonstrates the importance of correctly incorporating this risk into economic policy analysis. It highlights the importance of the interaction that exists between economic and environmental factors, such as the specification of the technologies that are utilized in agricultural production, the use of multidisciplinary knowledge in economic model development, the incorporation of heterogeneity across both production units and geographical locations, the consideration of multidimensionality in environmental problems, and the incorporation of different sources of variability.

The risk-generation model was presented to formally develop the risk processes involved in health risk generation, and the cost-minimizing safety rule approach was used to capture the uncertainty associated with limiting this risk. The importance of the risk-generation methodology was shown to lie in its ability to allow for policies to intervene at various points in the risk-generation chain, thereby allowing policymakers greater flexibility when designing environmental policy to achieve greater risk reduction at lower overall costs.

The strategies undertaken by farmers to manage risk in agriculture have important implications that influence environmental quality. Risk management strategies can alter resource allocations and, thus, negatively impact the environment through the externalities that are generated during production, through the depletion of the natural resources that are used in production, and also through the diversion of environmental amenities that would otherwise provide habitat for various flora and fauna.

Important results from this analysis demonstrate that the use of risk management strategies can directly influence environmental quality and that the use of a variable input that is either risk-increasing or risk-reducing is an important consideration in determining the direction and the magnitude of these effects. Also examined are the factors influencing the adoption of modern "green" technologies, and it is shown that the rate of adoption for these new technologies is dependent on the average profit as well as on the variance of profit generated by the technology.

There are many strains of literature examining the effects of risk and uncertainty on environmental policy in agriculture: this chapter only scratches the surface as to the implications of risk. It is clear that economic analyses need to incorporate an understanding of many issues in order to quantify the trade-offs from risk management and become a useful tool in the development of environmental policy under risk. The risk issues discussed in the other chapters of this volume can be used as inputs in this process and help to create a better understanding of the implications of risk and risk management for the environment.

REFERENCES

Babcock, B., E. Lichtenberg, and D. Zilberman. 1992. "Impact of Damage Control and Quality of Output: Estimating Pest Control Effectiveness." *American Journal of Agricultural Economics* 74: 163-172.

Bogen, K. 1990. *Uncertainty in Environmental Health Risk Assessment*. New York: Garland.

Byrd, D.M. 1997. "Goodbye Pesticides? The Food Quality Protection Act of 1996." *Regulation* (Fall): 57-62.

Caswell, M., E. Lichtenberg, and D. Zilberman. 1990. "The Effects of Pricing Policies on Water Conservation and Drainage." *American Journal of Agricultural Economics* 72: 883-890.

Caswell, M., and D. Zilberman. 1986. "The Effects of Well Depth and Land Quality on the Choice of Irrigation Technology." *American Journal of Agricultural Economics* 68: 798-811.

Cropper, M., W. Evans, S. Berardi, M. Ducla-Soares, and P. Portney. 1992. "The Determinants of Pesticide Regulation: A Statistical Analysis of EPA Decision Making." *Journal of Political Economy* 100: 175-197.

Crouch, E., and R. Wilson. 1981. "Regulation of Carcinogens." *Risk Analysis* 1: 47-57.

Feder, G. 1977. "The Impact of Uncertainty in a Class of Objective Functions." *Journal of Economic Theory* 16: 504-512.

Feder, G., and G. O'Mara. 1982. "On Information and Innovation Diffusion: A Bayesian Approach." *American Journal of Agricultural Economics* 64: 145-147.

Hanemann, M., E. Lichtenberg, and D. Zilberman. 1989. "Conservation versus Cleanup in Agricultural Drainage Control." Working Paper No. 88-37, Department of Agricultural and Resource Economics, University of Maryland, College Park.

Harper, C.R., and D. Zilberman. 1992. "Pesticides and Worker Safety." *American Journal of Agricultural Economics* 74: 68-78.

Hochman, E., and D. Zilberman. 1979. "Two-Goal Environmental Policy: An Integration of Micro and Macro ad hoc Decision Rules." *Journal of Environmental Economics and Management* 6: 152-174.

Hueth, D., and U. Regev. 1974. "Optimal Agricultural Pest Management with Increasing Pest Resistance." *American Journal of Agricultural Economics* 56: 543-552.

Innes, R. 2000. "The Economics of Livestock and Its Regulation." *American Journal of Agricultural Economics* 82: 97-117.

Just, R., and R. Pope. 1978. "Stochastic Specification of Production Functions and Economic Implications." *Journal of Econometrics* 7: 67-86.

Just R., and D. Zilberman. 1988. "A Methodology for Evaluating Equity Implications of Environmental Policy Decisions in Agriculture." *Land Economics* 64: 37-52.

Kataoka, S. 1963. "A Stochastic Programming Model." *Econometrica* 31: 181-196.

Khanna, M., and D. Zilberman. 1996. "Incentives, Precision Technology, and Environmental Protection." *Ecological Economics* 23: 25-43.

Lichtenberg, E. Forthcoming. "Agriculture and the Environment." In B. Gardner and G. Rausser, eds., *Handbook of Agricultural Economics.* Amsterdam: Elsevier.

Lichtenberg, E., and D. Zilberman. 1986. "The Welfare Economics of Price Supports in U.S. Agriculture." *American Economic Review* 76: 1135-1141.

___. 1988. "Efficient Regulation of Environmental Health Risks." *Quarterly Journal of Economics* 103: 167-178.

___. 1989. "Regulation of Marine Contamination under Environmental Uncertainty: Shellfish Contamination in California." *Marine Resource Economics* 4: 211-225.

Lichtenberg, E., D. Zilberman, and K. Bogen. 1989. "Regulating Environmental Health Risks Under Uncertainty: Groundwater Contamination in California." *Journal of Environmental Economics and Management* 17: 22-34.

Marra, M., and G. Carlson. 1987. "The Role of Farm Size and Resource Constraints in the Choice Between Risky Technologies." *Western Journal of Agricultural Economics* 12: 109-118.

Metcalfe, M. 2000. "Environmental Regulation and Implications for the U.S. Hog and Pork Industries." Ph.D. Dissertation, Department of Agricultural and Resource Economics, North Carolina State University.

Metcalfe, M., and D. Zilberman. 2001. "The Environmental Implications of Risk Management in Agriculture." Working Paper, Department of Agricultural and Resource Economics, University of California, Berkeley.

National Research Council (NRC). 2000. "The Future Role of Pesticides in U.S. Agriculture." Washington, D.C.: National Academy Press.

Rawlins, S. 1998. "Overview of the Food Quality Protection Act of 1996." American Farm Bureau Federation, FQPA Briefing Book (July).

Sandmo, A. 1971. "On the Theory of the Competitive Firm Under Price Uncertainty." *American Economic Review* 61: 65-73.

Schierow, L. 2000. "FQPA: Origin and Outcome." *Choices* (third quarter): 18-20.

Spear, R.C. 1978. "Organophosphate Residue Poisoning Among Agricultural Workers in the U.S.: Towards a Strategy for a Long Term Solution." CRES Report AS-R-18, Australian National University, Sidney.

Sunding, D., and J. Zivin. 2000. "Insect Population Dynamics, Pesticide Use, and Farmworker Health." *American Journal of Agricultural Economics* 82: 527-540.

Tinbergen, J. 1952. *On the Theory of Macroeconomic Policy.* Amsterdam: North-Holland.

U.S. Environmental Protection Agency. 1999. "EPA State Compendium: Programs and Regulatory Activities Related to Animal Feeding Operations." U.S. EPA (August).

Wu, J., and B. Babcock. 1996. "Purchase of Environmental Goods from Agriculture." *American Journal of Agricultural Economics* 78: 935-945.

Zivin, J., and D. Zilberman. Forthcoming. "Optimal Environmental Health Regulations with Heterogeneous Populations: Treatment versus Tagging." *Journal of Environmental Economics and Management.*

Part 6

CONCLUSIONS REGARDING THE SIGNIFICANCE OF RISK RESEARCH FOR AGRICULTURE

Chapter 24

HOW MUCH DOES RISK REALLY MATTER TO FARMERS?

Wesley N. Musser and George F. Patrick
University of Maryland and Purdue University

INTRODUCTION

Agriculture is inherently risky. Farm production is a biological process subject to unpredictable weather, diseases, and biological pests. In addition, farming is spatially dispersed on heterogeneous soils. Weather and spatial dispersion particularly affect crops and grazing livestock. In contrast, confinement production of animals partially controls production risk. This biological uncertainty is a fundamental cause of agricultural price uncertainty. Price uncertainty in crops is mirrored in livestock and poultry feed price uncertainty, causing output price uncertainty even for confined livestock and poultry production. Over the past 40 years, a number of changes in the U.S. agricultural sector have further increased risk. Substitution of income payments for high price supports in federal commodity programs resulted in major crop prices being determined with markets rather than policy price floors. In addition, increasing international trade in U.S. agricultural commodities in the 1970s resulted in output prices being subject to supply and demand shifts throughout the world, as well as in this country. Increases in size of farm operations and the pace of technological change have increased the managerial complexity of farming and the need for farmers to manage risk. Urbanization of farming areas has increased conflicts with new neighbors. A final change is increased environmental and other regulations in farming that influence risk management and introduce other sources of risk. These themes of termination of commodity programs, globalization of markets, increased managerial complexity, increased neighbor conflicts, and increased governmental regulations are fundamental shifts in the risk environment of U.S. farmers over the past 50 years.

Given the above characteristics, it seems naïve, if not silly, to ask if risk matters to producers. Coping with risk is and always has been an important part of farm management. Nevertheless, this characteristic does not tell us how risk affects farm management or how important it is. Risk obviously affects the choice set of farmers, but is risk an important part of the decision criteria of farmers? Even if risk aversion exists, does risk require active management? Or, does risk get controlled as expected profits are maximized or increased? These issues are the concerns of this chapter.

We do not provide complete answers to these issues. We will review evidence in the agricultural economics literature and suggest our views. We will proceed as follows. First, we review sources of risk and management responses to these risks. Next, we review empirical evidence on the importance of these sources and modern farm management responses. In the third section, we evaluate the implications of this evidence for the importance of different sources of risk in decision objectives and the choice set. Given these implications, we then briefly review models and methods used in past risk research and their relation to our inferences in previous sections of the chapter. We then conclude with some suggestions for future directions.

SOURCES OF AND RESPONSES TO RISK

Different classifications of sources of farm risk are available. For example, Barry et al. (1995) list the following types: (1) production and yield, (2) market and prices, (3) severe casualties and disasters, (4) social and legal, (5) human management and labor, (6) technological change and obsolescence, and (7) finance. Alternatively, Baquet, Hambleton, and Jose (1997) have identified five major sources of risk: (1) production, (2) marketing, (3) financial, (4) legal and environmental, and (5) human resources. We will use the second list in this section as it has fewer sources. Each source of risk is briefly explained, and then some possible management responses are summarized.

Production risk concerns variations in crop yields and in livestock and poultry production due to weather, diseases, and pests. Some common risk responses are the following:

- Diversification, which includes combinations of products, spatial dispersion, different varieties of crops, and different stages of production for animals and perennials
- Input use includes irrigation, pesticides, fertilizers, and veterinary medicine
- Management information such as soil tests, scouting, feed nutrient testing, and production records
- Crop insurance

Marketing risk concerns variations in commodity prices and/or quantities that can be marketed. Among the responses are the following:

- Marketing plans
- Forward marketing techniques such as futures, options, and cash forward contracts
- Sequential marketing – marketing several times per year
- Marketing contracts that guarantee price and quantity
- Direct sales to consumers
- Vertical integration

Financial risk is the ability to pay bills when due, to have money to continue farming, and to avoid bankruptcy. Risk responses are the following:

- Farm records
- Cash flow planning
- Controlling and limiting debt
- Pacing investments over time
- Maintaining financial/credit reserves
- Off-farm work and investments
- Controlling family consumption

Legal and environmental risk concerns the possibility of lawsuits initiated by other businesses or individuals, disputes over contractual agreements, and changes in government regulations of pollution and other farm practices. Following are some risk responses:

- Liability insurance
- Nutrient management plans
- Retaining legal counsel
- Good neighbor relations
- Road vehicle maintenance

Human resources risk is the threat that owners, family, and employees will not be available for farm labor and management. Useful risk responses are the following:

- Strategic business planning
- Estate and ownership succession planning
- Modern personnel practices and human resource management
- Health, disability, and life insurance
- Pre-nuptial agreements and marriage counseling
- Using management consultants

EMPIRICAL EVIDENCE

Agricultural economists have conducted several surveys on sources of and responses to risk among farmers. The first two parts of this section summarize parts of these surveys. Some standard USDA surveys and other data also have implications that are summarized in the third part of this section. The final part reviews some survey data on willingness to take risk.

Sources of Risk

Commodity prices, weather, and other factors affecting production consistently receive the highest ratings as the primary sources of risk faced by farmers and ranchers. In a 1983 survey of a limited number of producers in 12 states (Patrick et al. 1985), these factors received ratings of over 4 on a 5-point scale. A broad range of sources of risk including safety and health, family plans, credit cost, and credit availability were in the list of sources of risk, but with the exception of health and safety, their ratings were substantially lower.

Table 1 summarizes the ratings of importance of a number of risk sources for participants in the Top Farmer Crop Workshop at Purdue University in 1993, 1997, and 1999 (Patrick and Collins 2000). A different number of sources was used each year. For example, 16 were used in 1999. Table 1 includes the nine highest-rated sources in 1999. Participants in this workshop have larger acreages, more education, and are younger than most farmers in the Eastern Corn Belt. Commodity prices and yields are highly rated in all three years. Injury, illness, or death of the operator was the most highly rated source of risk in 1993 at 4.35, while by 1999 at 3.82 it was the lowest rated of the nine sources of risk in Table 1. Although changes over time in the ratings of other sources of risk were statistically significant, the change for illness, injury, or death of the operator was not. In part, this lack of significance is due to the large standard deviation, indicating that respondents varied widely in their assessment of its importance in their individual situations. Research procedures can also influence ratings of sources of risk. Business arrangements with output purchasers was added as a source of risk in 1997 and was the third most highly rated source of risk. Sources of risk that were not highly rated by producers included changes in land rents, family labor force, family relationships, and credit availability. The average rating of these sources of risk had no statistically significant change in averages in the 1993 to 1999 period in part because their standard deviations are larger than the sources of risk in Table 1.

The 1999 survey of crop producers in Indiana, Nebraska, Mississippi, and Texas (Coble et al. 1999) confirms the importance of crop price variability (4.55) and crop yield variability (4.07) as sources of risk faced by crop producers. However, only six sources of risk were included in the mail survey. In contrast, limited-resource farmers in Mississippi rated changes in input

Table 1. Averages and Standard Deviations (in parentheses) of Ratings of Importance* of Highly Rated Sources of Risk for Top Farmer Crop Workshop Participants**

Source of Risk	1993 N=73	1997 N=41	1999 N=28
Crop price variability	4.12[b]	4.61[a]	4.61[a]
	(0.87)	(0.63)	(0.63)
Crop yield variability	4.08	4.49[a]	4.32[ab]
	(0.78)	(0.68)	(0.77)
Business arrangements with output purchasers	NA	4.12[a]	4.18[a]
	–	(0.75)	(0.86)
Cost of capital items	3.79[a]	3.95[a]	4.11[a]
	(0.89)	(0.89)	(0.92)
Government commodity programs	3.62[ab]	3.20[b]	4.00[a]
	(1.04)	(0.88)	(1.05)
Technology	3.86[a]	3.80[a]	4.00[a]
	(0.95)	(0.81)	(0.72)
Input costs	3.93[a]	3.90[a]	3.89[a]
	(0.82)	(0.80)	(0.97)
Injury, illness, or death of operator	4.35[a]	4.10[a]	3.82[a]
	(0.94)	(1.16)	(1.16)
Environmental regulations	4.17[a]	3.73[b]	3.82[ab]
	(0.77)	(0.78)	(1.06)

* Importance was evaluated on a Likert-type scale of 1 (not very important) to 5 (very important).

** Averages in different years with the same superscript are not statistically different.

costs and credit availability higher than crop yields and prices as sources of risk (Coble et al. 2000b).

Preliminary tabulations of a survey of cow-calf operations in Nebraska and Texas found the highest rated sources of risk that had the largest potential to affect income were cattle price variability, severe drought, and pasture yield variability (Hall 2000). Hog price variability was also the highest rated source of risk for Indiana and Nebraska hog producers (Patrick et al. 2000). Diseases and changes in environmental regulations were rated second and third. Limited resource farmers with livestock in Mississippi rated availability of livestock markets and livestock price variability as the sources of risk with the greatest potential to affect their income (Coble et al. 2000b).

Responses to Risk

Producers' responses to risk have also changed. In the 1983 multi-state survey, enterprise diversification, market information, and pacing investments were the highest rated responses to production, marketing, and financial risk, respectively (Patrick et al. 1985). Somewhat surprisingly, only 2 of the 21 risk responses were rated above 3.0 on the 4-point scale.

Table 2 summarizes the ratings of importance of the 8 of 20 responses to risk rated 3.85 or higher on a 5-point scale in 1999 by Top Farmer Crop Workshop participants in 1993, 1997, and 1999. Being a low-cost producer and having liability insurance have been the top-rated responses over time. It is interesting to note the change in importance associated with participating in the government farm program. Forwarding contracting and hedging are also highly rated. Production, marketing, financial, legal, and human resource sources of risk all have highly rated risk responses. Ratings of other risk responses, not included in Table 2, were in 1999 maintaining financial/credit reserves, 3.74; having a written marketing plan, 3.68; diversification of enterprises and using a marketing consultant, 3.64; buy/sell agreements, 3.36; crop or revenue insurance, 3.27; and prenuptial agreements, 2.56.

Table 2. Averages and Standard Deviations (in parentheses) of Ratings of Importance* of Risk Management Responses by Top Farmer Crop Workshop Participants**

Risk Management Response	1993 N=70	1997 N=41	1999 N=28
Being a low-cost producer	4.40^{b}	4.15^{b}	4.79^{a}
	(0.79)	(0.96)	(0.50)
Liability insurance	4.40^{a}	4.54^{a}	4.57^{a}
	(0.62)	(0.87)	(0.69)
Government program participation	3.86^{b}	3.49^{b}	4.32^{a}
	(1.04)	(1.12)	(0.86)
Forward contracting the selling price of crops	4.14^{a}	4.32^{a}	4.18^{a}
	(0.79)	(0.72)	(0.90)
Using production techniques which work under a variety of conditions	4.35^{a}	4.10^{ab}	3.93^{b}
	(0.66)	(0.74)	(0.83)
Hedging selling price of crops	3.62^{a}	3.78^{a}	3.93^{a}
	(1.22)	(0.94)	(1.05)
Life insurance for operator/key personnel	3.64^{a}	3.98^{a}	3.86^{a}
	(1.09)	(0.96)	(1.33)
Debt-leverage management	3.81^{a}	3.66^{a}	3.85^{a}
	(1.08)	(1.11)	(0.92)
Average of 20 responses to risk	3.47^{a}	3.54^{a}	3.56^{a}
	(0.51)	(0.44)	(0.49)

* Importance was evaluated on a Likert-type scale of 1 (not very important) to 5 (very important).

** Average values for the importance of a risk management response in different years with the same superscript are not statistically different.

The four-state survey of crop producers found the following ratings of risk responses: being a low-cost producer, 3.64; maintaining financial/credit reserves, 3.58; and diversification of farming enterprises, 3.27. Forward pricing and crop/revenue insurance were rated at 3.12 and 2.93, respectively (Coble et al. 1999). Maintaining good animal health and being a low-cost producer were the highest rated responses to risk for both types of livestock producers (Patrick et al. 2000 and Hall 2000).

Evidence from Other Sources

Data from other sources provide more information on human resources risk. Drury and Tweeten (1997) reported a divorce rate from survey responses from 1972 to 1993 of about 13 percent for farmers compared to 28.6 to 38.3 percent in metropolitan areas. Table 3 has census data on hired labor and farm-related injuries and deaths for farms with $100,000 or more in sales in Indiana and Maryland. About 60 percent of the farms in Indiana and 70 percent in Maryland have hired labor: these farms have to deal with human resource problems related to hired labor. The number of farms with injuries and deaths are much lower. More than 3 percent of the farms have injuries and fewer than one in 1,000 has a death. It must be stressed that these are only farm-related problems: off-farm or personal injuries and deaths would give a higher incidence when added to the totals.

Table 3. Human Resources Characteristics of Farmers with Sales of $100,000 or More in Indiana and Maryland

	Number of Farms	
Characteristic	Indiana	Maryland
Farm-related injuries		
operators and family members	207	40
hired workers	167	63
Farm-related deaths		
operators and family members	8	0
hired workers	2	0
Hired labor expenses	7,537	1,823
Farms	12,395	2,597

Source: USDA-NASS 1997, Part 14, Chapter 1, Indiana state-level data, Table 46; and Part 20, Chapter 1, Maryland state-level data, Table 4

Information about important risk responses is also in secondary data. Table 4 has census tabulations on numbers of farms with different crops. In Indiana, corn and soybeans dominate, with almost equal numbers of farms having both crops. Farms with both these crops are more than 80 percent of total farms. Obviously, many of these farms have a corn-soybean rotation. About 40 percent of the farms also have wheat, some of which is double-cropped with soybeans. Farms with at least two of these three crops have very limited diversification. Common rotations in Maryland are not as clear from the data. About half the farms have soybeans and about 35 percent have corn and wheat. Many dairy farms produce corn but are not among the farms marketing it in Table 4. Larger proportions of these farms have wheat and corn silage than in Indiana. A smaller percentage of farmers grow corn and soybeans in Maryland, so the proportion having a corn-soybean rotation may not be as high as in Indiana. However, the number of other crops being quite low suggests there is not much diversification in Maryland.

Table 4. Numbers of Farms with Different Crops for Farms with Sales of $100,000 or More in Indiana and Maryland, 1997

	Number of Farms	
Crop	**Indiana**	**Maryland**
Corn	10,224	962
Wheat	5,252	900
Soybeans	10,738	1,290
Oats	216	34
Hay and silage	1,672	283
Vegetables	315	238
Fruit and nuts	69	78
Total farms	12,395	2,597

Source: USDA–NASS 1997, Vol. 1, Part 14, Chapter 1, Indiana state-level data, Table 46; and Vol. 1, Part 20, Chapter 1, Maryland state-level data, Table 46

Other relevant data concern farm finance. Among the aggregate farm ratios that USDA reports, a debt-to-asset ratio of 15.8 percent is a measure of solvency, and a debt service coverage of 2.22 is a measure of liquidity (USDA-ERS 2001). Standard benchmark values for strong values are 40 percent or less for the debt-asset ratio and debt service coverage of 1.10 or more. Thus, farmers as a whole have considerable financial reserves for solvency and liquidity.

Producers' Willingness to Pay

The risk-return trade-off is fundamental to risk research and producers' risk management. However, some evidence indicates that some producers are unwilling to accept reduced average returns to reduce variability. For yield risk, 33.6 percent of cotton producers and 36.8 percent of corn producers indicated that if a new method of growing the crop was developed with no change in production costs, they would be unwilling to give *any* of their current average yield to get the same yield every year. A majority of the crop producers surveyed, 51.3 percent, did not agree with the statement, "My primary marketing goal is to reduce risks rather than raise my net sales price" (Coble et al. 2000a). More than two-thirds of hog producers surveyed (67.3 percent) did not agree with that statement (Patrick et al. 2000). Only 34.3 percent of crop producers indicated agreement with the statement, "I am willing to accept a lower price to reduce price risk." Of the hog producers, only 32.1 percent agreed with that statement. Among the limited-resource producers surveyed in Mississippi, only 10.9 percent agreed with the statement (Coble et al. 2000b).

RISK AVERSION OR EXPECTED PROFIT MAXIMIZATION?

When agricultural economists began actively studying farm risk, the expected utility model was the theoretical foundation of this research. Theories and methods were being developed in finance and economics to analyze risk problems. The most systematic effort had been in finance. Copeland and Weston list six theories about risk that they considered the foundation of modern finance theory: (1) utility theory, (2) state-preference theory, (3) mean-variance theory and the capital asset-pricing model, (4) arbitrage pricing theory, (5) option pricing theory, and (6) the Modigliani-Miller theories. Articles outlining these theories were published at least by 1977. Advances in the theory of the firm under risk were also being made – Sandmo in 1971 and Batra and Ullah in 1974 were particularly influential in agricultural economics. The theory of stochastic dominance was also being developed (Hanoch and Levy 1969, Hadar and Russell 1969, Meyer 1977).

These emerging economic theories were accompanied by a surge of literature on risk in agricultural economics. Eidman, Dean, and Carter in 1967 and Carlson in 1970 were early applications of decision theory in farm management. Halter and Dean completed an early textbook on risk analysis in farm management in 1971. Hazell also published his article on a linear programming approximation to quadratic programming in that year. In 1974, the influential articles by Just and by Lin, Dean, and Moore appeared. In 1975, Hazell and Scandizzo incorporated risk in an aggregate mathematical programming model. Articles on risk management by Barry and Willmann, and Barry and Fraser, appeared in 1976; Baquet, Halter, and Conklin also published an article on the value of information under risk in that year. Activity further accelerated in 1977. Anderson, Dillon, and Hardaker's treatise on risk appeared in that year. Robison and Barry presented and implemented a theoretical framework to analyze changes in risk, Boehlje and Trede had an article published on risk management, and Walker and Nelson released their survey of risk management literature in that year. The above review is just a sample of the surge in agricultural economics literature during this time period, all the articles of which were based on expected utility theory.

Expected profit maximization did receive some attention during this period. Just (1975) had an article on this topic, which was largely ignored.

Much of the firm growth literature used some form of net present value, which concerns profit maximization over time (WRCC-16 1977, Barry and Baker 1971). While it can be argued that the discount rate as the cost of capital incorporates risk aversion, risk aversion is not explicit. However, there was some consideration of multiple goals, including risk, and satisficing models (Patrick and Eisgruber 1968).

One must also recognize the contribution of McCarl. His proposition was that risk aversion is not necessary to explain behavior if the risk choice set is fully specified. Brink and McCarl (1978) was one of the first articles to make this point. Subsequent articles are reviewed in Musser, McCarl, and Smith (1986). The number of these studies is fewer than those having risk in the objective function.

It is ironic that the firm growth literature focused on finance, while the early risk aversion research, with the exception of Barry and Willmann (1976), focused largely on production. Our review of empirical evidence indicates that finance is one of the sources of risk where risk aversion is clearly present. Aggregate financial ratios indicate that many farmers are holding liquidity and solvency reserves that are much larger than what is commonly considered as adequate, indicating a risk-returns trade-off. The importance of liability insurance also indicates a risk-returns trade-off in the legal source. Some 97 percent of Top Farmer Crop Workshop participants in 1999 reported having liability insurance (Patrick and Collins 2000). Segerson (1986) also notes that environmental policies can require firms to share the environmental risk. Human resources is obviously also a source of risk, but we have very limited evidence about farmers' responses – Tauer (1985) shows that risk-averse individuals often prefer life insurance over installment payments to finance intergenerational transfer of property, and Tauer (1995) documents the change in efficiency of farm managers at different ages.

Logic and empirical evidence suggests that finance, legal, and human resources are types of sources of risk where risk aversion is fundamentally related to survival. A firm with liquidity and solvency problems, legal problems, or human casualties is seriously threatened with failure. Risk responses for these sources can assist in ameliorating these threats. Now, these responses mostly have risk-returns trade-offs, but survival seems to be the motive here. Patrick, Loehman, and Fernandez (1984) documented that survival is an important goal for farmers. Depending on their goal orientation, farmers were willing to give up $638 to $4,264 of income to avoid a one percent increase in the possibility of bankruptcy. Foster and Rausser (1991) found evidence that threat of failure could be related to behavior. This idea is related to safety-first models. Safety-first formulations of risk choice have considerable intuitive appeal despite theoretical limitations. Target MOTAD, developed by Tauer (1983), was popular largely because it had solutions consistent with stochastic dominance, which was widely used in this period. Does this literature suggest that risk aversion is related to survival rather than being a fundamental goal itself? What about production and marketing risk?

Even though the survey data indicated that production was an important source of risk, the behavioral data was not necessarily related to risk aversion in the previous section. This outcome is consistent with the above reasoning. Production risk will not usually cause firm failure if the appropriate financial responses are in place – agricultural firms regularly survive droughts, frost, and other phenomena that regularly cause complete crop failure or liquidation of livestock. Why do many production responses reduce risk? The key here is that many also increase expected profits rather than have a risk-returns trade-off. The standard E-V frontier can demonstrate increasing expected returns with constant or decreasing risk. In Figure 1, moving from D to C increases expected returns but does not change risk. Going from D to B increases expected returns and decreases risk. While C would be preferred for an expected profit maximizer, researchers and farmers may know the technology to get to B but not C.

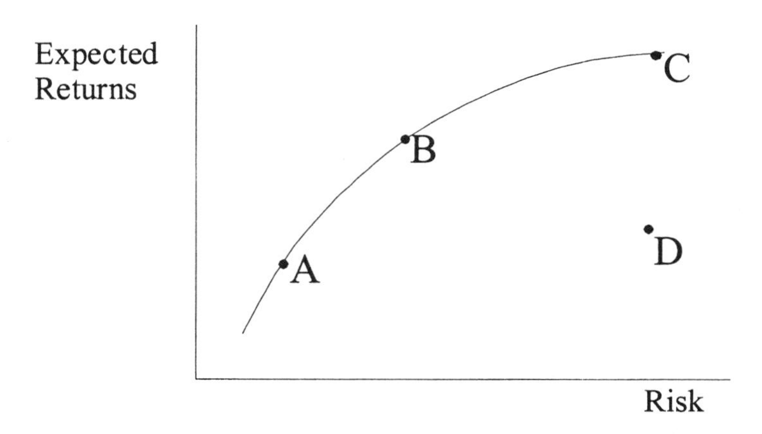

Figure 1. Expected Returns-Risk Trade-Off

Using the Just-Pope (1979) formulation of technology, we can further investigate this production risk response. Assuming that $y = f(x) + g(x)e$, then derivatives of the functions with respect to x_i are $\partial f / \partial x_i > 0$ and $\partial g / \partial x_i < 0$ when risk is decreased while returns are increased. Many of the input and information responses to production risk have these characteristics. Only if $\partial g / \partial x_i > 0$, such as for fertilizer (Just and Pope 1979), does risk increase with expected returns. In the more common cases, both risk-neutral and risk-averse individuals can ignore risk.

In these common situations, expected returns are increased because the lower part of the yield probability density function is truncated. For example, irrigation eliminates low yields caused by dry weather, extra machinery capacity allows planting and harvesting to be accomplished more quickly and offsets potential losses associated with inclement weather, pest scouting identifies pest populations that would cause yield losses, and feed testing precludes feeding animals a ration with inadequate nutrients. These responses have fixed costs that exist in all states of nature, while benefits only occur in part of the states of nature. In some cases, such as scouting, the fixed costs are low, while in others, such as irrigation, the fixed costs are much higher. The magnitudes of yield increases often are in direct proportion to these fixed costs. However, these technologies can decrease expected profits if the fixed costs are too high – too much machinery and an irrigation system on silt soils in humid areas are examples. In these cases, a risk-returns trade-off emerges. Many producers can avoid this trade-off by just maximizing expected profits. It should be noted that Carlson and Wetzstein (1993) and Boggess, Lacewell, and Zilberman (1993) review studies that indicate the importance of risk management in pest control and irrigation. Thus, the above logic must be considered a hypothesis for further study.

Diversification can also maximize expected profits. Product diversification seems to be limited, as discussed in the past section. The diversification that does occur often results in limited risk reduction and increases in expected profit. For example, let us examine soybeans and corn, which together form the most common rotation for crop farmers in the Midwest, the mid-Atlantic, and some areas in the South, as was discussed above. Musser and Stamoulis (1981) reported a correlation coefficient between returns for these crops in Georgia of 0.68, which may be higher in the Midwest. Corn and soybeans have several joint production relationships – this rotation largely eliminates problems with corn rootworms so the pesticide application on continuous corn is eliminated, soybeans can provide a limited amount of residual nitrogen for corn, and yields of both rotated crops are higher. In addition, fixed machinery and labor inputs can be more fully utilized with a rotation because planting and harvest times are different for the two crops. Thus, this rotation increases expected profits because yields are increased while insecticides, nitrogen fertilizer, machinery, and labor costs are reduced. While some rotations may have a risk-returns trade-off, this one does not. When farmers add wheat to this rotation, it may have more diversification effects but may also have further joint production relationships – planting and harvesting of small grains is at different times than planting and harvesting of row crops, so that labor and machinery are again more efficiently used. If wheat is double-cropped with soybeans, land use intensity and potential profit is increased. One cannot infer that rotations always are for risk reduction; if they were, farmers would not be so specialized.

The other risk response that has not been discussed is crop insurance. Insurance is usually considered a risk-averse response. However, the problems of adverse selection in this form of insurance could result in only those farmers who can increase expected profits buying crop insurance. Just, Calvin, and Quiggin (1999) found this result. The premium subsidy is now much higher today. With a 50 percent premium subsidy for 2001, almost all farmers should increase expected profits with this insurance.

Consideration of marketing risk can start with forward contracting, which has a large literature. Johnson (1960) and Stein (1961) developed the basic conceptual foundation for this research around 1960. An extensive literature has followed focusing on the optimal hedge ratio. However, surveys of farmers have found they have a much lower percent forward price than implied by these models – for example, see Goodwin and Schroeder (1994) and Musser, Patrick, and Eckman (1996). One suggested explanation for these results is that farmers perceive that forward contracting reduces expected returns (Anderson and Mapp 1966, Brorsen and Irwin 1996). Patrick, Musser, and Eckman (1996) found that farmers perceived that cash forward contracts provided price enhancement. While they also thought that these contracts provided price risk protection, the percent reporting this goal was lower than for price enhancement. While the evidence is mixed, farmers do not follow predictions of the risk-aversion models, and some evidence exists that they use them to enhance profits and prices.

Less work has been done on the other marketing responses. Goodwin and Kasten (1996) found that the mean number of times that Kansas farmers market is between three and four, depending on the crop; Patrick, Musser, and Eckman (1996) found that the majority of farmers in the Eastern Cornbelt attending the Top Farmer Crop Workshop marketed crops more than six times per year. However, the effect on risk and returns is not known. Marketing contracts and integration have been documented as reducing risk (Knoeber and Thurman 1995, Johnson and Foster 1994). Although Schrader (1986) and Rhodes (1995) proposed that these arrangements would likely increase efficiency and expected returns due to market coordination, empirical evidence is mixed. Johnson and Foster (1994) found reduced returns, while Zering and Beals (1990) found positive returns. Direct sales to consumers can increase profits (Govindasamy, Hossain, and Adelaja 1999), but their effect on risk is unclear. Overall, the effects of marketing responses on expected profits and risk are not as clear as for production responses, at least for the authors. However, limited evidence is available that marketing responses reduce risk and expected returns.

MODELS AND METHODS

Mapp and Helmers (1984) identify two overall approaches that have been used to model farm firm risk – mathematical programming and simulation. While both these approaches are used at the industry or national level, a third, the econometric approach, is more common in that area. This section will briefly discuss all three of these approaches.

Programming models have been used considerably in studying risk – Boisvert and McCarl (1990) review these applications. They have the advantage of being optimizing models so that comparative statics of a decision can be easily evaluated. Programming models are most useful when decision variables are continuous. They have been used primarily to evaluate the impact of risk in the decision criteria on optimal behavior. While discrete stochastic programming can model risk in the choice set, limited numbers of risky production activities or states of nature can be included before the model becomes too large. Thus, this method has limitations in analysis of risk choice sets that are important in production and marketing. This same problem is a limitation for multi-period analyses of finance decisions in which linkages between periods are stochastic.

Johnson and Rausser (1977) note that simulation is particularly useful for problems that involve risk, are dynamic, and have discrete decision variables. The first two are of obvious relevance in risk decisions, particularly involving finance. Many production and marketing decisions, such as crop acreage and amounts to hedge, are continuous choice variables, and simulation would require a search over all possible combinations of choice variables in order to be an effective method. However, many choice variables are discrete – purchasing more land or new machinery and adoption of new technology are usually discrete choices. In addition, simulation is useful in cases where data are limited. The use of biophysical simulators for firm and resource interactions was quite widespread in the early 1980s. These models had the advantage in situations with dynamic choice sets within a production period and when relevant probability distributions were unknown. Irrigation and pest control are examples of such decisions (Musser and Tew 1984). Compared to programming models, this approach does not have the optimizing relationships with economic theory, and ad hoc behavioral relations can be included.

Econometric analysis is useful in evaluating the effects of risk on decisions. One disadvantage is that it is limited to existing or easily collectible data on decisions. Such data sets are largely limited to static production and marketing decisions, which this chapter argues are not the decisions in which risk may be most important. Just (2000) has advocated development of a panel data set, which would facilitate research on other sources of risk. In addition, these uses have all the common limitations of econometrics. One that is most salient is specification error. As in programming models, risk aversion is often modeled as a residual variable to other common variables under risk neutrality. If an important physical or behavioral variable is not included, risk aversion can be supported if its measure is correlated with the omitted variable. For example, a correlation between a risk variable and a weather influence can falsely support risk aversion.

Where are we left with this modeling issue? All three of the methods have advantages and disadvantages. However, econometrics is being used the most today (Just 2000). Given our need to study finance, legal, and human resources risks, simulation needs to be re-evaluated. These decisions are dynamic, many are discrete, and most have very limited data, if any, available. Unfortunately, many younger discipline members have limited experience with this method and would have to do some retooling. Spreadsheet programs with @RISK add-ons may be easier to access than simulators previously used; these programs are an innovation that makes simulation more feasible.

WHERE DO WE GO FROM HERE?

Just (2000) recently stated that agricultural economists may have been examining risk in management and policy where it is of less importance. We tend to agree with him, but still do not want to downplay the importance of risk. Agriculture is inherently a risky industry. Our challenge is to delineate the importance of this risk for decisions and behavior. One distinction that is important is the impact of risk on the choice set and the decision criteria. The importance of risk indicates that risk does influence the choice set. One of our challenges is to fully specify risk in the choice set; the farm management and policy literature has many examples where failure to fully consider all the stochastic influences in the choice set biased the outcomes of the research. One bias is support of risk aversion rather than just risk neutrality. This bias is an obvious case of where risk aversion does not matter when the risk in the choice set is fully modeled. Besides the traditional case of enterprise diversification, adoption of technologies such as irrigation or larger machinery size may be desirable with expected profit maximization, but not profit maximization at mean response levels.

Another situation in which risk aversion does not matter is the case in which a choice can increase expected returns and decrease risk. Then, a risk-averse decision maker would likely make the same choice as a risk-neutral decision maker, so the outcome can be evaluated with expected profit maximization. We also gave examples of such a relationship with production inputs, production information, and enterprise diversification.

As the above discussion indicates, it is difficult to identify situations where risk matters for production. Agricultural economists have spent considerable time on research on enterprise diversification. However, joint production relations and allocation of fixed inputs can explain better than risk the limited set of crops that are produced. A similar situation seems to exist for responses to marketing risk. The literature on how farmers do not use optimal hedges is extensive. While the evidence is limited, it seems that producers are attempting to enhance price through their actions. A similar conclusion can be made about most of the other marketing responses.

The situation is different for the other three sources of risk. In finance, many situations support the importance of risk aversion – differences in returns among alternative securities and the term interest rate structure are two examples. In farming, the extremely low debt-to-asset ratios and the high liquidity ratios are consistent with risk aversion. In the legal area, use of liability insurance provides some evidence. While empirical examples of risk-averse behavior other than insurance are limited, the human resources area also seems to be an area where risk aversion is important. The lower rate of divorce for farm families may reflect the economic reality of farming rather than any real difference in attitudes. These three sources of risk are more important for firm survival than are production and marketing risk. While finance has received attention in the literature, research on risk in the other two areas is very limited. Compared to production and marketing, these areas mostly involve multi-period decisions, and firm or industry data are not as readily available (Just 2000).

With the importance of these three sources, how do we begin to increase emphasis on these areas? The emphasis on production and marketing risk is understandable. They are single-period decisions that can be modeled as short-run decisions, which is much easier than long-run decisions. In addition, price and production data are readily available. Finance has been given considerable attention, but lack of data has provided limitations to testing hypotheses, as in the production and marketing areas. Most of what we know about firm decisions in finance has come from use of mathematical programming and simulation, neither of which are in high favor these days. In farm management and production economics, the attention to legal and human resources risk has been even more limited. However, some literature can be used to begin. Environmental economists have been studying risks associated with different pollution policies. Collaboration with environmental economists or work in the firm risk area by production-resource economists seems like a logical beginning. For the human resources source, examining the impact of risks in the life cycle of a firm may be promising.

As Just (2000) notes, theoretical advances and more data are also necessary. As he suggests, farm panel data would be helpful. However, we might need very large panels to examine some of the legal and human resources risks that do not have a high incidence of manifestation. In those cases, some case studies might be useful in providing more knowledge. Purposeful identification of firms with legal or human resources problems for close study would provide some hypotheses. As with all case studies, identification of the impact of the event of interest compared with other ongoing economic events is difficult. Programming or simulation models may be helpful in isolating the effect of the risk event. In general, simulation and programming models deserve renewed attention and acceptance from the profession.

As for theory, it is difficult to be too optimistic. Intertemporal decisions are complex simply because so many decision variables exist. Aggregation is then likely – the standard theory of consumption over time has total annual consumption rather than individual commodities as choice variables. As one moves towards empirical application, more aggregation is likely to occur – the aggregated inputs used in dual estimation are an example familiar to many production economists. Another outcome is the use of restrictive assumptions to obtain theoretical results. Finding a multi-period risk adjusted discount rate is an example in finance (Copeland and Weston 1988). Then one has to be concerned if the theoretical results are sufficiently robust for broad empirical application. More concentrated attention to these issues may prove us wrong.

In conclusion, we see considerable challenging work that can be done on risk decisions in the future. Hopefully, the efforts can be redirected to areas more promising than those emphasized in the past. In moving research forward, we must emphasize that our statements and conclusions are largely hypotheses rather than firm conclusions. We are anxious for you to test some of these ideas.

REFERENCES

Anderson, J.R., J.L. Dillon, and J.B. Hardaker. 1977. *Agricultural Decision Analysis.* Ames, IA: Iowa State University Press.

Anderson, K.B., and H.P. Mapp. 1966. "Risk Management Education Programs in Extension." *Journal of Agricultural and Resource Economics* 21: 31-38.

Baquet, A.E., A.N. Halter, and F.S. Conklin. 1976. "The Value of Frost Forecasting: A Bayesian Appraisal." *American Journal of Agricultural Economics* 58: 511-520.

Baquet, A., R. Hambleton, and D. Jose. 1997. *Introduction to Risk Management.* U.S. Department of Agriculture, Risk Management Agency.

Barry, P.J., and C.B. Baker. 1971. "Reservation Prices on Credit Use: A Measure of Response to Uncertainty." *American Journal of Agricultural Economics* 53: 222-228.

Barry, P.J., P.N. Ellinger, J.A. Hopkins, and C.B. Baker. 1995. *Financial Management in Agriculture* (5th edition). Danville, IL.: Interstate Publishers, Inc.

Barry, P.J., and D. Fraser. 1976. "Risk Management in Agricultural Production." *American Journal of Agricultural Economics* 58: 286-295.

Barry, P.J., and D.R. Willmann. 1976. "A Risk Programming Analysis of Forward Contracting with Credit Constraints." *American Journal of Agricultural Economics* 58: 62-70.

Batra, R.N., and A. Ullah. 1974. "Competitive Firm and the Theory of Input Demand Under Price Uncertainty." *Journal of Political Economy* 82: 537-548.

Boehlje, M.D., and L.D. Trede. 1977. "Risk Management in Agriculture." *Journal of American Society of Farm Managers and Rural Appraisers* 41: 20-27.

Boggess, W., R. Lacewell, and D. Zilberman. 1993. "Economics of Water Use in Agriculture." In G.A. Carlson, D. Zilberman, and J.A. Miranowski, eds., *Agricultural and Environmental Resource Economics.* New York: Oxford University Press.

Boisvert, R.N., and B.A. McCarl. 1990. "Agricultural Risk Modeling Using Mathematical Programming." Southern Cooperative Series Bulletin 346, Cornell University.

Brink, L., and B. McCarl. 1978. "The Trade-Off Between Expected Return and Risk Among Corn Belt Farmers." *American Journal of Agricultural Economics* 60: 259-263.

Brorsen, B.W., and S.H. Irwin. 1996. "Improving the Relevance of Research on Price Forecasting and Marketing Strategies." *Agricultural and Resource Economics Review* 25: 68-75.

Carlson, G.A. "A Decision Theoretic Approach to Crop Disease Prediction and Control." *American Journal of Agricultural Economics* 52: 216-223.

Carlson, G.A., and M.E. Wetzstein. 1993. "Pesticides and Pest Management." In G.A. Carlson, D. Zilberman, and J.A. Miranowski, eds., *Agricultural and Environmental Resource Economics."* New York: Oxford University Press.

Coble, K.H., T.O. Knight, G.F. Patrick, and A.E. Baquet. 1999. "Crop Producer Risk Management Survey: A Preliminary Summary of Selected Data." Information Report 99-001, Department of Agricultural Economics, Mississippi State University (available at www.agecon.msstate.edu/riskedu).

Coble, K.H., T.O. Knight, G.F. Patrick, and A.E. Baquet. 2000a. "Farmer Risk Perceptions and Demand for Risk Management Education." Paper presented at the Extension RME Workshop, St. Louis, June 6-7 (available at www.agecon.msstate.edu/riskedu).

Coble, K.H., O. Vergara, T.O. Knight, G.F. Patrick, and A.E. Baquet. 2000b. "Understanding Limited Resource Farmers Risk Management Decision Making Summary and Preliminary Analysis." Unpublished paper, Department of Agricultural Economics, Mississippi State University.

Copeland, T.E., and J.F. Weston. 1988. *Financial Theory and Corporate Policy* (3rd edition). Reading, MA: Addison-Wesley Publishing Company.

Drury, R., and L. Tweeten. 1997. "Have Farmers Lost Their Uniqueness?" *Review of Agricultural Economics* 19: 58-90.

Eidman, V.R., G.W. Dean, and H.O. Carter. 1967. "An Application of Statistical Decision Theory to Commercial Turkey Production." *Journal of Farm Economics* 49: 852-868.

Foster, W.E., and G. Rausser. 1991. "Farmer Behavior Under Risk of Failure." *American Journal of Agricultural Economics* 73: 276-288.

Goodwin, B.K., and T.L. Kasten. 1996. "An Analysis of Marketing Frequency by Kansas Grain Producers." *Review of Agricultural Economics* 18: 575-584.

Goodwin, B.K., and T.C. Schroeder. 1994. "Human Capital, Producer Education Programs, and the Adoption of Forward Pricing Methods." *American Journal of Agricultural Economics* 76: 936-947.

Govindasamy, R., F. Hossain, and A. Adelaja. 1999. "Income of Farmers Who Use Direct Marketing." *Agricultural and Resource Economics Review* 28: 76-83.

Hadar, J., and W.R. Russell. 1969. "Rules for Ordering Uncertain Prospects." *American Economics Review* 59: 25-34.

Hall, D. 2000. "Beef Cattle Producer Risk Management Survey." Unpublished paper, Department of Agricultural Economics, Texas A&M University.

Halter, A.M., and G.W. Dean. 1971. *Decisions Under Uncertainty and Research Applications.* Cincinnati: South-Western Publishing.

Hanoch, G., and H. Levy. 1969. "The Efficiency Analysis of Choices Involving Risk." *Review of Economics Studies* 36: 335-346.

Hazell, P.B.R. 1971. "A Linear Alternative to Quadratic and Semivariance Programming for Farm Planning Under Uncertainty." *American Journal of Agricultural Economics* 53: 53-62.

Hazell, P.B.R., and P.L. Scandizzo. "Market Intervention Policies When Production is Risky." *American Journal of Agricultural Economics* 57: 641-649.

Johnson, C.S., and K.A. Foster. 1994. "Risk Preferences and Contracting in the U.S. Hog Industry." *Journal of Agricultural and Applied Economics* 26: 393-405.

Johnson, L.I. "The Theory of Hedging and Speculation in Commodities." *Review of Economic Studies* 27: 139-151.

Johnson, S.R., and G.C. Rausser. 1977. "Systems Analysis and Simulation: A Survey of Applications in Agricultural and Resource Economics." In G.G. Judge et al., eds., *A Survey of Agricultural Economics Literature* (Vol. 2). Minneapolis: University of Minnesota Press.

Just, R.E. 1974. "An Investigation of the Importance of Risk in Farmers' Decisions." *American Journal of Agricultural Economics* 56: 14-25.

___. 1975. "Risk Aversion Under Profit Maximization." *American Journal of Agricultural Economics* 57: 347-352.

___. 2000. "Some Guiding Principles for Empirical Production Economics Research." *Agricultural and Resource Economics Review* 29: 138-158.

Just, R.E., L. Calvin, and J. Quiggin. 1999. "Adverse Selection in Crop Insurance: Actuarial and Asymmetric Incentives." *American Journal of Agricultural Economics* 81: 834-849.

Just, R.E., and R. Pope. 1979. "Production Function Estimation and Related Risk Considerations." *American Journal of Agricultural Economics* 61: 276-284.

Knoeber, C.R., and W.N. Thurman. 1995. "'Don't Count Your Chickens...': Risk and Risk Shifting in the Broiler Industry." *American Journal of Agricultural Economics* 77: 486-496.

Lin, W.G., G.W. Dean, and C.V. Moore. 1974. "An Empirical Comparison of Utility versus Profit Maximization." *American Journal of Agricultural Economics* 56: 497-508.

Mapp, Jr., H.P., and G.A. Helmers. 1984. "Methods of Risk Analysis for Farm Firms." In P.J. Barry, ed., *Risk Management in Agriculture.* Ames, IA: Iowa State University Press.

Meyer, J. 1977. "Choice Among Distributions." *Journal of Economic Theory* 14: 326-336.

Musser, W.N., B.A. McCarl, and G.S. Smith. 1986. "A Case Study of Deletion of Constraints from Risk Programming Models." *Southern Journal of Agricultural Economics* 17: 147-154.

Musser, W.N., G.F. Patrick, and D.T. Eckman. 1996. "A Risk and Grain Marketing Behavior of Large-Scale Farmers." *Review of Agricultural Economics* 18: 1-13.

Musser, W.N., and K.G. Stamoulis. 1981. "Evaluating the Food and Agricultural Act of 1977 with Firm Quadratic Risk Programming." *American Journal of Agricultural Economics* 63: 447-456.

Musser, W.N., and B.V. Tew. 1984. "Use of Biophysical Simulation in Production Economics." *Southern Journal of Agricultural Economics* 16: 79-86.

Patrick, G.F., A.E. Baquet, K.H. Coble, and T.O. Knight. "Hog Risk Management Survey: Summary and Preliminary Analysis." Staff Paper No. 00-9, Department of Agricultural Economics, Purdue University (available at www.agecon.msstate.edu/riskedu).

Patrick, G.F., and K. Collins. 2000. "Producers' Adjustments to 'Freedom to Farm'." Purdue Agricultural Economics Report, September, pp. 13-16 (available at www.agecon.purdue.edu/extensio/paer/0900).

Patrick, G.F., and L.M. Eisgruber. 1968. "The Impact of Managerial Ability and Capital Structure on Farm Firm Growth." *American Journal of Agricultural Economics* 50: 491-507.

Patrick, G.F., E.T. Loehman, and A. Fernandez. 1984. "Estimation of Risk-Income and Labor-Income Trade-Offs with Conjoint Analysis." *North Central Journal of Agricultural Economics* 6: 151-156.

Patrick, G.F., W.N. Musser, and D.T. Eckman. 1998. "Forward Marketing Practices and Attitudes of Large Scale Midwestern Grain Producers." *Review of Agricultural Economics* 20: 38-53.

Patrick, G.F., P.N. Wilson, P.J. Barry, W.G. Boggess, and D.L. Young. "Risk Perceptions and Management Responses: Producer-Generated Hypotheses for Risk Modeling." *Southern Journal of Agricultural Economics* 17: 231-238.

Rhodes, V.J. 1995. "The Industrialization of Hog Production." *Review of Agricultural Economics* 17: 107-118.

Robison, L.J., and P.J. Barry. 1977. "Portfolio Adjustments: An Application to Rural Banking." *American Journal of Agricultural Economics* 59: 311-320.

Sandmo, A. 1971. "On the Theory of the Competitive Firm Under Price Uncertainty." *American Economics Review* 61: 65-73.

Schrader, L. 1986. "Responses to Forces Shaping Agricultural Markets: Contracts." *American Journal of Agricultural Economics* 68: 161-166.

Segerson, K. 1986. "Risk Sharing in the Design of Environmental Policy." *American Journal of Agricultural Economics* 68: 1261-1265.

Stein, J.L. 1961. "The Simultaneous Determination of Spot and Future Prices." *American Economics Review* 51: 1012-1025.

Tauer, L.W. 1983. "Target MOTAD." *American Journal of Agricultural Economics* 65: 606-610.

___. 1985. "Use of Life Insurance to Fund the Farm Purchase from Heirs." *American Journal of Agricultural Economics* 67: 60-69.

___. 1995. "Age and Farmer Productivity." *Review of Agricultural Economics* 17: 61-69.

U.S. Department of Agriculture, Economic Research Service. 2001. "Data: Farm Business Balance Sheet and Financial Ratios." Available at http://www.ers.usda.gov/Data/Farm BalanceSheet.

U.S. Department of Agriculture, National Agricultural Statistics Service. 1997. *1997 Census of Agriculture* (Vol. I). Available at http://www.nass.usda.gov/census/census97.

Walker, O.L., and A.G. Nelson. 1977. *Agricultural Research and Education Related to Decision Making Under Uncertainty: An Interpretative Review of Literature.* Research Report No. P-747, Oklahoma State University (March).

WRCC-16. 1977. "Economic Growth of the Agricultural Firm." Technical Bulletin No. 86, College of Agriculture Research Center, Washington State University.

Zering, K., and A. Beals. 1990. "Financial Characteristics of Swine Production Contracts." *J. American Society of Farm Managers and Rural Appraisers* 54: 43-53.

Chapter 25

PAST PROGRESS AND FUTURE OPPORTUNITIES FOR AGRICULTURAL RISK RESEARCH

Richard E. Just and Rulon D. Pope
University of Maryland and Brigham Young University

INTRODUCTION

The purpose of this book is to provide an assessment of the role of risk in U.S. agriculture and offer a critical evaluation of the current state of agricultural risk research and prospects for further improvements. We commend the contributors for fulfilling that purpose. Collectively, the chapters of this book provide a very useful assessment of the state of empirical research on agricultural risk. Many excellent insights are offered about weaknesses in current research and a number of insightful possibilities are suggested for further research.

After all the research on agricultural risk to date, the treatment of risk in economic models of agriculture is far from harmonious. Many competing risk models have been proposed. Some new methodologies are largely untested. Some of the leading empirical methodologies in agricultural economic research are poorly suited for problems where risk is important. Many important areas of risk research appear to remain. While cursory assessments of the existing wide variety of research were offered recently by Hallam (forthcoming) with a retrospective view and by Just (forthcoming) with a prospective view, the chapters of this book go into far greater detail and offer a far more comprehensive assessment and critical evaluation.

We highlight eight major themes that emerge from this book:[1]

1. Uniformity of measurement and stylized facts that transfer among models and applications cannot be expected to emerge without careful

[1] This summary chapter is not an attempt to summarize the other chapters in terms of the authors' views but rather to summarize the contributions from our perspective for purposes of drawing joint implications from them.

consistency in defining the arguments of utility functions.

2. Risk models can be improved by considering the psychology of risk assessment and decision making under risk including human abilities of identification, comprehension, and information processing.
3. The basic risks in agricultural models are endogenous depending on information and other farmer choices in ways that have not yet been appropriately incorporated in economic modeling.
4. Risk behavior is likely most important for long-term and intertemporal decision making because instabilities cannot be mitigated by standard risk management tools as they can in short-term problems. Much remains to be understood about how decision maker preferences weigh consumption versus wealth accumulation and how they affect long-term decisions related to capital acquisition, financial planning, debt management, consumption versus investment decisions, etc.
5. Models are needed to represent the role of risks that are unknown prior to experience or observation, to reflect decisions to acquire such information once such risks are identified, and to reflect how such information affects risk assessments and production behavior as it is acquired.
6. Some have argued that risk is of secondary importance in agricultural production and marketing, but risk research has (i) focused on the short-run decisions, (ii) addressed them with aggregate data where risk is least likely to appear as important, and (iii) aggregated observations and decisions over time intraseasonally so that risk implications for decision making are obscured.
7. Data availability is the most important constraint in expanding risk research to address these concerns. Data on allocation of inputs among crops, conditionally over time during the growing season, and on wealth are particularly lacking and are key to uncovering risk behavior and revealed preferences.
8. After all of the anomalies and variations suggested over the past two decades, the expected utility model with proper representations of the realities of decision making still seems to be the most promising model for empirical analyses of agricultural problems. Nevertheless, generalizations should be investigated on an ongoing basis.

IMPERFECT MARKETS AND AGRICULTURE

Chavas and Bouamra-Mechemache (this volume) provide a careful and elegant characterization of general equilibrium and welfare with transaction and informational costs. As such, they motivate the entire topic of this book but especially the discussion in Just and Rausser (this volume) on the microeconomics of information acquisition. Making use of Luenberger's benefit function in a state contingent framework, risk aversion, efficiency, and competitive equilibria are defined and characterized. Pareto efficiency with transaction costs and information costs is considered. As expected, convexity is required to ensure that a competitive equilibrium exists. Further, reducing transaction costs expands the feasible set and, thus, can enhance efficiency. Similarly, greater information (or a reduction in the cost of information) enlarges the feasible set and can lead to efficiency gains as the Pareto utility frontier is shifted outward. Because information is costly, information will not in general be perfect.

Chavas and Bouamra-Mechemache conclude by indicating how diverse information affects efficiency and comment on the proper role of government in the creation and dispersal of information. Both the amount and distribution of information is crucial to the existence and efficiency of markets. Clearly, the scope for government provision or regulation of information is enhanced when there are typically informational externalities or public goods, and when there are economies of scale. When the numbers of individuals wishing to exchange standard goods is small, then contracting can facilitate exchange. Thus contracts and contract design are an inherently important aspect of agricultural exchange (Hueth and Hennessy, this volume; Ligon, this volume).

THE ARGUMENTS OF THE UTILITY FUNCTION

Jack Meyer's chapter (this volume) provides an excellent overview of the evolvement of accepted thinking regarding the expected utility (EU) approach. While many criticisms were raised regarding the expected utility approach beginning in the 1980s, he argues that the EU model has emerged from the myriad of challenges as the leading tool for addressing the broad spectrum of risk problems facing economic research. He provides an excellent review of the state of understanding of stylized facts regarding the structure of expected utility based both on theory and empirical work.

Some of the most important stylized "facts" developed in the literature thus far include the following:

1. Utility is increasing at a decreasing rate in wealth and profit.
2. Decreasing absolute risk aversion is widely supported by both theory and empirical work.
3. Increasing relative risk aversion is widely supported by theory but less by empirical work.
4. Relative risk aversion less than or equal to one is suggested by theory but empirical work has reached widely differing conclusions based on the context of application.

While the first "fact" is largely undisputed, the primary model used for agricultural risk analysis, the mean-variance expected utility model, is at variance with the second. Without doubt, one reason for the wide use of mean-variance models is convenience. While a small and growing literature has begun to examine more general models, the implications for agricultural risk research of more general models are largely unexplored.

Perhaps the most important implication of Meyer's chapter (this volume) for further assessment of the role of risk in agriculture has to do with inconsistent empirical practices. Meyer points out that the definition of the variable included as the argument of the utility function has fundamental implications for measurement of properties of the utility function. Common practices in agricultural risk analysis for empirical purposes consider utility as a function of wealth, true profit, accounting profit, consumption, and various definitions of partial profit. Even though a decision maker has a unique set of preferences, the associated specification of utility can differ dramatically among these alternatives. Especially relevant for agriculture is the difference in true profit and accounting profit because of the prevalence of family labor as an input on family farms, and the difference in wealth and consumption where much of a farmer's wealth is tied up in productive land and capital. Meyer makes the case that taking account of these differences could explain most of the empirical results that seem to find relative risk aversion greater than one, and that these differences could also explain widely varying empirical findings about whether relative risk aversion is increasing.

This important theme developed in detail by Meyer is reinforced by Buschena (this volume), who discusses the importance of reference points, by D.R. Just (this volume), who underscores the sensitivity of Arrow-Pratt measures of risk aversion, by Holt and Chavas (this volume), who emphasize a need to incorporate conditioning on wealth in econometric measurement of risk response, and by Roe and Randall (this volume), who discuss the importance of context and reference effects. We conclude that developing a uniform reference point for measuring risk preferences will be a major key for future empirical risk research. Uniform comparison to such a reference point will provide opportunity to find empirical support for stylized facts currently entertained mostly on theoretical grounds, and may facilitate opportunity for developing further stylized facts. As explained by Just and Pope (1999, 2001), research from multiple studies becomes additive as it relates to stylized facts, regardless of whether evidence is positive or negative, because such constructs facilitate comparability across research studies. One reason that many studies of agricultural markets and policies have not incorporated risk is that risk behavior is not well identified. For the case of single-argument utility functions, the underlying reference point of obvious choice is ex post wealth because the basic theoretical literature is so couched.

Conclusion 1. *For purposes of comparison, studies using single-argument utility functions to describe risk preferences should convert estimated parameterizations into relationships describing the dependence of absolute and relative risk aversion on wealth, whether or not utility is actually estimated as a function of true profit, accounting profit (excluding family labor), consump-*

tion, or some other partial profit measure. Where necessary data are lacking, ex post sensitivity analysis based on plausible hypothetical data can at least preserve the integrity of the literature.

PSYCHOLOGY AND HUMAN CAPACITY FOR RISK ASSESSMENT AND INFORMATION PROCESSING

David Buschena's chapter (this volume) provides an excellent overview of the non-expected utility literature and its implications for assessment of the role of risk in agriculture. He delineates the four major areas of anomalies in which the non-EU criticisms fall: (i) violations of independence of irrelevant alternatives, (ii) intransitivities, (iii) relevance of alternative reference points, and (iv) probability miscalibrations. After summarizing the literature that evidences these violations, he makes the point that practical non-EU alternatives also do not consider intransitivities and that miscalibrations may be due largely to posing experimental alternatives in contexts unfamiliar to respondents. Context-dependent anomalies are unlikely to be a problem for risk research using revealed preference data on typical agricultural risk problems, although they may be important when new risks arise.

Consistent with Meyer's view that EU models have emerged as the favored approach, Buschena argues that the remaining anomalies are better addressed in EU models by sufficient consideration of (1) heteroscedastic error structures, (2) reference points, and (3) information processing. With respect to (1), Buschena and Zilberman (1995, 2000) show that accounting for similarity of alternatives and heteroscedasticity of choices can cause EU models to perform as well as or better than the leading generalized expected utility models of Quiggin (1991) and Machina (1989). With respect to (2), while perhaps the most basic step in establishing reference points for EU models is proposed by Meyer as discussed above, other considerations may also be important. For example, in the same way that the permanent income hypothesis leads to money illusion in consumer models, wealth illusion could lead to differences in risk premiums depending on recent experience. For example, the same absolute change in wealth may be valued differently depending on whether it is a loss or a gain because of dependence on recent experience (see Roe and Randall, this volume, for further discussion). Such additional hypotheses about the importance of reference points require further empirical exploration.

With respect to (3), the potential for incorporating human limits on information processing and comprehension in EU models is explored in David Just's chapter (this volume). He reviews the various objections to EU models and concludes that they can be adequately addressed with EU models based on modified Bayesian updating after incorporating weightings reflecting learning and recall. A significant literature has focused on models where risk aversion is due to both curvature in the utility function and modification of probabilities due, for example, to misperceptions. The latter source of risk behavior is based on judgment bias and consists of both resolution and calibration bias. Psychologists have identified consistent patterns of (i) overconfidence when faced with wide distributions, (ii) availability bias associated with events that are not notable and thus less likely to be recalled, (iii) the law of small numbers associated with overweighting limited new information, (iv) representative bias associated with underweighting prior information when new information is less diffuse, and (v) conservatism associated with overweighting prior information when new information is more difficult to incorporate.

As analytic representation of these biases has been explored, some have suggested representing these biases by inappropriate weighting of prior versus likelihood information in the Bayesian updating model. Noting that these weights must depend on circumstances (Grether 1980), David Just (this volume) proposes a model of the form

$$p_{t+1}(x) = \frac{p_t(x)^{R(l,p,z)} l(\theta \mid x)^{L(l,p,z)}}{\int_{-\infty}^{\infty} p_t(x)^{R(l,p,z)} l(\theta \mid x)^{L(l,p,z)} dx}, \quad (1)$$

where p_t is the density representing beliefs in period t, l is the likelihood function representing new information in period t, z indexes individual circumstances, and R and L are weights representing recall and learning ($R = 1-L$), respectively. In this model, R and L can represent the difficulty of recalling prior experience depending on its complexity and diffusion, the difficulty of incorporating new information depending on its diffusion and the difficulty of processing it, and the effect of factors in z such as education that affect information processing abilities. In related work, D.R. Just (forthcoming) shows that this model can explain each of the numerical anomalies noted by Kahneman and Tversky (1979) as well as other "non-EU" results.

While a number of non-EU models have been developed in the context of experimental situations that are perhaps context-dependent, many so-called non-EU results seem to carry through to revealed preference, real-world data. Roe and Randall (this volume), for example, conclude that context, reference, availability, and process effects apply to problems with real payouts. As concluded by both Meyer (this volume) and Buschena (this volume), the standard EU model seems to be the most satisfactory model for addressing agricultural risk at this point. However, a generalization that includes the standard EU model as a special case and yet addresses the major anomalies would be more desirable. Marra and Carlson (this volume) note that information acquisition and learning is particularly critical for problems of technology adoption. Generalizations to EU that add psychological realities may be particularly advantageous and even necessary as untested products derived through biotechnology enter agricultural markets more rapidly, with each accompanied by uncertain consumer acceptance, international markets, productivity, vulnerability, pending policy controls, etc.

In the context of these developments, human abilities to collect, recall, and process information are likely to become increasingly limiting. Thus, further exploration with models that account for learning and recall (or, equivalently, representativeness and conservatism) as in equation (1) appear promising. However, we suggest that such models will be more useful if they embrace the EU model for standard problems where it performs well, while enhancing it to address well-recognized psychological limitations of decision makers. Because it incorporates these possibilities, a model such as equation (1) appears highly promising, particularly if stylized facts about the properties of the recall and learning functions can be developed through empirical applications. As David Just (this volume) points out, considerations of human learning and recall capabilities could even lead to policy issues about how public information should be produced so as to minimize problems with human comprehension, recall, and information processing.

Conclusion 2. *Expected utility models need to be expanded to reflect limitations of human abilities to comprehend, recall, and process information. The psychology literature and observed human behavior suggest minor modifications of standard Bayesian decision models that can potentially greatly improve the realistic potential of economics models.*

ENDOGENEITY OF RISK DUE TO COSTS OF INFORMATION ACQUISITION AND PROCESSING

Just and Rausser, Hueth and Hennessy, and Ligon add other perspectives about imperfect information in agriculture. Just and Rausser (this volume) motivate why information is not processed and utilized uniformly by decision makers (see also Chavas and Bouamra-Mechemache, this volume). They demonstrate that decision makers faced with a tradeoff between quality and cost of information (including information processing costs) will choose less information in periods of stability and more information in periods of volatility, and that decision makers with higher information gathering and processing costs will choose less information. The endogeneity of risk is also underscored by Hueth and Hennessy (this volume). They show how optimal risk arrangements in contracting problems depend on the circumstances of the problem, which are subject to change. These results add additional dimensions to the endogeneity of risk suggested in the original work of Just and Pope (1978) on risk implications of input use and the ensuing work by Antle (1983) and his work with others as summarized by Antle and Capalbo (this volume).

Just and Rausser (this volume) additionally demonstrate an optimality in relying on imperfect information about remote events because of the high cost of acquiring and processing such information. These results add a potential economic justification for imperfect use of information observed in the psychological literature based on individual human circumstances such as education. They also explain why decision makers faced with their own inabilities to comprehend, recall, and process information do not simply pay for services that compensate for their individual limitations.

The added complication that arises when the concept of rational expectations is expanded to account for information costs is that the choice of expectations mechanism becomes endogenous. This complicates empirical estimation because the typical separation assumption (whereby expectations formation is treated separately from optimizing behavior) fails. As noted by Nerlove and Bessler (2001), virtually all empirical work on expectations formation is based on this assumption. Just and Rausser (this volume) also suggest that existence of and convergence to equilibrium are greatly complicated because heterogeneity of information acquisition and processing costs generates expectations heterogeneity, which in turn causes heterogeneity of risks. Such considerations have not been incorporated in empirical models of agriculture to date.

Markets and contracts are institutional responses to facilitate exchange. Ligon (this volume) considers another institutional response that can reduce the cost of information and enhance efficiency – grading. Grading is an attempt to communicate information about particular product characteristics and presumably reduces the amount of heterogeneous information relevant to the market. Grading may benefit producers, consumers, or various third parties involved in producing and marketing a product. At one extreme is the case where the quality of foodstuffs is perfectly observed by all parties yielding perfect markets while at the other extreme no party observes quality. Only in the second case is there uncertainty to the consumer. Because firms have no incentive to produce a particular quality, it will tend to be low in order to reduce costs. Intermediate are cases with grading uncertainty. Ligon finds that whether or not producers or consumers bear risk depends crucially on the uncertainties facing intermediaries and consumers. For example, only when the intermediary does not observe the investments (characteristics) of the producer does the producer bear risk associated with product quality and grading. In this case, contingent contracts based upon quality will be observed rather than flat price or wage schedules. This contracting arrangement is ubiquitous in California fruit and vegetable production.

All of the chapters discussed in this section suggest that the marginal costs and benefits of information acquisition are crucial in understanding efficiency, behavior, and institutional responses. These costs and benefits are heterogeneous across individual consumers and producers and, in the case of grading and contracting, are likely different between the two groups. One of the general themes found in these chapters is the following:

Conclusion 3. *Applications of expected utility models should consider cost-consistent rational expectations that balance the marginal benefits of information (used to form expectations) with the marginal costs (of obtaining that information). The question of optimality of grades and contracts arises because information is costly.*

LONG-TERM AND INTERTEMPORAL RISK

Modeling intertemporal agricultural choice provides a major challenge for agricultural economists. Biological and behavioral dynamics are complex and stochastic. Representations of preferences must be developed accordingly. Such representations of incentives and preferences must consider risk aversion as well as intertemporal substitution and the timing of the resolution of risk. Typical models of intertemporal choice represent these problems with additive probabilities, such as expected utility, and additive utility over time (R.E. Just, forthcoming). Though risk aversion and temporal substitution are seemingly independent concepts, it is well known that these additive models imply a very restrictive relationship (Weil 1990). Decision makers may have a clear preference for early resolution of risk compared to later resolution of risk (Kreps and Porteus 1978). However, operational examples of general representations of risk and temporal substitution are possible (Epstein and Zin 1989, Weber 2000). These approaches have yet to be exploited productively by agricultural economists. Because of the predominance of intertemporal risk issues, agriculture should provide a very laboratory for studying intertemporal risk preferences. And certainly, the study of the role of risk in agriculture will not be complete without considerable research along these lines. Additionally, intertemporal problems imply that risk is inherently endogenous.

The endogeneity of risk has been emphasized in the dynamic setting of typical agricultural production by Antle and Capalbo (this volume). They emphasize that much of the reality of dealing with risk by farmers has been missed because of inappropriate temporal and spatial aggregation of data. Biophysical constraints, which are largely unique to agricultural production, generate temporal and spatial variability that contribute to localized dynamic risk responses. Annual crop models, for example, must be considered as a temporal aggregation of a number of sequential risky decision problems where each successive decision is conditioned on the risks that have been resolved by that point in the growing season. Similarly, interseasonal production problems depend heavily on dynamic factors such as accumulated soil moisture, soil nutrients, pest populations and resistance, etc.

As Antle and Capalbo emphasize, these factors have significant implications for risk that fail to be reflected in typical data sets. Lack of data forces estimated relationships to take on reduced form with spatial and temporal aggregation. Antle and Capalbo find for a Montana data set that adding risk considerations to their model does not improve predictive performance under these conditions, even though statistical significance can be found on risk factors. They conclude that the reason why risk considerations have not improved predictive performance is that the spatial and temporal realities of agricultural decision making have been ignored. Until the spatial and temporal nature of agricultural production are better specified, the benefits of risk research cannot be gained.

The importance of better representing the dynamic aspects of farm decision making is emphasized also by Barry, Marra and Carlson, Musser and Patrick, Roberts and Key, Roe and Randall, and Taylor and Zacharias. To explain rates of adoption of long-term investments in agriculture that are slower than neoclassical predictions, Marra and Carlson (this volume) emphasize the necessity of adding stochastic control considerations in dynamic programming and control models, and of considering real option value as an explanation for observed hysteresis. Upon considering the full range of risks faced in long-term adoption decisions, such as the risk of arrival of improved technologies during the life of an investment, optimal behavior can be quite different than when current risks are viewed myopically. Based on a broad survey of this literature, Marra and Carlson conclude that, even though risk is important, behavior once thought to be the effects of risk aversion is explained as well or better with other factors such as credit constraints, learning, or asset fixity. Their conclusion is in harmony with Antle and Capalbo's suggestion (this volume) that the benefits of risk research are likely not to be obtained until the dynamic aspects of risk problems are properly characterized.

Barry (this volume) considers the effects of agricultural finance on risk behavior in agriculture, which are inherently dynamic in nature. He argues that traditional tools such as the Capital Asset Pricing Model do not apply well to agriculture because risks cannot be well diversified and because equities are not readily marketable. Alternatively, in the realities of agriculture, lenders impose a significant set of risk management constraints on farmers that may well explain significant departures from the behavior implied by various other optimization models under risk. Roberts and Key (this volume) report empirical findings that demonstrate the risk-related importance of liquidity constraints. Barry's summary as well as Roberts and Key's results suggest that correct dynamic characterization of risk problems may not be possible until the risk issues that arise in finance have been well linked or integrated into other analyses of agricultural risk. Thus, again, the conclusion is that the benefits of risk research may not be attained until dynamic representations are improved.

Conclusion 4. *Risk behavior is intimately linked to a variety of dynamic considerations. Thus, the benefits of studying risk behavior in agriculture are unlikely to be captured until the dynamic aspects of agricultural production and finance are better discovered and incorporated into risk models both intraseasonally and interseasonally.*

UNCERTAINTY VERSUS RISK

Goodwin and Ker (this volume) consider the estimation of the probability distributions of yields, prices, and revenue. They provide an excellent overview of appropriate techniques for measurement that includes time-series parametric and nonparametric estimation with both Bayesian and classical approaches, giving the strengths and limitations of each. Particularly, because of the demands for accurate knowledge of yield distributions from crop insurance policy, much progress has been made. However, a significant limitation remains due to the paucity of yield data.

Taylor and Zacharias (this volume) begin their chapter with a reference to Frank Knight's (1921) distinction between risk and uncertainty and emphasize that many problems in agriculture do not lend themselves to standard frameworks of risk where all possible outcomes can be identified and neatly quantified by subjective probabilities associated with specific outcomes (Knight's case of risk). Alternatively, Knight argued that cases of pure uncertainty are not analyzable because probabilities of purely uncertain events are not measurable. Taylor and Zacharias suggest that reality is actually somewhere between Knight's polar cases of risk and uncertainty, and that fuzzy set theory provides a framework in which to think about these issues.

As Taylor and Zacharias suggest, agricultural production and marketing indeed seem to be characterized by significant uncertainty (in Knight's terminology). Farmers sometimes realize unforeseen outcomes associated with unanticipated farm programs, unforeseen changes in macroeconomic variables and their effects, unexpected changes in the structure of agriculture, failure of contracts, retaliation or intimidation by integrators, and unexpected changes in environmental regulations. As the potential range of technological developments and their implications expands with genetic engineering and biotechnology in the twenty-first century, and as the Internet promises to speed dissemination of information about both new technologies and perceptions of associated consumer risk (whether correctly or falsely founded), the frequency and magnitude of such unexpected outcomes may increase.

Just and Rausser (this volume) discuss the problem of Knightian uncertainty from the standpoint of the cost of acquiring information about unanticipated events. They suggest that models of risk and uncertainty in agriculture need to be expanded to take explicit account of unanticipated outcomes. One implication of the psychological literature reviewed by Buschena (this volume) and David Just (this volume) is that decision makers tend to ignore the possibility of events about which they are uninformed. On the other hand, once a particular unanticipated event occurs, decision makers gain some information with which to consider it in the future. With this somewhat practical and realistic approach, actions that have possible outcomes characterized partially by risk and partially by uncertainty (in the Knightian sense) become tractable. Furthermore, choosing to remain uninformed or partially informed is a rational decision whenever experience suggests that the marginal benefits from becoming informed about certain types of outcomes are less than the costs. Just and Rausser suggest that this approach seems to explain many of the observed anomalies of expected utility models by simply expanding the definition of rationality in a plausible way.

Taylor and Zacharias (this volume) suggest that, unless Knight's case of uncertainty is addressed, researchers are forced to ignore some of the most important applied problems and the most intellectually challenging problems in agriculture. To expand analysis to the case of risk and uncertainty in the Knightian sense, one of the main objectives becomes appropriate characterization of the degree to which decision makers are informed. Casual observation suggests that the process by which decision makers become informed about previously unanticipated outcomes is based on experience – basically, realization of a previously unanticipated outcome causes a decision maker to take it into account. Furthermore, some experience, however thin, gives some basis for subjective characterization of its likelihood, which is necessary to consider it with the Bayesian probability calculus. Thus, contrary to Knight's approach of regarding uncertain outcomes as not subject to analysis, the objective becomes tracking the rate at which decision makers become aware of unanticipated outcomes. In other words, the focus of research is turned toward modeling discovery.

Modeling discovery empirically is a difficult challenge because data are typically unavailable until an outcome has become important. To be ready to address unanticipated outcomes with empirical models, economists must ask "what if" questions. One area in which considerable experience has been developed along this line is in contingent valuation of unobserved environmental outcomes. While contingent valuation has essentially not been used in analyzing problems of risk and uncertainty, contingent valuation approaches may hold promise for early quantification of emerging issues when decision makers are beginning to become aware of previously unanticipated outcomes. Roe and Randall (this volume) discuss the possibilities for using contingent valuation for assessing risk problems. They characterize the contingent valuation literature as largely overlapping with the experimental economics literature from which EU anomalies have arisen. In spite of the anomalies, they underscore the need for experimental methods to address problems with previously unobserved phenomena such as reactions to new policies, new market developments, and new technologies.

Conclusion 5. *Analysis of risk and uncertainty should represent the role of outcomes that are unanticipated possibilities ("uncertain," in Knight's terminology) prior to experience or observation, the resulting adjustments in economic behavior that occur when they are realized, and decisions to acquire further information once such potential outcomes are identified.*

IS RISK OF SECONDARY IMPORTANCE?

Some critics of risk research in agriculture claim that risk is of secondary importance and argue that agricultural economists should focus more on sorting out first-order effects. Others argue that while policymakers cite risk as an important reason for agricultural policy, careful analyses of the effects reveal that agricultural policies are not designed accordingly.[2] One objective of this book is to address the question of whether risk is of crucial importance to the understanding of agriculture.

Several of the chapters in this book suggest a secondary importance of risk. For example, Hueth and Hennessy (this volume) argue that risk does not explain the tendency to contract although it is an important consideration in designing optimal contracts once decisions to contract are made. Antle and Capalbo (this volume) conclude that the predictive power of agricultural production models is not improved by risk considerations until the dynamic aspects of production are modeled correctly.

In contrast, Larson, English, and Roberts (this volume) conclude from their empirical work on variable rate technologies that risk management rather than profit maximization explains observed behavior. Marra and Carlson (this volume) observe that technology adoption behavior once thought to be the effect of risk aversion is explained as well or better with other factors such as credit constraints, learning, or asset fixity. Nevertheless, they conclude that risk is important in explaining behavior and that more work is needed especially on long-term investment models and applications.

Robison and Myers (this volume) offer a more pragmatic explanation for why many regard risk as of secondary importance. They suggest that the apparent inability to measure risk aversion explains the perception that risk is not important. If this is true, then comparison of risk preferences with more careful attention to the arguments of the utility function following the suggestions of Meyer (this volume) as discussed above might lead to greater clarity in the empirical literature and thus a change in the perceived importance of risk.

Musser and Patrick (this volume) devote their entire chapter to addressing the question of whether risk is of secondary importance. They conclude that some of the standard problems of the agricultural risk literature that involve crop diversification and use of marketing instruments such as hedging are explained quite well by risk-neutral models because farmers are not willing to trade off much short-run expected profit for reductions in short-run risk. Just

---

[2] For example, Just, Calvin, and Quiggin (1999) show that crop insurance participation is explained essentially by opportunities for adverse selection. Summarizing the implications of this and related literature, Glauber and Collins (this volume) conclude that the government is out of touch with the realities of crop insurance. In a similar discussion at the meetings where the chapters in this book were discussed, Brian Wright (who served as a discussant) underscored more broadly the fallacy of regarding risk as supporting agricultural policy.

(2000) presents a brief example explaining why short-run risk may not have serious consequences.

Musser and Patrick conclude, however, that other risk problems are likely to cause a much greater departure from risk-neutral behavior and may be largely impossible to analyze without considering risk. These include risks related to agricultural finance, legal liability, and human resources. These are cases of longer-term risk where survival of the firm may be at stake or the lifetime cycle of the firm is a critical issue. R.E. Just (forthcoming) argues that risk research has focused on the short-run decisions and addressed them with aggregate data where risk is least likely to be important.

In contrast, for longer-term problems, such as in the agricultural finance literature, the importance of risk considerations appears to be a foregone conclusion. Studies typically do not test for the importance of risk behavior but simply assume it. In fact, empirical work on longer-term risk problems has tended to use a very different methodology than used for short-term risk research. Barry (this volume) comments on this dichotomy of research approaches and notes that traditional simulation models are preferred to optimization models for multi-period, stochastic farm-level risk models.

Musser and Patrick, in fact, call for more simulation models to capture dynamics of agricultural decision making. By comparison, relatively few studies uncover direct positive evidence of decision maker preferences regarding intertemporal risk considerations. In this sense, the study by Roberts and Key (this volume) finds important evidence of preferences for liquidity. Barry (this volume) characterizes the strengths of simulation in terms of flexibility in accommodating empirical characteristics, inter-relationships among model components, varying shapes of probability distributions, and a broader range of decision maker objectives.

We suggest, however, that the short-run and longer-run methodologies need to be merged and that statistical evidence should be compiled to confirm or reject model interrelationships, alternative probability distributions, and broader decision maker preferences if they apply. Only then can a coherent understanding of risk preferences and their relative prevalence in agriculture be acquired and only then can the investigations of short-run and longer-run risk behavior in agriculture become synergistic.

Conclusion 6. *Risk behavior is likely most important for long-term intertemporal decision making because instabilities cannot be mitigated by standard risk management tools as in short-term problems. Much empirical work remains to be done on such problems, including investigation of how decision maker preferences weigh consumption versus wealth accumulation and how those preferences affect long-term decisions related to capital acquisition, financial planning, debt management, consumption versus investment decisions, etc.*

THE LIMITATIONS OF DATA AVAILABILITY ON RISK RESEARCH

Of course, a major obstacle to better understanding risk problems is that data are typically inadequate. Many of the chapters in this book refer to the data limitations of risk research. Antle and Capalbo (this volume) cite lacking midseason growth data on crops as a reason why crop production models have not been very successful in understanding the role of risk in crop production. Holt and Chavas (this volume) argue that disaggregate data are needed to understand risk problems because risk is inherently a disaggregate issue. Marra and Carlson (this volume) conclude that panel data are needed to understand issues of risk in technology adoption and investment. Barry (this volume), Chambers and Quiggin (this volume), and Coble and Knight (this volume) also cite the need for improved data to sort out risk issues in agriculture. Larson, English, and Roberts (this volume) conclude that even data within fields is necessary to determine the risk implications of variable rate technologies.

Agricultural economists probably have an insatiable appetite for data. Clearly, fully detailed site-specific data within farms is unlikely to be collected for general purposes. However, some careful consideration of alternatives is needed. One approach to lacking data is explored by Roe and Randall (this volume). Contingent valuation methods have found wide use in environmental problems where many hypothetical issues arise. They suggest that, due to lacking data, risk preferences for unobserved contingencies may be uncovered by such survey methods. Similarly, the dynamics of decision making under risk may be better understood with such survey responses designed to uncover how decision makers think through such problems. Certainly, lacking sufficient data for understanding the temporal and spatial complexities of dynamic risk problems suggested by Antle and Capalbo (this volume), direct interviews may permit better understanding of human limitations in addressing complex intertemporal decision problems.

In addition to such survey approaches and the potential of modeling site-specific, within-farm behavior with one-time data sets, however, some enhancement of widely accessible data is needed if risk considerations are ever to have benefits in policy modeling and analysis. In several recent papers, we have called for enhancement of public data by reporting higher moments of survey data than the mean and for reporting of cross moments, and we have called for the development of a broadly accessible general panel data set on agriculture (Just and Pope 1999 and 2001, Just 2000). The former could be made available at low reporting cost because additional sampling is not required, while the latter could facilitate additive and comparable debate in risk research.

Conclusion 7. *Data availability is the most important constraint in risk research. Data on allocation of inputs among crops, conditionally over time* *during the growing season, and on wealth are particularly lacking and are critical in uncovering risk behavior and revealed preferences. A mix of approaches is needed to relax data limitations, including more effective reporting of data already collected publicly (e.g., reporting of higher moments); development of a widely available panel data set reflecting allocations over time and crops, wealth and financial data, and age distributions of capital; and continued development of one-shot data sets that permit investigation of emerging within-farm and even within-field risk issues.*

HOW TO MODEL RISK PROBLEMS

One of the objectives of this book is to assess the relative strengths of methodologies used to investigate risk behavior. Several contrasts are of interest:

Econometrics versus Programming versus Simulation. Holt and Chavas (this volume) explore the strengths and weaknesses of the econometric approach, in contrast to Taylor and Zacharias (this volume), who explore the programming approach. Barry (this volume) and Musser and Patrick (this volume) further discuss the strengths of simulation. Clearly, each approach has its strengths, which has led to mutual use in the literature. We note in practice, however, that simulation models can be essentially econometric models, essentially programming models, or a mix of the two. Thus, we restrict our comments here to the former two. With respect to programming versus econometric approaches, we find the comments of Holt and Chavas (this volume) that point to a merging of the two methodologies most appealing. Couched in the discussion of Pope and Saha (this volume), one of the weaknesses of the econometric approach has been its 20-30 year preoccupation with dual approaches developed under certainty. Programming approaches have perhaps become relatively appealing for addressing risk problems, as suggested by Taylor and Zacharias (this volume), because they reveal the primal structure of problems necessary to investigate allocations and dynamic relationships where risk dependence is most likely to be manifested. But we also find appealing the emphasis of Holt and Chavas (this volume) on possibilities for investigating the importance of risk terms directly in first-order conditions using generalized method of moments (GMM) estimation. Redirection of econometric modeling toward primal structures and GMM estimation can allow inference in econometric models that does not preclude risk behavior a priori, while use of more boot strapping and statistical testing will move programming methodology closer to econometric methodology.

Primal versus Dual Modeling. Pope and Saha (this volume) investigate the relative strengths of traditional dual approaches compared to primal methods. Models based on moments of the distribution of wealth are found to offer convenience for both direct (primal) as well as indirect (dual) methods. This is particularly true if one insists on developing coherent curvature restrictions based upon parameters of the problem. Pope and Saha argue that many conditional expected utility of wealth models such as expected profit maximization offer attractive alternatives to unconditional expected utility maximization. However, they conclude that sufficient progress has been made to allow agricultural researchers to choose among conditional or unconditional models in either direct or indirect form – particularly in mean-variance models. Thus, the choice is largely dictated by taste, econometric convenience, and the consideration of specification errors. They argue that at this stage, researchers will often find it cumbersome to impose all of the appropriate restrictions from theory when both production and price uncertainty are considered. Pope and Saha conclude by arguing that many researchers will find it useful to begin with methods that do not require the development of the implications of duality under uncertainty but use instead direct first-order conditions using empirical distribution functions. Generalized method of moments estimation is generally couched in these terms and has had widespread application throughout the economics of uncertainty. It has a simple and natural application to the primal, which allows the researcher to concentrate on identifying the proper specification of the problem. This might be a natural starting point for agricultural researchers based upon the simplicity of implementation and on recent work by Mundlak (1996). If the direct approach's shortcomings are empirically evident, then researchers can move to dual methods (which are essentially inverses of the primal) that conventionally have different error structures and endogenous issues than do primal systems. A more careful approach would test for which error structure is the most appropriate. Holt and Chavas (this volume) also conclude in favor of using the primal approach for risk analyses.

State-contingent versus Parametric Distributional Modeling. Chambers and Quiggin (this volume) present the strengths of the state-contingent approach. On a conceptual level, their approach is highly general and overcomes the traditional weaknesses of dual models for problems of uncertainty. Furthermore, it provides a natural opportunity to apply state-contingent generalized utility models. From the standpoint of the purpose of this book, however, the remaining issue is how useful the state-contingent approach is for determining the role of risk empirically, e.g., in U.S. agriculture. Assuming expected utility provides an adequate characterization of the decision paradigm, the critical question is whether stochastic production technology can be better characterized by a tractable number of states versus a tractable number of moments in the context of a parametric structure (see Just 2000). We see this as the crucial issue for agricultural risk research if expected utility provides an appropriate representation of the decision problem compared to state-contingent utility.

Expected Utility versus Non-expected Utility. On this issue, we find the arguments of Meyer (this volume) and Buschena (this volume) to be convincing. They conclude that, in spite of the non-expected utility results in the literature, the expected utility model still explains a greater share of observed behavior than its alternatives, and that more realistic consideration of heteroscedasticity, reference points, and psychological factors in decision making can potentially explain most of the anomalies that have been identified.

Conclusion 8. *The expected utility model with proper representation of the realities of decision making under risk appears to be the most promising model for risk research in agriculture. Traditional dual models are inadequate for risk modeling but both primal risk models and state-contingent dual models offer substantial challenges for empirical modeling. Progress is most likely through increasing convergence of econometric and programming models and GMM estimation and testing of flexible representations of first-order conditions.*

RISK, POLICY, AND CONCLUDING REMARKS

While this short concluding chapter identifies some of the main themes of this book, and perhaps imposes some of our thinking on the interpretation of those themes, the chapters of this book identify many other fruitful areas for further risk research in agriculture. These include application of the new institutional economics (Barry), analysis of risk considerations in contracting (Hueth and Hennessy) and grading (Ligon), exploring risk considerations in mechanism design, and application of the production implications of environmental risk (Metcalfe, Sunding, and Zilberman).

Perhaps the true test of the importance of risk research in agriculture will be whether risk researchers can quantify and forecast policy issues related to risk and explain them to policymakers well enough to influence policy. Taylor and Zacharias raise the issue of whether subsidizing risk increases risk taking. This is one of the central issues considered by Coble and Knight as they examine crop insurance as a policy tool for farm firms to manage price and yield risk. During the last twenty years, the U.S. crop program has grown sevenfold with very poor actuarial results. This raises the question of whether the insurability of crops can be quantified and forecasted. Incentives to purchase insurance are due to risk aversion, subsidies, adverse selection, and moral hazard. Separately, these effects are difficult to measure and disentangle empirically. However, due to the relatively large magnitudes of the combined effects, a number of alternatives have been considered. These include increased efficiency in the rating of and economic evaluation of the existing program and products. As well, new alternatives such as area yield and crop revenue insurance have been proposed and studied. They conclude with an analysis of the future directions that insurance policy might sensibly take. The effectiveness of any insurance policy depends crucially on understanding farm-level heterogeneity. Coble and Knight argue that only panel data is likely to substantially enlighten policy analysis.

Metcalfe, Sunding, and Zilberman argue that the Coase Theorem, which has been widely propounded by economists to policymakers over the past few decades, fails when risk is important. Glauber and Collins review convincing evidence that federal crop insurance increases acreage planted to covered crops and thus reduces market price. Thus, the premium subsidy may well be offset entirely by the reduction in price that might occur. This confers a large negative pecuniary externality to the uninsured. They conclude that government is largely out of touch with the real risk issues related to crop insurance. And, as Gardner concludes, the government acts as if it is largely unaware of the policy risk it imposes on farmers. In this context, David Just argues that education of policymakers about risk, taking into account human capacities to comprehend risk problems, may be a worthwhile pursuit of economic research.

In conclusion, a rich and promising research agenda remains for uncovering understanding of the role of risk in agriculture. We hope this book is a useful contribution to establishing that agenda.

REFERENCES

Antle, J.M. 1983. "Testing the Stochastic Structure of Production: A Flexible Moment-Based Approach." *Journal of Business and Economic Statistics* 1: 192-201.

Antle, J.M., and S.M. Capalbo. 2001. "Agriculture As a Managed Ecosystem: Implications for Econometric Analysis of Production Risk." In R.E. Just and R.D. Pope, eds., *A Comprehensive Assessment of the Role of Risk in U.S. Agriculture.* Boston, MA: Kluwer Academic Publishers.

Barry, P.J. 2001. "Finance and Risk Bearing in Agriculture." In R.E. Just and R.D. Pope, eds., *A Comprehensive Assessment of the Role of Risk in U.S. Agriculture.* Boston, MA: Kluwer Academic Publishers.

Buschena, D.E. 2001. "Non-Expected Utility: What Do the Anomalies Mean for Risk in Agriculture?" In R.E. Just and R.D. Pope, eds., *A Comprehensive Assessment of the Role of Risk in U.S. Agriculture.* Boston, MA: Kluwer Academic Publishers.

Buschena, D.E., and D. Zilberman. 1995. "Performance of the Similarity Hypothesis Relative to Existing Models of Risky Choice." *Journal of Risk and Uncertainty* 11: 233-262.

___. 2000. "Generalized Expected Utility, Heteroscedastic Error, and Path Dependence in Risky Choice." *Journal of Risk and Uncertainty* 20: 67-88.

Chambers, R.G., and J. Quiggin. 2001. "Dual Approaches to State-Contingent Supply Response Systems Under Price and Production Uncertainty." In R.E. Just and R.D. Pope, eds., *A Comprehensive Assessment of the Role of Risk in U.S. Agriculture.* Boston, MA: Kluwer Academic Publishers.

Chavas, J.-P., and Z. Bouamra-Mechemache. "The Significance of Risk Under Incomplete Markets." In R.E. Just and R.D. Pope, eds., *A Comprehensive Assessment of the Role of Risk in U.S. Agriculture.* Boston, MA: Kluwer Academic Publishers.

Coble, K.H., and T.O. Knight. 2001. "Crop Insurance As a Tool for Price and Yield Risk Management." In R.E. Just and R.D. Pope, eds., *A Comprehensive Assessment of the Role of Risk in U.S. Agriculture.* Boston, MA: Kluwer Academic Publishers.

Epstein, L.G., and S.E. Zin. 1989. "Substitution, Risk Aversion, and the Temporal Behavior of Consumption and Asset Returns: A Theoretical Framework." *Econometrica* 55: 251-276.

Gardner, B.L. 2001. "Risk Created by Policy in Agriculture." In R.E. Just and R.D. Pope, eds., *A Comprehensive Assessment of the Role of Risk in U.S. Agriculture.* Boston, MA: Kluwer Academic Publishers.

Glauber, J.W., and K.J. Collins. 2001. "Risk Management and the Role of the Federal Government." In R.E. Just and R.D. Pope, eds., *A Comprehensive Assessment of the Role of Risk in U.S. Agriculture.* Boston, MA: Kluwer Academic Publishers.

Goodwin, B.K., and A.P. Ker. 2001. "Modeling Price and Yield Risk." In R.E. Just and R.D. Pope, eds., *A Comprehensive Assessment of the Role of Risk in U.S. Agriculture.* Boston, MA: Kluwer Academic Publishers.

Grether, D.M. 1980. "Bayes Rule as a Descriptive Model." *Quarterly Journal of Economics* 95: 537-557.

Hallam, J.A. Forthcoming. "Risk Research in Agricultural Economics: From the Agent to the Market." *Agricultural Systems* (special issue: "Advances in Risk Impacting Agriculture and the Environment").

Holt, M.T., and J.-P. Chavas. 2001. "The Econometrics of Risk." In R.E. Just and R.D. Pope, eds., *A Comprehensive Assessment of the Role of Risk in U.S. Agriculture.* Boston, MA: Kluwer Academic Publishers.

Hueth, B., and D.A. Hennessy. 2001. "Contracts and Risk in Agriculture: Conceptual and Empirical Foundations." In R.E. Just and R.D. Pope, eds., *A Comprehensive Assessment of the Role of Risk in U.S. Agriculture.* Boston, MA: Kluwer Academic Publishers.

Just, D.R. 2001. "Information, Processing Capacity, and Judgment Bias in Risk Assessment." In R.E. Just and R.D. Pope, eds., *A Comprehensive Assessment of the Role of Risk in U.S. Agriculture.* Boston, MA: Kluwer Academic Publishers.

___. Forthcoming. "Learning and Information." Unpublished Ph.D. dissertation, Department of Agricultural and Resource Economics, University of California, Berkeley.

Just, R.E. 2000. "Some Guiding Principles for Empirical Production Economics Research." *Agricultural and Resource Economics Review* 29: 138-158.

___. Forthcoming. "Risk Research in Agricultural Economics: Opportunities and Challenges for the Next Twenty-Five Years." *Agricultural Systems* (special issue: "Advances in Risk Impacting Agriculture and the Environment").

Just, R.E., L. Calvin, and J. Quiggin. 1999. "Adverse Selection in Crop Insurance: Actuarial and Asymmetric Information Incentives." *American Journal of Agricultural Economics* 81: 834-849.

Just, R.E., and R.D. Pope. 1978. "Stochastic Specification of Production Functions and Economic Implications." *Journal of Econometrics* 7: 67-86.

___. 1999. "Implications of Heterogeneity for Theory and Practice in Production Economics." *American Journal of Agricultural Economics* 81: 711-718.

___. 2001. "The Agricultural Producer: Theory and Statistical Measurement." In B. Gardner and G.C. Rausser, eds., *Handbook of Agricultural Economics.* Amsterdam: Elsevier-North-Holland.

Just, R.E., and G.C. Rausser. 2001. "Conceptual Foundations of Expectations and Implications for Estimation of Risk Behavior." In R.E. Just and R.D. Pope, eds., *A Comprehensive Assessment of the Role of Risk in U.S. Agriculture.* Boston, MA: Kluwer Academic Publishers.

Kahneman, D., and A. Tversky. 1979. "Prospect Theory: An Analysis of Decision under Risk." *Econometrica* 47: 263-292.

Knight, F.H. 1964. *Risk, Uncertainty and Profit.* Silver Lake Publishing Company, New York: Sentry Press. (Originally published in 1921 by the Houghton Mifflin Company, Boston.)

Kreps, D.M., and E.L. Porteus. 1978. "Temporal Resolution of Uncertainty and Dynamic Choice Theory." *Econometrica* 46: 185-200.

Larson, J.A., B.C. English, and R.K. Roberts. 2001. "Precision Farming Technology and Risk Management." In R.E. Just and R.D. Pope, eds., *A Comprehensive Assessment of the Role of Risk in U.S. Agriculture.* Boston, MA: Kluwer Academic Publishers.

Ligon, E. 2001. "Quality and Grading Risk." In R.E. Just and R.D. Pope, eds., *A Comprehensive Assessment of the Role of Risk in U.S. Agriculture.* Boston, MA: Kluwer Academic Publishers.

Machina, Mark J. 1989. "Comparative Statics and Non-Expected Utility Preferences." *Journal of Economic Theory* 47: 393-405.

Marra, M.C., and G.A. Carlson. 2001. "Agricultural Technology and Risk." In R.E. Just and R.D. Pope, eds., *A Comprehensive Assessment of the Role of Risk in U.S. Agriculture.* Boston, MA: Kluwer Academic Publishers.

Metcalfe, M., D. Sunding, and D. Zilberman. 2001. "Risk Management and the Environment." In R.E. Just and R.D. Pope, eds., *A Comprehensive Assessment of the Role of Risk in U.S. Agriculture.* Boston, MA: Kluwer Academic Publishers.

Meyer, J. 2001. "Expected Utility as a Paradigm for Decision Making in Agriculture." In R.E. Just and R.D. Pope, eds., *A Comprehensive Assessment of the Role of Risk in U.S. Agriculture.* Boston, MA: Kluwer Academic Publishers.

Mundlak, Y. 1996. "Production Function Estimation: Reviving the Primal." *Econometrica* 64: 431-438.

Musser, W.N., and G.F. Patrick. 2001. "How Much Does Risk Really Matter to Farmers?" In R.E. Just and R.D. Pope, eds., *A Comprehensive Assessment of the Role of Risk in U.S. Agriculture.* Boston, MA: Kluwer Academic Publishers.

Nerlove, M., and D.A. Bessler. 2001. "Expectations, Information and Dynamics." In B. Gardner and G.C. Rausser, eds., *Handbook of Agricultural Economics.* Amsterdam: Elsevier-North-Holland.

Pope, R.D., and A. Saha. 2001. "Can Indirect Approaches Represent Risk Behavior Adequately?" In R.E. Just and R.D. Pope, eds., *A Comprehensive Assessment of the Role of Risk in U.S. Agriculture.* Boston, MA: Kluwer Academic Publishers.

Quiggin, J. 1991. "Comparative Statics for Rank-Dependent Expected Utility Theory." *Journal of Risk and Uncertainty* 4: 339-350.

Roberts, M.J., and N. Key. 2001. "Does Liquidity Matter to Agricultural Production?" In R.E. Just and R.D. Pope, eds., *A Comprehensive Assessment of the Role of Risk in U.S. Agriculture.* Boston, MA: Kluwer Academic Publishers.

Robison, L.J., and R.J. Myers. 2001. "Ordering Risky Choices." In R.E. Just and R.D. Pope, eds., *A Comprehensive Assessment of the Role of Risk in U.S. Agriculture.* Boston, MA: Kluwer Academic Publishers.

Roe, B., and A. Randall. 2001. "Survey and Experimental Techniques As an Approach for Agricultural Risk Analysis." In R.E. Just and R.D. Pope, eds., *A Comprehensive Assessment of the Role of Risk in U.S. Agriculture.* Boston, MA: Kluwer Academic Publishers.

Taylor, C.R., and T.P. Zacharias. "Programming Methods for Risk-Efficient Choice." In R.E. Just and R.D. Pope, eds., *A Comprehensive Assessment of the Role of Risk in U.S. Agriculture.* Boston, MA: Kluwer Academic Publishers.

Weber, C.E. 2000. "'Rule-of-Thumb' Consumption, Intertemporal Substitution, and Risk Aversion." *Journal of Business and Economic Statistics* 18: 497-502.

Weil, P. 1990. "Nonexpected Utility in Macroeconomics." *Quarterly Journal of Economics* 105: 29-42.

INDEX